D1605503

VAN WYLEN LIBRARY

AUG 31 2005

WITHDRAWN

Hope College
Holland, Michigan

Invasive Plant Species of the World

A Reference Guide to Environmental Weeds

To My Father

Invasive Plant Species of the World

A Reference Guide to Environmental Weeds

Ewald Weber

Geobotanical Institute
Swiss Federal Institute of Technology
Zurich
Switzerland

CABI Publishing

CABI Publishing is a division of CAB International

CABI Publishing
CAB International
Wallingford
Oxon OX10 8DE
UK

CABI Publishing
44 Brattle Street
4th Floor
Cambridge, MA 02138
USA

Tel: +44 (0)1491 832111
Fax: +44 (0)1491 833508
E-mail: cabi@cabi.org
Website: www.cabi-publishing.org

Tel: +1 617 395 4056
Fax: +1 617 354 6875
E-mail: cabi-nao@cabi.org

© CAB International 2003. All rights reserved. No part of this publication may be reproduced in any form or by any means, electronically, mechanically, by photocopying, recording or otherwise, without the prior permission of the copyright owners.

A catalogue record for this book is available from the British Library, London, UK.

Library of Congress Cataloging-in-Publication Data
Weber, Ewald, 1960-
  Invasive plant species of the world : a reference guide to environmental weeds / Ewald Weber.
    p. cm.
Includes bibliographical references (p.   ).
  ISBN 0-85199-695-7 (alk. paper)
  1. Invasive plants. 2. Weeds. 3. Invasive plants--Control. 4. Weeds--Control. I. Title.
  SB613.5.W43 2004
  632'.5--dc21
                              2003010034

ISBN 0 85199 695 7

Printed and bound in the UK by Biddles Ltd, King's Lynn, from copy supplied by the author.

Contents

Foreword	vi
Introduction	1
Abbreviations and symbols	10
Species accounts	11
References	461
List of synonyms	537
Glossary	545

Foreword

Organisms from distant lands, which have breached the ancient biogeographic barriers that once delimited their distribution, can do enormous damage to their new home if they become established and spread. They can disrupt the ecosystems upon which we depend for food, fibre, water and biological diversity. Of course introductions, both purposeful and inadvertent, have been occurring since humans developed the capacity for intercontinental transport, however, as the tempo of transport is increasing, so too, are new introductions and invasions.

Those organisms that are successful invaders represent every taxonomic group, and they come from everywhere. The one thing they do share in common is that they are self-replicating. This is why they pose a special problem. It is not like an oil spill that can be cleaned up, or even like global warming, where we know that by reducing our input of greenhouse gases to the atmosphere we can solve the problem.

Once established, invasive organisms can be difficult, or if not caught early, virtually impossible to eradicate. Controlling them to low, non-harmful population sizes may be possible, but only with constant attention and expenditure of time and money. That is why the best, and most economical and prudent, strategy is to keep out those species that can do damage. However, the catch is that, in general, we do not have sufficient knowledge, based on species traits, to be able to predict fully which introduced species will become established and invasive. The best predictor actually is whether it is invasive in other foreign lands where it has been previously introduced.

Unfortunately we do not have an established global early warning system to tell us which species are a growing threat in one region to alert those in other regions to a potential menace. We of course, know the dangers of those invasives that are already widespread globally. For these, the battle has already been lost and enormous effort is directed to controlling them to reasonable levels, such as what happens with water hyacinth. Further, and crucially, we do not have a global system for early recognition and containment of potentially dangerous invasive species, other than the highly effective Centers for Disease Control and the World Health Organization. Comparable organizations need to be developed for organisms that, although they may be non-pathogenic to humans, are none the less often a threat to human livelihood and well being.

This book, by Ewald Weber, on the invasive plant species of the world, focuses on non-agricultural species, and provides such an early warning, since it shows clearly global distributions of selected plant species and where they are native as well as where they are already invasive. There is already available an abundance of information on agricultural weeds. Over 400 species are covered in this book, a selection of many of the worst global problem species, but unfortunately for all of us, only a fraction of the number of species that are becoming problems one place of another.

The great merit and value of the book is in the uniform treatment of each of the species covered. In a one-page compact form, information is noted on the growth form, synonymy, commercial use, global distribution (both natural and introduced), the kinds of habitats invaded and the ecology and control methods for the species, as well as the primary references relevant to the plant. The information is compact but comprehensive, and thus, very useful. The literature covered is extensive and can lead the reader to more detailed information if needed.

Weber no doubt struggled over which species to include in this monumental effort. But what is included represents many significant invaders and an important foundation for whatever follows. It will certainly contribute to the global digital database for invasive species that will surely develop in the not too distant future.

An important analysis in the book is the notation of economic uses of plants that subsequently became invasive, utilizing a database of over 700 species. This analysis builds on others that have shown that ornamental species are the largest pool for species that subsequently become invasive. Hopefully these studies will bring a more precautionary approach to importation of new and untested ornamentals in various regions of the world.

H.A. Mooney
Department of Biological Sciences
Stanford University
Stanford
California
USA

Introduction

Invasive plant species – a threat to our natural heritage

In the last few decades, ecologists and natural resource managers have recognized that the spread of non-native or alien organisms poses a serious threat to the conservation of natural and semi-natural habitats, and that such invaders can have a tremendous impact on the native faunal and floral communities. The spread of alien plants is a lasting and pervasive threat as the invaders proliferate and continue to spread, even if their introduction has ceased or ecosystems are not longer under the influence of disturbances and pollution. Indeed, the ever increasing number of alien species threatens our native diversity in many ways, and this threat is just an addition to other threats caused by pollution, fragmentation, and climatic change. The problem of biological invasions has become a central issue in the conservation of our biological diversity, and their control and management became costly and labour intensive (Pimentel 2002).

Whereas most of the introduced plant species that became naturalized do not cause significant problems as they are confined to highly disturbed sites or are not expanding their range, some few species do. These are called invasive species. The International Conservation Union (IUCN) defines alien invasive species as (McNeely 2001):

An alien species which becomes established in natural or seminatural ecosystems or habitats, is an agent of change, and threatens native biological diversity.

Thus, invasive plant species are distinguished from weeds growing in agroecosystems or highly disturbed man-made habitats. In the literature, this distinction is rarely made and the term 'invasive' is understood in different ways (Pysek 1995, Richardson et al. 2000). The reasons are that it is often difficult to clearly separate natural or seminatural habitats from man-made habitats, and that species may occur in all of these habitats. However, for practical conservation purposes and for developing an ecological theory on the nature of plant invasions, it is necessary to separate invasive plant species from weeds on arable land. A commonly used term for invasive plants is 'environmental weeds'.

Weeds in agroecosystems differ in their ecology from invasive plants because the invaded ecosystems are different. Agroecosystems are highly artificial and represent simple, species poor habitats with environmental homogeneity and predictable disturbance regimes. In contrast, natural habitats are mostly species rich, environmentally heterogeneous and often unpredictable. Plants invading agroecosystems represent mostly herbaceous species, often adapted to the crop system, whereas plants invading natural habitats comprise the full range of life forms.

On the origins of invasive plants

According to the above definition, invasive plant species are nonnative in the area where they are invasive. There are exceptions as some few native plants become invasive within their native range, e.g. *Phragmites australis* in central Europe. The vast majority of plant invaders are, however, aliens. They originate either from intentionally introduced or from accidentally introduced species. The former comprises species that are cultivated as ornamentals, for timber production and other economic uses. The latter comprises species that are introduced as contaminants of seeds, soil, agricultural produce or packing material. In both cases, the species establishes self-reproducing populations outside the area of cultivation or point of introduction and spreads into natural communities.

Within the last 100 years, the number of established alien plant species increased rapidly in many regions of the world as a result of increasing trade volume and travel around the globe (Rejmánek and Randall 1994). In California, for example, the number of naturalized alien plant species increased from 150 to 1000 between 1900 and 1990 (Rejmánek and Randall 1994). The number of naturalized plant species in a region is related to the number of invasive plants: the more aliens are present, the higher the likelihood that invasive plants are under them. An important part of current management practice of plant invasions is therefore the prevention of introduction of potentially invasive plants, which requires risk assessment tools to recognize such species (Groves *et al.* 2001, Pheloung *et al.* 1999, Reichard and Hamilton 1997, Zamora *et al.* 1989).

Considering the economic uses of plant species that are invasive in various parts of the world reveals some interesting patterns (Fig. 1). Most of the invaders that have been intentionally introduced are ornamentals. Indeed, horticulture has been and still is an important source of alien plants that become invasive (Reichard and White 2001). In USA for example, 82% of woody plant species colonizing areas outside of cultivation have been introduced for landscape use (Reichard and Hamilton 1997). Other significant sources of invasive plants are species used for erosion control or timber production. These figures make clear that there is a need to involve horticulture and other plant production industries in dealing with the problem of invasive plant species. Many serious plant invasions we are witnessing today could have been avoided if the species had not been introduced and planted.

Impacts caused by invasive plants

Invasive plants that spread into natural or seminatural ecosystems can have various effects on the native fauna and flora, and these effects can occur on different scales. Most general, effects of invasive plants can be grouped into direct and indirect ones. Direct effects include

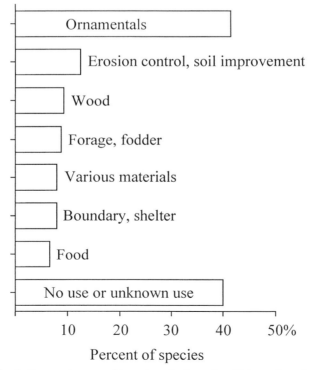

Fig. 1. Economic uses of invasive plant species. Data are based on 774 species that are considered as being invasive in one or more regions.

competition for space, nutrients, water and light, resulting in crowding out native species, displacing them with populations of the invader and preventing the establishment of native plants. Ecologically, these effects are due to competition between the invader and its neighbouring native plants, with the invader having a superior competitive ability. Indirect effects include changing soil water relations, nutrient cycling, light conditions, disturbance regimes, and affecting habitats of wildlife. Hybridization between an invader and closely related native species may lead to genetic changes in the populations of the native species. Most invaders have several effects and it is not always easy to reveal the ecological effects of invasive plant species. A general pattern is that the spread of invasive plant species leads to replacement of species rich and diverse communities with a single species stand of the invader, because most invasive plants grow vigorously and form extensive patches that tolerate only very few native species. The insect and bird fauna may change in its composition as an exotic plant expands and replaces the original vegetation. Such changes could also be caused by shading effects of an invader, as is the case with *Salix fragilis* in New Zealand. This tree colonizes extensively river and stream sides, reducing light levels and thereby affecting the macroinvertebrate fauna.

The process of an invasion

Biological invasions are dynamic processes during which several stages can be distinguished. An invasion starts with the *arrival* of propagules in an area. It has been shown that propagule pressure, e.g. the frequency and abundance of propagules entering an area is positively related to the invasion success of a species (Williamson 1996). In the subsequent *establishment* phase, one or several self-reproducing founder populations are formed. It is a commonly accepted theory that disturbances of an ecosystem favour the establishment of invasive species. Such disturbances can be fires, storm damage, logging, floods or any other external force leading to openings in the vegetation. Even invaders that seemingly enter undisturbed habitats have most likely profited from disturbances during their establishment phase. Once an invasive species has established, it reproduces and spreads. In this *spread* phase, propag-ules are dispersed and more populations are built up. Successful spread depends on the species ability to disperse and on the competitive nature between the invader and its surrounding native neighbours. A successful invader must have a higher competitive ability than the native species and this is often manifested in a high relative growth rate. The next phase may be called *range expansion*, where the species spreads not only locally but across the landscape, colonizing new suitable habitats. Indeed, such range expansions can be rapid in the case of invasive plants. *Impatiens glandulifera*, for example, spread at a rate of *c.* 645 km^2 $year^{-1}$ in Great Britain (Perrins *et al.* 1993). Successful spread and range expansion is facilitated if the species has a wide ecological amplitude and is able to cope with a wide range of new environmental conditions.

The management of plant invasions

The above explanations make clear that a proper management of invasive plants and invaded communities is necessary to reduce the impact of the invaders and to prevent their spread to new yet uninfested areas. A complete eradication of an exotic species, after it has become naturalized, is virtually impossible, and control methods must aim at minimizing the impacts where an invader has established, and preventing its spread to new sites. Controlling invasive plants is quite different from controlling weeds on arable lands because it has to take place in natural, often species rich and fragile communities that are of considerable conservation value. The intensity and frequency of control measures depend on the size of the infested area, the local abundance of the invader, and the type of vegetation (Fig. 2).

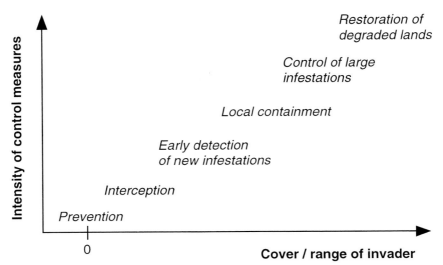

Fig. 2. Different approaches to control and manage plant invasions depend on the cover and range of the invasive plant. Modified after Westbrooks and Eplee (1996).

Control techniques
Control of invasive plants can be accomplished by basically three techniques: biological control, chemical control, physical control. Any combination of these methods are the basis for integrated pest management. The environmentally safest control techniques are those that have the greatest effect on the invader but do only little harm to the environment. Considerable damage to the environment can be caused as a result of control measures themselves, e.g. soil disturbance, opening of the vegetation, burning. In many cases, prescribed fires are used as part of control measures. Chemical control involves applying herbicides to the leaves, cut stumps, or stems. Residues of the compounds may have adverse effects on the habitats. However, chemical control is often the only practical method with long-lasting results and may be advantageous over physical methods. Applying herbicides needs to be carefully planned as the herbicide may adversely affect native plant and animal species. There are various application techniques and compounds whose use depend on the species and the time of year. Since herbicides are toxic to plants as well as to the people working with them, proper protection gear and experience are necessary. In any case, the use of herbicides needs to be carefully evaluated before using them in natural plant communities. Some compounds are extremely toxic such as 2,4,5-T and are therefore banned in many countries. Other compounds may remain active in the soil for many years such as picloram, which might not be desirable in natural communities. Further details on the use of herbicides are found in Dunham (1965), Parsons (1995), Parsons and Cuthbertson (2001) and Anon. (1983). A good overview of herbicide application techniques is given in Cronk and Fuller (2001) and Muyt (2001). In the case of woody species, the most common ones include:
Frilling: notching the base of the stem into a frill of bark slivers, behind which the herbicide is applied to the moist sapwood.
Stem injection: drilling holes onto the stem that penetrate below the bark but not into the heartwood, and injecting the herbicide into these holes.
Basal bark application: herbicide is painted or sprayed all round onto the bark at the base of stems. The herbicide is often mixed with a light oil such as diesel.

Cut stump application: cutting stems at ground level and immediately treating the cut surfaces with herbicide. This is a common method for shrubs and trees that are capable of resprouting.

Vines are often treated by painting stem bunches with herbicides. Bunches of stems are gathered up, cut and the cut surfaces painted with herbicide. In the case of vines that produce aerial tubers, cutting may lead to a mass release of tubers that establish on the floor. A herbicide may be applied to a scrape on the side of the vine stems, so that the exposed sapwood gets in contact with the herbicide. For herbaceous species, common herbicide applying techniques include foliar sprays, where the herbicide is applied during the phase of active growth. This can be done from spot spraying to treating large infestations by aircraft. Soil application involves usually scattering picloram granules around the invasive plants. The herbicide remains in the soil for a long period of time, preventing any seedling establishment.

Classical biological control
Classical biological control means that a natural enemy of the invasive plant (mostly herbivorous insects, fungi) is collected in the native range of the species and released in the introduced range to control the plant. The selection of a biocontrol agent is done in such a way that only specific control agents are released, in order to prevent native plants from being attacked. The success of biological control depends on the climate, size of population of the control agent, and the community where the invasive plant grows. Biological control clearly has an advantage over the other techniques because it is cheap (once a control agent has been found and its host range has been evaluated), it does not require disturbing the community for containment of the invasive plant, and because large populations of an invasive plant may be successfully controlled within a rather short time. On the other hand, non-target effects can be minimized by thorough pre-release studies, but they cannot be completely excluded. Therefore, the decision for implementation of classical biological control should be based on a careful risk-benefit assessment, considering also the risks and benefits of alternative control strategies. An overview of currently approved biocontrol agents together with their target species is given by Julien and Griffiths (1998).

Where to get information on invasive plants

Within the last few decades, a vast amount of literature on plant invasions has accumulated, and even more is available on the internet. An excellent overview on the ecology and management of plant invasions is provided by Cronk and Fuller (2001) and Luken and Thieret (1997). General textbooks on the ecology of biological invasions include Cox (1999), Drake *et al.* (1989), Elton (1958), Mooney and Hobbs (2000), Williamson (1996), and Wittenberg and Cock (2001). Internet sites on invasive plants are almost enumerable. Some few general sites are listed below:

Aquatic, wetland, and invasive plant particulars: http://aquat1.ifas.ufl.edu/
California exotic pest plant council: http://www.caleppc.org/
Center for invasive plant management: http://www.weedcenter.org/
Ecology and management of invasive plants program: http://www.invasiveplants.net
Florida exotic pest plant council: http://www.fleppc.org/
Hawaiian alien plant studies: pest plants of Hawaiian native ecosystems: http://www.botany.hawaii.edu
Hawaiian ecosystems at risk project: http://www.hear.org
Invaders database system: http://invader.dbs.umt.edu/Noxious_Weeds/
Invasive alien plants: http://www.lboro.ac.uk/research/cens/invasives/
Invasive and exotic species of North America: http://www.invasive.org
Invasive species specialist group: http://www.issg.org/

NBII Invasive species: http://www.invasivespecies.gov/
New Zealand weeds: http://weeds.massey.ac.nz/
The Nature Conservancy stewardship abstracts: http://tncweeds.ucdavis.edu

GRIN taxonomy database: http://www.ars-grin.gov/npgs/tax/index.html
Flora Europaea on the web: http://rbg-web2.rbge.org.uk/FE/fe.html
Flora of North America: http://hua.huh.harvard.edu/FNA/
Plant database: http://plants.usda.gov
The plant autecology pages: http://www.atkinsonm.demon.co.uk/autecol/

Scope of this book

The aims of this work are threefold. First, it gives an overview on the wide variety of plant species that are currently regarded as invasive plant species in one or more regions of the globe. Second, it provides information on the species' origin, distribution, ecology, and management. Third, an extensive list of references should point the reader to the relevant literature on these species.

The book does not cover species that are solely agricultural weeds. For these, several excellent reference books exist (Holm *et al.* 1977, 1997, Parsons and Cuthbertson 2001, Randall 2002), and there is a wealth of literature on their biology and control.

There are, of course, limitations as well. The list of invasive plant species covered in this book is certainly not complete; such a list can never be complete because new invasive species continue to appear in many regions, and the number of invasive plants is too large to be included in a single book. Any selection is arbitrary to some extent, and different individuals will have different opinions on which species to include or not. All species included in this book are considered to be invasive in natural areas and represent a good sample of the most serious plant invaders.

For many invasive plant species, only scarce information on their ecology and control is available. These gaps of knowledge are an important message and should stimulate research on those species. Many plant species have been recognized as being highly invasive in certain areas, but the biology and ecology of these invasions have not yet been investigated in detail. There is an urgent need to fill these gaps.

Source of information

A large amount of literature has been screened for this work. The main sources for compiling the species list for this book include the following: Blood (2001), Bossard *et al.* (2000), Carr *et al.* (1992), Cronk and Fuller (2001), Henderson (2001), Humphries *et al.* (1991), Lange-land and Craddock Burks (1998), Muyt (2001), Stone *et al.* (1992), Weidema (2000), White *et al.* (1993), Williams (1997), and various other publications in edited books and scientific journals. These citations are found in the reference section (pp. 461-536). Additional information was extracted from common floras and from the internet. The geographic distribution was taken from floras and from the GRIN taxonomy database (http://www.ars-grin.gov/ npgs/ tax/index.html).

Organization of the book

Geographic distribution and status

The geographic distribution of the species is summarized in the form of a table containing predefined geographic areas. A species' presence in one or more of these areas does not allow conclusions to be drawn on its abundance or distribution within this area. It merely indicates that the species is present. The status of a species is assigned to three categories: (i) native; (ii) invasive; or (iii) introduced but not invasive, or status as being invasive or not is unclear. Where a native species is also considered as invasive, both labels are provided. The status given to each species and geographic area is based on the literature mentioned above.

Most of the geographic areas are within political boundaries and hence clearly circumscribed. The areas covering several countries or states are defined below:

Northern Europe – Scandinavia: Finland, Norway, Sweden.
Central Europe – Austria, Benelux states, Denmark, France without the Mediterranean border, Germany, Switzerland.
Southern Europe – Southern France, Greece, Italy, Spain.
Eastern Europe – All European states east of Austria, Germany, Italy, European part of Russia, European part of Turkey.
Northern Africa – Algeria, Egypt, Libya, Morocco.
Tropical Africa – All African states between northern and southern Africa.
Southern Africa – Lesotho, Namibia, South Africa, Swaziland.
Western USA – States of continental USA bordering the Pacific Ocean.
Southeastern USA – States of continental USA bordering the Gulf of Mexico: Alabama, Florida, Louisiana, Mississippi, eastern Texas.
Remaining USA – All states of continental USA between western and southeastern USA.
Tropical South America – All states of continental South America beyond Mexico and except Argentina, Chile, Uruguay.
Temperate Asia – Middle East, China, Japan, Asian part of Russia
Tropical Asia – India, Sri Lanka, all states east of India and bordering the south of China, Malaysia, Indonesia, Philippines.

Selection of species

The decision as to which species to include in this book was based on the available information and the species' distribution. Each selection is arbitrary to some extent, and the species covered in this book present only a fraction of all invasive plant species. However, since only species were considered that are referred to as seriously problematic in the respective sources, the list of species of this book can be regarded as a set of environmental weeds that have significant negative effects on the native fauna and flora where they are introduced and invasive. Thus, these species should be regarded as species that deserve attention in terms of monitoring and control. Nomenclature follows the GRIN taxonomy database (http:// www.ars-grin.gov/npgs/tax/index.html).

Description, ecology and control

Two short sections provide information on the growth habit and morphology of the species, on their ecology and what the common methods are to control the species. In many instances, however, only scarce information is available on the ecology and control of invasive plant species, even if they are listed as being invasive in national weed lists. These gaps of knowledge should stimulate research on these species towards a better understanding of their invasion biology and control. Biological control agents are not listed as a successful biocontrol programme depends on various biotic and abiotic factors of the place where biocontrol agents are to be released. The reader is referred to the global overview of biocontrol agents provided by Julien and Griffiths (1998). Papers dealing with biological control of particular species are included in the reference section.

References

A list of selected references is given for each species, and all references are listed at the end of the book. References include original research papers, book chapters, books, and other published material, and were chosen to provide information on the ecology, biogeography, and control methods for the species. Floras from which species descriptions and data on their geographic distribution were extracted are not referenced. References of local bulletins, conference abstracts and literature that are not widely available are not referenced. These citations can easily be found by consulting the references given in this book.

Synonyms
Commonly used synonyms are listed at the end of the book. The list is far from complete but should help the reader to find additional literature and to find the species in common floras and textbooks with different taxonomic treatments.

Literature cited

Anon. (1983) *Herbicide handbook*. Weed Science Society of America, Champaign, Illinois, USA.

Blood, K. (2001) *Environmental weeds: a field guide for SE Australia*. CRC Weed Management Systems, Mt Waverley, 228 pp.

Bossard, C.C., Randall, J.M., and Hoshovsky, M.C. (2000) *Invasive plants of California's wildlands*. University of California Press, Berkeley, California, 360 pp.

Carr, G.W., Yugovic, J.V., and Robinson, K.E. (1992) *Environmental weed invasions in Victoria. Conservation and management implications.* Department of Conservation and Environment, Melbourne, Victoria, Australia.

Cox, G.W. (1999) *Alien species in North America and Hawaii: impacts on natural ecosystems*. Island Press, Washington DC, 387 pp.

Cronk, Q.C.B. and Fuller, J.L. (2001) *Plant invaders: the threat to natural ecosystems*. Earthscan Publications Ltd., London, 241 pp.

Drake, J.A., Mooney, H.A., di Castri, F., Groves, R.H., Kruger, F.J., Rejmánek, M., and Williamson, M. (eds) (1989) *Biological invasions: a global perspective*. John Wiley & Sons, Chichester, 525 pp.

Dunham, R.S. (1965) *Herbicide manual for noncropland weeds*. Agricultural Research Service, US Department of Agriculture, Washington DC.

Elton, C. (1958) *The ecology of invasions by animals and plants*. The University of Chicago Press, Chicago, 181 pp. *(reprint 2000)*

Groves, R.H., Panetta, F.D., and Virtue, J.G. (eds) (2001) *Weed risk assessment*. CSIRO Publishing, Collingwood, Australia, 244 pp.

Henderson, L. (2001) *Alien weeds and invasive plants: a complete guide to declared weeds and invaders in South Africa*. Agricultural Research Council, Cape Town, 300 pp.

Holm, L.G. *et al.* (1977) *The world's worst weeds*. University of Hawaii Press, Honolulu.

Holm, L.G., Doll, J., Holm, E., Pancho J., and Herberger, J. (1997) *World weeds*. John Wiley & Sons, New York.

Humphries, S.E., Groves, R.H., and Mitchell, D.S. (1991) *Plant invasions of Australian ecosystems: a status review and management directions*. Kowari 2. Australian National Parks and Wildlife Service. Canberra, Australia.

Julien, M.H. and Griffiths, M.W. (1998) *Biological control of weeds: a world catalogue of agents and their target weeds*. CAB International, Wallingford, 223 pp.

Langeland, K.A. and Craddock Burks, K. (eds) (1998) *Identification and biology of non-native plants in Florida's natural areas*. University of Florida, Gainesville.

Luken, J.O. and Thieret, J.W. (1997) *Assessment and management of plant invasions*. Springer, New York.

McNeely, J.A. (2001) *The great reshuffling: human dimensions of invasive alien species*. IUCN, Gland, Switzerland and Cambridge, UK, 242 pp.

Mooney, H.A. and Hobbs, R.J. (2000) *Invasive species in a changing world*. Island Press, Washington DC, 457 pp.

Muyt, A. (2001) *Bush invaders of south-east Australia*. R.G. and F.J. Richardson, Meredith, 304 pp.

Parsons, J.M. (ed.) (1995) *Australian weed control handbook*. Inkata Press, Melbourne.

Parsons, W.T. and Cuthbertson, E.G. (2001) *Noxious weeds of Australia*. CSIRO Publishing, Collingwood, Australia.

Perrins, J., Fitter, A., and Williamson, M. (1993) Population biology and rates of invasion of three introduced *Impatiens* species in the British Isles. *Journal of Biogeography* 20, 33-44.

Pheloung, P.C., Williams, P.A., and Halloy, S.R. (1999) A weed risk assessment model for use as a biosecurity tool evaluating plant introductions. *Journal of Environmental Management* 57, 239-251.

Pimentel, D. (2002) *Biological invasions: economic and environmental costs of alien plant, animal, and microbe species.* CRC Press, Boca Raton, 369 pp.

Pysek, P. (1995) On the terminology used in plant invasion studies. In: Pysek, P., Prach, K., Rejmánek, M., and Wade, M. (eds) *Plant invasions: general aspects and special problems.* SPB Academic Publishing, pp. 71-81.

Randall, R.P. (2002) *A global compendium of weeds.* R.G. and F.J. Richardson, Australia.

Reichard, S.H. and Hamilton, C.W. (1997) Predicting invasions of woody plants introduced into North America. *Conservation Biology* 11, 193-203.

Reichard, S.H. and White, P. (2001) Horticultural introductions of invasive plant species: a North American perspective. In: McNeely, J.A. (ed.) *The great reshuffling: human dimensions of invasive alien species.* IUCN, Gland, pp. 161-170.

Rejmánek, M. and Randall, J. (1994) Invasive alien plants in California: 1993 summary and comparison with other areas in North America. *Madroño* 41, 161-177.

Richardson, D., Pysek, P., Rejmánek, M., Barbour, M.G., Panetta, F.D., and West, C.J. (2000) Naturalization and invasion of alien plants: concepts and definitions. *Diversity and Distributions* 6, 93-107.

Stone, C.P., Smith, C.W., and Tunison, J.T. (eds) (1992) *Alien plant invasions in native ecosystems of Hawaii: management and research.* University of Hawaii Cooperative National Park Resources Studies Unit, Honolulu.

Weidema, I.R. (ed.) (2000) *Introduced species in the Nordic countries.* Nordic Council of Ministers, Copenhagen, 242 pp.

Westbrooks, R. and Eplee, R.E. (1996) Strategies for preventing the world movement of invasive plants. In: Sandlund, O.T., Schei, P.J., and Viken, A. (eds) *Proceedings of the Norway/UNEP conference on alien species.* Norwegian Institute for Nature Research (NINA), Trondheim, Norway, pp. 148-154.

White, D.J., Haber, E., and Keddy, C. (1993) *Invasive plants of natural habitats in Canada.* Environment Canada, Ottawa, 121 pp.

Williams, P.A. (1997) *Ecology and management of invasive weeds.* Conservation Sciences Publication no. 7. Department of Conservation, Wellington, New Zealand.

Williamson, M. (1996) *Biological invasions.* Chapman & Hall, London, 244 pp.

Wittenberg, R. and Cock, M.J.W. (eds) (2001) *Invasive alien species: a toolkit of best prevention and management practices.* CAB International, Wallingford, 228 pp.

Zamora, D.L., Thill, D.C., and Eplee, R.E. (1989) An eradication plan for plant invasions. *Weed Technology* 3, 2-12.

Acknowledgements

The following people are acknowledged for providing valuable information and giving generous help during various stages of the manuscript: P. Binggeli, C.C. Daehler, C.M. D'Antonio, H. Dietz, T. Dudley, P.J. Edwards, M. Lonsdale, W. Roschi, U. Schaffner, J. Stöcklin, E. Underwood, F. Weber. CABI Publishing is acknowledged for printing this work and T. Hardwick gave helpful advice during the production of the manuscript.

The preparation of this work was financially supported by a grant from the Huber-Kudlich foundation of the Swiss Federal Institute of Technology, Zurich.

Abbreviations and symbols

LF	Life form
CU	Commercial use
SN	Synonyms
●	Invasive in natural areas and not native to the area.
N	Native (indigenous)
X	Introduced, e.g. not native to the area, but not invasive in natural areas, or solely a weed of agroecosystems, or status as being invasive or not is unknown.

Abrus precatorius L. — Fabaceae

LF: Evergreen climber CU: Ornamental
SN: *Abrus abrus* (L.) Wight, *Glycine abrus* L.

Geographic Distribution

Europe	*Australasia*	*Atlantic Islands*
Northern	N Australia	Cape Verde
British Isles	New Zealand	Canary, Madeira
Central, France		Azores
Southern	*Northern America*	South Atlantic Isl.
Eastern	Canada, Alaska	
Mediterranean Isl.	● Southeastern USA	*Indian Ocean Islands*
	Western USA	N Mascarenes
Africa	Remaining USA	N Seychelles
Northern	X Mexico	N Madagascar
Tropical		
Southern	*Southern America*	*Pacific Islands*
	X Tropical	N Micronesia
Asia	X Caribbean	Galapagos
N Temperate	Chile, Argentina	X Hawaii
N Tropical		N Melanesia, Polynesia

Invaded Habitats
Tropical hammocks, open forests, pine rockland, disturbed sites.

Description
A slender twining or trailing woody vine with herbaceous branches and stems up to 9 m long. Leaves are pinnately compound and have 5-15 pairs of oval to oblong leaflets of 7-27 mm length and 3-10 mm width. White, yellow, pink or purple flowers of 10-15 mm length are borne in dense clusters in the axils of leaves. Fruits are oblong pods of 2-5 cm length and 1-1.5 cm width, splitting to release 3-8 shiny oval seeds of 6-7 mm length and 2-3 mm diameter. They are bright scarlet and have black bases.

Ecology and Control
This deep rooting plant has trailing and climbing shoots that smother native shrubs and small trees with a dense curtain of branches, impeding their growth and reproduction. The plant establishes well in disturbed sites and spreads rapidly after fires. It produces seeds prolifically which are dispersed by birds. The seeds are extremely poisonous to livestock and humans.

Once established, the plant is difficult to control. Small plants may be hand pulled or dug out. Larger infestations may be cut and painted with herbicide. Cutting before fruit ripen prevents seed dispersal.

References
715.

Acacia baileyana Muell.   Fabaceae

LF: Evergreen tree, shrub   CU: Ornamental, erosion control
SN: *Racosperma baileyanum* (Muell.) Pedley

Geographic Distribution

Europe	*Australasia*	*Atlantic Islands*
Northern	● N Australia	Cape Verde
British Isles	X New Zealand	Canary, Madeira
Central, France		Azores
Southern	*Northern America*	South Atlantic Isl.
Eastern	Canada, Alaska	
Mediterranean Isl.	Southeastern USA	*Indian Ocean Islands*
	X Western USA	Mascarenes
Africa	Remaining USA	Seychelles
Northern	Mexico	Madagascar
Tropical		
● Southern	*Southern America*	*Pacific Islands*
	Tropical	Micronesia
Asia	Caribbean	Galapagos
Temperate	Chile, Argentina	Hawaii
Tropical		Melanesia, Polynesia

Invaded Habitats
Grassland, riparian habitats, mediterranean climate scrub, heath- and woodland.

Description
A small unarmed, erect or spreading tree or bush of 3-8 m height, with a brown and smooth bark. The greyish or silvery-blue leaves are bipinnately compound, 2-5 cm long and have 2-6 pairs of pinnae. Glands at the base of each pair of pinnae are common. Pinnae are 1-2.5 cm long, each consisting of 12-24 pairs of linear to oblong leaflets of 5-9 mm length. The globular flowerheads consist of 20-25 bright yellow flowers, and 8-30 flowerheads form an extended axillary inflorescence with an axis of 5-10 cm length. Peduncles are 4-7 mm long. Fruits are greyish brown to black pods, straight to slightly curved, generally flat, 4-10 cm long and 8-12 mm wide. Each seed has a filiform aril at one end.

Ecology and Control
This tree grows in stony hills and in shrub communities where native. It is invasive because it forms dense thickets that compete for space, water and nutrients with native species, thereby replacing native vegetation. The fast growing tree is nitrogen-fixing and increases soil fertility. Reproduction is by seed, and it spreads rapidly after disturbances. Seeds are dispersed by birds and ants.
    Seedlings and small plants are hand pulled or sprayed with herbicides. Older trees are ringbarked or cut down, herbicide application is usually not necessary as the trees do not resprout. Fire is used to kill trees and stimulate seed germination. Seedlings are then treated with herbicides.

References
121, 215, 924, 931, 1089.

Acacia cyclops Cunn. ex Don — Fabaceae

LF: Evergreen shrub, tree
SN: *Acacia cyclopsis* Don
CU: Ornamental, erosion control

Geographic Distribution

Europe	*Australasia*	*Atlantic Islands*
Northern	N Australia	Cape Verde
British Isles	New Zealand	Canary, Madeira
Central, France		Azores
X Southern	*Northern America*	South Atlantic Isl.
Eastern	Canada, Alaska	
Mediterranean Isl.	Southeastern USA	*Indian Ocean Islands*
	● Western USA	Mascarenes
Africa	Remaining USA	Seychelles
X Northern	Mexico	Madagascar
Tropical		
● Southern	*Southern America*	*Pacific Islands*
	Tropical	Micronesia
Asia	Caribbean	Galapagos
Temperate	Chile, Argentina	Hawaii
Tropical		Melanesia, Polynesia

Invaded Habitats
Grassland, riparian habitats, coastal scrub and cliffs, coastal dunes.

Description
A dense, spreading large shrub or small tree of 1.5-6 m height with a brownish and fissured bark, without spines and with bright green phyllodes instead of true leaves. Phyllodes are leathery, 4-9 cm long and 6-12 mm wide, and have 3-5 longitudinal veins. Yellow flowers are borne in globular flowerheads of 4-6 mm diameter that are arranged in short racemes. Fruits are reddish-brown pods of 4-10 cm length and 8-12 mm width that are curved and become twisted when ripe. Fruits persist after seeds are shed. Seeds are encircled by orange to red and fleshy seed stalks.

Ecology and Control
In the native range, this tree grows in open scrub and rarely forms dense stands. Where invasive, it forms dense and impenetrable thickets that crowd out native vegetation; in South Africa, it forms a species poor dune scrub. Litter production is high, leading to increased soil nitrogen content. The tree reproduces by seeds which are dispersed by birds, ants and small mammals. Germination is enhanced after fire; the seedlings are intolerant of shade. The tree rarely resprouts after fire damage or felling.

Mechanical control can be achieved by cutting stems close to the ground. Clearing and burning stands of the tree are used to reduce the soil seed bank.

References
137, 284, 452, 465, 582, 583, 585, 586, 587, 638, 687, 795, 880, 881, 1428.

Acacia dealbata Link — Fabaceae

LF: Evergreen tree, shrub   CU: Ornamental, erosion control
SN: *A. decurrens* var. *dealbata* (Link) Muell., *Racosperma dealbatum* Pedley

Geographic Distribution

	Europe		*Australasia*		*Atlantic Islands*
	Northern	N	Australia		Cape Verde
	British Isles	●	New Zealand		Canary, Madeira
	Central, France			X	Azores
●	Southern		*Northern America*		South Atlantic Isl.
X	Eastern		Canada, Alaska		
X	Mediterranean Isl.		Southeastern USA		*Indian Ocean Islands*
		X	Western USA		Mascarenes
	Africa		Remaining USA		Seychelles
	Northern		Mexico	X	Madagascar
	Tropical				
●	Southern		*Southern America*		*Pacific Islands*
			Tropical		Micronesia
	Asia		Caribbean		Galapagos
	Temperate		Chile, Argentina		Hawaii
	Tropical				Melanesia, Polynesia

Invaded Habitats
Grassland, riparian habitats.

Description
A shrub or tree of 2-10 m height, occasionally reaching >25 m, without spines, and with a greyish green to black, usually smooth bark. Branchlets are slightly ribbed and pubescent. The bipinnately compound leaves are 4-10 cm long and have 8-20 pairs of pinnae, each consisting of 20-40 pairs of leaflets. These are linear-oblong and 2-5 mm long. A raised gland is present at each junction of pinnae pairs. Inflorescences form large racemes or panicles with globular flowerheads of 5-6 mm diameter, each having 25-35 bright yellow flowers. Fruits are greyish to brown pods, usually flat, 5-9 cm long and 6-12 mm wide, and slightly constricted between seeds.

Ecology and Control
Where native, this plant grows as a tall tree in mountain forests and along watercourses and in dry sclerophyll forests, remaining shrubby under dry conditions. It vigorously resprouts from stumps and is a prolific seed producer. It forms dense thickets that suppress native vegetation, disrupt water flow and increase erosion along streambanks. The plant is nitrogen-fixing and increases soil fertility.

Mechanical control is achieved by ringbarking or digging out, chemical control by basal stem treatments, stump treatments, or foliar spray with herbicides. After clearing large infestations, a follow-up programme is necessary to remove emerging seedlings and to prevent coppice regrowth. Stumps need to be treated with herbicides in order to prevent resprouting, and it is recommended to keep stumps lower than 15 cm.

References
206, 210, 549, 1073, 1089, 1415.

Acacia longifolia (Andr.) Willd. — Fabaceae

LF: Evergreen tree, shrub CU: Ornamental, erosion control
Syn: *Racosperma longifolium* (Andr.) Martius

Geographic Distribution

Europe	Australasia	Atlantic Islands
Northern	● N Australia	Cape Verde
British Isles	● New Zealand	Canary, Madeira
Central, France		Azores
● Southern	*Northern America*	South Atlantic Isl.
Eastern	Canada, Alaska	
X Mediterranean Isl.	Southeastern USA	*Indian Ocean Islands*
	X Western USA	Mascarenes
Africa	Remaining USA	Seychelles
Northern	Mexico	Madagascar
Tropical		
● Southern	*Southern America*	*Pacific Islands*
	Tropical	Micronesia
Asia	Caribbean	Galapagos
● Temperate	Chile, Argentina	Hawaii
Tropical		Melanesia, Polynesia

Invaded Habitats
Riparian habitats, woodland, grassland, coastal dunes and scrub.

Description
A spreading and unarmed shrub or small tree, 2-10 m tall, with a smooth and grey bark, and with bright green, flat phyllodes instead of leaves. Phyllodes are linear-lanceolate to obovate, 8-20 cm long and 1-2.5 cm wide, and have 2-5 prominent longitudinal veins. Bright yellow flowers appear in axillary, cylindrical flowerheads of 2-5 cm length and *c.* 7 mm width. Fruits are pale brown pods of 5-15 cm length and 3-6 mm width, more or less straight and cylindric, and constricted between seeds. Each contains 6-8 seeds having a thick aril.

Ecology and Control
A nitrogen-fixing tree common in coastal forests in the native range. Where invasive, it forms dense thickets that reduce native invertebrate and plant species richness. The plant produces large amounts of litter increasing nitrogen and phosphorus content of the soil. It accumulates large quantities of seeds in the soil that may remain dormant for many years. In South Africa, fires usually stimulate only a small proportion of seeds in the soil to germinate; seeds buried in deeper soil layers survive the fire and remain dormant.

Since the tree does not sprout after cutting, control of large plants might be feasible. If only a moderate number of seedlings emerge after fires, these can be removed by weeding or with herbicides. Seedling removal is best combined with re-establishment of native plants to reduce soil erosion problems. Seedlings need to be controlled for several years after clearance.

References
137, 215, 284, 322, 794, 795, 844, 880, 881, 924, 1025, 1026, 1027, 1129, 1415.

Acacia mearnsii De Wild.   Fabaceae

LF: Evergreen tree   CU: Ornamental, erosion control
SN: *Acacia decurrens* var. *mollis* Lindl., *Racosperma mearnsii* (De Wild.) Pedley

Geographic Distribution

	Europe		Australasia		Atlantic Islands
	Northern	N	Australia	X	Cape Verde
	British Isles	X	New Zealand	X	Canary, Madeira
	Central, France				Azores
X	Southern		*Northern America*		South Atlantic Isl.
	Eastern		Canada, Alaska		
X	Mediterranean Isl.		Southeastern USA		*Indian Ocean Islands*
			Western USA		Mascarenes
	Africa		Remaining USA		Seychelles
	Northern		Mexico		Madagascar
	Tropical				
•	Southern		*Southern America*		*Pacific Islands*
			Tropical		Micronesia
	Asia	X	Caribbean		Galapagos
	Temperate		Chile, Argentina	X	Hawaii
	Tropical				Melanesia, Polynesia

Invaded Habitats
Riparian habitats, coastal scrub, dry to mesic forests, grassland.

Description
An unarmed tree growing 5-20 m tall, with densely pubescent branchlets, a dark olive-green foliage, and a smooth and dark bark. Leaves are bipinnately compound with 9-20 pairs of pinnae, each pinna having 20-60 pairs of leaflets 1.5-4 mm long. Numerous nectar glands are present along the main axis of the leaves. Pale yellow flowers appear in globose flowerheads of c. 5 mm diameter that are arranged in large irregularly formed leafy panicles. Fruits are dark brown pods of 6-15 cm length and 5-9 mm width, and constricted between the seeds. Seeds are ellipsoid and flattened, 4-7 mm long and 3-6 mm in diameter.

Ecology and Control
In Australia, the tree often forms the understorey vegetation of eucalypt forests. Where invasive, it forms dense impenetrable thickets that displace native vegetation and reduce species richness. The tree is fast growing, nitrogen-fixing, and its high litter production leads to increased soil nitrogen levels. It is a prolific seed producer, and seeds are dispersed by small mammals, birds and water. The seed bank may contain up to 20,000 seeds m^{-2}. Fire stimulates germination and basal resprouting.

     Mechanical control includes removal of roots or cutting the stems as low as possible. Chemical control methods include spraying seedlings and saplings with glyphosate, and cutting larger trees followed by treating the stumps with herbicide. Older trees do not usually coppice from stumps.

References
120, 210, 549, 736, 1089, 1129, 1151.

Acacia melanoxylon R.Br.   Fabaceae

LF: Evergreen tree, shrub   CU: Ornamental, erosion control
SN: *Racosperma melanoxylon* (R. Br.) Mart.

Geographic Distribution

	Europe		*Australasia*		*Atlantic Islands*
	Northern	N	Australia		Cape Verde
X	British Isles	X	New Zealand		Canary, Madeira
X	Central, France			X	Azores
●	Southern		*Northern America*		South Atlantic Isl.
	Eastern		Canada, Alaska		
	Mediterranean Isl.		Southeastern USA		*Indian Ocean Islands*
		X	Western USA		Mascarenes
	Africa		Remaining USA		Seychelles
X	Northern		Mexico		Madagascar
	Tropical				
●	Southern		*Southern America*		*Pacific Islands*
			Tropical		Micronesia
	Asia		Caribbean		Galapagos
	Temperate	X	Chile, Argentina	X	Hawaii
	Tropical				Melanesia, Polynesia

Invaded Habitats
Forest edges and gaps, grass- and heathland, scrubland, riparian habitats.

Description
A tree of 8-30 m height, with a dark grey and furrowed bark and a dense, pyramidal crown. The tree has no spines and no leaves but dark green, straight or slightly curved phyllodes. These are smooth, oblong lanceolate, 4-13 cm long and 7-25 mm wide, and have a gland close to the base. True leaves often persist on young plants. Pale yellow to creamy flowers are borne in globular flowerheads of 8-10 mm diameter, arranged in branched racemes. Fruits are reddish-brown pods, 3-12 cm long and 5-10 mm wide, slightly constricted between seeds and becoming twisted. Seeds are black, 3-5 mm long and 1.5-3 mm wide, and encircled by pinkish-red seed stalks.

Ecology and Control
In the native range, the plant grows as a shrub or tree in rainforest margins and on stream banks. Where invasive, it forms dense thickets, competing for water and light and replacing native vegetation. The large amounts of litter produced increase the soil nitrogen content. The species vigorously regenerates from the soil seed bank after clearing or burning; fire stimulates germination of the seeds in the soil, and a copious seedling recruitment contributes to its invasiveness. The tree coppices after damage and frequently suckers from roots.
    Seedlings and small plants are hand pulled or dug out, roots must be removed. Larger plants are cut and the cut stumps treated with herbicide to prevent regrowth.

References
402, 549, 880, 881, 1016, 1036, 1089.

Acacia nilotica (L.) Willd. ex Delile   Fabaceae

LF: Deciduous tree   CU: Fuelwood, ornamental
SN: *A. arabica* (Lam.) Willd.

Geographic Distribution

Europe	*Australasia*	*Atlantic Islands*
Northern	● Australia	X Cape Verde
British Isles	New Zealand	Canary, Madeira
Central, France		Azores
Southern	*Northern America*	South Atlantic Isl.
Eastern	Canada, Alaska	
Mediterranean Isl.	Southeastern USA	*Indian Ocean Islands*
	Western USA	Mascarenes
Africa	Remaining USA	Seychelles
N Northern	Mexico	Madagascar
N Tropical		
N Southern	*Southern America*	*Pacific Islands*
	X Tropical	Micronesia
Asia	X Caribbean	X Galapagos
N Temperate	Chile, Argentina	Hawaii
N Tropical		Melanesia, Polynesia

Invaded Habitats
Grassland, savanna.

Description
A spreading shrub or small tree reaching 10 m height, branching almost from the base, with a dark, fissured bark and a deep taproot. The bipinnately compound leaves are glabrous to tomentose, have 3-10 pinnae of *c.* 4 cm length, each having 10-25 pairs of narrowly oblong leaflets of *c.* 6 mm length and 1.5 mm width. A petiolar gland is present between the two lowest pairs of pinnae. Flowers are bright yellow, 2.5-3.5 mm long, and are arranged in globular heads of *c.* 12 mm diameter. Inflorescences are clusters of 2-6 flowerheads. Fruits are indehiscent and flat pods of 6-25 cm length and 1-1.5 cm width, containing 8-15 seeds of 6-7 mm diameter.

Ecology and Control
This variable species with nine subspecies within the native range grows in grassland and savannas. *A. nilotica* subsp. *indica* (Benth.) Brenan spread rapidly over large areas of Australian grasslands and transformed them to thorny scrubland. Dense stands may have >900 trees ha^{-1}. Seeds are dispersed by mammals. Establishment depends on sufficient precipitation but 1-year-old plants are drought and fire resistant, and tolerate grazing. The tree easily resprouts from the base if damaged and is nitrogen-fixing.
     Chemical control methods include basal barking, cut stump treatment, stem injection, or foliar application; commonly used herbicides are 2,4-D or triclopyr. Mechanical clearing requires a follow-up programme to control regrowth and emerging seedlings.

References
16, 151, 166, 217, 607, 608, 630, 701, 801, 820, 863, 1061, 1192, 1312.

Acacia pycnantha Benth.   Fabaceae

LF: Evergreen tree   CU: Ornamental
SN: -

Geographic Distribution

Europe	*Australasia*	*Atlantic Islands*
Northern	N Australia	Cape Verde
British Isles	New Zealand	Canary, Madeira
Central, France		Azores
Southern	*Northern America*	South Atlantic Isl.
Eastern	Canada, Alaska	
Mediterranean Isl.	Southeastern USA	*Indian Ocean Islands*
	X Western USA	Mascarenes
Africa	Remaining USA	Seychelles
Northern	Mexico	Madagascar
Tropical		
● Southern	*Southern America*	*Pacific Islands*
	Tropical	Micronesia
Asia	Caribbean	Galapagos
Temperate	Chile, Argentina	Hawaii
Tropical		Melanesia, Polynesia

Invaded Habitats
Heath- and shrubland, forests.

Description
An erect or spreading tree or shrub, 3-8 m tall, with a smooth or finely fissured, dark brown to greyish bark. Phyllodes are 6-20 cm long, 5-30 mm wide, and have a conspicuous midvein and 1-2 glands along the margin. Flowerheads contain 30-70 golden yellow florets each, and are arranged on extended axillary inflorescences. Fruits are brown, more or less straight and flat pods of 5-14 cm length and 5-8 mm width, and are slightly constricted between the seeds.

Ecology and Control
Within the native range, this plant grows commonly in dry sclerophyll forest and heath communities, mostly on sandy or stony soils. It is invasive because it can form extensive and dense stands that crowd out native vegetation and prevent the regeneration of native shrubs and trees. It is nitrogen-fixing and increases soil fertility levels.

    Seedlings and small plants can be hand pulled or dug out. Larger individuals are cut and the cut stumps treated with herbicide.

References
549, 576, 736, 1089, 1336.

Acacia saligna (Labill.) Wendl.   Fabaceae

LF: Evergreen shrub, tree   CU: Erosion control
SN: *Acacia cyanophylla* Lindl., *Mimosa saligna* Labill.

Geographic Distribution

Europe		*Australasia*		*Atlantic Islands*	
	Northern	●	N Australia	X	Cape Verde
	British Isles		New Zealand		Canary, Madeira
	Central, France				Azores
●	Southern	*Northern America*			South Atlantic Isl.
	Eastern		Canada, Alaska		
	Mediterranean Isl.		Southeastern USA	*Indian Ocean Islands*	
		X	Western USA		Mascarenes
Africa			Remaining USA		Seychelles
X	Northern		Mexico		Madagascar
	Tropical				
●	Southern	*Southern America*		*Pacific Islands*	
			Tropical		Micronesia
Asia			Caribbean		Galapagos
	Temperate		Chile, Argentina		Hawaii
	Tropical				Melanesia, Polynesia

Invaded Habitats
Heath- and grassland, coastal scrub and beaches, forests.

Description
A shrub or small tree, mostly 2-6 m tall, with a smooth and grey to red-brown bark becoming fissured with age. Instead of true leaves, the species has flattened, linear to lanceolate, dark-green phyllodes, 8-25 cm long and 3-30 mm wide, and with conspicuous midribs. The stalked flowerheads are globular and 5-10 mm in diameter, each containing 25-65 bright yellow flowers. Flowerheads appear in irregularly shaped axillary racemes of 2-5 cm length. Petals are 2-3 mm long. Fruits are linear pods of 8-12 cm length and 4-6 mm width that are slightly constricted between seeds. The dark brown to black seeds are oblong and 5-6 mm long.

Ecology and Control
In the native range, this freely suckering tree grows frequently along watercourses and also on coastal dune systems. Where invasive, the species forms large and impenetrable thickets that replace native vegetation and threaten endangered plant species. The copious litter production and the ability to fix nitrogen increases soil nitrogen content. The shrub accumulates a large seed bank and easily resprouts if cut or burnt. Seeds are dispersed by birds and water, are long-lived and germinate rapidly after fire.

    Control methods include cutting trees and treating the stumps with herbicides to prevent regrowth. Prescribed fires after clearing are used to stimulate seed germination; seedlings are then either hand pulled, chemically treated or killed by repeated burning.

References
137, 215, 284, 570, 582, 584, 586, 638, 795, 796, 880, 881, 922, 923, 1429.

Acer negundo L.   Aceraceae

LF: Deciduous tree   CU: Ornamental, wood
SN: -

Geographic Distribution

Europe		*Australasia*		*Atlantic Islands*	
	Northern	●	Australia		Cape Verde
X	British Isles	X	New Zealand		Canary, Madeira
●	Central, France				Azores
	Southern	*Northern America*			South Atlantic Isl.
●	Eastern	X	Canada, Alaska		
	Mediterranean Isl.	N	Southeastern USA	*Indian Ocean Islands*	
		N	Western USA		Mascarenes
Africa		N	Remaining USA		Seychelles
	Northern	N	Mexico		Madagascar
	Tropical				
	Southern	*Southern America*		*Pacific Islands*	
		N	Tropical		Micronesia
Asia			Caribbean		Galapagos
	Temperate		Chile, Argentina		Hawaii
	Tropical				Melanesia, Polynesia

Invaded Habitats
Riparian habitats and forests, woodland.

Description
A dioecious tree of 3-20 m height, usually branching close to the ground, and with a broad irregular crown. The bark is light grey and smooth at first but becomes fissured with age. Leaves are bright green, 20-30 cm long, pinnately compound with 3-5 ovate to elliptic leaflets, each 5-10 cm long and 5-8 cm wide. Margins are coarsely toothed. The tiny, greenish yellow flowers appear in early spring prior to leaf growth. Fruits are composed of two fused yellowish samaras of 3-4 cm length that have strongly veined wings. These have a length of 25-35 mm each and are diverging at an acute angle. Seeds are markedly wrinkled.

Ecology and Control
Where native, this fast growing tree is commonly found in riparian forests and floodplains, and at least seven varieties have been recognized. It is a shade tolerant tree that easily resprouts after damage and forms root suckers. The tree is a successful and persistent invader once established, displaces native shrubs and trees and prevents their regeneration. Fruit production is high and seeds are dispersed by wind. The tree tolerates flooding and also drought to some extent.

Control includes cutting trees at ground level and treating the cut stumps with herbicides. Seedlings and saplings can be hand pulled or dug out. 2,4-D is a very effective herbicide to control this tree.

References
121, 321, 854, 966, 1118.

Acer platanoides L.   Aceraceae

LF: Deciduous tree   CU: Ornamental, wood
SN: -

Geographic Distribution

Europe		*Australasia*	*Atlantic Islands*
N	Northern	Australia	Cape Verde
●	British Isles	New Zealand	Canary, Madeira
N	Central, France		Azores
N	Southern	*Northern America*	South Atlantic Isl.
N	Eastern	X Canada, Alaska	
	Mediterranean Isl.	Southeastern USA	*Indian Ocean Islands*
		Western USA	Mascarenes
Africa		● Remaining USA	Seychelles
	Northern	Mexico	Madagascar
	Tropical		
	Southern	*Southern America*	*Pacific Islands*
		Tropical	Micronesia
Asia		Caribbean	Galapagos
N	Temperate	Chile, Argentina	Hawaii
	Tropical		Melanesia, Polynesia

Invaded Habitats
Forests, woodland.

Description
A tree reaching >30 m height, with a smooth, greyish bark becoming dark and fissured with age. Winter buds are glabrous and dark red to reddish brown. The opposite leaves are dark green on both sides, shiny, and have long petioles of 5-15 cm length. The leaf blade is up to 12 cm long and 20 cm wide, and has 3-7 lobes, each having few but long teeth. Flowers are yellow green, arranged in erect corymbose inflorescences, and appear together with the leaves. Petals are 5-6 mm long. Fruits are a pair of winged samaras of 8-10 cm length each. The wings form an angle of *c.* 180°.

Ecology and Control
In the native range, this tree grows commonly in broad-leaved forests, often as a shrub, sometimes forming large continuous stands. More than 80 cultivars have been developed. It is invasive because it casts heavy shade that excludes growth and regeneration of native species under its canopies. The tree transforms native woodland communities into species poor stands. The species is shade tolerant at all life stages, and reproduces vigorously by seeds. Seedlings establish abundantly under its own canopies. It is also frost tolerant and grows well on poor soils. Seeds are wind dispersed.

Seedlings and saplings can be dug out or hand pulled, all roots must be removed to prevent regrowth. Mature trees should be cut down as close to the base as possible. Cut stumps can be treated with herbicide.

References
679, 829, 941, 1367, 1368, 1369, 1439.

Acer pseudoplatanus L.   Aceraceae

LF: Deciduous tree   CU: Ornamental, wood
SN: -

Geographic Distribution

Europe		*Australasia*		*Atlantic Islands*	
●	Northern	●	Australia		Cape Verde
X	British Isles	●	New Zealand	●	Canary, Madeira
N	Central, France				Azores
N	Southern	*Northern America*			South Atlantic Isl.
N	Eastern		Canada, Alaska		
	Mediterranean Isl.		Southeastern USA	*Indian Ocean Islands*	
			Western USA		Mascarenes
Africa			Remaining USA		Seychelles
	Northern		Mexico		Madagascar
	Tropical				
	Southern	*Southern America*		*Pacific Islands*	
			Tropical		Micronesia
Asia			Caribbean		Galapagos
N	Temperate	X	Chile, Argentina		Hawaii
	Tropical				Melanesia, Polynesia

Invaded Habitats
Forests, riparian habitats, woodland.

Description
A tree of 8-25 m height, occasionally reaching 40 m, and with leaves up to 20 cm length. The bark of young trees is smooth and yellowish grey, becoming brown and fissured with age. Winter buds are glabrous and green. Leaves have petioles of 5-15 cm length, and usually five coarsely toothed lobes. They are dark green and glabrous above and light green beneath. Leaf blades are 10-20 cm wide. Young leaves are usually pubescent beneath. The greenish flowers appear together with leaves or shortly after leaf growth, and are arranged in large hanging inflorescences. Petals are 2-4 mm long. Fruits are a pair of winged samaras, the wings forming an angle of somewhat more than 90°.

Ecology and Control
In the native range, this fast growing tree grows commonly in moist and rich soils of broad-leaved and mixed forests and forest edges. It is a gap colonizer, growing best in sunny or lightly shaded sites. The juvenile period is 20-25 years. It produces up to 10,000 seeds annually. The tree is shade tolerant and forms dense, species poor infestations that alter light levels and impede the growth of native species. The tree easily resprouts if damaged.

If cut, the stumps should be treated with a herbicide to prevent resprouting. Seedlings and saplings are pulled by hand or dug out, roots should be removed to prevent regrowth.

References
121, 215, 284, 426, 607, 608, 635, 924, 1415, 1418, 1422.

Adenanthera pavonina L.   Fabaceae

LF: Deciduous tree   CU: Ornamental, shelter
SN: -

Geographic Distribution

Europe	Australasia	Atlantic Islands
Northern	N Australia	Cape Verde
British Isles	New Zealand	Canary, Madeira
Central, France		Azores
Southern	*Northern America*	South Atlantic Isl.
Eastern	Canada, Alaska	
Mediterranean Isl.	● Southeastern USA	*Indian Ocean Islands*
	Western USA	Mascarenes
Africa	Remaining USA	● Seychelles
Northern	Mexico	Madagascar
X Tropical		
Southern	*Southern America*	*Pacific Islands*
	Tropical	X Micronesia
Asia	X Caribbean	Galapagos
N Temperate	Chile, Argentina	Hawaii
N Tropical		N Melanesia, Polynesia

Invaded Habitats
Forests and forest edges, woodland, coastal areas.

Description
An unarmed tree of 6-15 m height and with a dark brown to greyish bark. Slightly buttressed trunks are common in older trees. Leaves are bipinnately compound, with 2-6 opposite pairs of pinnae. Each pinna has 8-21 alternate oval-oblong leaflets of 2-5 cm length and 2-2.5 cm width. Leaflets have an asymmetric base, are dull green above and blue-green beneath. Flowers are borne in narrow spike-like and hanging racemes of 10-15 cm length at the end of branches. They are small, white or yellow and fragrant. Fruits are curved and hanging, dark pods of 15-25 cm length and *c.* 2 cm width, releasing 8-12 seeds. These are hard-coated, bright scarlet and 7-10 mm in diameter.

Ecology and Control
In the native range, the tree is common in forests throughout the lowland tropics and occurs mostly on neutral to slightly acidic soils. Sparse nodulation has been reported for Singapore, indicating the potential for nitrogen-fixation. Growth is slow at first but increases rapidly after the first year: annual growth rates of 23-26 mm in diameter and 2-2.3 m in height have been reported. Ripened pods remain on the tree for long periods. The tree rapidly spreads in secondary forests of the Seychelles and displaces native trees and shrubs, preventing their regeneration and reducing diversity.

    Control measures include cutting larger trees. Seedlings and saplings can be hand pulled or dug out. Specific control methods need to be developed.

References
84, 416, 417, 700.

Agapanthus praecox subsp. *orientalis* Leight.   Liliaceae

LF: Perennial herb
SN: *Agapanthus orientalis* Leight.
CU: Ornamental

Geographic Distribution

Europe		*Australasia*		*Atlantic Islands*	
	Northern	●	Australia		Cape Verde
X	British Isles	X	New Zealand	X	Canary, Madeira
	Central, France				Azores
	Southern	*Northern America*			South Atlantic Isl.
	Eastern		Canada, Alaska		
	Mediterranean Isl.		Southeastern USA	*Indian Ocean Islands*	
			Western USA		Mascarenes
Africa			Remaining USA		Seychelles
	Northern		Mexico		Madagascar
	Tropical				
N	Southern	*Southern America*		*Pacific Islands*	
			Tropical		Micronesia
Asia			Caribbean		Galapagos
	Temperate		Chile, Argentina		Hawaii
	Tropical				Melanesia, Polynesia

Invaded Habitats
Coastal sand dunes, grass- and heathland, riparian habitats, forests, rock outcrops.

Description
A herb with flowering stalks of 75-100 cm height and with a thick and branched rhizome. The broad to narrowly lanceolate leaves are 40-80 cm long, 3-5 cm wide and emerge from a rosette. Inflorescences are large single umbels at the end of tall and hollow scapes of 50-100 cm length. Umbels are dense and contain numerous blue or white flowers on pedicels of 5-8 cm length. The blue to white perianth is 4-5 cm long and has dark midveins on the six segments. Stamens are 4-4.5 cm long, the anthers are 3-5 mm long and yellow. The fruit is an elongated ovoid capsule of 25-33 mm length, containing numerous black seeds that have a membranous wing and measure 7-8 mm in length.

Ecology and Control
The species is highly variable and many cultivars are known. In Australia, two subspecies have become naturalized: subsp. *praecox* and subsp. *orientalis*. The latter has become invasive and forms dense populations that crowd out native vegetation and reduce species richness. The plant spreads both by seeds and by vegetative growth. After establishment, it quickly forms dense clumps, and scattered clumps collate into a single matrix dominated by this species.

Small infestations can be dug out. Larger infestations are sprayed, best before fruits develop.

References
215.

Agave americana L.   Agavaceae

LF: Perennial succulent   CU: Ornamental
SN: *Agave rasconensis* Trel. ex Standl.

Geographic Distribution

Europe		*Australasia*		*Atlantic Islands*
	Northern	X	Australia	Cape Verde
X	British Isles	X	New Zealand	● Canary, Madeira
X	Central, France			● Azores
●	Southern	*Northern America*		South Atlantic Isl.
X	Eastern		Canada, Alaska	
●	Mediterranean Isl.		Southeastern USA	*Indian Ocean Islands*
		N	Western USA	Mascarenes
Africa			Remaining USA	Seychelles
	Northern	N	Mexico	Madagascar
	Tropical			
●	Southern	*Southern America*		*Pacific Islands*
			Tropical	Micronesia
Asia			Caribbean	Galapagos
	Temperate		Chile, Argentina	Hawaii
	Tropical			Melanesia, Polynesia

Invaded Habitats
Coastal bluffs and cliffs, dunes, rocky places, savanna, disturbed sites.

Description
A freely suckering plant with a large basal rosette formed by thick succulent leaves. Rosettes are 1-2 m tall and 2-4 m wide. Leaves reach 2 m in length and are 15-25 cm wide, lanceolate, light green or greyish, glaucous, and end in a sharp terminal spine. The flowering stalk is 5-9 m tall, straight and slender, and has rather small bracts of triangular shape. Panicles are generally long oval and open, with 15-35 spreading branches bearing umbels at the end. The yellow flowers are 7-10 cm long and have 8-10 cm long stamens. Fruits are oblong short-beaked capsules of 4-5 cm length. Seeds are brown or shiny black, 7-8 mm long and 5-6 mm wide.

Ecology and Control
The exact origin of this taxon is uncertain, and there are many subspecies, varieties and cultivars. *A. americana* spp. *protamericana*, a putative progenitor of the cultivated and naturalized taxa, is found naturally in mountain forests, tropical deciduous forests and thorn forests. Where invasive, the plant persists and spreads by abundant suckering from the root crown. A single individual can form dense impenetrable stands that eliminate native vegetation. The species is drought resistant and shows a broad tolerance to different soil types.
   Specific control methods for this species are not available. Smaller rosettes may be dug out but all roots must be removed to prevent regrowth.

References
444, 549, 1089.

Agave sisalana Perrine — Agavaceae

LF: Perennial herb
SN: -
CU: Fibre, ornamental

Geographic Distribution

Europe	Australasia	Atlantic Islands
Northern	X Australia	Cape Verde
British Isles	New Zealand	Canary, Madeira
Central, France		Azores
Southern	*Northern America*	South Atlantic Isl.
Eastern	Canada, Alaska	
Mediterranean Isl.	Southeastern USA	*Indian Ocean Islands*
	Western USA	Mascarenes
Africa	Remaining USA	Seychelles
Northern	N Mexico	● Madagascar
Tropical		
● Southern	*Southern America*	*Pacific Islands*
	Tropical	Micronesia
Asia	X Caribbean	Galapagos
Temperate	Chile, Argentina	X Hawaii
Tropical		Melanesia, Polynesia

Invaded Habitats
Grass- and heathland, forests, rock outcrops, coastal beaches.

Description
A large succulent perennial with thick leaves in a basal rosette of 1.5-2 m height, elongated rhizomes and suckering from the base. The bright green, linear-lanceolate leaves are 90-130 cm long, 9-12 cm wide and have dark brown terminal spines of 2-2.5 cm length. The margins are smooth or with numerous prickles. Inflorescences form 5-6 m tall panicles of elliptic shape, with 10-20 lateral branches bearing umbels at the ends. Flowers are 55-65 mm long, greenish yellow, malodorous, with dark-spotted or reddish filaments up to 6 cm long, and yellow anthers of *c.* 25 mm length. The fruit is a capsule of *c.* 6 cm length and has a beak at the end.

Ecology and Control
The origin of this plant is uncertain and it is probably of hybrid origin. It is an important economic plant for fibre production and is cultivated in many places. It forms suckers abundantly from the base, enabling the species to spread without reproduction by seed. In addition, bulbils are usually formed in the axils of the bracteoles after flowering. Fruits are rarely formed. The weedy behaviour is due to vegetative growth, allowing the species to colonize large areas and replace the native vegetation with its rosettes. It is drought tolerant and withstands waterlogging.

Small plants and smaller patches may be dug out, but all roots and rhizomes must be removed. Effective herbicides are glyphosate or triclopyr plus picloram applied to cut plants.

References
444, 549, 607, 792, 1041, 1089.

Ageratina adenophora (Spreng.) King & Rob.   Asteraceae

LF: Perennial herb   CU: Ornamental
SN: *Eupatorium adenophorum* Spreng.

Geographic Distribution

Europe		*Australasia*		*Atlantic Islands*	
	Northern	●	Australia		Cape Verde
	British Isles	●	New Zealand	●	Canary, Madeira
	Central, France			●	Azores
●	Southern	*Northern America*			South Atlantic Isl.
	Eastern		Canada, Alaska		
	Mediterranean Isl.	X	Southeastern USA	*Indian Ocean Islands*	
		X	Western USA		Mascarenes
Africa			Remaining USA		Seychelles
	Northern	N	Mexico		Madagascar
X	Tropical				
X	Southern	*Southern America*		*Pacific Islands*	
			Tropical		Micronesia
Asia			Caribbean		Galapagos
X	Temperate		Chile, Argentina	X	Hawaii
●	Tropical				Melanesia, Polynesia

Invaded Habitats
Dry to wet forests, river banks, swampy sites, grassland, disturbed places.

Description
A multi-stemmed herb, occasionally woody at the base, and with often purplish stems of 0.5-2 m height. Leaves and stems are densely hairy. Leaves are opposite, long petiolated, triple-nerved, and have ovate and dark green blades of 5-10 cm length and 1-6 cm width. They are purplish and glandular-pubescent beneath. Flowerheads are *c.* 7 mm long, contain 10-16 florets, and are arranged in terminal clusters. The corollas are white to pinkish and *c.* 3 mm long. Fruits are achenes of *c.* 2 mm length, each with a pappus of 5-40 bristles and five ribs along the achene.

Ecology and Control
The trailing branches easily root at the nodes on contact with the soil, enabling the species to build up dense thickets consisting of many intertwined stems. It has a prolific asexual seed production due to apomixis, and the seed rain in dense stands of this plant may reach 60,000 m^{-2}. Dense stands eliminate native vegetation and prevent the regeneration of native plants.

    Mechanical control is done by slashing followed by ripping or ploughing, often combined with sowing other species after removal. Herbicides used to control this plant include glyphosate, 2,4-D amine, dicamba and MCPA, or triclopyr, applied in late summer when the plant is actively growing.

References
344, 682, 986, 1301, 1415, 1440.

Ageratina riparia King & Rob. Asteraceae

LF: Perennial herb
SN: *Eupatorium riparium* Regel
CU: Ornamental

Geographic Distribution

Europe	Australasia	Atlantic Islands
Northern	● Australia	Cape Verde
British Isles	● New Zealand	X Canary, Madeira
Central, France		Azores
Southern	*Northern America*	South Atlantic Isl.
Eastern	Canada, Alaska	
Mediterranean Isl.	Southeastern USA	*Indian Ocean Islands*
	Western USA	X Mascarenes
Africa	Remaining USA	Seychelles
Northern	N Mexico	Madagascar
Tropical		
X Southern	*Southern America*	*Pacific Islands*
	Tropical	Micronesia
Asia	Caribbean	Galapagos
Temperate	Chile, Argentina	X Hawaii
● Tropical		Melanesia, Polynesia

Invaded Habitats
Riparian habitats, forest edges and gaps, disturbed places.

Description
A herb of 40-60 cm height, sometimes exceeding 1 m, with numerous spreading or ascending stems. Leaves have long petioles, are opposite and simple, lanceolate to elliptic, and toothed at the margins of the upper half. The leaf blades are 3-12 cm long and 0.8-3 cm wide. Flower-heads are 5-6 mm in diameter, subtended by green bracts, and contain 15-30 white or cream florets and are arranged in terminal clusters at the ends of branches. The corolla is 3-3.5 mm long. Fruits are dark brown achenes of *c.* 2 mm length and have a pappus of white hairs. The plant has a short and thick root stock.

Ecology and Control
The plant reproduces by seed and spreads also vegetatively by stem layering. The rapidly growing seedlings may reproduce within 8-10 weeks. The sprawling growth pattern leads to dense mats that crowd out native vegetation. The species is shade tolerant but sensitive to frost. Mature plants may produce 10,000-100,000 seeds per year. Seeds are both wind and water dispersed, and germination has found to be light dependent.

Mechanical removal of infestations needs resowing of other plants to prevent re-establishment of this plant. Freshly cut stumps are treated with 2,4-D or dicamba. Seedlings and small plants can be hand pulled or dug out.

References
607, 608, 986, 1301, 1415, 1440, 1452.

Agrostis capillaris L.   Poaceae

LF: Perennial herb   CU: Ornamental, erosion control, fodder
SN: *Agrostis tenuis* Sibth., *Agrostis vulgaris*

Geographic Distribution

Europe		*Australasia*		*Atlantic Islands*	
N	Northern	●	Australia		Cape Verde
N	British Isles	●	New Zealand		Canary, Madeira
N	Central, France				Azores
N	Southern		*Northern America*		South Atlantic Isl.
N	Eastern		Canada, Alaska		
N	Mediterranean Isl.		Southeastern USA		*Indian Ocean Islands*
		X	Western USA	X	Mascarenes
Africa		X	Remaining USA		Seychelles
N	Northern		Mexico		Madagascar
	Tropical				
	Southern		*Southern America*		*Pacific Islands*
			Tropical		Micronesia
Asia			Caribbean		Galapagos
N	Temperate		Chile, Argentina		Hawaii
	Tropical				Melanesia, Polynesia

Invaded Habitats
Pastures, grassland, shrubland, disturbed sites.

Description
A tufted perennial grass, spreading by short rhizomes and stolons, and with erect or ascending culms of 10-70 cm length. They are slender and have 2-5 nodes. Leaves are glabrous, linear, 1-15 cm long and 1-5 mm wide, and finely pointed at the ends. Ligules are mostly shorter than broad and 0.5-2 mm long. The green to purplish panicles are 1-20 cm long and 1-12 cm wide, oblong to pyramidal, and very loose. The lanceolate to narrow-oblong spikelets measure 2-3.5 mm in length, have one floret each and persistent glumes as long as the spikelet. Fruits are ellipsoid caryopses.

Ecology and Control
In the native range, this grass grows in nutrient poor meadows, heathland, and in forest gaps. The species is calcifuge and highly variable with many cultivars. It invades areas of poor, acid soils and forms dense swards that crowd out native vegetation and prevent shrub recruitment. It grows quickly and establishes well after disturbances such as fire or clearing. Extensive colonies of this grass alter the local invertebrate communities in New Zealand. It replaces *Festuca* dominated native grasslands and slows the natural colonization by woody plants.

    Specific control methods for this species are not available. Scattered plants may be hand pulled or dug out, larger infestations sprayed with herbicides.

References
215, 1066, 1067, 1157, 1158.

Agrostis stolonifera L.   Poaceae

LF: Perennial herb   CU: Erosion control, ornamental
SN: -

Geographic Distribution

Europe		*Australasia*		*Atlantic Islands*	
N	Northern	●	Australia	N	Cape Verde
N	British Isles	X	New Zealand	N	Canary, Madeira
N	Central, France			N	Azores
N	Southern	*Northern America*		X	South Atlantic Isl.
N	Eastern		Canada, Alaska		
N	Mediterranean Isl.		Southeastern USA	*Indian Ocean Islands*	
		X	Western USA	X	Mascarenes
Africa			Remaining USA		Seychelles
N	Northern		Mexico		Madagascar
	Tropical				
X	Southern	*Southern America*		*Pacific Islands*	
			Tropical		Micronesia
Asia		X	Caribbean		Galapagos
N	Temperate	X	Chile, Argentina	X	Hawaii
N	Tropical				Melanesia, Polynesia

Invaded Habitats
Riparian habitats, seasonal freshwater wetlands, marshes, coastal scrub and beaches.

Description
A multi-stemmed tufted perennial grass with long aboveground stolons and erect or ascending culms of 35-100 cm height, rooting at the nodes. Leaves are glabrous and linear, 1-10 cm long and 0.5-5 mm wide, and rolled when young. The obtuse and membranous ligules are 1-6 mm long. Inflorescences are green, whitish or purplish panicles, 1-13 cm long and 1-3 cm wide, linear to lanceolate or oblong, open during flowering and contracted afterwards. The one-flowered spikelets are 2-3 mm long and have persistent glumes as long as the spikelet. Caryopses are ellipsoid, pale brown to reddish brown, and 0.5-0.8 mm long.

Ecology and Control
In the native range, this grass is commonly found in wet meadows and riparian habitats. It is is a polyploid complex in which clonal propagation is predominant. It is fast growing and invasive because it forms mat-like dense colonies due to vegetative growth, displacing native vegetation and reducing species richness. It is a threat to *Acaena magellanica* scrub on the sub-Antarctic Marion Island. Plants in Australia do not fruit and spread solely by stolons. Fragments of stolons easily root and regenerate to new plants.
    Infestations can be sprayed with grass-selective herbicides. Small patches may be dug out if soil disturbance is not critical.

References
215, 491, 492.

Ailanthus altissima (Mill.) Swingle   Simaroubaceae

LF: Deciduous tree   CU: Ornamental
SN: *Ailanthus glandulosa* Desf., *Rhus cacodendron* Ehrh.

Geographic Distribution

	Europe		*Australasia*		*Atlantic Islands*
	Northern		● Australia		Cape Verde
X	British Isles	X	New Zealand		Canary, Madeira
X	Central, France			●	Azores
●	Southern		*Northern America*		South Atlantic Isl.
●	Eastern	●	Canada, Alaska		
X	Mediterranean Isl.	X	Southeastern USA		*Indian Ocean Islands*
		●	Western USA		Mascarenes
	Africa	●	Remaining USA		Seychelles
X	Northern	X	Mexico		Madagascar
	Tropical				
X	Southern		*Southern America*		*Pacific Islands*
			Tropical		Micronesia
	Asia		Caribbean		Galapagos
N	Temperate	X	Chile, Argentina	X	Hawaii
	Tropical				Melanesia, Polynesia

Invaded Habitats
Grassland, forest gaps, riparian habitats, flood plains, rock outcrops, disturbed places.

Description
A dioecious tree ranging 5-25 m in height, with a grey and more or less smooth bark becoming fissured with age. Leaves are 30-100 cm long and pinnately compound with 10-40 leaflets. These are lanceolate, 4-13 cm long and 25-50 mm wide, have 2-4 rounded teeth near the base and often a round gland on the lower surface. Small and greenish flowers are borne in panicles of 6-12 cm length at the ends of branches. Fruits are dry, indehiscent, light brown to yellowish, winged samaras of 25-50 mm length and 6-10 mm width, containing one seed of 3-5 mm diameter in the centre.

Ecology and Control
A fast growing and light-demanding pioneer tree forming extensive thickets due to root suckering, thereby displacing native vegetation. It tolerates drought and airborne salt, and grows well on poor soils. Older trees are resistant to freezing temperatures. The tree produces up to one million wind dispersed seeds per year. Seedlings and ramets may persist in the forest waiting for a gap, and then rapidly growing into the canopy.

    Seedlings and saplings are hand pulled but root fragments must be removed. Cutting induces root suckering and resprouting and must be combined with herbicide treatments. Girdling is done when trees are growing, followed by herbicide application. Effective herbicides include glyphosate applied to foliage or stem cuttings, triclopyr applied to barks of young stems and stem cuttings, and picloram for treating cut stumps.

References
121, 127, 309, 512, 596, 685, 695, 696, 698, 726, 879, 985, 986, 1317.

Aira caryophyllea L.   Poaceae

LF: Annual herb
SN: -
CU: Forage

Geographic Distribution

Europe		*Australasia*		*Atlantic Islands*	
X	Northern	●	Australia		Cape Verde
N	British Isles	X	New Zealand	N	Canary, Madeira
N	Central, France			N	Azores
N	Southern	*Northern America*		X	South Atlantic Isl.
N	Eastern	X	Canada, Alaska		
N	Mediterranean Isl.		Southeastern USA	*Indian Ocean Islands*	
		X	Western USA		Mascarenes
Africa			Remaining USA		Seychelles
N	Northern	X	Mexico		Madagascar
N	Tropical				
X	Southern	*Southern America*		*Pacific Islands*	
		X	Tropical		Micronesia
Asia			Caribbean		Galapagos
N	Temperate	X	Chile, Argentina	X	Hawaii
	Tropical				Melanesia, Polynesia

Invaded Habitats
Coastal vegetation, heath- and woodland, riparian habitats, wetlands, rock outcrops.

Description
An annual or overwintering grass with few to many tufted thin culms of 3-50 cm height and bright green leaf blades. The glabrous leaves are 0.5-5 cm long and *c.* 0.3 mm wide. Ligules are membranous, toothed, and up to 5 mm long. Spikelets are silvery to purplish, 2.2-3.5 mm long and arranged in panicles of 1-12 cm height and about the same width. Glumes are persistent, 2.2-3.5 mm long and have one to three veins. Lemmas are narrowly ovate and have two teeth. Fruits are caryopses of 1-2 mm length.

Ecology and Control
In the native range, this grass is found in forests on sandy and generally acid soils that are not too nutrient-poor, in heaths, on rocks and dunes. It is adapted to nutrient-poor habitats and grows in dense colonies, eliminating native vegetation and preventing the establishment of native plants.

Small infestations and scattered plants may be hand pulled or dug out. Larger infestations can be sprayed with grass-specific herbicides.

References
215, 428.

Akebia quinata (Houtt.) Decne. Lardizabalaceae

LF: Deciduous liana
SN: -
CU: Ornamental

Geographic Distribution

Europe	*Australasia*	*Atlantic Islands*
Northern	Australia	Cape Verde
British Isles	X New Zealand	Canary, Madeira
X Central, France		Azores
Southern	*Northern America*	South Atlantic Isl.
Eastern	Canada, Alaska	
Mediterranean Isl.	● Southeastern USA	*Indian Ocean Islands*
	Western USA	Mascarenes
Africa	Remaining USA	Seychelles
Northern	Mexico	Madagascar
Tropical		
Southern	*Southern America*	*Pacific Islands*
	Tropical	Micronesia
Asia	Caribbean	Galapagos
N Temperate	Chile, Argentina	Hawaii
Tropical		Melanesia, Polynesia

Invaded Habitats
Forests, forest edges.

Description
A twining, deciduous or half-evergreen shrub climbing up to 10 m or more, with glabrous and long-petioled, pinnately compound leaves, each having five entire and ovate leaflets. These are obovate or elliptic to oblong, 3-6 cm long, and glaucous on the lowerside. Petioles are up to 12 cm long. Flowers appear in axillary and drooping racemes and have slender stalks. Female flowers are purplish brown, 25-30 mm in diameter, and have broad-elliptical sepals. Male flowers are rosy purple, much smaller, and appear on short stalks of *c.* 5 mm length. The fruit is a purplish-violet, flattened pod of 6-8 cm length containing numerous tiny, black seeds.

Ecology and Control
This liana grows either as a climber or as a vigorous ground cover. It grows extremely quickly, e.g. 6-12 m per year and forms dense curtains of intertwined stems that can kill off understorey vegetation and overtop canopy trees. Once established, the dense growth habit of the plant prevents germination and establishment of native species. The liana is shade and drought tolerant and spreads mainly by vegetative growth.

    Manual control of small infestations is achieved by cutting which needs to be repeated several times per year due to the rapid regrowth of the vine. A minimum measure is to cut stems back to the ground at the end of the summer. Large infestations are treated with systemic herbicides containing glyphosate or triclopyr.

References
1095, 1224.

Albizia julibrissin Durazz.

Fabaceae

LF: Deciduous tree
SN: -
CU: Ornamental

Geographic Distribution

Europe	*Australasia*	*Atlantic Islands*
Northern	Australia	Cape Verde
British Isles	New Zealand	Canary, Madeira
Central, France		Azores
Southern	*Northern America*	South Atlantic Isl.
Eastern	Canada, Alaska	
Mediterranean Isl.	● Southeastern USA	*Indian Ocean Islands*
	Western USA	Mascarenes
Africa	Remaining USA	Seychelles
Northern	Mexico	Madagascar
Tropical		
Southern	*Southern America*	*Pacific Islands*
	Tropical	Micronesia
Asia	Caribbean	Galapagos
N Temperate	Chile, Argentina	Hawaii
Tropical		Melanesia, Polynesia

Invaded Habitats
Forest edges, disturbed sites.

Description
A small, spreading and low-branched shrub or tree up to 16 m tall, with a broad crown and a short greyish trunk. Leaves are alternate, bipinnately compound, 10-35 cm long and up to 15 cm wide, with 6-25 pinnae each having 40-60 leaflets. These are oblong, 5-15 mm long and 2-5 mm wide, ciliate and sometimes pubescent on the midrib beneath. They are asymmetric with the middle nerve located along the upper edge. Light pink flowers appear in heads that are clustered at the end of branchlets. Fruits are yellow-green flattened pods up to 15 cm long and 2-3 cm wide. Seeds are *c.* 3.5 mm long and 1.8 mm wide and have a hard seed coat.

Ecology and Control
Where native, this tree grows in scrub and woodland on moist sites. A large seed production and the ability to resprout after damage make it a strong competitor. It forms dense stands that reduce light levels and nutrients and prevent the establishment of native plants. The tree is nitrogen-fixing, usually abundantly nodulated and well adapted to poor soils. It is fast growing, resistant to drought and tolerates moderate frosts.

    Cutting at ground level is most effective at the beginning of flowering. Repeated cutting or herbicide treatment is needed for emerging resprouts. Girdling is effective to kill large trees. Seedlings can be hand pulled but the entire root should be removed. Effective herbicides for treating cut stumps or seedlings and saplings are glyphosate or triclopyr.

References
765, 1446.

Albizia lebbeck (L.) Benth.   Fabaceae

LF: Deciduous tree   CU: Ornamental, shelter
SN: *Acacia lebbeck* (L.) Willd., *Mimosa lebbeck* L.

Geographic Distribution

Europe		*Australasia*		*Atlantic Islands*	
	Northern	N	Australia	X	Cape Verde
	British Isles		New Zealand		Canary, Madeira
	Central, France				Azores
	Southern	*Northern America*			South Atlantic Isl.
	Eastern		Canada, Alaska		
	Mediterranean Isl.	●	Southeastern USA	*Indian Ocean Islands*	
			Western USA		Mascarenes
Africa			Remaining USA		Seychelles
	Northern		Mexico		Madagascar
X	Tropical				
●	Southern	*Southern America*		*Pacific Islands*	
		X	Tropical		Micronesia
Asia		X	Caribbean		Galapagos
	Temperate		Chile, Argentina	X	Hawaii
N	Tropical			X	Melanesia, Polynesia

Invaded Habitats
Tropical hammocks, disturbed sites.

Description
A spreading, unarmed, low-branched tree up to 15 m tall, with a dark-grey and rough but not peeling bark. Young branches are hairy. Leaves are glabrous or pubescent, 15-40 cm long and bipinnately compound with 2-5 pairs of pinnae. Petioles are 7-30 cm long and have a large gland near the base, with additional interjugary glands above. Each pinna bears 10-24 light green, entire leaflets that are pubescent to glabrous beneath, oblong to elliptic, and 2-4 cm long. Whitish or yellow flowers appear on pedicels of 1.5-4.5 mm length and are arranged in axillary globose heads. Fruits are light brown to straw-yellow pods of 15-30 cm length and 3-5 cm width, flat and papery, containing a single row of 8-12 rather large seeds.

Ecology and Control
Native habitats of this nitrogen-fixing tree include sand river banks, savannas, forests and bushy places. When grown solitary, the trunk is rather short; in forests it is long and straight. The tree has a shallow root system with widely spreading roots. It is well adapted to poor soils and tolerates coastal salt spray. The tree produces large amounts of seeds, and the rapidly growing seedlings may reach high densities. It suckers from the roots and once established, it forms dense stands that shade out native vegetation.

Seedlings and saplings may be hand pulled or dug out. Larger trees are cut and the cut stumps treated with herbicides.

References
549, 663, 715, 779.

Alliaria petiolata (Bieb.) Cav. & Grande   Brassicaceae

LF: Biennial herb   CU: Flavour
SN: *A. officinalis* Andrz. ex Bieb.

Geographic Distribution

Europe		*Australasia*	*Atlantic Islands*
N	Northern	Australia	Cape Verde
N	British Isles	New Zealand	Canary, Madeira
N	Central, France		Azores
N	Southern	*Northern America*	South Atlantic Isl.
N	Eastern	● Canada, Alaska	
N	Mediterranean Isl.	● Southeastern USA	*Indian Ocean Islands*
		Western USA	Mascarenes
Africa		● Remaining USA	Seychelles
N	Northern	Mexico	Madagascar
	Tropical		
	Southern	*Southern America*	*Pacific Islands*
		Tropical	Micronesia
Asia		Caribbean	Galapagos
N	Temperate	Chile, Argentina	Hawaii
N	Tropical		Melanesia, Polynesia

Invaded Habitats
Forests and forest edges, grassland, riparian habitats.

Description
A herb ranging in height from 5-190 cm at flowering time, with dark green basal leaves and alternate stem leaves. Basal leaves are 6-10 cm in diameter and have scalloped edges; the triangular to deltoid stem leaves are alternate, sharply toothed and of decreasing size towards the top of the stem. They are 3-8 cm long and wide with pubescent petioles of 1-5 cm length. White flowers are borne in terminal racemes, sometimes in short axillary racemes, and have a diameter of 6-7 mm. Fruits are linear siliques of 2.5-6 cm length and *c.* 2 mm width on short and stout pedicels of *c.* 5 mm length. One fruit contains 10-20 black, cylindrical and ridged seeds of *c.* 3 mm length and 1 mm width.

Ecology and Control
Where native, this shade tolerant, obligate biennial is commonly found in riparian habitats, forests, open woods and is an indicator of nutrient rich soils. It is invasive because it can quickly become dominant in understorey vegetation and eliminates native species. Seedling density may exceed 5,000 plants m^{-2}. The plant accumulates a soil seed bank and populations establish rapidly after disturbance.

    Control is difficult once established. Plants should be removed before seed set. Cutting flowering plants at ground level is effective but plants need to be removed repeatedly over several years to exhaust the soil seed bank. An effective herbicide is glyphosate applied to dormant rosettes in late autumn or early spring.

References
38, 79, 225, 288, 360, 571, 743, 839, 942, 943, 944, 1095, 1248, 1396.

Allium triquetrum L. Liliaceae

LF: Perennial herb
SN: -
CU: Ornamental

Geographic Distribution

Europe	*Australasia*	*Atlantic Islands*
Northern	● Australia	Cape Verde
X British Isles	X New Zealand	N Canary, Madeira
Central, France		N Azores
N Southern	*Northern America*	South Atlantic Isl.
Eastern	Canada, Alaska	
N Mediterranean Isl.	Southeastern USA	*Indian Ocean Islands*
	Western USA	Mascarenes
Africa	Remaining USA	Seychelles
N Northern	Mexico	Madagascar
Tropical		
Southern	*Southern America*	*Pacific Islands*
	Tropical	Micronesia
Asia	Caribbean	Galapagos
Temperate	Chile, Argentina	Hawaii
X Tropical		Melanesia, Polynesia

Invaded Habitats
Coastal sand dunes and beaches, grassland, forests, freshwater wetlands, riparian habitats.

Description
A bulbiferous herb of 10-50 cm height with a few glabrous leaves as long as the shoot and flower shoots that are triangular in section. Leaves are alternate, have flat to somewhat concave blades, are strongly keeled on the lowerside, 30-40 cm long and *c.* 15 mm wide. Inflorescences are umbels with 3-15 white and drooping flowers each, with pedicels of 1-2.5 cm length. The perianth segments have prominent green median stripes. Fruits are green and globular, 3-6 mm in diameter, containing black and oblong seeds of 2-3 mm length. The plant forms somewhat globose bulbs of 1-2 cm diameter with numerous bulblets.

Ecology and Control
In the native range, this plant is frequently found in moist places such as ditches, along watercourses and in damp woodland. The species reproduces both by seed and bulbs which are spread by water and soil movement. In Australia, seeds are also dispersed by ants. It may form monospecific stands due to its dense growth habit, thus replacing native vegetation and preventing establishment of native species. Aerial parts die back in early summer, and regrowth occurs from germinating bulbs in autumn.

    Plants can be dug out but all bulbs must be removed. Large stands can be cut at ground level during early flowering. Herbicides include 2,4-D, dicamba, or picloram in mixture with 2,4-D.

References
121, 215, 924, 986.

Alstroemeria aurea Graham — Liliaceae

LF: Perennial herb
SN: *A. aurantiaca* Don
CU: Ornamental

Geographic Distribution

Europe		*Australasia*		*Atlantic Islands*	
	Northern	●	Australia		Cape Verde
X	British Isles	X	New Zealand		Canary, Madeira
	Central, France				Azores
	Southern		*Northern America*		South Atlantic Isl.
	Eastern		Canada, Alaska		
	Mediterranean Isl.		Southeastern USA		*Indian Ocean Islands*
			Western USA		Mascarenes
Africa			Remaining USA		Seychelles
	Northern		Mexico		Madagascar
	Tropical				
	Southern		*Southern America*		*Pacific Islands*
			Tropical		Micronesia
Asia			Caribbean		Galapagos
	Temperate	N	Chile, Argentina		Hawaii
	Tropical				Melanesia, Polynesia

Invaded Habitats
Forests and forest edges, riparian habitats, disturbed places.

Description
A glabrous herb with tuberous roots, long rhizomes, and many simple, erect stems reaching 1.2 m in height. Leaves are alternate, 7-11 cm long and 5-20 mm wide, elliptic-lanceolate, entire and with a short petiole. Inflorescences are simple or compound umbels at the ends of stems. Each umbel consists of 3-7 rays, each having 1-3 flowers. Flowers have six perianth segments of 4-6 cm length and 1-2 cm width, the outer ones are orange or yellow, broadly ovate, the inner ones narrower, orange or yellow with red spots and streaks. Fruits are ellipsoid, ribbed capsules of 15-30 mm length and 10-18 mm width, containing numerous seeds of 2-3 mm diameter.

Ecology and Control
The plant is a frequent native of Chilean *Nothofagus* open shrubland and forests; it colonizes frequently forest gaps and disturbed sites. It shows considerable variation in size and shape according to environmental conditions. In addition, numerous cultivars have been developed. It is invasive because it forms dense populations with up to 200 shoots m^{-2}, crowding out native vegetation and preventing the establishment of native plants. The plant recovers easily after cutting to ground level by the formation of new sterile shoots.

Plants can be hand pulled or dug out when the soil is moist, all rhizomes must be removed to prevent regrowth. Larger infestations are sprayed with herbicide.

References
86, 215, 1051, 1052.

Alternanthera philoxeroides (Mart.) Griseb.   Amaranthaceae

LF: Aquatic perennial herb
SN: *Buchholzia philoxeroides* Mart.
CU: Ornamental

Geographic Distribution

Europe
- Northern
- British Isles
- Central, France
- Southern
- Eastern
- Mediterranean Isl.

Africa
- Northern
- Tropical
- Southern

Asia
- Temperate
- ● Tropical

Australasia
- ● Australia
- ● New Zealand

Northern America
- Canada, Alaska
- ● Southeastern USA
- X Western USA
- Remaining USA
- Mexico

Southern America
- N Tropical
- X Caribbean
- N Chile, Argentina

Atlantic Islands
- Cape Verde
- Canary, Madeira
- Azores
- South Atlantic Isl.

Indian Ocean Islands
- Mascarenes
- Seychelles
- Madagascar

Pacific Islands
- Micronesia
- Galapagos
- Hawaii
- Melanesia, Polynesia

Invaded Habitats
Freshwater wetlands, lakes, riparian vegetation.

Description
A perennial aquatic or semi-terrestrial stoloniferous herb with much branched, thick and hollow stems rooting at the nodes and reaching several metres in length. Stems have a hairy groove on two opposite sides along the internodes. Leaves are entire, opposite, glabrous, *c.* 9 cm long and 1.5 cm wide, oblong to narrowly obovate, and have short petioles of 1-6 mm length. White flowers are borne in axillary or terminal flowerheads on stalks of *c.* 5 cm length. Bracteoles are shiny and white, glabrous and 4-8 mm long.

Ecology and Control
The plant grows best in eutrophic conditions and can grow either as a floating aquatic plant or terrestrially. It forms dense mats as a result of vegetative growth and proliferates rapidly under favourable conditions. A single plant can cover several square metres, and infestations impair water flow and crowd out native species. The plant is salt-tolerant and can adapt to low light conditions, e.g. up to 12 % of full light.

Mechanical control and removal is difficult as even small stem fragments can produce new plants. The species is resistent to many common herbicides, although dicamba, triclopyr, and bentazone are used to control this plant. Excavation of terrestrial infestations must include removal of all roots to prevent regrowth.

References
23, 121, 170, 581, 645, 646, 647, 757, 804, 924, 986, 1214, 1234, 1415.

Alternanthera pungens Kunth   Amaranthaceae

LF: Perennial herb   CU: -
SN: *Alternanthera repens* (L.) Link., *A. achyrantha* (L.) R.Br., *Achyranthes repens* L.

Geographic Distribution

	Europe		*Australasia*		*Atlantic Islands*
	Northern	●	Australia	X	Cape Verde
	British Isles		New Zealand		Canary, Madeira
X	Central, France				Azores
X	Southern		*Northern America*		South Atlantic Isl.
	Eastern		Canada, Alaska		
	Mediterranean Isl.		Southeastern USA		*Indian Ocean Islands*
		X	Western USA		Mascarenes
	Africa		Remaining USA		Seychelles
X	Northern		Mexico		Madagascar
X	Tropical				
●	Southern		*Southern America*		*Pacific Islands*
		N	Tropical		Micronesia
	Asia		Caribbean		Galapagos
	Temperate	N	Chile, Argentina	X	Hawaii
X	Tropical				Melanesia, Polynesia

Invaded Habitats
Grassland, sandy soils, disturbed places.

Description
A prostrate, creeping perennial with trailing stems up to 60 cm long that are covered with soft and silky hairs. Stems are somewhat woody at the base and root at nodes. Leaves are opposite, 2-4 cm long and 0.5-1.5 cm wide, oval to ovate, glabrous or pubescent, and have short petioles. Flowers are small and inconspicuous, white, surrounded by sharply pointed bracts, and appear in axillary spikes of 5-15 mm length. Fruits are prickly burrs of *c.* 10 mm length. Seeds are yellowish and glabrous, and 1-2 mm in diameter. The plant has a deep woody taproot and is branching at the base.

Ecology and Control
The plant quickly colonizes bare or disturbed ground and once established, it forms dense and persisting infestations that exclude almost all other vegetation and prevent the regeneration of native species. Aboveground stems die back in the autumn and new shoots grow each spring.
     Manual removal can give effective control if the bulk of the taproot is removed, or if roots are cut well below the soil surface. Seedlings can be treated with herbicides containing dicamba. Established plants are treated with amitrole or picloram herbicides, best applied before flowering occurs.

References
985, 986, 1089.

Ammophila arenaria (L.) Link   Poaceae

LF: Perennial herb
CU: Erosion control
SN: *Calamagrostis arenaria* (L.) Roth, *Psamma arenaria* (L.) Roem & Schult.

Geographic Distribution

Europe
- N Northern
- N British Isles
- N Central, France
- N Southern
- N Eastern
- N Mediterranean Isl.

Africa
- N Northern
-    Tropical
-    Southern

Asia
- N Temperate
-    Tropical

Australasia
- ● Australia
- ● New Zealand

Northern America
-    Canada, Alaska
-    Southeastern USA
- ● Western USA
-    Remaining USA
-    Mexico

Southern America
-    Tropical
-    Caribbean
-    Chile, Argentina

Atlantic Islands
-    Cape Verde
-    Canary, Madeira
-    Azores
-    South Atlantic Isl.

Indian Ocean Islands
-    Mascarenes
-    Seychelles
-    Madagascar

Pacific Islands
-    Micronesia
-    Galapagos
- X Hawaii
-    Melanesia, Polynesia

Invaded Habitats
Coastal sand dunes and beaches.

Description
A grass reaching 120 cm in height, and with long horizontally and vertically spreading rhizomes. Aerial shoots appear in dense tufts and are produced mainly along the vertical rhizomes. Leaves are up to 60 cm long and up to 6 mm wide, sharply pointed and usually reflexed at the margins. Ligules are up to 25 mm long and acuminate. Inflorescences are dense, narrowly oblong to lanceolate-oblong panicles of 7-15 cm length. Spikelets contain one floret each and measure 10-16 mm in length. Glumes are whitish and keeled. The lemma is 8-12 mm long, lanceolate, and has two short points at the end and a short awn in between.

Ecology and Control
This grass is an important coastal plant where native. Well adapted to unstable sand, it tolerates burial of up to 1 m per year. It tolerates extreme exposure and drought. The dense clusters of tillers reduce native diversity and accumulate sand; the grass responds by rapid production of elongated stems with subsequent rhizome and root formation. It changes the topography of foredunes from low dunes with gentle slopes into dunes with steep walls. Rhizome fragments may be washed along the shore to form new populations.

Manual control includes repeated digging up of the rhizomes. In the first year, monthly treatment intervals are recommended. An effective herbicide is glyphosate applied as a foliar spray.

References
133, 149, 171, 215, 601, 602, 895, 994, 995, 1194, 1366, 1415.

Amorpha fruticosa L. Fabaceae

LF: Deciduous shrub
SN: *Amorpha virgata* Small.
CU: Ornamental

Geographic Distribution

Europe	*Australasia*	*Atlantic Islands*
Northern	Australia	Cape Verde
British Isles	New Zealand	Canary, Madeira
X Central, France		Azores
● Southern	*Northern America*	South Atlantic Isl.
● Eastern	N Canada, Alaska	
Mediterranean Isl.	N Southeastern USA	*Indian Ocean Islands*
	X Western USA	Mascarenes
Africa	N Remaining USA	Seychelles
Northern	N Mexico	Madagascar
Tropical		
Southern	*Southern America*	*Pacific Islands*
	Tropical	Micronesia
Asia	Caribbean	Galapagos
X Temperate	Chile, Argentina	Hawaii
Tropical		Melanesia, Polynesia

Invaded Habitats

Heath- and shrubland, riparian habitats, reed swamps, coastal estuaries.

Description

A bushy shrub up to 4 m tall, with alternate, pinnately compound leaves of 10-30 cm length. Leaflets number 9-35 per leaf and are oval, 1-5 cm long and 0.5-3 cm wide, and often densely pubescent. Violet-purple to deep blue flowers of 8-10 mm length are borne in dense elongated racemes up to 20 cm long, produced at the ends of shoots. Anthers are yellow. Fruits are short, indehiscent, somewhat curved pods of 7-9 mm length, mostly containing one seed. Seeds are *c.* 5 mm long and *c.* 1.8 mm wide.

Ecology and Control

Where native, the species is found in riparian habitats, open rich woods, and in floodplain forests. Where invasive, it forms extensive stands due to its dense growth habit, altering the vegetation structure and replacing native species. Seeds have an impermeable seed coat and show a high percentage of dormant seeds. The species produces retenone, a natural insecticide, which may explain the absence of phytophygous insects on this shrub in the introduced range. The shrub is nitrogen-fixing due to root nodulation. Seeds are dispersed by streams.

Specific control methods need to be developed. Seedlings and saplings may be hand pulled or dug out. Larger plants are cut and the cut stumps treated with herbicide.

References

456, 1268, 1453.

Ampelopsis brevipedunculata (Maxim.) Trautv.   Vitaceae

LF: Deciduous vine   CU: Ornamental
SN: *Ampelopsis heterophylla* (Thunb.) Sieb., *Vitis brevipedunculata* (Maxim.) Dippel

Geographic Distribution

Europe
   Northern
   British Isles
   Central, France
   Southern
   Eastern
   Mediterranean Isl.

Africa
   Northern
   Tropical
   Southern

Asia
 N Temperate
   Tropical

Australasia
   Australia
   New Zealand

Northern America
   Canada, Alaska
● Southeastern USA
● Western USA
   Remaining USA
   Mexico

Southern America
   Tropical
   Caribbean
   Chile, Argentina

Atlantic Islands
   Cape Verde
   Canary, Madeira
   Azores
   South Atlantic Isl.

Indian Ocean Islands
   Mascarenes
   Seychelles
   Madagascar

Pacific Islands
   Micronesia
   Galapagos
   Hawaii
   Melanesia, Polynesia

Invaded Habitats
Forest edges and gaps, riparian habitats, shorelines.

Description
A climbing vine reaching 6 m height or more, with alternate leaves and climbing by means of tendrils that grow opposite the leaves. The simple leaves are heart-shaped and dark green. Leaf blades range 3-12 cm in length and 2-8 cm in width and have coarsely toothed margins, petioles of 2-6 cm length, and hairy veins. Flowers are rather inconspicuous, small and greenish-white, 1-2 mm in diameter, and are borne in cymose inflorescences. Fruits are small berries of *c.* 5 mm diameter, ranging in colours from bright blue to yellow or purplish. Each berry contains 3-4 smooth, triangular-ovoid seeds of *c.* 3.5 mm length.

Ecology and Control
Once established, the vine grows and spreads quickly, overwhelming and shading out smaller plants and outcompeting them for resources. It smothers shrubs and small trees, impeding their growth and regeneration. Seeds are dispersed by birds, mammals and water.
    Manual control includes pulling the vines from trees and ground before fruit set, thus preventing seed dispersal. Hand-pruning in the autumn or spring will prevent the formation of flower buds in the following season. Repeated cutting or mowing reduces vigour. Effective herbicides are triclopyr or glyphosate. A basal bark application can be used to kill larger vines.

References
448, 1095.

Andropogon gayanus Kunth
Poaceae

LF: Perennial herb
SN: -
CU: Forage, erosion control

Geographic Distribution

Europe	Australasia	Atlantic Islands
Northern	● Australia	N Cape Verde
British Isles	New Zealand	Canary, Madeira
Central, France		Azores
Southern	*Northern America*	South Atlantic Isl.
Eastern	Canada, Alaska	
Mediterranean Isl.	Southeastern USA	*Indian Ocean Islands*
	Western USA	Mascarenes
Africa	Remaining USA	Seychelles
Northern	Mexico	Madagascar
N Tropical		
N Southern	*Southern America*	*Pacific Islands*
	● Tropical	Micronesia
Asia	Caribbean	Galapagos
Temperate	Chile, Argentina	Hawaii
Tropical		Melanesia, Polynesia

Invaded Habitats
Grassland, savanna.

Description
A large tufted rhizomatous grass forming dense tussocks up to 1 m in diameter and with culms 50-300 cm tall and 4-10 mm wide. The linear-lanceolate and acute leaf blades reach 40 cm in length and 15 mm in width, and are pubescent on both sides. The midrib has a cream to white keel and forms a sharp point at the end of the leaf. Ligules consist of a fringe of white hairs up to 7 mm long. Rhizomes have a diameter of *c.* 5 mm and very short branches, forming a compact mass. Inflorescences are panicles with racemes either on pedicels or almost sessile, each containg *c.* 16 pairs of spikelets. Awns measure 1-2 cm in length. Caryopses are *c.* 3 mm long and 0.8 mm wide.

Ecology and Control
In the native range, this drought and fire tolerant grass occurs frequently in shady places of savannas and open woodland, and four varieties (var. *bisquamulatus*, var. *gayanus*, var. *squamulatus*, var. *tridentatus*) are distinguished. The grass grows on a wide range of soils including those of low fertility. It is invasive because it forms dense and large tussocks, displacing native vegetation. The grass produces many fruits, and seeds remain viable for 2-3 years.

Burning removes both dead and living leaves, but the rhizomes and roots will soon resprout after a fire. The species is tolerant of moderate grazing. Smaller patches can be sprayed with grass-specific herbicides. Individual tufts may be hand pulled or dug out.

References
144, 145, 146, 1033, 1190.

Andropogon virginicus L. — Poaceae

LF: Perennial herb
SN: -
CU: -

Geographic Distribution

Europe	*Australasia*	*Atlantic Islands*
Northern	● Australia	Cape Verde
British Isles	X New Zealand	Canary, Madeira
Central, France		Azores
Southern	*Northern America*	South Atlantic Isl.
Eastern	N Canada, Alaska	
Mediterranean Isl.	N Southeastern USA	*Indian Ocean Islands*
	X Western USA	Mascarenes
Africa	N Remaining USA	Seychelles
Northern	N Mexico	Madagascar
X Tropical		
Southern	*Southern America*	*Pacific Islands*
	N Tropical	Micronesia
Asia	N Caribbean	Galapagos
Temperate	Chile, Argentina	● Hawaii
Tropical		Melanesia, Polynesia

Invaded Habitats
Grassland, scrubland, woodland.

Description
A perennial bunchgrass of 50-100 cm height, with branching, light-green to reddish brown stems. Leaf-sheaths have long hairs on the margins and a tuberculate surface. Ligules are yellow to brownish and membranous. Leaf blades reach 40 cm in length and are 2-5 mm wide, rough and have spathes of 3-5 cm length. Inflorescences are racemes of 2-4 cm length containing spikelets of 3-4 mm length. Flowers are either sessile and bisexual or stalked and male.

Ecology and Control
In the native range, this grass occurs mainly in prairies and is an indicator of acid soils. The grass invades extremely nutrient poor soils in Australia, burnt areas and grassland and has a low requirement for phosporus. It is also found on more fertile soils but abundance will decrease if competition from other species becomes too strong. On Hawaii, it forms dense stands in bogs, open mesic and dry habitats. The species is highly flammable due to accumulation of dead plant material and thus affects fire regimes by increasing fire intensity as well as the acreage burnt. The species is fire-stimulated, and its cover increases with each fire. Seeds are dispersed by wind.

    Chemical control methods include spraying bromacil, hexazinone, tebuthiuron, or a mixture of bromacil and diuron. Individual plants can be hand pulled or dug out.

References
284, 500, 600, 677, 908, 986, 1010, 1080, 1199.

Annona glabra L.   Annonaceae

LF: Semi-evergreen tree, shrub   CU: -
SN: *Annona palustris* L.

Geographic Distribution

Europe	*Australasia*	*Atlantic Islands*
Northern	● Australia	Cape Verde
British Isles	New Zealand	Canary, Madeira
Central, France		Azores
Southern	*Northern America*	South Atlantic Isl.
Eastern	Canada, Alaska	
Mediterranean Isl.	N Southeastern USA	*Indian Ocean Islands*
	Western USA	Mascarenes
Africa	Remaining USA	Seychelles
Northern	N Mexico	Madagascar
N Tropical		
Southern	*Southern America*	*Pacific Islands*
	N Tropical	Micronesia
Asia	N Caribbean	Galapagos
Temperate	Chile, Argentina	Hawaii
X Tropical		Melanesia, Polynesia

Invaded Habitats
Swampy places of lowland rainforests, mangrove swamps.

Description
A late-deciduous to evergreen small tree or shrub up to 16 m height, densely branched and with a taproot. Trunks are frequently buttressed at the base. Leaves are alternate, simple and entire, shiny green and ovate to elliptic. The leaf blade has a length of 5-15 cm and a width of 6-8 cm, with a broadly cuneate to rounded base. Leaves have petioles of 10-20 mm length. Solitary flowers of *c.* 25 mm diameter are borne on short stalks in the leaf axils of new shoots. Petals are cream-white to pale-yellow, six in number, the outermost being 16-30 mm in length. Inner petals have deep purple bases. Stamens are linear and 3-4 mm long. Sepals are glabrous and 5-6 mm long. The fruit is a fleshy syncarp of 5-13 cm length and 7-8 cm width, dull yellow with brown spots, smooth and ovoid, containing several seeds. These are ellipsoid to obovoid, 10-15 mm long and *c.* 8 mm wide, and have a short wing.

Ecology and Control
Where native, the shrub commonly grows in swamp forests of coastal regions and on riverbanks, edges of freshwater ponds and in wet hammocks. The tree is well adapted to standing waters and floods. It is invasive because it forms dense thickets. The dense foliage reduces light levels and shades out native shrubs and trees by preventing their establishment and growth. Species richness is reduced in stands of this tree.
    Specific control methods for this species are not available. Seedlings and saplings can be hand pulled, larger trees cut and the stumps treated with herbicide.

References
607, 608, 1462.

Anredera cordifolia (Ten.) Steenis — Basellaceae

LF: Perennial vine
SN: *Boussingaultia cordifolia* Ten.
CU: Ornamental

Geographic Distribution

Europe		*Australasia*		*Atlantic Islands*	
	Northern	●	Australia		Cape Verde
	British Isles	●	New Zealand	X	Canary, Madeira
	Central, France			X	Azores
X	Southern	*Northern America*			South Atlantic Isl.
	Eastern		Canada, Alaska		
	Mediterranean Isl.		Southeastern USA	*Indian Ocean Islands*	
		X	Western USA		Mascarenes
Africa			Remaining USA		Seychelles
	Northern		Mexico		Madagascar
	Tropical				
X	Southern	*Southern America*		*Pacific Islands*	
		N	Tropical		Micronesia
Asia			Caribbean		Galapagos
X	Temperate	N	Chile, Argentina	X	Hawaii
	Tropical				Melanesia, Polynesia

Invaded Habitats
Coastal sand dunes and beaches, forests, riparian habitats.

Description
A slender vine with tuber-bearing rhizomes and glabrous stems of 3-6 m length, and with additional small, axillary stem tubers. The alternate leaves are simple and entire, ovate to lanceolate, petioled and 1-10 cm long. Inflorescences are branching racemes, up to 30 cm long and have many white and fragrant flowers of *c.* 6 mm diameter on short stalks of *c.* 2 mm length. Sepals are two in number and petals five. Fruits are indehiscent globose nutlets, *c.* 1 mm in diameter, and crowned by the enlarged style base.

Ecology and Control
A shade tolerant vine reproducing mostly from the numerous aerial tubers. These fall to the ground and can remain viable for more than 2 years. In California, no fruit set has been observed. The vine is invasive because it climbs into the canopies of trees and shrubs, and smothers them completely. It forms dense infestations of tangled stems and trees can be crushed by the weight of the vines and tubers, as a result of which light enters the gap and breaks dormancy of the tubers in or on the soil. By time, invaded forests become degraded and displaced with vine growth, preventing the establishment of native trees.

Manual removal must include the removal of all tubers. Effective chemical control is done by applying glyphosate or fluoroxypyr. Repeated applications are necessary to achieve long-term control.

References
121, 215, 607, 608, 924, 1045, 1260, 1415.

Anthoxanthum odoratum L.   Poaceae

LF: Perennial herb   CU: Essential oils
SN: -

Geographic Distribution

Europe		*Australasia*		*Atlantic Islands*	
N	Northern	●	Australia		Cape Verde
N	British Isles	X	New Zealand	N	Canary, Madeira
N	Central, France			N	Azores
N	Southern	*Northern America*			South Atlantic Isl.
N	Eastern		Canada, Alaska		
N	Mediterranean Isl.		Southeastern USA	*Indian Ocean Islands*	
		X	Western USA	●	Mascarenes
Africa			Remaining USA		Seychelles
X	Northern Tropical		Mexico		Madagascar
X	Southern	*Southern America*		*Pacific Islands*	
		X	Tropical	●	Micronesia
Asia		X	Caribbean		Galapagos
N	Temperate Tropical	●	Chile, Argentina	●	Hawaii
					Melanesia, Polynesia

Invaded Habitats
Coastal dunes, grass- and heathland, forests, freshwater wetlands, riparian habitats.

Description
A tufted rhizomatous grass with a strong smell of coumarin. The unbranched and stiff culms are 10-100 cm tall and erect to spreading. The glabrous or loosely hairy leaves are variable in size, 1-12 cm long, 1.5-7 mm wide, linear and finely pointed at the end. The membranous ligules are 1-5 mm long. Inflorescences are dense green to purplish panicles of 1-12 cm length and 6-15 mm width. Spikelets are 6-10 mm long, lanceolate and compressed, containing three florets; they fall entirely at maturity. Glumes are persistent and keeled, finely pointed at the apex and loosely hairy. Caryopses are ellipsoid and *c.* 2 mm long.

Ecology and Control
In the native range, this grass is frequently found in poorer soils of dry fields, meadows, forest edges and disturbed places. Where introduced, it invades primarily nutrient poor soils and is common on soils with a low phosphorus content. The species is highly competitive to other grass species because it grows rapidly and flowers earlier than native species. The grass reproduces by seeds and does not spread vegetatively. A single plant may produce >1,000 seeds per year.
    Manual control includes hand pulling of individual plants and mowing. Mowing should be carried out early in the season, before seeds are ripened. The herbicide dalapon has proved to be effective for control of this grass.

References
121, 215, 665, 924, 1127, 1278.

Aponogeton distachyos L.f.   Aponogetonaceae

LF: Aquatic perennial herb   CU: Ornamental, food
SN: -

Geographic Distribution

	Europe		*Australasia*		*Atlantic Islands*
	Northern	●	Australia		Cape Verde
X	British Isles	X	New Zealand		Canary, Madeira
X	Central, France				Azores
	Southern		*Northern America*		
	Eastern		Canada, Alaska		South Atlantic Isl.
	Mediterranean Isl.		Southeastern USA		*Indian Ocean Islands*
		X	Western USA		Mascarenes
	Africa		Remaining USA		Seychelles
	Northern		Mexico		Madagascar
	Tropical				
N	Southern		*Southern America*		*Pacific Islands*
			Tropical		Micronesia
	Asia		Caribbean		Galapagos
	Temperate		Chile, Argentina		Hawaii
	Tropical				Melanesia, Polynesia

Invaded Habitats
Freshwater wetlands, ponds, riparian habitats.

Description
An aquatic herb with a large and tuberous rootstock. Leaves are all basal and emerge from the end of the rootstock. The petioles are long and the oblong-elliptic leaf blades float on the water surface; they are 6-25 cm long and 2-7 cm wide. Inflorescences are showy and forked spikes of *c.* 6 cm length that appear at the ends of emergent peduncles. Each arm of the fork bears 12-14 white to pinkish flowers with perianth segments up to 15 mm long. Anthers are dark purplish. The spikes appear close to the water surface and have a deciduous spathe at the base. The fruit is a group of follicles. Seeds are smooth, irregular in shape, and up to 1 cm long.

Ecology and Control
In the native range, this plant occurs in slow streams, shallow water, and ephemeral lakes. If the habitats dry up in summer, the plants die back aboveground but regrow from the rhizomes in spring. Where invasive, the plant forms a dense cover over the water body affecting its environmental conditions and native diversity. The species reproduces both by seeds and tubers. Ripe fruits become detached and float on the water for a while before seeds are released. The plant is a prolific seed producer, and seedlings may reach high densities.

    Specific control methods for this species are not available. Small patches may be removed by pulling plants out of the ground, the rootstocks must be removed to prevent regrowth.

References
215, 1003.

Araujia sericifera Brot.　　　　　　　　　　　　　　　Asclepiadaceae

LF: Evergreen climber　　　　CU: Ornamental
SN: -

Geographic Distribution

Europe		*Australasia*		*Atlantic Islands*	
	Northern	●	Australia		Cape Verde
	British Isles	●	New Zealand		Canary, Madeira
	Central, France			X	Azores
●	Southern		*Northern America*		South Atlantic Isl.
	Eastern		Canada, Alaska		
X	Mediterranean Isl.		Southeastern USA		*Indian Ocean Islands*
		X	Western USA		Mascarenes
Africa			Remaining USA		Seychelles
	Northern		Mexico		Madagascar
	Tropical				
●	Southern		*Southern America*		*Pacific Islands*
		N	Tropical		Micronesia
Asia			Caribbean		Galapagos
	Temperate	X	Chile, Argentina		Hawaii
	Tropical				Melanesia, Polynesia

Invaded Habitats
Riparian habitats, woodland, rainforest margins, disturbed sites.

Description
A somewhat woody, climbing, evergreen vine reaching >10 m length and containing a milky juice. The opposite leaves are ovate-oblong, dark green and glabrous above, pale green and hairy below, 5-10 cm long and 2-3 cm wide. The bell- or funnel-shaped flowers are white, pale pink or creamish and have corollas of 2-3 cm diameter. Fruits are deeply grooved follicles, green if young, up to 12 cm long and 6 cm wide. They contain numerous black seeds of 7-8 mm length, each with a tuft of silky hairs of *c.* 25 mm length.

Ecology and Control
This vigorously growing vine is invasive because it has a dense foliage and smothers native shrubs and trees. Dense infestations prevent regeneration of native overstorey species. The heavy weight of fruiting vines can break branches of trees. The vine produces large quantities of wind dispersed seeds.

Seedlings and smaller plants can be hand pulled or dug out, roots should be removed to prevent regrowth. Larger stems are cut at ground level, and the cut stumps treated with herbicide. Large infestations may be foliar sprayed. A follow-up programme is necessary to control regrowth and seedlings.

References
121, 924, 1415.

Arctotheca calendula (L.) Levyns   Asteraceae

LF: Annual, perennial herb   CU: Ornamental
SN: *Arctotis calendula* L., *Cryptostemma calendulaceum* (L.) R. Br.

Geographic Distribution

Europe		*Australasia*		*Atlantic Islands*	
	Northern	●	Australia		Cape Verde
X	British Isles	X	New Zealand		Canary, Madeira
	Central, France			X	Azores
●	Southern	*Northern America*			South Atlantic Isl.
	Eastern		Canada, Alaska		
X	Mediterranean Isl.		Southeastern USA	*Indian Ocean Islands*	
		X	Western USA		Mascarenes
Africa			Remaining USA		Seychelles
	Northern		Mexico		Madagascar
	Tropical				
N	Southern	*Southern America*		*Pacific Islands*	
			Tropical		Micronesia
Asia			Caribbean		Galapagos
	Temperate		Chile, Argentina		Hawaii
	Tropical				Melanesia, Polynesia

Invaded Habitats
Coastal sand dunes and beaches, grass- and heathland, forests.

Description
An annual or perennial herb up to 40 cm tall, with pinnately lobed leaves of 5-25 cm length. Leaves are pubescent above and densely hairy to white-woolly underneath. Flowerheads are *c.* 6 cm in diameter and are borne solitary on short stalks that elongate during ripening of the fruits to a length of 15-20 cm. Flowerheads contain less than 20 ray florets and numerous bisexual disk florets. Ligules are pale yellow on the upperside and purplish on the lowerside. Achenes are densely covered with long hairs and have a pappus of distinct scales. Without hairs, achenes measure 1-1.5 mm in length.

Ecology and Control
In the native range, this plant is found on coastal dunes and on rich soils off the coast. The plant grows rapidly and spreads vigorously by creeping stems that root at the nodes and may reach 2 m in length. The sprawling growth habit and vegetative growth leads to the formation of impenetrable mats that eliminate native vegetation. Stem fragments with nodes easily root if left in the soil. Seeds are wind dispersed. Plants are annual in Britain, Spain, and South Africa.

    If removed manually, all crowns and roots should be removed as intact as possible. Covering infestations with light-impenetrable fabric is effective but needs at least 1.5 years to cause sufficient mortality. Chemical control of large infestations is done by repeated application of glyphosate.

References
215, 364, 427, 557, 558, 1434.

Ardisia crenata Sims — Myrsinaceae

LF: Evergreen shrub
SN: *A. crenulata* Vent.
CU: Ornamental

Geographic Distribution

Europe	*Australasia*	*Atlantic Islands*
Northern	Australia	Cape Verde
British Isles	New Zealand	Canary, Madeira
Central, France		Azores
Southern	*Northern America*	South Atlantic Isl.
Eastern	Canada, Alaska	
Mediterranean Isl.	● Southeastern USA	*Indian Ocean Islands*
	Western USA	● Mascarenes
Africa	Remaining USA	● Seychelles
Northern	Mexico	Madagascar
Tropical		
● Southern	*Southern America*	*Pacific Islands*
	Tropical	Micronesia
Asia	Caribbean	Galapagos
N Temperate	Chile, Argentina	X Hawaii
N Tropical		Melanesia, Polynesia

Invaded Habitats
Forests, tropical hammocks, riverbanks.

Description
A low-growing, evergreen shrub up to 2 m tall and with reddish, glandular or papillate branchlets. The simple and alternate leaves have petioles of 6-10 mm length, are elliptic-lanceolate to narrowly lanceolate, dark green, 6-20 cm long and 2-4 cm wide. Leaf margins are entire but somewhat undulate. 12-18 lateral veins are present on each side of the midrib. Small white or pink flowers of 4-8 mm diameter are borne in umbels or cymose inflorescences at the ends of lateral branches. Sepals are *c.* 1.5 mm long and oblong-ovate. The fruit is a globose and red drupe of 5-10 mm diameter, containing one seed.

Ecology and Control
A very variable species. In the native range, the shrub is commonly found in closed forests, valleys, and dark damp places from 100-2,400 m elevation. Where invasive, it becomes dominant in the understorey and forms dense stands with >100 plants m^{-2}, reducing light and native species richness, and preventing the growth of native tree seedlings. Mature plants are often surrounded by a dense cover of their own seedlings. The plant grows best in moist soils and vigorously resprouts after cutting or fire. It produces fruits abundantly and seeds are dispersed by birds and small mammals.

Seedlings and saplings can be hand pulled if soil disturbance is not a problem. Glyphosate can be used to spray dense seedling populations. Larger trees are cut and the stumps treated with herbicide, or treated with a basal bark application of triclopyr.

References
715, 770, 1188, 1239.

Ardisia elliptica Thunb. — Myrsinaceae

LF: Evergreen shrub, tree
SN: *A. humilis* Vahl.
CU: Ornamental

Geographic Distribution

Europe	*Australasia*	*Atlantic Islands*
Northern	Australia	Cape Verde
British Isles	New Zealand	Canary, Madeira
Central, France		Azores
Southern	*Northern America*	South Atlantic Isl.
Eastern	Canada, Alaska	
Mediterranean Isl.	● Southeastern USA	*Indian Ocean Islands*
	Western USA	● Mascarenes
Africa	Remaining USA	● Seychelles
Northern	Mexico	Madagascar
Tropical		
Southern	*Southern America*	*Pacific Islands*
	Tropical	Micronesia
Asia	X Caribbean	Galapagos
N Temperate	Chile, Argentina	● Hawaii
N Tropical		Melanesia, Polynesia

Invaded Habitats
Freshwater wetlands, marshes and swamps, tropical hammocks, forest edges.

Description
A glabrous shrub of 1-2 m height and with angular, longitudinally ridged branchlets. Leaves are alternate, simple, and have petioles of 5-10 mm length. They are elliptic to oblong, 6-15 cm long and 3-6 wide, and have a dull green and densely punctate lower surface. There are 12-34 lateral veins on each side of the midrib. White or pinkish flowers of 6-8 mm diameter are borne in clusters in leaf axils. The densely black dotted pedicels are 1-2 cm long. Fruits are almost globose drupes, red at first and becoming black, 5-8 mm in diameter, and contain one large and round seed each.

Ecology and Control
In the native range, this rapidly growing shrub occurs in disturbed places, edges of fields, scrub, and along sandy and muddy coasts. The shrub is shade-tolerant and forms dense monotypic stands, preventing establishment and regeneration of all other species due to the shade casting dense foliage. Frugivorous birds are the principal dispersal agents of seeds. The shrub is probably not resistant to fire.

In areas where the soil is already disturbed, seedlings can be hand pulled. Infestations consisting of a dense cover of seedlings can be sprayed with a glyphosate herbicide. Mature plants are treated with a basal application of a triclopyr herbicide mixed with an oil diluent.

References
715, 1239.

Argemone ochroleuca Sweet — Papaveraceae

LF: Annual herb
CU: -
SN: *A. mexicana* var. *ochroleuca* L., *A. subfusiformis* Ownbey

Geographic Distribution

Europe
- Northern
- British Isles
- Central, France
- Southern
- Eastern
- Mediterranean Isl.

Africa
- X Northern
- ● Tropical
- ● Southern

Asia
- Temperate
- X Tropical

Australasia
- X Australia
- X New Zealand

Northern America
- Canada, Alaska
- Southeastern USA
- Western USA
- Remaining USA
- N Mexico

Southern America
- N Tropical
- Caribbean
- Chile, Argentina

Atlantic Islands
- X Cape Verde
- X Canary, Madeira
- Azores
- South Atlantic Isl.

Indian Ocean Islands
- X Mascarenes
- Seychelles
- Madagascar

Pacific Islands
- Micronesia
- Galapagos
- Hawaii
- Melanesia, Polynesia

Invaded Habitats
Grassland, stream beds and alluvial flats of ephemeral streams, disturbed places.

Description
A large and sparingly prickly annual herb with 30-100 cm tall stems. Stems are covered with stiff yellowish prickles. Stem leaves are alternate and sessile, basal leaves have short stalks. Leaves are variable in size, 6-20 cm long and 3-8cm wide, deeply lobed with 7-12 lobes, and have prickly margins. On the lower leaf surface, veins are prickly, on the upper surface usually unarmed. The large flowers are 3-7 cm across and have lemon yellow, rarely dark yellow petals. The fruit is an ellipsoid and prickly capsule of 25-50 mm length and *c.* 20 mm width (exclusive prickles), containing dark brown to black and globular seeds of *c.* 1.5 mm diameter.

Ecology and Control
In the native range, this plant occurs frequently in disturbed sites. Where invasive, it forms large and dense populations that affect wildlife movement and crowd out native vegetation. Seeds are dormant and have an after-ripening period of a few weeks to several months. A single plant can produce 4,000-30,000 seeds per year. Seedlings are poor competitors if perennial species are present, and establishment of this plant takes place on disturbed sites and bare ground.

Control should aim at preventing seed formation. Plants can hand pulled, grubbed or cut before fruits ripen. Seedlings can be sprayed or mowed. Effective herbicides are 2,4-D ester, 2,4-D amine, or glyphosate.

References
549, 580, 986, 1089.

Arrhenatherum elatius (L.) Beauv.   Poaceae

LF: Perennial herb   CU: Erosion control, forage
SN: -

Geographic Distribution

Europe		*Australasia*		*Atlantic Islands*	
N	Northern	X	Australia		Cape Verde
N	British Isles	●	New Zealand	N	Canary, Madeira
N	Central, France			N	Azores
N	Southern	*Northern America*			South Atlantic Isl.
N	Eastern		Canada, Alaska		
N	Mediterranean Isl.		Southeastern USA	*Indian Ocean Islands*	
		●	Western USA	X	Mascarenes
Africa			Remaining USA		Seychelles
N	Northern		Mexico		Madagascar
	Tropical				
X	Southern	*Southern America*		*Pacific Islands*	
		X	Tropical		Micronesia
Asia			Caribbean		Galapagos
N	Temperate	X	Chile, Argentina	X	Hawaii
X	Tropical				Melanesia, Polynesia

Invaded Habitats
Grassland, prairies, disturbed sites.

Description
A loosely tufted grass with yellowish roots and culms of 50-180 cm height that are erect or slightly spreading. The green and linear leaves are 10-40 cm long and 4-10 mm wide, and finely pointed at the end. Sheaths are rounded on the back, and the membranous ligules are 1-3 mm long. The inflorescence is a green to purplish and lanceolate to oblong panicle of 10-30 cm length, with clustered branches and pedicels of 1-10 mm length. Spikelets usually have two florets and are 7-11 mm long. Spikelets fall entirely at maturity. Glumes are membranous and finely pointed at the end. Fruits are ellipsoid, hairy caryopses of *c.* 1.5 mm length.

Ecology and Control
This grass is frequent in lightly grazed or mown grasslands in the native range, and occurs also on gravel banks. It mostly grows on soils of moderate to high fertility. Vegetative regeneration occurs by regrowth from the lowest nodes. In var. *bulbosum*, the lowest stem internodes form bulbs that can regenerate the plant. This variety is difficult to eradicate due to dispersal of bulbs. Seeds germinate readily as soon as they are ripe. The grass competes with native forbs and grasses and forms dense swards that displace the original vegetation.

Repeated cutting reduces vitality. In the western USA, a late spring mowing with removal of cut material is recommended over a period of at least 3 years. Smaller patches can be sprayed with grass-selective or non-selective herbicides.

References
507, 1013, 1277, 1415, 1424.

Arundo donax L. Poaceae

LF: Perennial herb
SN: -
CU: Ornamental, fibre

Geographic Distribution

Europe	*Australasia*	*Atlantic Islands*
Northern	● Australia	X Cape Verde
X British Isles	● New Zealand	X Canary, Madeira
Central, France		● Azores
● Southern	*Northern America*	South Atlantic Isl.
N Eastern	Canada, Alaska	
Mediterranean Isl.	Southeastern USA	*Indian Ocean Islands*
	● Western USA	Mascarenes
Africa	Remaining USA	Seychelles
X Northern	X Mexico	Madagascar
Tropical		
● Southern	*Southern America*	*Pacific Islands*
	Tropical	Micronesia
Asia	X Caribbean	Galapagos
N Temperate	X Chile, Argentina	X Hawaii
N Tropical		Melanesia, Polynesia

Invaded Habitats
Floodplains, riparian habitats, damp places.

Description
A large grass ranging in height from 2-9 m and growing in many-stemmed tussocks. Stems are somewhat woody and rarely branching, glabrous, tough and hollow, and 1-4 cm in diameter. The grass spreads from horizontal rootstocks. Leaves are alternate, pale green to bluish-green, up to 70 cm long and 8 cm wide, and have large basal ear lobes. The leaf sheaths are smooth, glabrous, and persistent. Inflorescences are compact, cream to brown, and 30-60 cm long. Spikelets have a length of 12-18 mm and contain 4-5 florets.

Ecology and Control
This grass grows best in moist soils and spreads mainly vegetatively by stem and rhizome fragments. In North America, no viable achenes are formed. Even small rhizome fragments can regrow and form new plants; rhizome fragments are carried by rivers and streams. Dead shoots are highly flammable and the grass resprouts quickly after burning. The plant forms species poor clones that may cover hundreds of acres and displace native vegetation and exclude associated wildlife species.

Smaller plants are hand pulled or dug out, but all rhizomes must be removed. Cutting the stems close to the ground or burning does not kill the rhizome system. Effective herbicides are glyphosate or fluaziprop, applied as a foliar spray or cut stem treatment. The best time of application is after flowering.

References
101, 133, 423, 549, 1089, 1415.

Asparagus asparagoides (L.) Druce   Liliaceae

LF: Perennial herb   CU: Ornamental
SN: *Myrsiphyllum asparagoides* (L.) Willd., *M. scandens* (Thunb.) Oberm.

Geographic Distribution

Europe	*Australasia*	*Atlantic Islands*
Northern	● Australia	Cape Verde
British Isles	● New Zealand	X Canary, Madeira
Central, France		X Azores
X Southern	*Northern America*	South Atlantic Isl.
Eastern	Canada, Alaska	
X Mediterranean Isl.	Southeastern USA	*Indian Ocean Islands*
	X Western USA	Mascarenes
Africa	Remaining USA	Seychelles
Northern	Mexico	Madagascar
Tropical		
N Southern	*Southern America*	*Pacific Islands*
	Tropical	Micronesia
Asia	Caribbean	Galapagos
Temperate	Chile, Argentina	Hawaii
Tropical		Melanesia, Polynesia

Invaded Habitats
Woodland, grass- and heathland, riparian habitats, coastal dunes, rock outcrops.

Description
A somewhat woody erect or climbing herb reaching 3 m in height, with slender, twining and branching stems. Leaves are reduced to small bract-like scales, instead, the plant has alternate, broadly ovate and glossy green cladodes of 1-7 cm length and 8-30 mm width. Greenish white flowers of 8-9 mm diameter appear solitary or in clusters in the axils of leaves, with flower stalks being bent. Fruits are red and sticky, globose berries of 6-10 mm diameter. The black and shiny seeds are ovoid to globose and 3-4 mm in diameter. The plant has a short and thick rhizome.

Ecology and Control
This shade tolerant plant competes with native species by forming thick root mats and dense canopies. It eliminates all native herbs and shrubs and prevents overstorey regeneration. The species produces tubers and rhizomes abundantly. Seedling growth is slow until the root system is well developed, but stems of established plants elongate rapidly. Seeds seem to be dispersed by birds and streams.

    Mechanical control is done by carefully digging out single plants and destroying them; all parts of the tuberous root system must be removed to prevent regeneration. Control by herbicides include spotspraying with glyphosate, leading to eradication if applied in several consecutive years. Other herbicides used are paraquat or metsulfuron-methyl.

References
121, 213, 215, 343, 484, 607, 986, 1014, 1032, 1047, 1415.

Asparagus densiflorus (Kunth) Jessop — Liliaceae

LF: Evergreen shrub CU: Ornamental
SN: *A. aethiopicus* L., *A. sprengeri* Regel, *Protasparagus densiflorus* (Kunth) Oberm.

Geographic Distribution

Europe	*Australasia*	*Atlantic Islands*
Northern	● Australia	Cape Verde
British Isles	X New Zealand	Canary, Madeira
Central, France		Azores
Southern	*Northern America*	South Atlantic Isl.
Eastern	Canada, Alaska	
Mediterranean Isl.	● Southeastern USA	*Indian Ocean Islands*
	Western USA	Mascarenes
Africa	Remaining USA	Seychelles
Northern	Mexico	Madagascar
Tropical		
N Southern	*Southern America*	*Pacific Islands*
	Tropical	Micronesia
Asia	X Caribbean	Galapagos
Temperate	Chile, Argentina	Hawaii
Tropical		Melanesia, Polynesia

Invaded Habitats
Coastal sand dunes and beaches, coastal cliffs and scrub, forests, riparian habitats.

Description
A sprawling, stiff shrub up to 2 m tall, with spines and cladodes instead of true leaves. The axillary spines reach 5-10 mm in length, cladodes are borne in axillary clusters of 2-5, are flattened, linear, 15-25 mm long and 2-3 mm wide. The inflorescence is a raceme of 4-10 cm length and has numerous flowers on stalks of 5-9 mm length. The pale pinkish white flowers are 4-6 mm in diameter and have orange anthers. The fruit is a red berry of 5-8 mm diameter, containing one to few seeds of 3-4 mm diameter.

Ecology and Control
A variable species widely used in horticulture, and several cultivars are known. Where invasive, the plant forms extensive and dense colonies that displace native understorey and ground cover species; it overtops shrubs and young trees. The plant grows in both low and high light conditions and produces a dense mass of somewhat fleshy roots, bearing numerous tubers. Once established, it grows vigorously and forms a dense crown of 2-3 m diameter, shading out other species. It fruits profusely and seeds are dispersed by birds.

Control should aim at removing the crown of plants and all fruits. Seedlings are hand pulled or sprayed. A follow-up programme is necessary to control seedlings and regrowth. Regrowth and seedlings can be effectively controlled with dicamba.

References
147, 715, 986.

Asparagus officinalis L.   Liliaceae

LF: Perennial herb   CU: Food
SN: *Asparagus longifolius* Fisch. ex Steud.

Geographic Distribution

Europe		*Australasia*		*Atlantic Islands*	
X	Northern	●	Australia		Cape Verde
X	British Isles	X	New Zealand		Canary, Madeira
N	Central, France				Azores
N	Southern	*Northern America*			South Atlantic Isl.
N	Eastern		Canada, Alaska		
N	Mediterranean Isl.		Southeastern USA	*Indian Ocean Islands*	
		X	Western USA		Mascarenes
Africa			Remaining USA		Seychelles
N	Northern		Mexico		Madagascar
	Tropical				
	Southern	*Southern America*		*Pacific Islands*	
			Tropical		Micronesia
Asia			Caribbean	X	Galapagos
N	Temperate	X	Chile, Argentina		Hawaii
	Tropical				Melanesia, Polynesia

Invaded Habitats
Grassland, riparian habitats, coastal sand dunes and beaches, salt meadows, swamps.

Description
A herb of 30-200 cm height, with much-branched shoots and needle-like cladodes of 5-20 mm length instead of true leaves. Cladodes occur in clusters of 3-10 in the axils of membranous scale-like strongly reduced leaves. Bell-shaped, hanging flowers of *c.* 5 mm length are borne solitary or in clusters of two, have long and thin stalks, and are whitish to pale yellow. The plant is usually dioecious, but hermaphroditic forms are also found. Fruits are red to brownish-red berries of 8-10 mm diameter, each containing 2-6 black and round seeds of 3-4 mm diameter. The plant has a thick rhizome.

Ecology and Control
In the native range, this plant is commonly found on disturbed and dry soils such as roadsides, sandy and grassy places, and on coastal cliffs and scrub. The natural distribution is obscure due to cultivation of this species in many parts of the world. New above-ground shoots appear every year from the crowns. The plant spreads mostly by rhizome fragmentation. Where invasive, it forms dense stands that eliminate native vegetation and prevent the regeneration of native shrubs and trees. Seeds are dispersed by birds.

Specific control methods for this species are not available. The same methods as for the previous species may apply.

References
215.

Asphodelus fistulosus L.   Liliaceae

LF: Annual, perennial herb   CU: Ornamental
SN: -

Geographic Distribution

	Europe			*Australasia*			*Atlantic Islands*
	Northern		●	Australia			Cape Verde
X	British Isles		X	New Zealand		N	Canary, Madeira
	Central, France					N	Azores
N	Southern			*Northern America*			South Atlantic Isl.
	Eastern			Canada, Alaska			
N	Mediterranean Isl.			Southeastern USA			*Indian Ocean Islands*
				Western USA			Mascarenes
	Africa			Remaining USA			Seychelles
N	Northern			Mexico			Madagascar
	Tropical						
	Southern			*Southern America*			*Pacific Islands*
				Tropical			Micronesia
	Asia			Caribbean			Galapagos
N	Temperate		X	Chile, Argentina			Hawaii
X	Tropical						Melanesia, Polynesia

Invaded Habitats
Grass- and heathland, coastal sand dunes and beaches, disturbed sites.

Description
A mostly perennial herb, occasionally annual, 15-70 cm tall, and with basal leaves up to 50 cm long and 4 mm wide. The rigid and hollow stems are branched in the upper part. Leaves are numerous, glabrous, round and hollow. The white to pink flowers are 15-20 mm in diameter and are arranged alternately along the branches of panicles up to 70 cm long. Each petal has one brown to reddish stripe, the anthers are orange. The fruit is a globular and wrinkled capsule of 4-7 mm diameter, divided into three segments with 1-2 seeds each. Seeds are brown to black, triangular, somewhat wrinkled, and 3-4 mm long.

Ecology and Control
In the native range, this plant is commonly found on pastures and grasslands. Seedlings develop slowly, and establishment occurs best in disturbed situations where competition from other species is low. Established plants, however, form dense and persisting tufts crowding out native species and preventing their regeneration. Scattered patches collate into a single matrix dominated by this plant.

Manual removal should be done before seed set. If cut below the soil surface, the roots are unable to regrow. Effective chemical control is achieved by applying amitrole, paraquat or picloram in the preflowering stage.

References
215, 985, 986.

Atriplex semibaccata R. Br. — Chenopodiaceae

LF: Semi-deciduous shrub
SN: -
CU: Erosion control

Geographic Distribution

Europe	*Australasia*	*Atlantic Islands*
Northern	N Australia	Cape Verde
British Isles	New Zealand	Canary, Madeira
Central, France		Azores
Southern	*Northern America*	South Atlantic Isl.
Eastern	Canada, Alaska	
Mediterranean Isl.	Southeastern USA	*Indian Ocean Islands*
	● Western USA	Mascarenes
Africa	Remaining USA	Seychelles
Northern	Mexico	Madagascar
Tropical		
Southern	*Southern America*	*Pacific Islands*
	Tropical	Micronesia
Asia	Caribbean	Galapagos
Temperate	Chile, Argentina	Hawaii
Tropical		Melanesia, Polynesia

Invaded Habitats
Coastal estuaries, salt marshes, coastal dunes and scrub, woodland.

Description
A perennial herb or small shrub of 15-35 cm height, with several spreading to ascending stems of 30-100 cm length, often covered with white scales. Leaves are oblong to narrowly elliptic, 8-30 mm long, and entire or toothed. Inflorescences with female flowers contain only few flowers. Bracts are fleshy, reddish and 3-6 mm long in fruit. Seeds are 1.5-2 mm long.

Ecology and Control
A ground-spreading plant that displaces native species. In the native range, it is a pioneer species of eroded soils. Once established, it persists and may become dominant over large areas. Stands of this plant are species poor. Seeds are produced in large numbers and dispersed by birds.

Plants can easily be hand pulled, best before seeds develop. Chemical control for closely related taxa (*Kochia scoparia*, *Salsola tragus*) include 2,4-D, dicamba, or picloram plus 2,4-D.

References
133.

Avena fatua L.   Poaceae

LF: Annual herb
SN: -
CU: Fodder, forage

Geographic Distribution

Europe
- N   Northern
- X   British Isles
- N   Central, France
- N   Southern
- N   Eastern
- N   Mediterranean Isl.

Africa
- N   Northern
- X   Tropical
- X   Southern

Asia
- N   Temperate
- N   Tropical

Australasia
- ●   Australia
- X   New Zealand

Northern America
- X   Canada, Alaska
-        Southeastern USA
- X   Western USA
- X   Remaining USA
-        Mexico

Southern America
- X   Tropical
-        Caribbean
- X   Chile, Argentina

Atlantic Islands
-        Cape Verde
- X   Canary, Madeira Azores
-        South Atlantic Isl.

Indian Ocean Islands
-        Mascarenes
-        Seychelles
-        Madagascar

Pacific Islands
-        Micronesia
-        Galapagos
- X   Hawaii
- X   Melanesia, Polynesia

Invaded Habitats
Grass- and woodland, riparian habitats, coastal sand dunes and beaches.

Description
A tufted or solitary grass with erect to ascending culms of 30-150 cm height. The green, linear-lanceolate leaves are 10-45 cm long and 3-15 mm wide, finely pointed at the end, and have sheaths that are rounded on the back. Ligules are membranous and up to 6 mm long. Inflorescences are panicles of 10-40 cm length and up to 20 cm width, nodding, green and pyramidal, with widely spreading branches. Spikelets are 18-25 mm long, hanging, with 2-3 florets each, and persistent glumes of 20-30 mm length. Glumes are lanceolate and have 7-11 veins. Lemmas are narrowly oblong-lanceolate, 14-20 mm long, and have 2-4 short teeth at the end as well as stiff hairs on the lower half. The caryopsis is oblong, sometimes silky-hairy, and 6-8 mm long.

Ecology and Control
A grass that prefers temperate and cool climates, and grows best in moist soils. The grass has a strong ability to emerge from deeper soil layers due to a great ability for elongation of the first internode. Reproduction may occur within 2-3 months after germination. Seed production is high; individual plants may produce up to 500 seeds. The grass forms dense swards that displace native grass and forb species.

Plants normally do not reproduce from vegetative parts, and mechanical removal may control the plant. Burning can reduce the number of seeds on or near the soil surface. Emerging seedlings can be killed by soil-applied herbicides such as trifluralin, barban or asulam.

References
65, 215, 249, 580, 1099, 1163, 1164.

Azolla filiculoides Lam. Salviniaceae

LF: Aquatic fern
SN: *Azolla rubra* R.Br.

CU: Ornamental, soil improvement

Geographic Distribution

Europe		Australasia		Atlantic Islands	
	Northern	X	Australia		Cape Verde
X	British Isles	X	New Zealand		Canary, Madeira
X	Central, France				Azores
X	Southern	*Northern America*			South Atlantic Isl.
	Eastern	N	Canada, Alaska		
	Mediterranean Isl.		Southeastern USA	*Indian Ocean Islands*	
		N	Western USA		Mascarenes
Africa			Remaining USA		Seychelles
	Northern	N	Mexico		Madagascar
	Tropical				
•	Southern	*Southern America*		*Pacific Islands*	
		N	Tropical		Micronesia
Asia			Caribbean		Galapagos
	Temperate	N	Chile, Argentina	N	Hawaii
•	Tropical				Melanesia, Polynesia

Invaded Habitats
Freshwater wetlands, riparian habitats.

Description
A small perennial, mat-forming and free-floating fern of *c.* 25 mm length. The small horizontal stems reach 25-35 mm length and bear silvery-green to reddish-brown, ovate to circular leaves of 1-1.5 mm length. Leaves are alternate and overlapping. Roots are borne singly and hang into the water. The plant has a short and branched rhizome. Minute fruiting bodies are borne in the axils of leaves. Fertile plants are somewhat ascending and have internodes of less than 1 mm length; immature plants are prostrate with internodes up to 5 mm long. Plants often become red or magenta during winter.

Ecology and Control
Azolla ferns grow in a symbiotic association with nitrogen-fixing cyanobacteria, enabling the plants to grow in waters with a low nitrogen content. The fern spreads rapidly by fragmentation of the plant. Sexual reproduction is common as well, and the spores are extremely resistant to desiccation. Vegetative growth leads to dense mats; in South Africa, mats of 5-30 cm thickness have been found, thereby reducing water quality, aquatic biodiversity, and increasing siltation.

Mechanical control includes removing plants by fine meshed nets. However, the plant can re-establish from spores, and manual removal may be impractical for large infestations. Herbicides used include glyphosate, paraquat, or diquat. A follow-up programme is needed to control new plants germinating from spores.

References
508, 565, 1089, 1355.

Baccharis halimifolia L.   Asteraceae

LF: Semi-deciduous shrub   CU: -
SN: -

Geographic Distribution

Europe		Australasia		Atlantic Islands	
	Northern	●	Australia		Cape Verde
X	British Isles	X	New Zealand		Canary, Madeira
	Central, France				Azores
●	Southern		*Northern America*		South Atlantic Isl.
	Eastern		Canada, Alaska		
	Mediterranean Isl.	N	Southeastern USA		*Indian Ocean Islands*
			Western USA		Mascarenes
Africa		N	Remaining USA		Seychelles
	Northern	N	Mexico		Madagascar
	Tropical				
	Southern		*Southern America*		*Pacific Islands*
			Tropical		Micronesia
Asia		N	Caribbean		Galapagos
	Temperate		Chile, Argentina		Hawaii
	Tropical				Melanesia, Polynesia

Invaded Habitats
Coastal swamps, coastal forests, disturbed places.

Description
A freely branching, evergreen shrub or small tree up to 4 m in height. The alternate leaves are pale green, 4-7 cm long and 1-4 cm wide, thick, with margins often being widely serrate, occasionally entire. Flowers are small, greenish to dull white. Flowerheads are arranged in peduncled clusters of 1-5, and bracts of the involucre are oblong-ovate, obtuse and glutinous. Fruits are winged achenes of 1-2 mm length. The pappus is bright white, much exceeding the involucre, and has 1-2 series of capillary bristles. The plant is dioecious.

Ecology and Control
In the native range, this shrub is found mostly in coastal habitats, e.g. salt marshes and tidal rivers, sandy places, but also on disturbed places far off the coast. In the cooler parts within the native range, the shrub is deciduous. Seeds are dispersed by wind and water, and germinate easily if sufficient soil moisture is available. It is a prolific seed producer and seeds are formed even in dense shade. The shrub tolerates a high level of soil salinity. It is invasive because it forms dense and impenetrable thickets that crowd out native vegetation and prevent the establishment of other plant species.

If plants are removed manually, the roots should be cut well below the surface to prevent resprouting. Chemical control is done by spraying herbicides containing 2,4-D, dicamba plus MCPA, glyphosate, or picloram plus 2,4-D.

References
607, 608, 699, 974, 975, 986, 1391.

Berberis darwinii Hook.   Berberidaceae

LF: Evergreen shrub   CU: Ornamental
SN: -

Geographic Distribution

Europe		*Australasia*		*Atlantic Islands*
	Northern	●	Australia	Cape Verde
X	British Isles	●	New Zealand	Canary, Madeira
	Central, France			Azores
	Southern		*Northern America*	South Atlantic Isl.
	Eastern		Canada, Alaska	
	Mediterranean Isl.		Southeastern USA	*Indian Ocean Islands*
		X	Western USA	Mascarenes
Africa			Remaining USA	Seychelles
	Northern		Mexico	Madagascar
	Tropical			
	Southern		*Southern America*	*Pacific Islands*
			Tropical	Micronesia
Asia			Caribbean	Galapagos
	Temperate	N	Chile, Argentina	Hawaii
	Tropical			Melanesia, Polynesia

Invaded Habitats
Forests and forest edges.

Description
A spiny shrub, 1-3 m tall, with a brown and densely tomentose bark. Stems have long primary and short axillary shoots. Branchlets are reddish-brown and pubescent, with slender spines of 3-7 mm length. Leaves are simple, with short petioles of 1-3 mm length and obovate, 1-veined leaf blades of 15-30 mm length and 9-12 mm width. Leaves are thick and rigid, the margins being reflexed, toothed or lobed, and have terminal spines of *c.* 1.5 mm length. The upper leaf surface is dark green, the lower one light green. The rather dense and racemose inflorescences are 3-4 cm in diameter, borne in the axils of leaves, and contain 10-20 bright orange flowers each. Fruits are dark purple globose berries of 6-8 mm diameter, each containing 2-7 seeds.

Ecology and Control
In the native range, the shrub is a prominent understorey plant of montane forests. Where invasive, it persists once established and replaces native understorey plants. The shrub is shade tolerant, resprouts from the base, and forms dense, spiny thickets. Occasionally, it grows as a climber up to 10 m height. Seedling density can be high, e.g. >200 m^{-2}. Seeds are dispersed by birds.
    Seedlings can be hand pulled or dug out. Larger plants are cut and the cut stumps treated with herbicide.

References
18, 20, 21, 121, 1415, 1457.

Berberis glaucocarpa Stapf — Berberidaceae

LF: Semi-evergreen shrub
SN: *Berberis chitria* Lindley
CU: Ornamental

Geographic Distribution

Europe		*Australasia*		*Atlantic Islands*	
	Northern		Australia		Cape Verde
X	British Isles	●	New Zealand		Canary, Madeira
	Central, France				Azores
	Southern		*Northern America*		South Atlantic Isl.
	Eastern		Canada, Alaska		
	Mediterranean Isl.		Southeastern USA		*Indian Ocean Islands*
			Western USA		Mascarenes
Africa			Remaining USA		Seychelles
	Northern		Mexico		Madagascar
	Tropical				
	Southern		*Southern America*		*Pacific Islands*
			Tropical		Micronesia
Asia			Caribbean		Galapagos
	Temperate		Chile, Argentina		Hawaii
N	Tropical				Melanesia, Polynesia

Invaded Habitats
Forests and forest margins, scrub, disturbed sites.

Description
A spiny, semi-evergreen or deciduous shrub up to 3 m tall, with a spreading habit and pale yellow stems. Leaves are oblanceolate to obovate, sparsely serrated or entire, bright green and 3-8 cm long. Pale yellow flowers are borne in stiff but often hanging panicles of up to 15 cm length. Fruits are bluish-black berries covered with a dense white bloom.

Ecology and Control
The seedlings of this shrub are shade tolerant and establish well in forests. The shrub grows vigorously and shades out native plants with its dense foliage, thereby preventing regeneration of native shrubs and trees. Older shrubs may reproduce vegetatively from roots. Seeds are dispersed by birds.

Specific control methods for this species are not available. Seedlings and young plants may be hand pulled or dug out, larger plants cut and the cut stumps treated with herbicide to prevent regrowth.

References
1415.

Berberis thunbergii DC.   Berberidaceae

LF: Evergreen tree, shrub   CU: Ornamental
SN: -

Geographic Distribution

Europe
    Northern
    British Isles
● Central, France
    Southern
    Eastern
    Mediterranean Isl.

Africa
    Northern
    Tropical
    Southern

Asia
N   Temperate
    Tropical

Australasia
    Australia
    New Zealand

Northern America
X   Canada, Alaska
● Southeastern USA
    Western USA
● Remaining USA
    Mexico

Southern America
    Tropical
    Caribbean
    Chile, Argentina

Atlantic Islands
    Cape Verde
    Canary, Madeira
    Azores
    South Atlantic Isl.

Indian Ocean Islands
    Mascarenes
    Seychelles
    Madagascar

Pacific Islands
    Micronesia
    Galapagos
    Hawaii
    Melanesia, Polynesia

Invaded Habitats
Forests and forest gaps, floodplains, grassland, woodland.

Description
A glabrous and thorny shrub with a purple to brown bark, reaching heights from 0.3-3 m, and thorns of *c.* 12 mm length. Bud scales are 1-2 mm long and deciduous. Leaves are simple and entire; petioles, if present, are up to 8 mm long. The leaf blade is obovate to spatulate, 12-24 mm long and 3-10 mm wide. Pale yellow flowers are borne in umbel-like inflorescences with 1-5 flowers each. Flowers range from 10-15 mm in diameter and have membranous bracts. Fruits are bright red, ellipsoid to globose berries of 8-10 mm diameter, containing seeds of 5-6 mm length.

Ecology and Control
The shrub grows well in full sun or in more shady places. Once established, it may quickly form dense and thorny thickets that shade out all other understorey species and displace native vegetation. New shoots arise from stolons and rhizomes, and stems touching the ground may become rooted. Soil pH and the amount of available nitrate in the soil increase under established stands of this shrub. Seeds are dispersed by birds.

    Manual removal includes hand pulling or digging out before seed set. The root system needs to be removed as far as possible because the shrub easily resprouts from roots and stem fragments. Cutting the shrub at the base may require treating resprouts with a glyphosate herbicide. Glyphosate applied in early spring proved to be very effective. Prescribed fire has been used to reduce populations in oak savannas.

References
311, 375, 376, 571, 693, 694, 1176, 1431.

Bidens pilosa L.   Asteraceae

LF: Annual herb   CU: -
SN: *Bidens chinensis* Willd., *Bidens leucantha* Willd.

Geographic Distribution

Europe
-     Northern
- X   British Isles
- X   Central, France
- X   Southern
- X   Eastern
-     Mediterranean Isl.

Africa
-     Northern
- X   Tropical
- ●   Southern

Asia
- X   Temperate
- X   Tropical

Australasia
- ●   Australia
- X   New Zealand

Northern America
-     Canada, Alaska
- X   Southeastern USA
- X   Western USA
-     Remaining USA
- X   Mexico

Southern America
- N   Tropical
- X   Caribbean
- X   Chile, Argentina

Atlantic Islands
- X   Cape Verde
- X   Canary, Madeira
- X   Azores
-     South Atlantic Isl.

Indian Ocean Islands
- X   Mascarenes
- X   Seychelles
-     Madagascar

Pacific Islands
- X   Micronesia
-     Galapagos
- X   Hawaii
-     Melanesia, Polynesia

Invaded Habitats
Grass- and heathland, forests, disturbed sites.

Description
A glabrous or soft-hairy herb of 30-180 cm height and stems that are quadrangular in cross section. Leaves are opposite, petioled and compound with 3-5 leaflets each. The lanceolate to ovate leaflets are 2-6 cm long, acute and have serrated margins. Flowerheads with 40-50 florets each appear on peduncles of 1-9 cm length. Involucres have a diameter of 7-8 mm and 7-9 outer phyllaries of 4-5 mm length. Ray florets are absent or strongly reduced. The yellow disk florets have a corolla of *c.* 2 mm length. Fruits are black achenes with 2-4 barbed awns, covered with short hairs, and 4-16 mm long.

Ecology and Control
A significant agricultural weed of many crop systems. In natural areas, it may form dense stands that cover large areas and eliminate native vegetation, preventing the establishment of native species. It quickly invades burned areas. Seed production is abundant: a single plant may produce 3,000-6,000 seeds per year. Seeds are widely dispersed by adhering easily to clothes, or animal fur. Many seeds germinate readily at maturity, allowing the species to have 3-4 generations per year under favourable conditions. Germination is best in light and if good aeration of the soil is present. Seeds may remain viable for several years.

    Scattered plants can be hand pulled or dug out if soil disturbance is not a problem. Larger infestations are mown or sprayed with herbicide.

References
63, 248, 407, 580, 1070, 1089, 1133, 1461.

Bischofia javanica Blume   Euphorbiaceae

LF: Evergreen tree   CU: Ornamental
SN: *Andrachne trifoliata* Roxb., *Bischofia trifoliata* (Roxb.) Hook

Geographic Distribution

Europe	*Australasia*	*Atlantic Islands*
Northern	N Australia	Cape Verde
British Isles	New Zealand	Canary, Madeira
Central, France		Azores
Southern	*Northern America*	South Atlantic Isl.
Eastern	Canada, Alaska	
Mediterranean Isl.	● Southeastern USA	*Indian Ocean Islands*
	Western USA	Mascarenes
Africa	Remaining USA	Seychelles
Northern	Mexico	Madagascar
Tropical		
Southern	*Southern America*	*Pacific Islands*
	Tropical	N Micronesia
Asia	Caribbean	Galapagos
N Temperate	Chile, Argentina	X Hawaii
N Tropical		N Melanesia, Polynesia

Invaded Habitats
Forests, tropical hammocks.

Description
A large glabrous tree up to 22 m tall, with a shady oval crown and a deep green foliage, turning red before falling. The thick bark is dark grey to brown, smooth or rough. Leaves are alternate, trifoliate with large and leathery, elliptic, and acuminate leaflets that are 10-15 cm long and 4-7 cm wide. Leaves have a petiole of 7-20 cm length and toothed margins. Flowers are greenish-yellow and *c.* 3 mm in diameter. The fruit is a brown, orange, reddish or blue-black, globose drupe of 9-10 mm diameter with three indehiscent cells, each containing 1-2 seeds of *c.* 5 mm length and 3 mm width. The tree is dioecious, and female trees produce large quantities of hanging fruits.

Ecology and Control
Native habitats of this tree are shady ravines in evergreen, mixed deciduous and swamp forests. The tree is fast growing and suckers easily from roots. In dry places, it grows as a stunted tree or shrub of 4-6 m height. Where invasive, the tree establishes dense stands, casting dense shade and crowding out native trees, displacing them and altering the vegetation structure. Seedlings establish well under shady forest canopies and grow rapidly into the canopy if a gap occurs. Once established, the tree will become abundant throughout the understorey. Seeds are easily dispersed by birds.

    Female trees should be removed first to prevent seed dispersal. Seedlings may be hand pulled; older trees treated basally with a triclopyr herbicide mixed with an oil diluent.

References
648, 649, 715, 1443.

Brassica tournefortii Gouan   Brassicaceae

LF: Annual herb   CU: -
SN: -

Geographic Distribution

	Europe		*Australasia*		*Atlantic Islands*
	Northern	●	Australia		Cape Verde
X	British Isles	X	New Zealand		Canary, Madeira
	Central, France				Azores
N	Southern		*Northern America*		South Atlantic Isl.
	Eastern		Canada, Alaska		
N	Mediterranean Isl.		Southeastern USA		*Indian Ocean Islands*
		●	Western USA		Mascarenes
	Africa		Remaining USA		Seychelles
N	Northern		Mexico		Madagascar
	Tropical				
X	Southern		*Southern America*		*Pacific Islands*
			Tropical		Micronesia
	Asia		Caribbean		Galapagos
N	Temperate		Chile, Argentina		Hawaii
N	Tropical				Melanesia, Polynesia

Invaded Habitats
Grassland, heathland, desert scrub, coastal sites.

Description
A highly variable species with stems 10-100 cm tall and leaves usually in a basal rosette. Basal leaves are deeply lobed, 7-30 cm long, and with the lobes being toothed. Stem leaves are much smaller. Inflorescences consist of racemes with 6-20 flowers each. The pale yellow flowers have petals of *c.* 15 mm width and 5-7 mm length, and pedicels of 10-20 mm length that elongate during fruit ripening. Fruits are dehiscent siliques of 3-7 cm length and 2-3 mm width, and have a beak of 10-20 mm length. Seeds are arranged in two rows of 7-15 seeds each. They are globose, reddish seeds and *c.* 1 mm in diameter.

Ecology and Control
In the native range, this plant grows in rocky and sandy places, and is a common weed in disturbed sites. The species is invasive because it forms dense stands, accumulates large quantities of dried plant material and thus increases fire hazard. It establishes well after fire from the soil seed bank. The species is fast growing, fruiting within 2-3 months after germination, and is a prolific seed producer; a large plant may form 800-9,000 seeds per year. Establishment of this species depends on disturbance. If wet, seeds are covered with a sticky gel, allowing dispersal over long distances by animals and vehicles.

    Scattered plants can be hand pulled. Larger patches are treated with herbicide. Cutting before seed set prevents dispersal of seeds.

References
133, 215.

Briza maxima L.   Poaceae

LF: Annual herb   CU: -
SN: -

Geographic Distribution

Europe		*Australasia*		*Atlantic Islands*	
	Northern	●	Australia		Cape Verde
X	British Isles	X	New Zealand	N	Canary, Madeira
	Central, France			N	Azores
N	Southern	*Northern America*			South Atlantic Isl.
N	Eastern		Canada, Alaska		
N	Mediterranean Isl.		Southeastern USA	*Indian Ocean Islands*	
		X	Western USA		Mascarenes
Africa			Remaining USA		Seychelles
N	Northern		Mexico		Madagascar
	Tropical				
X	Southern	*Southern America*		*Pacific Islands*	
		X	Tropical		Micronesia
Asia		X	Caribbean		Galapagos
N	Temperate	X	Chile, Argentina	X	Hawaii
X	Tropical				Melanesia, Polynesia

Invaded Habitats
Grass- and heathland, forests, riparian habitats, coastal beaches.

Description
A grass with loosely tufted or solitary culms of 10-60 cm height and 2-4 nodes each. The linear and glabrous leaves are 5-20 cm long and 3-8 mm wide, flat and finely pointed at the ends, and have rounded sheaths. Ligules are membranous and 2-5 mm long. The inflorescence is a loose and nodding panicle of 3-10 cm length, containing 1-12 spikelets. Pedicels are 6-20 mm long; the spikelets are 14-25 mm long and 8-15 mm wide, ovate to oblong, and contain 7-20 pale green and silvery flowers. Glumes are persistent, broadly rounded and membranous, 5-7 mm long, and have 5-9 veins. Lemmas measure 6-8 mm in length, and are closely overlapping and rounded on the back. Fruits are pale brown caryopses of *c.* 2.5 mm length that are narrowly ellipsoid.

Ecology and Control
Native habitats include hillslopes, coastal scrub and disturbed sites. It forms dense, species poor swards where invasive that impede the growth and regeneration of native plants. Seeds are abundantly produced and large populations can build up in a short time.
    Plants are easy to hand pull. Mowing before seeds are ripe prevents seed dispersal and kills the plant. Burning before flower open kills the grass and destroys seeds on the soil. Chemical control is done by spraying herbicides, best before the flowering stems emerge.

References
121, 215, 755, 849, 924.

Bromus inermis Leyss.   Poaceae

LF: Perennial herb
SN: -
CU: Forage

Geographic Distribution

Europe		Australasia	Atlantic Islands
X	Northern	Australia	Cape Verde
X	British Isles	New Zealand	Canary, Madeira
N	Central, France		Azores
N	Southern	*Northern America*	South Atlantic Isl.
N	Eastern	● Canada, Alaska	
	Mediterranean Isl.	Southeastern USA	*Indian Ocean Islands*
		X Western USA	Mascarenes
Africa		Remaining USA	Seychelles
	Northern	Mexico	Madagascar
	Tropical		
	Southern	*Southern America*	*Pacific Islands*
		Tropical	Micronesia
Asia		Caribbean	Galapagos
N	Temperate	Chile, Argentina	Hawaii
	Tropical		Melanesia, Polynesia

Invaded Habitats
Grassland, tallgrass prairies, old pastures.

Description
A perennial grass with erect and almost glabrous stems of 30-150 cm height and long rhizomes. Leaves are flat, glabrous and 6-10 mm wide. Inflorescences are erect and spreading panicles of 10-25 cm length, with 3-7 ascending or stiff branches per node. Spikelets have 5-10 florets and are 15-30 mm long and 3-5 mm wide. The awnless lemmas are 8-10 mm long.

Ecology and Control
In the native range, the grass is found on riverbanks, in woods, pastures and disturbed places. The grass is highly variable and numerous agricultural varieties exist. It has a deeply penetrating root system, and due to its rhizomes, it forms dense mats excluding other species and reducing native diversity. In the Rocky Mountains, it hybridizes with the native *B. pumpellianus*. The species grows best in full sun, and is moderately salt tolerant. Rhizome growth is important for the colonization ability of the grass, and rhizome formation starts as early as three weeks after germination.

    Effective control is achieved by mowing or burning in boot stage, e.g. when the flowering stems still are being enclosed within the sheaths. Mowing in spring is often combined with a later prescribed burning. An effective chemical control is applying picloram. Other herbicides used are based on glyphosate or atrazine. In Canada, excellent control was achieved with spring burning followed by spraying glyphosate when the grass is in the boot stage.

References
119, 335, 502, 1101, 1396, 1423.

Bromus rubens L.   Poaceae

LF: Annual herb   CU: Forage, revegetation
SN: *Bromus purpurascens* Delile, *Bromus madritensis* subsp. *rubens* (L.) Husnot

Geographic Distribution

	Europe		*Australasia*		*Atlantic Islands*
	Northern	●	Australia		Cape Verde
X	British Isles		New Zealand		Canary, Madeira
	Central, France			N	Azores
N	Southern		*Northern America*		South Atlantic Isl.
	Eastern		Canada, Alaska		
	Mediterranean Isl.		Southeastern USA		*Indian Ocean Islands*
		●	Western USA		Mascarenes
	Africa		Remaining USA		Seychelles
	Northern		Mexico		Madagascar
	Tropical				
	Southern		*Southern America*		*Pacific Islands*
			Tropical		Micronesia
	Asia		Caribbean		Galapagos
	Temperate		Chile, Argentina	X	Hawaii
	Tropical				Melanesia, Polynesia

Invaded Habitats
Deserts and desert scrub, coastal scrub, disturbed sites.

Description
A caespitose grass with culms of 5-50 cm height, generally pubescent and unbranched above. Leaf blades are flat, linear, 2-12 cm long and 2-3 mm wide, densely pilose on both sides. Inflorescences are elliptic, strongly contracted racemes or panicles of 5-20 cm length. Fruits are long elliptical, 8.5-10 mm long and 1-1.3 mm wide.

Ecology and Control
The plant germinates in the autumn, grows slowly during winter and rapidly in early spring. It forms dense stands and competes for nutrients and water. The grass enhances the potential for the start and spread of fires because the dead and dry stems persist for long times. Once established, it remains dominant and may thus reduce the soil seed bank of native annuals. The rather short-lived seeds are abundantly produced. The sharp florets become easily dislodged and have caused eye infections in hawks and wildlife.

    Scattered plants are easy to remove by hand pulling. Mulching is used to reduce seedling emergence, prescribed burning in late autumn to kill seedlings. Larger stands are treated with herbicides such as atrazine or glyphosate.

References
87, 133, 180, 215, 611, 1138.

Bromus tectorum L. Poaceae

LF: Annual herb
SN: *Anisantha tectorum* (L.) Nevski
CU: Forage

Geographic Distribution

Europe		*Australasia*		*Atlantic Islands*	
N	Northern	X	Australia		Cape Verde
X	British Isles	X	New Zealand	N	Canary, Madeira
N	Central, France				Azores
N	Southern		*Northern America*		South Atlantic Isl.
N	Eastern		Canada, Alaska		
N	Mediterranean Isl.		Southeastern USA		*Indian Ocean Islands*
		●	Western USA		Mascarenes
Africa		X	Remaining USA		Seychelles
N	Northern		Mexico		Madagascar
	Tropical				
	Southern		*Southern America*		*Pacific Islands*
			Tropical		Micronesia
Asia			Caribbean		Galapagos
N	Temperate		Chile, Argentina	X	Hawaii
N	Tropical				Melanesia, Polynesia

Invaded Habitats
Grassland, scrubland, rangelands.

Description
A 5-70 cm tall grass often growing in large tufts, with a fibrous root system and only few main roots. Stems are erect and sometimes pubescent; leaves are 4-16 cm long and 2-4 mm wide, light green and pubescent. The membranous ligules are *c.* 3 mm long. Inflorescences are rather dense and drooping panicles of 5-20 cm length, with slender and pubescent branches. The nodding spikelets have 2-8 florets and are 2-4 cm long. Lower glumes are 5-8 mm long, upper glumes 7-11 mm, and the toothed lemmas reach 9-12 mm length. Lemmas have long and soft hairs. Awns are slender, straight, and 12-14 mm long.

Ecology and Control
In western North America, this grass has become dominant on millions of acres previously dominated by native perennial grasses. The grass is highly variable and a prolific seed producer. It accumulates a large soil seed bank and seeds remain viable for several years. Due to the shallow root system, it extracts soil moisture from the upper soil layers, thus preventing the establishment of other species. The early maturation and accumulation of dead and highly inflammable material greatly increases fire hazards. Establishment and subsequent spread is enhanced by disturbances.

Mowing within one week after flowering is an effective control method for preventing seed formation. Herbicides, e.g. paraquat, glyphosate, are best applied in early spring before native perennial grasses have emerged.

References
133, 356, 535, 603, 611, 674, 799, 800, 859, 940, 1007, 1022, 1023, 1024, 1081, 1170, 1235, 1320, 1400, 1403, 1404.

Bryophyllum pinnatum (Lam.) Oken   Crassulaceae

LF: Perennial herb   CU: Ornamental
SN: *Bryophyllum calycinum* Salisb., *Kalanchoe pinnata* Pers.

Geographic Distribution

	Europe		*Australasia*		*Atlantic Islands*
	Northern	●	Australia	X	Cape Verde
	British Isles	X	New Zealand	X	Canary, Madeira
	Central, France			X	Azores
	Southern		*Northern America*		South Atlantic Isl.
	Eastern		Canada, Alaska		
	Mediterranean Isl.		Southeastern USA		*Indian Ocean Islands*
			Western USA	X	Mascarenes
Africa			Remaining USA	X	Seychelles
	Northern		Mexico	N	Madagascar
	Tropical				
	Southern		*Southern America*		*Pacific Islands*
			Tropical		Micronesia
Asia		X	Caribbean	●	Galapagos
X	Temperate		Chile, Argentina	X	Hawaii
	Tropical			X	Melanesia, Polynesia

Invaded Habitats
Forest margins, coastal heath, rock outcrops.

Description
A glabrous somewhat succulent herb with erect stems of 50-200 cm height and up to 2 cm width, woody at the base. Leaves are opposite, the lower ones simple, the upper ones compound with 3-5 elliptic leaflets of 5-20 cm length and 2-10 cm width. Petioles are 2-10 cm long. Flowers are pale green mottled with red or deep red, borne in terminal cymes of 10-80 cm length, and hanging on pedicels of 1-2.5 cm length. The flower tubes are 28-35 mm long. Seeds are *c.* 1 mm long.

Ecology and Control
The plant is suckering from the base and spreads vegetatively by forming young plantlets on the leaf margins. It is a drought tolerant species that often forms dense stands and displaces native species. The plant spreads rapidly due to vegetative growth.

     Plants can be hand pulled or dug out, the root crown should be removed to prevent regrowth. Effective chemical control is done by applying 2,4-D together with a surfactant.

References
521, 725, 835, 1212, 1313.

Buddleja davidii Franch. Buddlejaceae

LF: Deciduous shrub
SN: *Buddleja variabilis* Hemsl.
CU: Ornamental

Geographic Distribution

Europe		*Australasia*		*Atlantic Islands*	
X	Northern	●	Australia		Cape Verde
●	British Isles	●	New Zealand		Canary, Madeira
●	Central, France				Azores
●	Southern	*Northern America*			South Atlantic Isl.
X	Eastern		Canada, Alaska		
	Mediterranean Isl.		Southeastern USA	*Indian Ocean Islands*	
		X	Western USA		Mascarenes
Africa			Remaining USA		Seychelles
	Northern		Mexico		Madagascar
	Tropical				
	Southern	*Southern America*		*Pacific Islands*	
			Tropical		Micronesia
Asia			Caribbean		Galapagos
N	Temperate		Chile, Argentina		Hawaii
	Tropical				Melanesia, Polynesia

Invaded Habitats
Riparian habitats, grassland, quarries, forest edges.

Description
A much-branched shrub up to 5 m tall with long, arching branches and opposite leaves. Leaves are lanceolate to narrowly ovate, 10-25 cm long, white tomentose beneath, and have serrated margins and petioles of 2-5 mm length. Inflorescences are long pyramidal and dense panicles of 10-25 cm length. The pale lilac to deep violet flowers are 8-13 mm long and have corollas with four lobes and an orange centre. Fruits are capsules of *c.* 10 mm length that contain numerous seeds. Seeds are long-winged at both ends and 2-4 mm long.

Ecology and Control
A frequently used ornamental of which many cultivars exist with varying flower colours. It is invasive because it quickly displaces primary native colonizers on fresh alluvial plains and accelerates succession to forests. The fast growing and rapidly spreading shrub reaches high densities, e.g. >2,000 plants ha^{-1}. Initial growth rate is high, and it quickly suppresses herbaceous pioneer species. The shrub tolerates burial by alluvium deposition by producing adventitious roots and shoots; damaged shrubs can resprout. Seeds are produced in large quantitites and dispersed by wind and water.

If established plants are removed, the stumps should be either removed as well or treated with a glyphosate herbicide, because the shrub can regrow from the roots if cut. Seedlings and smaller plants can be hand pulled or dug out.

References
121, 153, 215, 605, 606, 967, 968, 1195, 1415.

Caesalpinia decapetala (Roth) Alston   Fabaceae

LF: Evergreen shrub, vine   CU: Ornamental
SN: *Caesalpinia japonica* Sieb. & Zucc., *Caesalpinia sepiaria* Roxb.

Geographic Distribution

Europe		*Australasia*		*Atlantic Islands*	
	Northern	X	Australia		Cape Verde
	British Isles	●	New Zealand		Canary, Madeira
	Central, France				Azores
	Southern	*Northern America*			South Atlantic Isl.
	Eastern		Canada, Alaska		
	Mediterranean Isl.		Southeastern USA	*Indian Ocean Islands*	
			Western USA		Mascarenes
Africa			Remaining USA		Seychelles
	Northern		Mexico		Madagascar
X	Tropical				
●	Southern	*Southern America*		*Pacific Islands*	
			Tropical		Micronesia
Asia			Caribbean		Galapagos
N	Temperate		Chile, Argentina	X	Hawaii
N	Tropical			X	Melanesia, Polynesia

Invaded Habitats
Forest margins and gaps, grassland, streambanks.

Description
A robust and sprawling shrub or climber of 0.5-10 m height, with numerous straight to hooked thorns on the stems. Leaves are dark green above, paler beneath, up to 30 cm long and bipinnately compound. They have deciduous stipules of 8-20 mm length. The leaf rachis is armed with downwardly hooked prickles. Each leaf consists of 3-15 pairs of pinnae, each pinna having 5-12 pairs of leaflets. The elliptic-oblong to ovate leaflets are rounded at the end, 10-22 mm long and 4-11 mm wide. Pale yellow flowers of 25-30 mm diameter are borne in axillary and terminal racemes of 10-40 cm length. Petals are 10-15 mm long and 8-15 mm wide. Fruits are dehiscent pods of 6-11 cm length and 2-3 cm width, each containing 4-9 black ellipsoid, flattened and black seeds of 8-12 mm length and 6-8 mm width.

Ecology and Control
Where native, this plant is commonly found in woodland and grassy places. The species is invasive because it forms dense impenetrable thickets and climbs over shrubs and trees, impeding their growth and regeneration. Seeds are dispersed by birds and small mammals, as well as water. Despite belonging to the Fabaceae, the plant is not nitrogen-fixing.

Mechanical control of established plants is difficult due to the long and sharp thorns of the plant. Chemical control is done by foliar application of herbicides.

References
549, 792, 1089, 1415.

Calluna vulgaris (L.) Hull — Ericaceae

LF: Evergreen shrub
SN: -
CU: Erosion control, ornamental

Geographic Distribution

Europe
N   Northern
N   British Isles
N   Central, France
N   Southern
N   Eastern
N   Mediterranean Isl.

Africa
N   Northern
    Tropical
    Southern

Asia
N   Temperate
    Tropical

Australasia
X   Australia
●   New Zealand

Northern America
X   Canada, Alaska
    Southeastern USA
X   Western USA
    Remaining USA
    Mexico

Southern America
    Tropical
    Caribbean
    Chile, Argentina

Atlantic Islands
    Cape Verde
N   Canary, Madeira
N   Azores
    South Atlantic Isl.

Indian Ocean Islands
    Mascarenes
    Seychelles
    Madagascar

Pacific Islands
    Micronesia
    Galapagos
    Hawaii
    Melanesia, Polynesia

Invaded Habitats
Heathland, tussock grassland, bog edges.

Description
A densely-branched, low shrub reaching 80 cm in height and with stems branching from the base. Leaves are minute and needle-like; those on long shoots are widely spaced and 3-4 mm long, those on short shoots closely spaced and only 1-2 mm long. They are glabrous, linear and sessile, and have reflexed margins. Purplish to white flowers with short stalks are borne solitary in the axils of leaves. They are slightly zygomorphic and 1-2 mm in diameter. Fruits are capsules containing 20-30 minute seeds of *c.* 0.6 mm length and 0.4 mm width.

Ecology and Control
This shrub is a frequent species of heath communities in the native range, and occurs mostly on acid sandy soils in moors and on peat, in bogs, on fixed sand dunes, on gravel, and also in woodland. The species is light demanding and optimal growth and abundance are achieved in the open. The plant easily forms adventitious roots, especially if the soil is covered by a moist moss layer. The shrub can form large and dense stands: a productivity of 14,400 kg ha^{-1} has been estimated in Norway. In New Zealand, it outcompetes native tussock grasses by shading them. Seeds are wind dispersed.

Heavy grazing by sheep can strongly reduce the abundance. Moderate burning will encourage new growth but strong fires can completely kill plants of all ages. However, the plant will recolonize from seed.

References
231, 453, 454, 890, 1401, 1415.

Calophyllum antillanum Britt. Hypericaceae

LF: Evergreen tree
SN: *Calophyllum calaba* Jacq.
CU: Ornamental

Geographic Distribution

Europe	*Australasia*	*Atlantic Islands*
Northern	Australia	Cape Verde
British Isles	New Zealand	Canary, Madeira
Central, France		Azores
Southern	*Northern America*	South Atlantic Isl.
Eastern	Canada, Alaska	
Mediterranean Isl.	● Southeastern USA	*Indian Ocean Islands*
	Western USA	Mascarenes
Africa	Remaining USA	Seychelles
Northern	N Mexico	Madagascar
Tropical		
Southern	*Southern America*	*Pacific Islands*
	N Tropical	Micronesia
Asia	N Caribbean	Galapagos
Temperate	Chile, Argentina	X Hawaii
Tropical		Melanesia, Polynesia

Invaded Habitats
Coastal areas, mangrove forests, tropical hammocks.

Description
A tree of 12-20 m height, with a straight trunk, and young stems that are green, minutely hairy, and becoming gray with age. The opposite, simple and shiny leaves are elliptic, 10-15 cm long, with numerous lateral veins on the surface, with petioles and entire margins. Small, white and fragrant flowers are borne in axillary racemes with a few flowers each. They have ten yellow stamens. Fruits are brown, indehiscent, globose drupes of *c.* 2.5 cm diameter, each containing one seed.

Ecology and Control
A native of woodland, closed forests and river banks in areas of higher rainfall. Where invasive, the species forms dense stands crowding out native species and preventing their establishment. Seedlings and saplings reach high densities at the edge of mangrove swamps in Florida. The tree withstands inundation and brackish conditions, and produces large quantities of fruits.

Specific control methods for this species are not available. Seedlings and saplings can be hand pulled. Larger trees can be cut and the stumps treated with herbicide.

References
715.

Calotropis procera (Aiton) Aiton   Asclepiadaceae

LF: Evergreen shrub, tree   CU: -
SN: *Asclepias procera* Ait.

Geographic Distribution

Europe
   Northern
   British Isles
   Central, France
   Southern
   Eastern
   Mediterranean Isl.

Africa
N   Northern
N   Tropical
   Southern

Asia
N   Temperate
N   Tropical

Australasia
●   Australia
   New Zealand

Northern America
   Canada, Alaska
   Southeastern USA
   Western USA
   Remaining USA
   Mexico

Southern America
X   Tropical
X   Caribbean
   Chile, Argentina

Atlantic Islands
N   Cape Verde
   Canary, Madeira
   Azores
   South Atlantic Isl.

Indian Ocean Islands
   Mascarenes
   Seychelles
   Madagascar

Pacific Islands
   Micronesia
   Galapagos
   Hawaii
   Melanesia, Polynesia

Invaded Habitats
Riparian habitats, forest edges, sandy places, pastures, disturbed sites.

Description
A large shrub or small tree of 2-4 m height, with a white latex and smooth, grey-green stems and a thick, soft bark. The plant has a deep taproot of 3-4 m length. The simple and opposite leaves are 8-25 cm long, 4-14 cm wide, ovate, thick and waxy. They have a short pointed tip at the end and a heart-shaped base partly clasping the stem. The white and purple flowers have five lobes, are more or less tubular, and 2-3 cm in diameter. Fruits are grey-green, fleshy or dry capsules of 8-12 cm length and 6-8 cm width. They contain numerous small, brown and flattened seeds of 8-10 mm length and *c.* 5 mm width, with long white hairs attached at one end.

Ecology and Control
In the native range, this species occurs in savannas but becomes weedy in many disturbed sites. The shrub reproduces both by seeds and by suckering. It has a strong ability to regrow when damaged, and quickly forms adventitious shoots. Seeds are dispersed by wind and water, and are tolerant to high salinity levels. Due to root suckering, the species can form dense and large thickets that displace native shrubs and trees.

   Manual removal should aim at removing as much of the taproot and lateral roots as possible to prevent resprouting. Actively growing seedlings and larger plants can be treated with a mixture of 2,4-D and picloram. In the case of mature plants, herbicides may also be applied to the basal bark.

References
607, 608, 986, 1432.

Cardiospermum grandiflorum Sw.   Sapindaceae

LF: Evergreen climber   CU: Ornamental
SN: *Cardiospermum barbicule* Baker, *Cardiospermum hirsutum* Willd.

Geographic Distribution

Europe	*Australasia*	*Atlantic Islands*
Northern	● Australia	Cape Verde
British Isles	New Zealand	X Canary, Madeira
Central, France		Azores
Southern	*Northern America*	South Atlantic Isl.
Eastern	Canada, Alaska	
Mediterranean Isl.	Southeastern USA	*Indian Ocean Islands*
	Western USA	Mascarenes
Africa	Remaining USA	Seychelles
Northern	N Mexico	Madagascar
N Tropical		
● Southern	*Southern America*	*Pacific Islands*
	N Tropical	X Micronesia
Asia	N Caribbean	Galapagos
Temperate	N Chile, Argentina	X Hawaii
Tropical		Melanesia, Polynesia

Invaded Habitats
Forests, riparian habitats, rocky places.

Description
A tall liana reaching 10 m, with ribbed and branched stems. Leaves are 6-16 cm long, twice ternately compound with ovate to lanceolate, toothed leaflets of 2.5-6 cm length and 1.5-4 cm width. Leaves are often pubescent on the veins of the lower surface. Flowers are fragrant, cream to pale yellow, 8-10 mm long and borne on peduncles of 7-15 cm length. Petals are 7-9 mm long. Fruits are ellipsoid to ellipsoid-ovoid capsules of 5-7 cm length, green first then becoming dry, releasing round and black, winged seeds of 6-7 mm diameter.

Ecology and Control
This fast growing plant forms large and dense smothering curtains of tangled stems that impede the growth of supporting vegetation, eventually killing trees by the heavy weight. Seedlings of native shrubs and trees are unable to establish under the stands of this plant. The plant tolerates periodic inundation and the vigorous growth destroys riparian forests in Australia. It is a prolific seed producer and seeds are dispersed by water and wind.

    Seedlings and smaller plants can be pulled or dug out, the taproot must be removed to prevent regrowth. Larger vines are cut and the taproot dug out, cut stumps can be treated with herbicide. Follow-up programmes are necessary to control seedlings and regrowth.

References
549, 607, 608, 1089.

Carduus acanthoides L.   Asteraceae

LF: Annual, biennial herb   CU: -
SN: -

Geographic Distribution

Europe		Australasia		Atlantic Islands	
N	Northern		Australia		Cape Verde
N	British Isles	X	New Zealand		Canary, Madeira
N	Central, France				Azores
N	Southern		*Northern America*		South Atlantic Isl.
N	Eastern	X	Canada, Alaska		
N	Mediterranean Isl.		Southeastern USA		*Indian Ocean Islands*
		X	Western USA		Mascarenes
Africa		●	Remaining USA		Seychelles
	Northern		Mexico		Madagascar
	Tropical				
	Southern		*Southern America*		*Pacific Islands*
			Tropical		Micronesia
Asia			Caribbean		Galapagos
N	Temperate	X	Chile, Argentina		Hawaii
	Tropical				Melanesia, Polynesia

Invaded Habitats
Grassland and prairies, disturbed sites.

Description
A 20-150 cm tall, sparsely hairy herb with a well developed basal rosette and stems that are branched above and have spiny wings extending to the flowerheads. Leaves are narrowly elliptic to oblong, deeply lobed with the lobes having 1-3 points each ending in a spine. Flowerheads are terminal, solitary or clustered on young branches, 12-16 mm in diameter, and have numerous phyllaries each tapering into a spine of 1.5-2 mm length. The usually purple florets have tubes of 7-9 mm length. Fruits are compressed, light brown achenes of 2.5-4 mm length and have a pappus of 11-13 mm length.

Ecology and Control
This species grows best in soils of high fertility and the plant is found in the same habitats as *Carduus nutans*. It produces a large number of seeds that contribute to its invasion success. The species spreads rapidly after disturbances and can form dense species poor stands with 90,000 plants ha^{-1} that prevent the establishment of native forbs and grasses, displacing the invaded areas with thorny thickets.

An effective chemical control is applying a 2,4-D herbicide during periods of active growth. Seed production is strongly reduced if herbicides are applied just prior to flowering.

References
219, 330, 405, 406, 571.

Carduus nutans L. Asteraceae

LF: Annual, biennial herb
SN: -
CU: -

Geographic Distribution

Europe	*Australasia*	*Atlantic Islands*
N Northern	X Australia	Cape Verde
N British Isles	X New Zealand	Canary, Madeira
N Central, France		Azores
N Southern	*Northern America*	South Atlantic Isl.
N Eastern	X Canada, Alaska	
N Mediterranean Isl.	● Southeastern USA	*Indian Ocean Islands*
	X Western USA	Mascarenes
Africa	● Remaining USA	Seychelles
Northern	Mexico	Madagascar
Tropical		
Southern	*Southern America*	*Pacific Islands*
	X Tropical	Micronesia
Asia	Caribbean	Galapagos
N Temperate	X Chile, Argentina	Hawaii
X Tropical		Melanesia, Polynesia

Invaded Habitats
Grass- and woodland, pastures, rangeland.

Description
A biennial or annual herb of 20-200 cm height, with much branched stems and a long tap-root. Stems are erect and have spiny wings. Basal leaves are elliptic to lanceolate, 15-30 cm long, and pinnately lobed with each lobe ending in a spine. Stem leaves are smaller and simple. Large globose flowerheads of 2-7 cm diameter appear solitary at the ends of branches or branchlets. Phyllaries are numerous, glabrous or pubescent, 9-27 mm long and tapering to a spine of 2-5 mm length. The flowers are pink to purple and have a corolla tube of 10-14 mm length. Filaments are hairy. Fruits are achenes of 3.5-4 mm length.

Ecology and Control
A pioneer species growing best in fertile soils but also occuring in saline and low-fertility soils. It is taxonomically a complex group with several subspecies. Where invasive, it forms extensive stands with a density of up to 150,000 plants ha^{-1}. A single plant may produce up to 11,000 seeds. These are dispersed by wind, water and animals. The weed eliminates native vegetation and prevents the regeneration of native plants.

Effective herbicides are picloram or 2,4-D, applied during active growth of seedlings or rosettes. Mowing flowering plants is effective, but in populations with large phenological differences among plants, mowing more than once per season may be necessary. Cutting is most effective if done at the top of the root crown to destroy both terminal and lateral buds.

References
219, 330, 347, 506, 571, 581, 728, 852, 853, 1040, 1072, 1183, 1204.

Carduus pycnocephalus L. Asteraceae

LF: Annual, biennial herb CU: -
SN: -

Geographic Distribution

Europe		Australasia		Atlantic Islands	
X	Northern	●	Australia		Cape Verde
X	British Isles	X	New Zealand	N	Canary, Madeira
N	Central, France			N	Azores
N	Southern		*Northern America*		South Atlantic Isl.
N	Eastern		Canada, Alaska		
N	Mediterranean Isl.		Southeastern USA		*Indian Ocean Islands*
		●	Western USA		Mascarenes
Africa		X	Remaining USA		Seychelles
N	Northern		Mexico		Madagascar
	Tropical				
X	Southern		*Southern America*		*Pacific Islands*
		X	Tropical		Micronesia
Asia			Caribbean		Galapagos
N	Temperate	X	Chile, Argentina	X	Hawaii
N	Tropical				Melanesia, Polynesia

Invaded Habitats
Grassland, scrub, woodland, disturbed places.

Description
A tall herb of 20-200 cm height with erect, winged and branched stems, and a stout taproot. The plant has a grey to whitish felty pubescence. Leaves are prickly, oblong, with 2-5 pairs of palm-like lobes and a spine of up to 12 mm length at the end. The oblong-cylindrical flowerheads are 15-20 mm long, purplish, and borne in clusters of 2-5 at the ends of branches. Involucral bracts are slightly recurved or erect. Florets vary from pink to purple and have corolla lobes of 4-5 mm length. Fruits are golden to brown achenes of 4-6 mm length with 20 veins and an early deciduous pappus of 10-20 mm length.

Ecology and Control
In the native range, this highly variable species grows on stony hillslopes, in coastal scrub and disturbed places. Once established where introduced, it can become dominant and exclude native species. The cover of overwintering rosettes prevents establishment of other plant species. In savannas, it can carry grass fires to tree canopies and thus increases fire hazards. Fruit production is prolific: a single plant can produce up to 20,000 seeds per year. The long-lived seeds may remain viable up to 10 years.

Hand pulling is possible, but the root must be cut at least 10 cm below ground level to prevent regrowth. Mowed plants can regrow fast and still produce seeds. Grazing by sheep or goats has been proved to be effective. Chemical control is done by herbicides containing glyphosate, picloram, or 2,4-D.

References
178, 215, 392, 530, 581, 657, 986, 1183.

Carpobrotus edulis (L.) N. E. Br.   Aizoaceae

LF: Succulent perennial   CU: Erosion control, ornamental
SN: *Mesembryanthemum edule* L.

Geographic Distribution

Europe	*Australasia*	*Atlantic Islands*
Northern	● Australia	X Cape Verde
● British Isles	X New Zealand	X Canary, Madeira
Central, France		● Azores
● Southern	*Northern America*	● South Atlantic Isl.
Eastern	Canada, Alaska	
● Mediterranean Isl.	X Southeastern USA	*Indian Ocean Islands*
	● Western USA	Mascarenes
Africa	Remaining USA	Seychelles
X Northern	Mexico	Madagascar
Tropical		
N Southern	*Southern America*	*Pacific Islands*
	Tropical	Micronesia
Asia	Caribbean	Galapagos
Temperate	Chile, Argentina	Hawaii
Tropical		Melanesia, Polynesia

Invaded Habitats
Coastal dunes and cliffs, salt marshes, coastal scrub.

Description
A large perennial herb with succulent, opposite leaves and long, trailing stems that root at nodes and branch frequently. Stems are 8-13 mm in diameter and up to 2 m long. The dull to bright shining green leaves are triangulate in cross section, 4-10 cm long, 5-12 mm wide, straight or slightly curved, and finely toothed in the upper part. The keel is usually reddish. The large pink or yellow flowers are 6-9 cm in diameter and have pedicels of 1-2 cm length. The fleshy fruits are globose to obovoid, 2.5-3 cm long, green at first, becoming purple red, and contain numerous black seeds of *c.* 1 mm length.

Ecology and Control
The extensive vegetative growth of this plant leads to the formation of extensive, impenetrable and species poor mats up to 50 cm thick that may cover large areas, displacing native beach vegetation and preventing the establishment of native plants. In California, the plant poses a threat to several rare and endangered plant species. Fruits are eaten by mammals which effectively disperse the seeds. The plant grows both in moist and dry sites. Soils under mats of this plant are becoming increasingly acid.

    Individual plants can be hand pulled, but buried stems should be removed to prevent resprouting. Re-establishment of seedlings can be reduced by replanting the area with native species or covering with mulch. Large mats are removed by pulling and rolling them like a carpet. Chemical control is done by spraying individual patches with glyphosate.

References
12, 13, 133, 215, 300, 301, 302, 303, 434, 1242, 1338, 1339, 1340, 1342, 1374, 1375, 1376, 1455.

Carrichtera annua (L.) DC.   Brassicaceae

LF: Annual herb   CU: -
SN: -

Geographic Distribution

Europe	*Australasia*	*Atlantic Islands*
Northern	● Australia	Cape Verde
British Isles	New Zealand	X Canary, Madeira
Central, France		Azores
N Southern	*Northern America*	South Atlantic Isl.
Eastern	Canada, Alaska	
N Mediterranean Isl.	Southeastern USA	*Indian Ocean Islands*
	Western USA	Mascarenes
Africa	Remaining USA	Seychelles
N Northern	Mexico	Madagascar
Tropical		
Southern	*Southern America*	*Pacific Islands*
	Tropical	Micronesia
Asia	Caribbean	Galapagos
Temperate	Chile, Argentina	Hawaii
Tropical		Melanesia, Polynesia

Invaded Habitats
Grassland, heath- and shrubland.

Description
A pubescent herb of 5-60 cm height, branched from the base, with erect or ascending stems and a deep taproot. The petiolated leaves are up to 10 cm long, deeply divided and have linear segments. Inflorescences are racemes of 5-15 cm length with 10-18 flowers each. Petals are pale yellow, 6-8 mm long, and have violet veins. Fruits are siliques of 5-8 mm length, borne on pedicels of 2-4 mm length.

Ecology and Control
A common plant of dry open and disturbed sites in the native range. Seeds develop a mucilaginous layer when wetted and adhere to the soil surface. Where invasive, the plant builds dense populations that compete for space, water and nutrients and shade out native forbs and grasses. Little is known on the ecology of this species.

Scattered plants can be hand pulled or dug out, the taproot must be removed. Larger patches are sprayed with herbicides.

References
215, 515, 1172, 1450.

Casuarina equisetifolia L. Casuarinaceae

LF: Evergreen tree
SN: -
CU: Erosion control, ornamental, wood

Geographic Distribution

Europe	*Australasia*	*Atlantic Islands*
Northern	N Australia	Cape Verde
British Isles	New Zealand	Canary, Madeira
Central, France		Azores
Southern	*Northern America*	South Atlantic Isl.
Eastern	Canada, Alaska	
Mediterranean Isl.	• Southeastern USA	*Indian Ocean Islands*
	Western USA	X Mascarenes
Africa	Remaining USA	Seychelles
Northern	Mexico	Madagascar
X Tropical		
• Southern	*Southern America*	*Pacific Islands*
	Tropical	Micronesia
Asia	• Caribbean	X Galapagos
Temperate	Chile, Argentina	X Hawaii
Tropical		N Melanesia, Polynesia

Invaded Habitats
Grassland, forests, coastal swamps and dunes.

Description
A semi-deciduous to evergreen, much-branched tree of 7-25 m height, with a conical shape and a rough dark grey bark. The minute and scale-like leaves are *c.* 1 mm long, borne in whorls of 6-8, and encircling the branchlets at regularly spaced nodes. Branchlets are very slender and resemble pine needles. Male flowers consist of scale-like perianth segments and one stamen, female flowers lack a perianth. Male spikes are 1-4 cm long, female flowerheads ovoid and 3-4 mm long. Cones are 1-2 cm long and 10-15 mm wide, releasing pale brown, winged samaras of 6-8 mm length.

Ecology and Control
In the native range, the tree grows in beach forests and on strands. It is a fast growing pioneer species, regenerating both by seeds and by resprouting. The tree is nitrogen-fixing, highly salt tolerant, and grows in sandy soils of the coast. It forms dense monospecific stands crowding out native vegetation and eliminating the food sources for wildlife. Its shallow root system inhibits nest building by sea turtles in Florida. The large amounts of litter hampers germination and growth of native species.

Seedlings and saplings can be hand pulled. Mature trees are effectively killed by spraying triclopyr in a band around the base of the trunk. If trees are cut, the stumps should be treated with triclopyr. In fire-tolerant communities, prescribed fires are used to eradicate this tree, but they are effective only in dense stands with sufficient dry fuel.

References
45, 684, 704, 715, 884, 927, 972, 1089.

Casuarina glauca Sieber ex Spreng. Casuarinaceae

LF: Evergreen tree
SN: -

CU: Ornamental, wood

Geographic Distribution

Europe
   Northern
   British Isles
   Central, France
   Southern
   Eastern
   Mediterranean Isl.

Africa
   Northern
   Tropical
   Southern

Asia
   Temperate
   Tropical

Australasia
  N Australia
      New Zealand

Northern America
   Canada, Alaska
  ● Southeastern USA
    Western USA
    Remaining USA
    Mexico

Southern America
    Tropical
  X Caribbean
    Chile, Argentina

Atlantic Islands
   Cape Verde
   Canary, Madeira
   Azores
   South Atlantic Isl.

Indian Ocean Islands
   Mascarenes
   Seychelles
   Madagascar

Pacific Islands
   Micronesia
   Galapagos
  ● Hawaii
   Melanesia, Polynesia

Invaded Habitats
Coastal beaches, sandy banks.

Description
A tree with a conical shape, 8-20 m tall, with a finely fissured and scaly, grey-brown bark. The spreading to drooping branchlets are up to 38 cm long and *c.* 1.5 mm thick. Leaves are minute and scale-like, *c.* 1 mm long, and are borne in whorls of 10-16, encircling the branchlets; whorls are 8-12 mm apart. The tree is dioecious with male spikes being 1-4 cm in length. Female inflorescences appear on short lateral branchlets of 3-12 mm length. Cones are ferruginous to white, pubescent at first and becoming glabrous. The cone body is 9-18 mm long and 7-9 mm in diameter. Fruits are samaras of 3.5-5 mm length.

Ecology and Control
In the native range, this tree occurs on saline swamp flats, on estuarine floodplains, wetland forests and along salt marshes. The tree is nitrogen-fixing and not as salt tolerant as the previous species. It frequently produces root suckers and coppices vigorously. It often forms pure stands that reduce native species diversity in the understorey. The tree produces large amounts of litter that prevent the growth and establishment of native species. Seeds are wind dispersed.

Specific control methods for this species are not available. Seedlings and saplings can be removed manually. For further control methods see the previous species.

References
254, 715, 927.

Cecropia peltata L.   Moraceae

LF: Evergreen tree   CU: Ornamental
SN: -

Geographic Distribution

Europe		*Australasia*		*Atlantic Islands*	
	Northern		Australia		Cape Verde
	British Isles		New Zealand		Canary, Madeira
	Central, France				Azores
	Southern	*Northern America*			South Atlantic Isl.
	Eastern		Canada, Alaska		
	Mediterranean Isl.		Southeastern USA	*Indian Ocean Islands*	
			Western USA		Mascarenes
Africa			Remaining USA		Seychelles
	Northern	N	Mexico		Madagascar
●	Tropical				
	Southern	*Southern America*		*Pacific Islands*	
		N	Tropical		Micronesia
Asia		N	Caribbean		Galapagos
	Temperate		Chile, Argentina	X	Hawaii
X	Tropical			X	Melanesia, Polynesia

Invaded Habitats
Forests and forest gaps, disturbed sites.

Description
A dioecious tree reaching 20 m height or more, with hollow stems and a thin spreading crown of a few stout branches. The bark is grey, smooth and thin. Leaves are peltate, clustered at the ends of branches, and have stout petioles of 30-50 cm length. The leaf blade is rounded, 30-75 cm in diameter, and has 7-11 lobes. Male flowers are borne in slender spikes of 3-5 cm length, arranged in clusters. Female flowers are borne in spikes of 4-6 cm length, enlarging in fruit and becoming slightly fleshy. Fruits are minute, numerous and contain one seed of *c.* 1.5 mm length each.

Ecology and Control
The tree is an early successional, light demanding, fast growing species that colonizes tree-fall gaps of rainforests in the native range. It produces large seed crops annually, and the seeds are dispersed by birds and bats. The small seeds accumulate to high densities in the soil. The tree quickly occupies gaps and forms dense stands that replace the tree *Musanga cecropioides* in Africa. It sprouts easily after damage.

Young trees and seedlings can easily be hand pulled. If larger trees are cut, the cut stumps should be treated with herbicide to prevent regrowth.

References
401, 418, 588, 650, 848, 1053, 1091, 1177, 1365.

Cedrela odorata L.   Meliaceae

LF: Deciduous tree   CU: Ornamental, wood
SN: *Cedrela mexicana* Roem.

Geographic Distribution

Europe	*Australasia*	*Atlantic Islands*
Northern	Australia	Cape Verde
British Isles	New Zealand	Canary, Madeira
Central, France		Azores
Southern	*Northern America*	South Atlantic Isl.
Eastern	Canada, Alaska	
Mediterranean Isl.	X Southeastern USA	*Indian Ocean Islands*
	Western USA	Mascarenes
Africa	Remaining USA	Seychelles
Northern	N Mexico	Madagascar
● Tropical		
Southern	*Southern America*	*Pacific Islands*
	N Tropical	Micronesia
Asia	N Caribbean	● Galapagos
Temperate	Chile, Argentina	Hawaii
Tropical		Melanesia, Polynesia

Invaded Habitats
Forests and forest edges.

Description
A large tree of 30-60 m height and a trunk diameter of up to 1.5 m, with a dense, rounded or somewhat flat crown, buttressed to *c.* 1 m height, and with a greyish brown to grey-black bark that is regularly and evenly furrowed or fissured. Leaves are alternate, 30-60 cm long, pinnately compound with 10-22 pairs of lanceolate to ovate, entire leaflets of 5-15 cm length. The small greenish white to yellowish flowers have an unpleasant odour and are borne in much branched thyrses. The fruit is a woody capsule of 2-4 cm length, opening from the end by five valves, and contains 40-50 winged seeds.

Ecology and Control
In the native range, this tree occurs in semi-deciduous to evergreen lowland and montane tropical forests. It grows rapidly and is a highly variable species in the native range. It is intolerant of shade and does not coppice or sucker from roots. Seeds are wind dispersed and germinate easily. High seedling densities are common. The tree shades out native plants with its large leaves, displacing them and building up species poor monospecific stands. Native shrubs and trees are unable to establish in these stands.

Specific control methods for this species are not available. Seedlings and saplings may be hand pulled or dug out. Larger trees can be killed by cutting at ground level.

References
253, 273, 711, 835, 1168, 1313.

Celastrus orbiculatus Thunb.   Celastraceae

LF: Deciduous vine   CU: Ornamental
SN: *Celastrus articulatus* Thunb.

Geographic Distribution

Europe	*Australasia*	*Atlantic Islands*
Northern	Australia	Cape Verde
X British Isles	● New Zealand	Canary, Madeira
Central, France		Azores
Southern	*Northern America*	South Atlantic Isl.
Eastern	Canada, Alaska	
Mediterranean Isl.	● Southeastern USA	*Indian Ocean Islands*
	Western USA	Mascarenes
Africa	Remaining USA	Seychelles
Northern	Mexico	Madagascar
Tropical		
Southern	*Southern America*	*Pacific Islands*
	Tropical	Micronesia
Asia	Caribbean	Galapagos
N Temperate	Chile, Argentina	Hawaii
Tropical		Melanesia, Polynesia

Invaded Habitats
Forests, riparian habitats, rocky places.

Description
A shrub or vine, climbing up to 12 m height, with twining young stems and a pair of spines at each bud. The oblong to obovate leaves are alternate, serrated at margins, 5-10 cm long, acute or acuminate, and have petioles of 3-8 cm length. Small greenish to whitish flowers of *c.* 4 mm diameter are borne in small axillary cymes, with the peduncles more or less as long as the pedicels. The fruit is a distinct orange-yellow, globose and dehiscent capsule of *c.* 8 mm diameter, containing 3-6 scarlet-coated seeds. The species has an extensive below-ground rhizome system.

Ecology and Control
This fast growing vine prolifically sprouts from below-ground rhizomes. It grows along the floor, into the canopies, and its impacts are blanketing and shading out the native vegetation, killing trees due to strangulation, and inceasing susceptibility to wind and ice damage of host trees, reducing forest regeneration in invaded areas. Seeds are dispersed by birds and mammals, and germinate even in low light. Seedlings are very shade tolerant; they respond to gaps by quickly increasing photosynthesis and growth rate.

    Control is difficult because root fragments may produce new plants. Manual control includes regular mowing, cutting, and hand pulling. Triclopyr herbicides applied to cut stems is best done late in season when the first frosts appear. Foliar sprays with 2,4-D plus triclopyr are also effective.

References
362, 363, 413, 489, 616, 788, 850, 1095, 1178, 1291, 1397.

Celtis sinensis Pers.   Ulmaceae

LF: Deciduous tree
SN: *Celtis japonica* Planch.
CU: Ornamental, shelter

Geographic Distribution

Europe
    Northern
    British Isles
    Central, France
    Southern
    Eastern
    Mediterranean Isl.

Africa
    Northern
    Tropical
X Southern

Asia
N Temperate
    Tropical

Australasia
● Australia
    New Zealand

Northern America
    Canada, Alaska
    Southeastern USA
    Western USA
    Remaining USA
    Mexico

Southern America
    Tropical
    Caribbean
    Chile, Argentina

Atlantic Islands
    Cape Verde
    Canary, Madeira
    Azores
    South Atlantic Isl.

Indian Ocean Islands
    Mascarenes
    Seychelles
    Madagascar

Pacific Islands
    Micronesia
    Galapagos
    Hawaii
    Melanesia, Polynesia

Invaded Habitats
Alluvial flats, moist gullies, riparian habitats.

Description
A tree up to 20 m tall with dark brown branchlets. Leaves are dark green, ovate-oblong to broad-ovate, almost glabrous above, entire in the lower half and serrate in the upper half. Leaf blades are 4-10 cm long, the petiole is 5-10 mm long. Flowers are dark orange and appear in axillary inflorescences with one to few flowers. The calyx has 4-5 lobes that are broadly ovate and *c.* 2 mm long. Stamens are slightly exceeding the perianth. Fruits are globose to ovoid drupes of 7-9 mm length, orange to black and with firm outer coats. They contain one pitted and ribbed stone seed each.

Ecology and Control
Where native, this tree often forms thickets along the coast. The tree is a relatively shade tolerant species that often becomes dominant in riparian forests. It fruits prolifically, and seeds are dispersed by birds and bats. Seeds are relatively short-lived in or on the soil and the species does not accumulate a persistent seed bank; recolonization is enhanced by the abundant fruiting. It outcompetes native shrubs and trees by forming dense stands and prevents their regeneration.

    Specific control methods for this species are not available. Seedlings and saplings may be pulled or dug out, larger trees cut and the cut stumps treated with herbicide.

References
607, 608, 979.

Cenchrus ciliaris L.   Poaceae

LF: Perennial herb
CU: Erosion control, forage
SN: *Pennisetum cenchroides* Rich., *Pennisetum ciliare* (L.) Link

Geographic Distribution

Europe
- Northern
- British Isles
- Central, France
- N Southern
- Eastern
- N Mediterranean Isl.

Africa
- N Northern
- N Tropical
- N Southern

Asia
- N Temperate
- N Tropical

Australasia
- ● Australia
- New Zealand

Northern America
- Canada, Alaska
- Southeastern USA
- ● Western USA
- Remaining USA
- X Mexico

Southern America
- X Tropical
- X Caribbean
- X Chile, Argentina

Atlantic Islands
- N Cape Verde
- X Canary, Madeira
- Azores
- South Atlantic Isl.

Indian Ocean Islands
- Mascarenes
- Seychelles
- Madagascar

Pacific Islands
- Micronesia
- Galapagos
- ● Hawaii
- Melanesia, Polynesia

Invaded Habitats
Grass- and shrubland, arid rangeland.

Description
A tufted or spreading, sometimes short-rhizomatous, deep rooting grass of 10-150 cm height, with erect or creeping and ascending culms. Leaf blades are usually glabrous, 3-25 cm long and 2-13 mm wide. The sheaths are keeled and the ligules consist of densely ciliate membranes of 0.5-2.5 mm length. Inflorescences are cylindrical to ovoid panicles of 2-14 cm length and 1-2.5 cm width, grey, purple, or yellowish. Spikelets are 2-5.5 mm long, surrounded by bristles, and contain two florets each. Glumes are distinct and unequal. The first glume is membranous and 1-3 mm long, the second glume longer. Caryopses are ovoid, 1.5-2 mm long and *c.* 1 mm in diameter.

Ecology and Control
A native of semi-arid areas, the grass grows mainly in open bush and grassland. It is very drought resistant but cannot stand flooding for more than a week. Several cultivars exist, with and without rhizomes. Where invasive, the grass forms dense mats or tussocks that displace native grassland communities and reduce their species richness. The dried shoots provide excellent fuel for fires and thus increase fire hazards. The cover of this grass may increase rapidly after a fire. Seeds are wind dispersed. Once established, the grass will withstand considerable grazing.

Chemical control is done by herbicides containing 2,2-DPA or paraquat applied before flowering. Tebuthiuron is an effective pre-emergence herbicide.

References
180, 607, 1068, 1190.

Centaurea calcitrapa L. Asteraceae

LF: Annual, biennial herb   CU: -
SN: *Centaurea myacantha* L.

Geographic Distribution

Europe		*Australasia*		*Atlantic Islands*	
	Northern	●	Australia	X	Cape Verde
X	British Isles	X	New Zealand	N	Canary, Madeira
N	Central, France				Azores
N	Southern	*Northern America*			South Atlantic Isl.
N	Eastern		Canada, Alaska		
N	Mediterranean Isl.		Southeastern USA	*Indian Ocean Islands*	
		●	Western USA		Mascarenes
Africa		X	Remaining USA		Seychelles
N	Northern		Mexico		Madagascar
	Tropical				
X	Southern	*Southern America*		*Pacific Islands*	
		X	Tropical		Micronesia
Asia			Caribbean		Galapagos
N	Temperate	X	Chile, Argentina		Hawaii
	Tropical				Melanesia, Polynesia

Invaded Habitats
Dry forests, grassland, woodland.

Description
A 40-80 cm tall herb with whitish to pale green and much branched stems, sparsely pubescent and without spines. The dark green leaves are deeply lobed, covered with short hairs, and sometimes toothed. Basal rosette leaves reach 25 cm length whereas stem leaves are much smaller and sessile. Numerous flowerheads appear solitary at the ends of branches or in the axils of upper leaves. They are mostly sessile and surrounded by numerous bracts each having a sharp white or yellowish spine of 15-30 mm length with 2-6 shorter spines at the base. Florets are mostly purple, occasionally pink or white. Fruits are whitish achenes with brown streaks, 3-4 mm long and *c.* 2 mm wide, smooth and without pappus. The plant develops a fleshy, thick taproot of 2-4 cm diameter.

Ecology and Control
This plant establishes in disturbed sites and builds up large populations that eliminate native vegetation by shading out and competing for space, water and nutrients. The large rosettes form a dense cover over large areas. Seeds lack a pappus and are not well adapted to wind dispersal. They remain viable in the soil for 2-3 years.

Isolated plants can be removed mechanically. Slashing is effective when done just before flowering and prevents seed dispersal. Plants are susceptible to 2,4-D and dicamba in the seedling and rosette stage. If plants are removed manually, as much of the taproot as possible should be removed.

References
133, 215, 985, 986.

Centaurea solstitialis L. — Asteraceae

LF: Annual herb
SN: -
CU: -

Geographic Distribution

Europe	*Australasia*	*Atlantic Islands*
Northern	X Australia	Cape Verde
X British Isles	X New Zealand	Canary, Madeira
X Central, France		Azores
N Southern	*Northern America*	South Atlantic Isl.
N Eastern	X Canada, Alaska	
N Mediterranean Isl.	Southeastern USA	*Indian Ocean Islands*
	● Western USA	Mascarenes
Africa	X Remaining USA	Seychelles
N Northern	Mexico	Madagascar
Tropical		
X Southern	*Southern America*	*Pacific Islands*
	Tropical	Micronesia
Asia	Caribbean	Galapagos
N Temperate	X Chile, Argentina	Hawaii
Tropical		Melanesia, Polynesia

Invaded Habitats
Grassland, rangeland, scrub, disturbed places.

Description
A mostly annual herb with several much-branched, winged stems of 15-200 cm height, and a deep taproot. The basal and lower stem leaves are up to 20 cm long and deeply lobed, the upper ones long, narrow, entire, and decurrent. Leaves become greyish green due to a dense cover of hairs. One to many solitary and spiny flowerheads with involucres of 13-17 mm height are produced at the ends of branches. The outer phyllaries have spiny appendages and a long central spine of 10-25 mm length. Florets are 13-20 mm tall and bright yellow. The inner seeds are 2-3 mm long, and have pappus hairs of *c.* 5 mm length, the outer seeds are dark brown, 3-4 mm long, and usually without a pappus.

Ecology and Control
The plant quickly forms dense infestations that may cover large areas, displace native vegetation and affect wildlife. It depletes soil moisture reserves in grasslands of western North America. The species is very variable in its growth. Large plants can produce up to 75,000 seeds in a year. Seeds may remain viable in the soil for more than 10 years.
    Physical removal by tillage is effective but exposes the soil to reinfestations. Mowing is best done when plants start to flower to minimize plant regrowth. In California, prescribed burning is used to kill the plant and to reduce the seed bank. Chemical control is done by applying 2,4-D or triclopyr to plants in the seedling or rosette stages. If plants have reached the bolting stage, glyphosate solution is effective.

References
103, 133, 195, 342, 537, 802, 803, 986, 1096, 1169, 1170.

Centaurea stoebe L. — Asteraceae

LF: Biennial, perennial herb
SN: *Centaurea maculosa* Lam.
CU: -

Geographic Distribution

Europe
    Northern
    British Isles
N   Central, France
N   Southern
N   Eastern
N   Mediterranean Isl.

Africa
    Northern
    Tropical
    Southern

Asia
    Temperate
    Tropical

Australasia
    Australia
X   New Zealand

Northern America
X   Canada, Alaska
●   Southeastern USA
X   Western USA
●   Remaining USA
    Mexico

Southern America
    Tropical
    Caribbean
    Chile, Argentina

Atlantic Islands
    Cape Verde
    Canary, Madeira
    Azores
    South Atlantic Isl.

Indian Ocean Islands
    Mascarenes
    Seychelles
    Madagascar

Pacific Islands
    Micronesia
    Galapagos
    Hawaii
    Melanesia, Polynesia

Invaded Habitats
Grassland, riparian habitats, rangeland, woodland.

Description
A herb of 30-100 cm height with erect or ascending, branched stems. Leaves are alternate and much divided, except the upper leaves which are linear. Numerous flowerheads of *c.* 6 mm diameter and 16-20 mm height are borne in corymbs or corymbose panicles. Bracts have a black-fringed tip of 1-2 mm length. Florets are purple, rarely white. Fruits are brownish achenes of *c.* 3 mm length, with a persistent pappus of 1-2 mm length.

Ecology and Control
Naturalized plants are referred to subsp. *stoebe*. The plant quickly establishes after disturbance and forms a dense cover in the form of rosettes, the density may exceed 400 plants m^{-2}. It displaces native perennial grasses, reduces native species richness, and causes degraded wildlife habitat. On rangelands, sites dominated by this plant show increased runoff and erosion. The plant is a prolific seed producer with up to 40,000 seeds m^{-2}. Seeds are not well adapted for long-distance dispersal but may adhere to animal furs. Germination takes place over a wide range of environmental conditions and under low light conditions.

    Scattered plants may be pulled or dug out. Slashing before flowering prevents seed dispersal. Effective chemical control include herbicides containing 2,4-D, dicamba, or picloram. Mowing bolting and fruiting stems can reduce the number of seed bearing stems.

References
128, 310, 537, 571, 653, 1082, 1213, 1314, 1315, 1363.

Cestrum diurnum L.   Solanaceae

LF: Evergreen shrub   CU: Ornamental
SN: -

Geographic Distribution

Europe	*Australasia*	*Atlantic Islands*
Northern	Australia	Cape Verde
British Isles	New Zealand	Canary, Madeira
Central, France		Azores
Southern	*Northern America*	South Atlantic Isl.
Eastern	Canada, Alaska	
Mediterranean Isl.	● Southeastern USA	*Indian Ocean Islands*
	Western USA	Mascarenes
Africa	Remaining USA	Seychelles
Northern	Mexico	Madagascar
Tropical		
Southern	*Southern America*	*Pacific Islands*
	N Tropical	Micronesia
Asia	N Caribbean	Galapagos
Temperate	Chile, Argentina	X Hawaii
X Tropical		Melanesia, Polynesia

Invaded Habitats
Tropical hammocks, coastal vegetation, disturbed places.

Description
A shrub of 2-5 m height or tree up to 10 m, with a whitish hard wood and multiple trunks reaching 8 cm diameter. The outer bark is smooth and grey, the inner yellow brown and slightly bitter. Twigs are slender, greenish grey and often covered with fine appressed hairs. The alternate leaves have light green petioles of 5-13 mm length, are 6-10 cm long and 2.5-3.5 cm wide, shiny green to yellow green above and whitish green beneath. The leaf blades are usually glabrous and entire. Greenish-white flowers of 13-16 mm length are borne in clusters in the axils of the uppermost leaves. The whitish green calyx is bell-shaped and the corolla has a narrow tube. Fruits are elliptic to globose, purplish black and juicy berries of 8-10 mm diameter, containing 4-14 brown and angled seeds of *c*. 3 mm length.

Ecology and Control
Where native, this shrub is common in thickets and woodland margins. A number of varieties have been described. The species is shade tolerant and also to some extent salt tolerant. It is mostly found on dry soils and it forms dense thickets that crowd out native vegetation and prevent the regeneration of native shrubs and trees. Seeds are dispersed by birds. The fruits are attractive to birds but poisonous to humans and other mammals.
    Seedlings and saplings can be hand pulled or dug out. Larger plants are cut and the cut stumps treated with herbicide. For effective herbicides, see next species.

References
715.

Cestrum laevigatum Schltdl.   Solanaceae

LF: Evergreen shrub, tree   CU: Ornamental
SN: -

Geographic Distribution

Europe	*Australasia*	*Atlantic Islands*
Northern	Australia	Cape Verde
British Isles	New Zealand	Canary, Madeira
Central, France		Azores
Southern	*Northern America*	South Atlantic Isl.
Eastern	Canada, Alaska	
Mediterranean Isl.	Southeastern USA	*Indian Ocean Islands*
	Western USA	Mascarenes
Africa	Remaining USA	Seychelles
Northern	Mexico	Madagascar
Tropical		
● Southern	*Southern America*	*Pacific Islands*
	N Tropical	Micronesia
Asia	Caribbean	Galapagos
Temperate	N Chile, Argentina	Hawaii
Tropical		Melanesia, Polynesia

Invaded Habitats
Riparian habitats, coastal dunes, forests, grassland.

Description
A sparsely hairy shrub of 1-2 m height or a tree up to 15 m tall. Leaves and stems have an unpleasant smell. Leaves are elliptic-lanceolate, up to 15 cm long and 5 cm wide. Flowers are greenish-yellow and borne in clusters in the axils of leaves. They have a long tube. Fruits are berries of *c.* 10 mm length, green at first, becoming purple-black when ripe.

Ecology and Control
The plant forms dense stands that shade out native plants and prevent the natural regeneration of shrubs and trees. Seeds are dispersed by birds. Young leaves and berries are poisonous to livestock. Little is known on the ecology of this species as an invasive plant.

    Seedlings and small plants are hand pulled or dug out. Larger plants are cut and the cut stumps treated with herbicide. Effective herbicides are picloram, triclopyr plus imazapyr, or 2,4-D plus picloram. Triclopyr can be used as a basal stem treatment.

References
325, 549, 1089.

Chamaecytisus prolifer (L.f.) Link   Fabaceae

LF: Evergreen shrub CU: Fodder
SN: *Cytisus prolifer* L., *Cytisus palmensis* (Christ) Hutch.

Geographic Distribution

Europe	*Australasia*	*Atlantic Islands*
Northern	● Australia	Cape Verde
British Isles	X New Zealand	N Canary, Madeira
Central, France		Azores
Southern	*Northern America*	South Atlantic Isl.
Eastern	Canada, Alaska	
Mediterranean Isl.	Southeastern USA	*Indian Ocean Islands*
	Western USA	Mascarenes
Africa	Remaining USA	Seychelles
Northern	Mexico	Madagascar
Tropical		
Southern	*Southern America*	*Pacific Islands*
	Tropical	Micronesia
Asia	Caribbean	Galapagos
Temperate	Chile, Argentina	Hawaii
Tropical		Melanesia, Polynesia

Invaded Habitats
Grass- and heathland, dry forests, riparian habitats, dry coastal hills.

Description
A variable shrub of 2-5 m height and with rounded, slightly angled twigs. The compound leaves have petioles of *c.* 1 cm length and three linear-lanceolate to elliptic leaflets of 2-4 cm length. White to cream, fragrant flowers of 16-25 mm length are borne in axillary clusters of 3-8; the pedicels are 7-13 mm long. The calyx is densely hairy. Fruits are grey to black, oblong pods of 3-8 cm length, each containing many shiny black, compressed and ellipsoid seeds of *c.* 5 mm length.

Ecology and Control
In the native range, this shrub is commonly found in pine forests and in evergreen laurel forests. Three subspecies and five varieties have been described, two of which have become naturalized in Australia: var. *palmensis* and var. *prolifer*. The plant is nitrogen-fixing and increases soil fertility levels. It is a prolific seed producer, and seeds may remain dormant in the soil for more than 10 years. Seeds are dispersed by birds and ants. Dense infestations shade out native species and prevent their regeneration.

Seedlings and small plants can be pulled or dug out. Larger shrubs are cut and the cut stumps treated with herbicide. Larger infestations are slashed or burned to stimulate seed germination. Seedlings are then sprayed with herbicide. Follow-up treatments over several years are necessary to control seedlings and regrowth.

References
215, 607, 608, 924.

Chasmanthe floribunda (Salisb.) N.E.Br. Iridaceae

LF: Perennial herb
SN: *Antholyza floribunda* Salisb.
CU: Ornamental

Geographic Distribution

Europe	*Australasia*	*Atlantic Islands*
Northern	• Australia	Cape Verde
British Isles	New Zealand	Canary, Madeira
Central, France		Azores
Southern	*Northern America*	South Atlantic Isl.
Eastern	Canada, Alaska	
Mediterranean Isl.	Southeastern USA	*Indian Ocean Islands*
	X Western USA	Mascarenes
Africa	Remaining USA	Seychelles
Northern	Mexico	Madagascar
Tropical		
N Southern	*Southern America*	*Pacific Islands*
	Tropical	Micronesia
Asia	Caribbean	Galapagos
Temperate	Chile, Argentina	Hawaii
Tropical		Melanesia, Polynesia

Invaded Habitats
Grass- and heathland, sclerophyll forests, riparian habitats.

Description
A perennial herb with a tuber and corms up to 6 cm in diameter, and annual leaves and flowering stems. Leaves are 25-80 cm long and 2-4 cm wide, and have a prominent midvein. Flower bearing stems are glabrous, few-branched and up to 1 m tall. The tubular and asymmetric flowers are orange-red to yellow, 6-8 cm long, and borne in spikes of 15-25 cm length. Fruits are depressed globose capsules of 10-15 mm diameter, releasing smooth and orange seeds.

Ecology and Control
The species grows best on well drained calcareous soils. Two varieties have become naturalized in Australia: var. *floribunda* with orange flowers and var. *duckittii* with yellow flowers. The flowers are attractive to nectar feeding birds, and the plant produces copious amounts of seed. Vegetative spread is possible due to the formation of daughter cormlets at the end of the growing season. The plant builds up large and dense populations that eliminate native vegetation by competing for space, water and nutrients.

   Specific control methods for this species are not available. The same methods as for *Watsonia* or *Moraea* species may apply.

References
121, 215, 655.

Chenopodium album L.   Chenopodiaceae

LF: Annual herb   CU: -
SN: -

Geographic Distribution

Europe		*Australasia*		*Atlantic Islands*	
X	Northern	X	Australia		Cape Verde
X	British Isles	X	New Zealand	X	Canary, Madeira
N	Central, France			X	Azores
X	Southern	*Northern America*			South Atlantic Isl.
N	Eastern	X	Canada, Alaska		
X	Mediterranean Isl.	X	Southeastern USA	*Indian Ocean Islands*	
		X	Western USA		Mascarenes
Africa		X	Remaining USA		Seychelles
X	Northern	X	Mexico		Madagascar
X	Tropical				
•	Southern	*Southern America*		*Pacific Islands*	
		X	Tropical		Micronesia
Asia		X	Caribbean		Galapagos
N	Temperate	X	Chile, Argentina	X	Hawaii
X	Tropical				Melanesia, Polynesia

Invaded Habitats
Grassland.

Description
An erect and rigid herb with green to reddish and grey-farinose stems of 10-200 cm height and a strong taproot. The alternate, simple leaves range from rhombic-ovate to lanceolate in shape, are 1-8 cm long and 3-55 mm wide, entire or weakly toothed at the margins, and greyish green below. Inflorescences are spike-like panicles in leaf axils or at the ends of stems and branches. Greenish and sessile, small flowers are borne in dense clusters. The somewhat keeled sepals have whitish and membranous margins, and nearly cover the mature fruit. A single fruit consists of the seed that is covered by the thin papery pericarp. Seeds are black, glossy, lens-shaped and 1.2-1.8 mm in diameter.

Ecology and Control
Taxonomically a highly variable and complex species whose exact native distribution and origin are uncertain. It is found from sea level up to 3,600 m elevation and is one of the most successful colonizing species. Growth is best on fertile soils but it tolerates a wide range of different soil types. It forms dense patches that crowd out native species. Seeds are abundantly produced and carried by waters or dispersed by animals. Large plants may produce more than 40,000 seeds per year. Establishment of this plant depends on disturbance.

The species is unable to withstand clipping, and cutting plants during early stages of growth will provide good control. An effective herbicide is MCPA.

References
81, 553, 580, 1089, 1412.

Chloris virgata Sw.   Poaceae

LF: Annual herb   CU: Fodder, forage
SN: *Chloris compressa* DC., *Chloris elegans* Kunth

Geographic Distribution

Europe
     Northern
X British Isles
     Central, France
     Southern
     Eastern
     Mediterranean Isl.

Africa
     Northern
X Tropical
     Southern

Asia
     Temperate
N Tropical

Australasia
● Australia
     New Zealand

Northern America
     Canada, Alaska
     Southeastern USA
     Western USA
X Remaining USA
     Mexico

Southern America
N Tropical
     Caribbean
     Chile, Argentina

Atlantic Islands
     Cape Verde
X Canary, Madeira
     Azores
     South Atlantic Isl.

Indian Ocean Islands
     Mascarenes
     Seychelles
     Madagascar

Pacific Islands
     Micronesia
     Galapagos
● Hawaii
     Melanesia, Polynesia

Invaded Habitats
Grass- and shrubland, forest edges.

Description
A tufted or stoloniferous, annual or short-lived perennial grass of 30-90 cm height with mostly erect culms, sometimes rooting at the lower nodes. Basal sheaths are strongly keeled, ligules are membranous and short. Leaf blades are flat, 10-30 cm long and 2-6 mm wide. Inflorescences are composed of 4-12 white to yellowish brown spikes of 2-10 cm length. Spikelets have 2-3 florets each. The lower lemma has an awn of 5-15 mm length. Fruits are golden yellow, cylindrical caryopses of 1.5-2 mm length.

Ecology and Control
A successful competitor to native grasses and forbs, this grass displaces native vegetation by forming dense swards. The species spreads mainly by vegetative growth. Little is known on the ecology of this species.
     Specific control methods for this species are not available. Scattered plants may be pulled or dug out, larger patches are cut to prevent seed formation or sprayed with herbicide.

References
76, 1190.

Chromolaena odorata (L.) King & Rob. — Asteraceae

LF: Evergreen shrub, climber CU: -
SN: *Eupatorium odoratum* L.

Geographic Distribution

Europe	*Australasia*	*Atlantic Islands*
Northern	X Australia	Cape Verde
British Isles	New Zealand	Canary, Madeira
Central, France		Azores
Southern	*Northern America*	South Atlantic Isl.
Eastern	Canada, Alaska	
Mediterranean Isl.	Southeastern USA	*Indian Ocean Islands*
	Western USA	X Mascarenes
Africa	Remaining USA	Seychelles
Northern	N Mexico	Madagascar
● Tropical		
● Southern	*Southern America*	*Pacific Islands*
	N Tropical	X Micronesia
Asia	N Caribbean	Galapagos
Temperate	Chile, Argentina	X Hawaii
● Tropical		X Melanesia, Polynesia

Invaded Habitats
Forests and forest gaps, riparian habitats, grass- and woodland, disturbed places.

Description
A spreading profusely branching shrub of 3-7 m height, with a deep stout taproot and yellowish stems. Young plants are herbaceous and become semi-woody with age. The opposite leaves are dark green, ovate to deltoid in shape, gradually tapering to a point, 6-12 cm long and 3-7 cm wide, and with toothed or entire margins. Flowerheads with 10-35 florets each are arranged in terminal and axillary peduncled clusters. Florets are pale blue to whitish. Fruits are linear, brown to black achenes of *c.* 5 mm length, and with stiff hairs on the angles. The pappus is white, *c.* 5 mm long and consists of rough bristles.

Ecology and Control
A nutrient-demanding, early successional species that is native in grasslands, savannas, and forest margins. Where invasive, it rapidly colonizes disturbed and cleared areas, and forms dense thickets that persist and prevent the establishment of all other species. It seriously degrades indigenous forests and savannas and also invades dune forests. The plant grows fast but cannot survive in shade. In fire prone regions, dry plants may be a fire hazard. Seed production is prolific and seeds are dispersed by wind. The shrub accumulates a soil seed bank.
  Control is difficult as the plant easily regenerates from its rootstock after damage. Repeated cutting and burning has been recommended for infested grasslands, combined with resowing of desired species. Chemical control is done by foliar sprays of 2,4-D esters, picloram, imazapyr, or 2,4,5-T.

References
326, 328, 381, 439, 470, 580, 624, 706, 792, 838, 920, 1006, 1089, 1113, 1135, 1430.

Chrysanthemoides monilifera (L.) Norl.   Asteraceae

LF: Evergreen shrub   CU: Ornamental
SN: -

Geographic Distribution

Europe	*Australasia*	*Atlantic Islands*
Northern	● Australia	Cape Verde
British Isles	● New Zealand	Canary, Madeira
Central, France		Azores
Southern	*Northern America*	X South Atlantic Isl.
Eastern	Canada, Alaska	
Mediterranean Isl.	Southeastern USA	*Indian Ocean Islands*
	Western USA	Mascarenes
Africa	Remaining USA	Seychelles
Northern	Mexico	Madagascar
Tropical		
N Southern	*Southern America*	*Pacific Islands*
	Tropical	Micronesia
Asia	Caribbean	Galapagos
Temperate	Chile, Argentina	Hawaii
Tropical		Melanesia, Polynesia

Invaded Habitats
Coastal dunes and heathland, riparian habitats, coastal woodland.

Description
A much branched, dark-green shrub up to 3 m height and with alternate leaves. The plant has shallow roots and no distinct taproot. Leaves are ovate to spathulate, glabrous, 3-9 cm long and 2-5 cm wide, and coarsely toothed at the margins. Flowerheads of 2-3 cm diameter with 5-8 large ray florets each appear in clusters at the ends of branches. Florets are bright yellow. Fruits are ovoid, contain one seed each and have a fleshy, green skin at first which becomes black and flakes off. Seeds are globular, light brown, 6-8 mm in diameter, and have a very hard seed coat.

Ecology and Control
Two subspecies have become naturalized in Australia: subsp. *monilifera* and subsp. *rotundata*. The latter forms extensive monospecific thickets that suppress native species in a range of vegetation types. The plant accumulates a large seed bank and regeneration after fire is extensive. Seeds may remain viable in the soil for more than 10 years. They are dispersed by birds and small mammals, and germinate easily once the hard seed coat is opened. The shrub vigorously resprouts if damaged.

    Physical control includes mulching stands of the shrub. Fire is commonly used to destroy seedlings and adult plants, and stimulates the seeds in the soil to germinate. The emerging seedlings can then be treated with herbicides such as bromoxyril, glyphosate, picloram, or amine 2,4-D.

References
121, 215, 607, 924, 985, 986, 1179, 1286, 1352, 1381, 1382, 1383, 1415.

Cinchona pubescens Vahl Rubiaceae

LF: Evergreen tree
CU: Flavouring, wood
SN: *Cinchona succirubra* Pav. ex Klotzsch

Geographic Distribution

Europe		*Australasia*		*Atlantic Islands*	
	Northern		Australia	X	Cape Verde
	British Isles		New Zealand		Canary, Madeira
	Central, France				Azores
	Southern	*Northern America*		X	South Atlantic Isl.
	Eastern		Canada, Alaska		
	Mediterranean Isl.		Southeastern USA	*Indian Ocean Islands*	
			Western USA		Mascarenes
Africa			Remaining USA		Seychelles
	Northern		Mexico		Madagascar
	Tropical				
	Southern	*Southern America*		*Pacific Islands*	
		N	Tropical	●	Micronesia
Asia		X	Caribbean	●	Galapagos
	Temperate		Chile, Argentina	●	Hawaii
	Tropical				Melanesia, Polynesia

Invaded Habitats
Forests, mesic sites.

Description
A tree of 4-10 m height and with opposite, broadly elliptic-ovate leaves of 10-20 cm length and 7-11 cm width. Leaves have 9-11 pairs of lateral veins and entire margins. Petioles are 1.5-4.5 cm long, stipules ovate and large. Numerous pink to red flowers are borne in panicles 20 cm long or more. The corolla tube is 10-12 mm long and appressed pubescent. Fruits are lanceolate to oblong capsules of 1-2 cm length. Seeds have a broad ciliate wing and are *c.* 2 mm long.

Ecology and Control
This tree is native in neotropical forests where it is usually not a dominant species. Where invasive, it becomes dominant and reaches high densities, e.g. >1,000 plants ha^{-1}. The tree shades out native plants with its large leaves. It grows rapidly and has a wide ecological tolerance. Damaged trees coppice and the plant can regrow from root remnants. It produces seeds abundantly that are wind dispersed. The tree is shade tolerant; seedlings establish well even in strong shade. On the Galapagos Islands, the tree displaces extensively the native tree *Miconia robinsoniana*.

Small individuals can be uprooted, larger individuals are removed by digging out. Roots must be removed as the tree can regrow from root remnants. Treating cut stumps with herbicides is effective and prevents resprouting.

References
284, 798, 835, 1313.

Cinnamomum camphora (L.) Presl — Lauraceae

LF: Evergreen tree
SN: *Laurus camphora* L.
CU: Essential oils

Geographic Distribution

Europe		Australasia		Atlantic Islands	
	Northern	●	Australia		Cape Verde
	British Isles		New Zealand	X	Canary, Madeira
	Central, France				Azores
X	Southern		*Northern America*		South Atlantic Isl.
	Eastern		Canada, Alaska		
	Mediterranean Isl.	●	Southeastern USA		*Indian Ocean Islands*
			Western USA		Mascarenes
Africa			Remaining USA		Seychelles
	Northern		Mexico	X	Madagascar
	Tropical				
●	Southern		*Southern America*		*Pacific Islands*
			Tropical		Micronesia
Asia			Caribbean		Galapagos
N	Temperate		Chile, Argentina		Hawaii
N	Tropical				Melanesia, Polynesia

Invaded Habitats
Rain forests, riparian habitats, bushland, wet sclerophyll forests.

Description
A stout 6-30 m tall tree with an enlarged base, and spreading branches forming a round crown of dense foliage. The bark is rough and greyish brown. Leaves are alternate, simple, light green, lanceolate, elliptic or obovate, 7-10 cm long and 3-5 cm wide, triplenerved. Petioles are slender and *c.* 2-3 cm long. Leaf blades are glabrous, 5-8 cm long and 2.5-4.5 cm wide, with small oil dots on the surface. Small white or yellowish flowers are borne in axillary clusters. The fruit is a small, black, indehiscent, globose berry of 8-10 mm length and width, containing a single seed of *c.* 7.5 mm diameter.

Ecology and Control
The tree has an excellent adaptation to disturbed sites and easily becomes naturalized where planted. It produces fruits abundantly and is able to reproduce vegetatively by root suckering. The tree forms single-dominant stands that delay or preclude native rainforest regeneration. Seeds are rather short-lived and dispersed by frugivore bird species. The tree does not accumulate a persistent seed bank. Initial seedling growth rate is rather slow. The wood is resistant to insect attack.

Control of this tree on cultivated ground has relied on intensive grazing. Seedlings can be hand removed. Cutting trees requires treating the cut stumps with herbicides because of rapid regeneration from stumps. Young plants can be treated with 2,4-D esters and 2,4,5-T. Larger plants can be killed by a basal bark or cut stump application of the same herbicides.

References
414, 549, 607, 608, 715, 924, 979.

Cinnamomum verum Presl   Lauraceae

LF: Evergreen shrub, tree   CU: Essential oils, flavouring
SN: *Cinnamomum zeylanicum* Blume, *Laurus cinnamomum* L.

Geographic Distribution

Europe	*Australasia*	*Atlantic Islands*
Northern	Australia	Cape Verde
British Isles	New Zealand	Canary, Madeira
Central, France		Azores
Southern	*Northern America*	South Atlantic Isl.
Eastern	Canada, Alaska	
Mediterranean Isl.	Southeastern USA	*Indian Ocean Islands*
	Western USA	Mascarenes
Africa	Remaining USA	• Seychelles
Northern	Mexico	Madagascar
Tropical		
Southern	*Southern America*	*Pacific Islands*
	Tropical	Micronesia
Asia	X Caribbean	Galapagos
Temperate	Chile, Argentina	Hawaii
N Tropical		Melanesia, Polynesia

Invaded Habitats
Rock outcrops, forests and forest gaps.

Description
A shrub or tree reaching 17 m height, with dark brown branches, a dense, rounded or cylindric crown, a greyish brown and rather smooth bark, and a strong smell of cinnamon. Leaves are mostly opposite, ovate to broadly ovate, 10-15 cm long and 4-8 cm wide, triple nerved, glabrous, and have stout petioles of *c.* 10 mm length. Flowers are grey pubescent and are borne in axillary inflorescences that are as long as or longer than the leaves. Fruits are ellipsoid berries of *c.* 10 mm length.

Ecology and Control
The tree has a vigorous growth and a prolific seed production. Seeds are dispersed by birds. The dense canopy shades out all other plants, creating species poor stands that may cover large areas. Once established, it becomes the dominant tree and eliminates the native forest.
    Specific control methods for this species are not available. The same methods as for the previous species may apply.

References
416.

Cirsium arvense (L.) Scop.   Asteraceae

LF: Perennial herb   CU: -
SN: -

Geographic Distribution

Europe
- N   Northern
- N   British Isles
- N   Central, France
- N   Southern
- N   Eastern
- N   Mediterranean Isl.

Africa
- X   Northern
- X   Tropical
- X   Southern

Asia
- N   Temperate
-     Tropical

Australasia
- X   Australia
- X   New Zealand

Northern America
- ●   Canada, Alaska
- X   Southeastern USA
- ●   Western USA
- ●   Remaining USA
- X   Mexico

Southern America
-     Tropical
-     Caribbean
- X   Chile, Argentina

Atlantic Islands
-     Cape Verde
-     Canary, Madeira
- X   Azores
-     South Atlantic Isl.

Indian Ocean Islands
- X   Mascarenes
-     Seychelles
-     Madagascar

Pacific Islands
-     Micronesia
-     Galapagos
- X   Hawaii
-     Melanesia, Polynesia

Invaded Habitats
Forests, grassland, riparian habitats, lakeshores and marshes, sand dunes.

Description
A perennial herb of 30-150 cm height, with much-branched and grooved stems arising from numerous buds on a rhizome, and far-creeping roots. Leaves are alternate, oblong to lanceolate, entire to deeply lobed, and have spiny-toothed margins. Upper leaves are sessile. Flowerheads are 20-25 mm in diameter with an involucre of 10-20 mm height, and appear in terminal and axillary clusters. Bracts are numerous and without spines. The pale-purple corollas are 10-18 mm long. Achenes are oblong, smooth and finely grooved, usually four-angled, and 2.5-4 mm long. The pappus consists of white to brownish hairs of *c.* 2 mm length. The plant has a deep taproot.

Ecology and Control
In the native range, this highly variable plant grows commonly in woodland, pastures and disturbed places. A single individual can build up large patches covering several square metres and crowding out native vegetation. The plant regenerates easily from root cuttings. Seed production is prolific, and seeds may remain viable in the soil for more than 20 years. Establishment of seedlings depends on disturbance, once established, the plant grows mainly vegetatively. The horizontally creeping, branching rhizomes may exceed 5 m length.

    Control methods include burning, spot mowing, and herbicide treatments. Repeated mowing will weaken the plants and prevent seed production. The most effective herbicides are glyphosate, 2,4-D, or picloram.

References
27, 29, 133, 390, 571, 580, 708, 897, 986, 1072, 1139, 1396, 1426.

Cirsium vulgare (Savi) Ten.   Asteraceae

LF: Annual, biennial herb   CU: -
SN: *Cirsium lanceolatum* (L.) Scop., *Carduus lanceolatus* L.

Geographic Distribution

Europe	*Australasia*	*Atlantic Islands*
Northern	● Australia	Cape Verde
British Isles	X New Zealand	Canary, Madeira
N Central, France		N Azores
N Southern	*Northern America*	South Atlantic Isl.
N Eastern	X Canada, Alaska	
Mediterranean Isl.	Southeastern USA	*Indian Ocean Islands*
	● Western USA	Mascarenes
Africa	● Remaining USA	Seychelles
X Northern	Mexico	Madagascar
X Tropical		
● Southern	*Southern America*	*Pacific Islands*
	Tropical	Micronesia
Asia	Caribbean	Galapagos
X Temperate	X Chile, Argentina	X Hawaii
Tropical		X Melanesia, Polynesia

Invaded Habitats
Grassland, rangeland, disturbed places.

Description
A robust and spiny biennial or short-lived perennial, with a deep and stout taproot, and winged stems of 30-300 cm height. Rosettes are up to 65 cm in diameter and have oblanceolate to elliptical leaves of *c.* 30 cm length. Stem leaves are decurrent, prickly-hairy above and rough beneath; the leaf blades range from 4-30 cm length. Flowerheads of up to 5 cm length appear solitary or in groups of a few at the ends of branches. The outer bracts of the involucre have spiny tips. The pale red to purple corollas are 25-35 mm long. Achenes are 3.5-5 mm long and *c.* 1.5 mm wide and have a white pappus of 2-3 cm length, consisting of soft and fluffy hairs.

Ecology and Control
In the native range, the plant is commonly found in many different habitats, e.g. grassland, coastal dunes, woodland, rock outcrops, and disturbed ground. It rarely grows on acid soils. The plant's large rosettes may cover large areas that outcompete native species and prevent their establishment. Seeds are dispersed by wind, water and soil movement. They show various degrees of dormancy and some germinate rapidly after imbibition. The plant cannot withstand deep shade and its establishment depends on disturbance.

    Mowing later in the season, e.g. just before seed dispersal, proved to be effective in reducing populations. Seedlings can be treated with herbicides containing MCPA, 2,4-D, dicamba or picloram.

References
215, 571, 581, 678, 985, 1064, 1089, 1183, 1271.

Citharexylum caudatum L. Verbenaceae

LF: Evergreen shrub, tree CU: Ornamental
SN: -

Geographic Distribution

Europe	*Australasia*	*Atlantic Islands*
Northern	Australia	Cape Verde
British Isles	New Zealand	Canary, Madeira
Central, France		Azores
Southern	*Northern America*	South Atlantic Isl.
Eastern	Canada, Alaska	
Mediterranean Isl.	Southeastern USA	*Indian Ocean Islands*
	Western USA	Mascarenes
Africa	Remaining USA	Seychelles
Northern	N Mexico	Madagascar
Tropical		
Southern	*Southern America*	*Pacific Islands*
	N Tropical	Micronesia
Asia	N Caribbean	Galapagos
Temperate	Chile, Argentina	● Hawaii
Tropical		Melanesia, Polynesia

Invaded Habitats
Mountain slopes, forests and forest edges, disturbed sites.

Description
A large shrub or small tree reaching 20 m height, with nearly terete and glabrous branches. The somewhat leathery glabrous leaves are oblong to obovate or elliptic-oblanceolate, 7-16 cm long, 3.5-5 cm wide, glossy above, and have petioles of 1-2 cm length. The white flowers are borne in narrow elongate-recurved racemes of 4-10 cm length. Pedicels are 2-5 mm long. The drupe-like fruits are orange to purplish-black, oblong to globose, 5-8 mm long, and splitting into two halves to release the seeds.

Ecology and Control
A rapidly spreading species that forms dense thickets and crowds out native vegetation due to the dense canopies. Seedlings of native plants are unable to grow under stands of this species. Seeds are dispersed by birds and are able to germinate in low light conditions. The plant spreads also vegetatively as stems touching the ground become rooted.

Specific control methods for this species are not available. Seedlings and small plants can be hand pulled. Larger stems are cut and the cut stumps treated with herbicide.

References
752.

Clematis vitalba L.   Ranunculaceae

LF: Deciduous climber   CU: Ornamental
SN: -

Geographic Distribution

Europe	*Australasia*	*Atlantic Islands*
X Northern	● Australia	Cape Verde
N British Isles	● New Zealand	Canary, Madeira
N Central, France		Azores
N Southern	*Northern America*	South Atlantic Isl.
N Eastern	Canada, Alaska	
N Mediterranean Isl.	Southeastern USA	*Indian Ocean Islands*
	X Western USA	Mascarenes
Africa	Remaining USA	Seychelles
N Northern	Mexico	Madagascar
Tropical		
Southern	*Southern America*	*Pacific Islands*
	Tropical	Micronesia
Asia	Caribbean	Galapagos
N Temperate	Chile, Argentina	Hawaii
Tropical		Melanesia, Polynesia

Invaded Habitats
Forests, heath- and shrubland, riparian forests.

Description
A large liana climbing up to 30 m height, with stems becoming woody and reaching 6 cm in diameter. Leaves are pinnately compound with usually five ovate to ovate-lanceolate leaflets of 3-10 cm length. The leaflets have stalks of 1-3 cm length. Greenish-white and fragrant flowers of *c.* 2 cm diameter are borne in axillary or terminal panicles of 7-12 cm length. The perianth is pubescent on both surfaces. Fruits are pubescent achenes and have the persistent styles attached; these are hairy and 2-3 cm long.

Ecology and Control
In the native range, this liana grows in riparian forests and forest edges, generally on rich and mesic soils, and predominates in early successional forests. It is invasive because it forms a dense smothering blanket over native trees, impeding their growth and increasing wind and ice damage. The vine rapidly climbs into the crown by its leaf tendrils. It invades forests from the edge or in canopy gaps, alters their structure and reduces the diversity of native understorey species. Seed production is high, a soil seed bank is produced, and seeds are easily wind dispersed. Seeds can remain on the vine over winter and well into summer.

    Small seedlings can be hand pulled, larger stems are cut. Roots must be removed as they can resprout. Cut stumps can be treated with herbicide such as 2,4,5-T. Aerial shoots touching the ground after cutting may become rooted.

References
57, 172, 173, 174, 284, 604, 924, 955, 1141, 1415, 1418.

Clidemia hirta (L.) Don   Melastomataceae

LF: Evergreen shrub   CU: Ornamental
SN: *Melastoma hirta* L., *Melastoma hirsutum* L.

Geographic Distribution

Europe	*Australasia*	*Atlantic Islands*
Northern	Australia	Cape Verde
British Isles	New Zealand	Canary, Madeira
Central, France		Azores
Southern	*Northern America*	South Atlantic Isl.
Eastern	Canada, Alaska	
Mediterranean Isl.	Southeastern USA	*Indian Ocean Islands*
	Western USA	X Mascarenes
Africa	Remaining USA	● Seychelles
Northern	N Mexico	X Madagascar
X Tropical		
Southern	*Southern America*	*Pacific Islands*
	N Tropical	X Micronesia
Asia	N Caribbean	Galapagos
Temperate	Chile, Argentina	● Hawaii
● Tropical		● Melanesia, Polynesia

Invaded Habitats
Forests and forest edges, grassland, disturbed sites.

Description
A densely-branching shrub growing 0.5-4 m tall, covered with straight bristles, and with rounded young branches. The opposite leaves are ovate to oblong-ovate, 5-16 cm long and 3-8 cm wide, with a sparsely strigose upper surface and a finely bristly lower surface. Leaves have 5-7 conspicuous, spreading veins. Petioles are 0.5-3 cm long. Flowers are borne in compact cymes or small panicles of 2-3 cm length in the upper leaf-axils. Each flower has 5-7 white or pinkish petals of 8-11 mm length. Fruits are dark-blue, ovoid berries of 6-9 mm length, covered with long patent bristles. Each berry contains 100 or more ovoid seeds of 0.5-0.8 mm length.

Ecology and Control
In the native range, this fast growing and shade-tolerant pioneer shrub grows in primary forests and along steep embankments. Where invasive, the tree establishes in forest gaps and other disturbed sites, forming dense and almost impenetrable thickets that shade out all native vegetation due to the large leaves. Fruits are abundantly produced and seeds dispersed mainly by birds but also may be carried by animals moving through the thickets. The species is probably not resistant to fire, however, it rapidly colonizes burned areas. The plant is a serious weed in mesic and wet environments on the Hawaiian Islands and on the Seychelles.

    Seedlings and saplings can be pulled or dug out. If cut, the cut stumps are treated with triclopyr to prevent coppicing. Glyphosate is an effective herbicide as a foliar spray.

References
75, 268, 446, 464, 969, 1011, 1037, 1201, 1390, 1444.

Colocasia esculenta (L.) Schott   Araceae

LF: Perennial herb   CU: Food, ornamental
SN: *Caladium esculentum* (L.) Vent., *Colocasia antiquorum* L.

Geographic Distribution

Europe		*Australasia*		*Atlantic Islands*	
	Northern		Australia		Cape Verde
	British Isles	X	New Zealand	X	Canary, Madeira
	Central, France			●	Azores
X	Southern	*Northern America*		X	South Atlantic Isl.
	Eastern		Canada, Alaska		
	Mediterranean Isl.	●	Southeastern USA	*Indian Ocean Islands*	
			Western USA		Mascarenes
Africa			Remaining USA		Seychelles
	Northern		Mexico		Madagascar
	Tropical				
X	Southern	*Southern America*		*Pacific Islands*	
			Tropical		Micronesia
Asia		X	Caribbean		Galapagos
	Temperate		Chile, Argentina		Hawaii
N	Tropical				Melanesia, Polynesia

Invaded Habitats
Riparian habitats, marshes.

Description
A large herb of 1-2.5 m height, with rhizomes and a stout tuber. The simple and alternate leaves are cordate-sagittate in shape, shining, have long petioles up to 1 m length or more, and blades up to 60 cm length and 35 cm width. Scapes are shorter than the leaves. The inflorescence is a terminal or axillary spike enfolded by a spathe of 20-45 cm length. The spathe is lanceolate and acuminate, with the basal part green and the upper part orange-yellow. The flower-bearing spadix is less than half as long as the spathe. Flowers are sessile, female ones are green and in the lower part of the spadix, male flowers creamy white and in the upper part, and sterile flowers in between.

Ecology and Control
Due to its edible rootstocks and wide cultivation, many cultivars have been developed. The plant spreads vegetatively, and plants in the introduced range rarely produce inflorescences. Vegetative growth leads to dense populations and its large leaves shade out native vegetation. Extensive stands of this herb alter the vegetational structure and dynamics of riparian plant communities. Rhizome fragments are easily carried by streams, and floods can dislodge bud-laden rhizomes from the banks. Growth seems to be best in the silty soils lining the river-banks.
   Specific control methods for this species are not available. Smaller plants may be pulled or dug out, larger plants cut and regrowth treated with herbicide.

References
10, 715.

Colubrina asiatica (L.) Brongn.   Rhamnaceae

LF: Evergreen shrub   CU: Ornamental
SN: *Ceanothus asiaticus* L.

Geographic Distribution

Europe
    Northern
    British Isles
    Central, France
    Southern
    Eastern
    Mediterranean Isl.

Africa
    Northern
N Tropical
    Southern

Asia
N Temperate
N Tropical

Australasia
N Australia
    New Zealand

Northern America
    Canada, Alaska
● Southeastern USA
    Western USA
    Remaining USA
    Mexico

Southern America
    Tropical
    Caribbean
    Chile, Argentina

Atlantic Islands
    Cape Verde
    Canary, Madeira
    Azores
    South Atlantic Isl.

Indian Ocean Islands
    Mascarenes
N Seychelles
    Madagascar

Pacific Islands
    Micronesia
    Galapagos
N Hawaii
N Melanesia, Polynesia

Invaded Habitats
Coastal beach and dune vegetation, coastal hammocks, mangrove swamps.

Description
A climbing or sprawling shrub or small tree up to 5 m tall, with a brown bark and long, trailing or spreading, thin and reddish-brown branches that are covered with a whitish bloom. The simple leaves are alternate, ovate, 4-9 cm long and up to 5 cm wide, shiny dark green above and paler below. There are 3-5 conspicuous main veins and several pairs of lateral veins. Leaves have slender petioles of 10-15 mm length. Small, cream to yellowish flowers of *c.* 4 mm diameter are borne in loose axillary heads or cymes. Flower stalks are up to 7 cm long. The fruit is a globose, reddish-brown and dehiscent capsule of *c.* 10 mm diameter, breaking into three parts.

Ecology and Control
Natural habitats of this fast growing plant include coastal sand dunes and littoral scrub. It is invasive because it climbs over native shrubs and small trees, smothering them and impeding their growth and regeneration.
    Specific control methods for this species are not available. Seedlings and saplings may be hand pulled or dug out. Larger shrubs can be cut and the cut stumps treated with herbicide.

References
715.

Conicosia pugioniformis (L.) N.E.Br.   Aizoaceae

LF: Succulent perennial   CU: Ornamental
SN: *Mesembryanthemum pugioniforme* L.

Geographic Distribution

Europe	*Australasia*	*Atlantic Islands*
Northern	Australia	Cape Verde
British Isles	New Zealand	Canary, Madeira
Central, France		Azores
Southern	*Northern America*	South Atlantic Isl.
Eastern	Canada, Alaska	
Mediterranean Isl.	Southeastern USA	*Indian Ocean Islands*
	● Western USA	Mascarenes
Africa	Remaining USA	Seychelles
Northern	Mexico	Madagascar
Tropical		
N Southern	*Southern America*	*Pacific Islands*
	Tropical	Micronesia
Asia	Caribbean	Galapagos
Temperate	Chile, Argentina	Hawaii
Tropical		Melanesia, Polynesia

Invaded Habitats
Coastal scrub and grassland, coastal dunes, woodland.

Description
A usually short-lived perennial with prostrate to ascending moderately branched shoots of 50-150 cm length. The grey-green to green leaves are narrowly linear to cylindrical, succulent, 15-20 cm long and 10-15 mm in diameter. Flowers are borne solitary on axillary peduncles of 5-12 cm length. Flowers are 5-8 cm in diameter and have numerous shiny yellow and linear petals. Fruits are dehiscent cylindrical capsules containing numerous smooth and roundish seeds of *c.* 1 mm length. The plant has a taproot of 50-100 cm length.

Ecology and Control
This plant is most abundant in open patches and in disturbed areas. The plant forms dense populations that expand laterally and inhibits regeneration of native species. The dense patches preclude the establishment of other vegetation. The plant produces many seeds and reproduction occurs only by seeds. Capsules separate from plants when dry and are readily moved by wind. The plant can resprout from the buried root crown.
    Small plants can be hand pulled, larger plants are easily killed by cutting the taproot. The underground root crown must be severed to prevent regrowth. An effective herbicide is glyphosate.

References
133.

Conium maculatum L.　　　　　　　　　　　　　　　　　　　　　Apiaceae

LF: Biennial herb　　　　　CU: Ornamental
SN: -

Geographic Distribution

Europe		*Australasia*		*Atlantic Islands*	
N	Northern	●	Australia		Cape Verde
N	British Isles	X	New Zealand		Canary, Madeira
N	Central, France			N	Azores
N	Southern	*Northern America*			South Atlantic Isl.
N	Eastern	X	Canada, Alaska		
N	Mediterranean Isl.		Southeastern USA	*Indian Ocean Islands*	
		●	Western USA		Mascarenes
Africa		X	Remaining USA		Seychelles
N	Northern		Mexico		Madagascar
	Tropical				
X	Southern	*Southern America*		*Pacific Islands*	
		X	Tropical	●	Micronesia
Asia			Caribbean		Galapagos
N	Temperate	X	Chile, Argentina		Hawaii
N	Tropical				Melanesia, Polynesia

Invaded Habitats
Grassland, forests, riparian habitats, freshwater wetlands.

Description
A biennial herb with a long taproot and 1-2 m tall hollow, ribbed and branched stems that have purple spots. The opposite leaves are ovate, 20-40 cm long, finely pinnately compound, and serrated at the margins. The petiole base tends to sheath the stem. Inflorescences are large terminal and lateral, compound umbels of 4-6 cm diameter, with 4-6 brown bracts. Flowers are small, white or yellowish, and have five petals. Fruits are ovate achenes of 2-3 mm width, with five distinct ribs.

Ecology and Control
This plant grows best in moist and fertile soils. A single plant may produce up to 38,000 seeds, and seeds may remain viable in the soil for several years. All plant parts contain toxic alkaloids and may be fatal to both livestock and wildlife. The plant can spread quickly in disturbed areas and is highly competitive, preventing the establishment of native grasses and forbs by shading and competing for space.

Hand pulling plants may be effective, especially prior to seed set. Spring mowing kills mature plants effectively; a second mow in late summer kills emerged seedlings and regrowth. Effective post-emergent herbicides include 2,4-D ester, 2,4-D amine, or glyphosate applied in late spring.

References
78, 133, 215, 468, 581, 888, 985, 986.

Coprosma repens L.   Rubiaceae

LF: Evergreen shrub, tree   CU: Ornamental, shelter
SN: *Coprosma baueri* Endl.

Geographic Distribution

Europe
    Northern
    British Isles
    Central, France
    Southern
    Eastern
    Mediterranean Isl.

Africa
    Northern
    Tropical
    Southern

Asia
    Temperate
    Tropical

Australasia
● Australia
N New Zealand

Northern America
    Canada, Alaska
    Southeastern USA
    Western USA
    Remaining USA
    Mexico

Southern America
    Tropical
    Caribbean
    Chile, Argentina

Atlantic Islands
    Cape Verde
    Canary, Madeira
    Azores
    South Atlantic Isl.

Indian Ocean Islands
    Mascarenes
    Seychelles
    Madagascar

Pacific Islands
    Micronesia
    Galapagos
    Hawaii
    Melanesia, Polynesia

Invaded Habitats
Coastal heath- and shrubland, forests, rock outcrops.

Description
A dioecious shrub or small tree up to 8 m tall, with thick, broadly ovate to oblong leaves that are shiny green above and dull green beneath. They are opposite, 2-10 cm long and 1-5 cm wide. Flowers of 4-5 mm length are borne in clusters in the axils of leaves. Fruits are obovoid yellow-orange drupes of *c.* 9 mm length.

Ecology and Control
In the native range, this plant often occurs as a shrub of not more than 90 cm height in rocky sites near the ocean. It is invasive because it forms dense and species poor thickets that shade out native vegetation. It impedes the growth and regeneration of native shrubs and trees. Seeds are dispersed by birds. The shrub is tolerant of salt spray and vigorously resprouts after damage. Branches touching the ground easily become rooted.
    Seedlings are easy to pull out, small plants can be dug out. Larger plants are cut and the cut stumps treated with herbicide. Initial control should aim at removing female plants to prevent seed dispersal.

References
121, 215, 607, 608, 924.

Cortaderia jubata (Lemoine) Stapf   Poaceae

LF: Perennial herb   CU: Erosion control, ornamental
SN: -

Geographic Distribution

Europe	*Australasia*	*Atlantic Islands*
Northern	● Australia	Cape Verde
British Isles	● New Zealand	Canary, Madeira
Central, France		Azores
Southern	*Northern America*	South Atlantic Isl.
Eastern	Canada, Alaska	
Mediterranean Isl.	Southeastern USA	*Indian Ocean Islands*
	● Western USA	Mascarenes
Africa	Remaining USA	Seychelles
Northern	Mexico	Madagascar
Tropical		
X Southern	*Southern America*	*Pacific Islands*
	N Tropical	Micronesia
Asia	Caribbean	Galapagos
Temperate	N Chile, Argentina	X Hawaii
Tropical		Melanesia, Polynesia

Invaded Habitats
Coastal scrub and dunes, grassland, riparian habitats, disturbed places.

Description
A large tussock grass, forming clumps up to 3 m in diameter, with a tufted base and flowering culms of 2-7 m height. The bright green leaf blades are 1-1.5 m long and 5-12 mm wide. Leaf sheaths are glabrous to densely hairy. The 30-100 cm long inflorescence is a dense, oblanceolate panicle with flexuous branches. Spikelets are 14-16 mm long and have 3-5 florets each. Lemmas are hairy and have an acuminate tip. The plant is dioecious, e.g. male and female flowers are borne on different individuals.

Ecology and Control
A native of the Andes, occurring from 2,800-3,400 m elevation in the native range. In California and Australia, only female plants are present that produce viable seeds asexually through apomixis. Each plume can contain more than 100,000 seeds which are wind dispersed. Seeds germinate easily on bare soil with sufficient moisture. The species forms large stands that displace native species and prevents forest re-establishment in forests that have been burned or clear cut.

     Seedlings can easily be pulled or grubbed. Mechanical removal of larger plants before flowering is effective but the entire crown and top sections of the roots must be removed to prevent resprouting. Good control is achieved by spot treatments with a glyphosate herbicide. Other post-emergence herbicides used are fluaziprop or imazapyr.

References
133, 215, 531, 765, 986, 1415.

Cortaderia selloana (Schult) Asch. & Graebn.   Poaceae

LF: Perennial herb
CU: Ornamental, soil stabilization
SN: *Cortaderia argentea* (Nees) Stapf, *Arundo selloana* Schult. & Schult.

Geographic Distribution

Europe	*Australasia*	*Atlantic Islands*
Northern	● Australia	Cape Verde
X British Isles	● New Zealand	X Canary, Madeira
Central, France		Azores
● Southern	*Northern America*	South Atlantic Isl.
Eastern	Canada, Alaska	
Mediterranean Isl.	Southeastern USA	*Indian Ocean Islands*
	● Western USA	Mascarenes
Africa	Remaining USA	Seychelles
Northern	Mexico	Madagascar
Tropical		
X Southern	*Southern America*	*Pacific Islands*
	N Tropical	Micronesia
Asia	Caribbean	Galapagos
Temperate	N Chile, Argentina	Hawaii
Tropical		Melanesia, Polynesia

Invaded Habitats
Grassland, heath- and shrubland, riparian habitats, freshwater wetlands, coastal dunes.

Description
A somewhat smaller tussock grass than the former species, with clumps of up to 3.5 m diameter and flowering culms of 1.5-4 m height. Leaf blades are markedly V-shaped in cross-section, glaucous-green, 80-180 cm long and 8-10 mm wide, and with glabrous to sparsely hairy sheaths. Inflorescences are obovate to ovate panicles of 30-130 cm length, with stiff branches. Spikelets are 15-17 mm long and have 5-7 florets each in female plants, and 2-4 each in hermaphrodites. The plant is dioecious or gyno-dioecious. Lemmas of female flowers are hairy, those of male flowers glabrous. Glumes are 8-15 mm long and glabrous.

Ecology and Control
In the native range, this grass is common on moist, sandy soils on grassy plains and slopes. Several distinct cultivars of this grass are in use. The vigorous growth and the large tussocks make it a strong competitor to indigenous species. It may also increase fire hazards in fire prone regions. Establishment depends on disturbance providing bare soil. The grass grows in a wide range of soils, and may reach a rooting depth of >3 m. Caryopses are dispersed by wind.

Small plants can be removed by hand, larger plants can be cut at ground level or burned. Regrowth is unlikely once the crown and top section of the roots have been removed. Effective herbicides are glyphosate, hexazinone, and imazapyr. Glyphosate has been widely used to control seedlings and mature plants of this grass.

References
121, 133, 215, 264, 986, 1415.

Cotoneaster divaricatus Rehder & Wilson   Rosaceae

LF: Deciduous shrub   CU: Ornamental
SN: -

Geographic Distribution

Europe	*Australasia*	*Atlantic Islands*
Northern	● Australia	Cape Verde
X British Isles	New Zealand	Canary, Madeira
Central, France		Azores
Southern	*Northern America*	South Atlantic Isl.
Eastern	Canada, Alaska	
Mediterranean Isl.	Southeastern USA	*Indian Ocean Islands*
	Western USA	Mascarenes
Africa	Remaining USA	Seychelles
Northern	Mexico	Madagascar
Tropical		
Southern	*Southern America*	*Pacific Islands*
	Tropical	Micronesia
Asia	Caribbean	Galapagos
N Temperate	Chile, Argentina	Hawaii
Tropical		Melanesia, Polynesia

Invaded Habitats
Dry forests, woodland, riparian habitats.

Description
A shrub up to 2 m tall with spreading branches. Leaves are elliptic to broad-elliptic, usually acute, 8-20 mm long, dark glossy green above and light green beneath. Pink flowers are borne on short pedicels and arranged in cymes of usually three flowers. The calyx is slightly pubescent. The fruit is a red and ellipsoid pome of *c.* 8 mm length and 6 mm width, containing 1-3 seeds.

Ecology and Control
The spreading growth habit of this shrub leads to dense thickets that may cover large areas and eliminate native vegetation by shading. Regeneration of native shrubs and trees is prevented, and the species richness is strongly reduced. Fruits are abundantly produced and seeds dispersed by birds.

Seedlings and small plants can be hand pulled or dug out. Larger plants are best sprayed with herbicide, or cut and the cut stumps treated with herbicide.

References
215, 1456.

Cotoneaster glaucophyllus Franch.   Rosaceae

LF: Semi-deciduous shrub   CU: Ornamental
SN: *Cotoneaster serotinus* Hutch.

Geographic Distribution

Europe
- Northern
- British Isles
- Central, France
- Southern
- Eastern
- Mediterranean Isl.

Africa
- Northern
- Tropical
- Southern

Asia
- N Temperate
- N Tropical

Australasia
- ● Australia
- X New Zealand

Northern America
- Canada, Alaska
- Southeastern USA
- Western USA
- Remaining USA
- Mexico

Southern America
- Tropical
- Caribbean
- Chile, Argentina

Atlantic Islands
- Cape Verde
- Canary, Madeira
- Azores
- South Atlantic Isl.

Indian Ocean Islands
- Mascarenes
- Seychelles
- Madagascar

Pacific Islands
- Micronesia
- Galapagos
- Hawaii
- Melanesia, Polynesia

Invaded Habitats
Forests, woodland, riparian habitats.

Description
A shrub of 3-4 m height, with an irregular growth habit. Leaves are elliptic to obovate, 3-8 cm long and 15-40 mm wide, glabrous above and white-tomentose beneath, with petioles of 5-10 mm length. Inflorescences are white-tomentose, and have numerous white flowers of *c.* 6 mm diameter. Fruits are slightly obovoid pomes of 7-8 mm diameter, scarlet to red, and contain 2-3 seeds each.

Ecology and Control
The shrub is a prolific fruit producer, and seeds are dispersed by birds. Damaged shrubs resprout vigorously from the base. The dense growth habit leads to dense thickets that shade out native plants and reduces growth and regeneration of overstorey plants.
    Seedlings and small plants can be hand pulled or dug out. All roots must be removed to prevent regrowth. Spraying herbicide is best done before fruits develop. Larger stems can be cut and the stumps treated with herbicide.

References
121, 215, 924.

Cotoneaster pannosus Franch.   Rosaceae

LF: Evergreen shrub   CU: Ornamental
SN: -

Geographic Distribution

Europe	*Australasia*	*Atlantic Islands*
Northern	● Australia	Cape Verde
British Isles	New Zealand	Canary, Madeira
Central, France		Azores
Southern	*Northern America*	South Atlantic Isl.
Eastern	Canada, Alaska	
Mediterranean Isl.	Southeastern USA	*Indian Ocean Islands*
	● Western USA	Mascarenes
Africa	Remaining USA	Seychelles
Northern	Mexico	Madagascar
Tropical		
X Southern	*Southern America*	*Pacific Islands*
	Tropical	Micronesia
Asia	Caribbean	Galapagos
N Temperate	Chile, Argentina	Hawaii
Tropical		Melanesia, Polynesia

Invaded Habitats
Grassland, coastal beaches and scrub.

Description
A shrub up to 2 m tall and with slender and arching branches. Young branchlets are densely pubescent. The elliptic leaves are acute, 1-2.5 cm long, dull green and glabrous above, and white-tomentose beneath. Petioles are 2-7 mm long. White flowers of *c.* 8 mm diameter are borne in dense corymbose inflorescences of 15-40 mm diameter and consisting of 6-20 flowers. Anthers are purple. The fruit is a globose to ellipsoid, dull red pome of *c.* 6 mm length and contains two seeds.

Ecology and Control
Seeds are distributed by birds. The species thrives in poor and droughty soils and shades out native sun-loving plant species. Eventually, large areas of native vegetation may become displaced by cotoneasters. The shrub has a strong and deep root system and branches profusely at ground level.

    Young plants can be hand pulled or uprooted with a weed wrench. Larger plants should be cut and the stumps treated with glyphosate to prevent resprouting. If a high cover of cotoneaster seedlings is present, they can be eliminated by smothering with mulch or black plastic, hand pulling or spraying.

References
133, 215, 1065.

Cotoneaster simonsii Baker Rosaceae

LF: Deciduous shrub
SN: -
CU: Ornamental

Geographic Distribution

Europe	*Australasia*	*Atlantic Islands*
X Northern	Australia	Cape Verde
X British Isles	• New Zealand	Canary, Madeira
X Central, France		Azores
Southern	*Northern America*	South Atlantic Isl.
Eastern	Canada, Alaska	
Mediterranean Isl.	Southeastern USA	*Indian Ocean Islands*
	X Western USA	Mascarenes
Africa	Remaining USA	Seychelles
Northern	Mexico	Madagascar
Tropical		
Southern	*Southern America*	*Pacific Islands*
	Tropical	Micronesia
Asia	Caribbean	Galapagos
Temperate	Chile, Argentina	Hawaii
N Tropical		Melanesia, Polynesia

Invaded Habitats
Forest margins, disturbed sites.

Description
A deciduous or semi-evergreen erect shrub up to 4 m height, with the young branches covered with a dense, brown wool. Leaves are obovate to broad elliptic, 15-30 mm long and 7-15 mm wide, dark green above, paler and pubescent beneath, and have short petioles of 2-3 mm length. White flowers are borne in cluster of 2-4. Fruits are ellipsoid to obovoid pomes of 5-10 mm length, shining scarlet, and contain 2-5 seeds each.

Ecology and Control
In the native range, this plant is found in forest edges and occurs up to 3,800 m elevation in Bhutan. The shrub is invasive because it forms dense thickets that impede the growth of native plants and prevent their regeneration. It can become dominant in the understorey of invaded forests, thereby changing the vegetation structure and reducing species richness.

Specific control methods for this species are not available. Seedlings and small plants may be hand pulled or dug out. Larger shrubs can be cut and the cut stumps treated with herbicide.

References
1415.

Cotula coronopifolia L. — Asteraceae

LF: Annual, perennial herb CU: -
SN: -

Geographic Distribution

Europe	Australasia	Atlantic Islands
Northern	● Australia	Cape Verde
X British Isles	X New Zealand	Canary, Madeira
X Central, France		Azores
● Southern	*Northern America*	X South Atlantic Isl.
Eastern	Canada, Alaska	
X Mediterranean Isl.	Southeastern USA	*Indian Ocean Islands*
	X Western USA	Mascarenes
Africa	Remaining USA	Seychelles
Northern	Mexico	Madagascar
Tropical		
Southern	*Southern America*	*Pacific Islands*
	X Tropical	Micronesia
Asia	Caribbean	Galapagos
Temperate	X Chile, Argentina	Hawaii
Tropical		Melanesia, Polynesia

Invaded Habitats
Saline and freshwater marshes, swamp edges, streambanks.

Description
A glabrous somewhat succulent herb with ascending stems reaching 50 cm height, rooting at the nodes. The alternate leaves are linear to lanceolate or oblong, entire to irregularly lobed or dissected, 2-7 cm long, the base forming a fused sheath. Flowerheads are 6-15 mm in diameter and are borne solitary on slender stalks. They are without ray florets but have bright yellow disc florets. Fruits are achenes of 1-2 mm length.

Ecology and Control
A pioneer plant of bare, wet and nutrient-rich soils, growing both in freshwater and brackish water habitats. It is a prolific seed producer, and seeds are mainly dispersed by water. Seeds remain viable for 1-2 years. The plant is able to build up dense populations that crowd out native vegetation.

    Specific control methods for this species are not available. Plants may be hand pulled or dug out. Larger patches can be sprayed with non-selective herbicides.

References
215, 1326, 1328.

Crassula helmsii (Kirk) Cockayne   Crassulaceae

LF: Perennial aquatic   CU: Ornamental
SN: *Tillaea helmsii* Kirk

Geographic Distribution

Europe
- Northern
- • British Isles
- X Central, France
- Southern
- Eastern
- Mediterranean Isl.

Africa
- Northern
- Tropical
- Southern

Asia
- X Temperate
- Tropical

Australasia
- N Australia
- N New Zealand

Northern America
- Canada, Alaska
- X Southeastern USA
- Western USA
- Remaining USA
- Mexico

Southern America
- Tropical
- Caribbean
- Chile, Argentina

Atlantic Islands
- Cape Verde
- Canary, Madeira
- Azores
- South Atlantic Isl.

Indian Ocean Islands
- Mascarenes
- Seychelles
- Madagascar

Pacific Islands
- Micronesia
- Galapagos
- Hawaii
- Melanesia, Polynesia

Invaded Habitats
Freshwater wetlands, ponds.

Description
An aquatic or semi-terrestrial perennial, with round stems of 10-130 cm length, floating or creeping, much branched and rooting at the nodes. The tips of stems are emerging above the water surface. The opposite and sessile leaves are 4-20 mm long, 0.7-1.6 mm wide, connate at the base, and succulent. Flowers are borne solitary on pedicels of 2-8 mm length in the axils of leaves, are 3-3.5 mm in diameter, and white or pinkish. Fruits contain 2-5 elliptical and smooth seeds of *c.* 0.5 mm length.

Ecology and Control
In the native range, this aquatic grows in shallow still water bodies, along streams, in salt marshes and on coastal rocks. On moist soils, it also grows terrestrially. In the introduced range, it grows in waters up to 3 m deep. Vegetative growth leads to dense mats of inter-mixed stems floating on the water surface that outcompete native emergent aquatic plants. Even small stem fragments are able to produce new plants and are easily dispersed by water. The plant is highly tolerant of anaerobic soil conditions and salt tolerant to some degree. The growth habit depends on water depth: fully submerged plants in deep water have thin leaves and long stems, plants in shallow water have much-branched and prostrate stems.

Control methods include covering infestations with a dark geotextile material to shade out the plants, and physical removal of infestations. All plants and plant fragments must be removed to prevent regrowth. The herbicide diquat has proved to be effective.

References
242, 284, 312, 314, 673, 705, 722, 1215.

Crataegus monogyna Jacq.   Rosaceae

LF: Deciduous shrub   CU: Ornamental
SN: -

Geographic Distribution

Europe		*Australasia*		*Atlantic Islands*	
N	Northern	•	Australia		Cape Verde
N	British Isles	•	New Zealand	X	Canary, Madeira
N	Central, France				Azores
N	Southern	*Northern America*			South Atlantic Isl.
N	Eastern		Canada, Alaska		
N	Mediterranean Isl.		Southeastern USA	*Indian Ocean Islands*	
		•	Western USA		Mascarenes
Africa			Remaining USA		Seychelles
N	Northern		Mexico		Madagascar
	Tropical				
X	Southern	*Southern America*		*Pacific Islands*	
			Tropical		Micronesia
Asia			Caribbean		Galapagos
N	Temperate		Chile, Argentina		Hawaii
	Tropical				Melanesia, Polynesia

Invaded Habitats
Grass- and woodland, damp forests, riparian habitats.

Description
A much-branched, erect shrub or small tree of 2-6 m height, with many branchlets ending in spines, and armed with thorns of 5-25 mm length. The bark is generally smooth. Leaves are alternate, triangular to ovate, deeply lobed and have coarsely serrated margins. They are very variable in shape and in the degree of lobing. The fragrant flowers are white, cream or pink, 8-12 mm in diameter and appear in clusters at the ends of branches. Fruits are globular and red berries of *c.* 8 mm diameter, each containing usually one ovoid and brown seed of 5-6 mm length. The plant has deep and spreading roots.

Ecology and Control
The plant easily suckers from roots, enabling it to form large and dense stands that crowd out almost all native understorey plants. The shrub prefers light and is tolerant to airborne salt in coastal areas. Fruits are frequently eaten by birds and other wildlife, spreading the seeds to new places. The plant tolerates both shade and sun. In the Pacific Northwest, it hybridizes with the native *C. douglasii*.

If removed manually, the top few centimetres of the root system should be removed to prevent resprouting. Cutting trunks close to the ground is often combined with treating the stumps with herbicides such as picloram, amine 2,4-D, triclopyr, or glyphosate. Smaller seedlings can be pulled by hand.

References
80, 133, 179, 215, 277, 607, 608, 691, 777, 924, 985, 1124, 1125, 1415, 1416.

Cryptostegia grandiflora Roxb. ex R.Br.　　　Asclepiadaceae

LF: Evergreen climber　　　CU: Ornamental
SN: -

Geographic Distribution

	Europe		Australasia		Atlantic Islands
	Northern	●	Australia		Cape Verde
	British Isles		New Zealand		Canary, Madeira
	Central, France				Azores
	Southern		*Northern America*		South Atlantic Isl.
	Eastern		Canada, Alaska		
	Mediterranean Isl.	X	Southeastern USA		*Indian Ocean Islands*
			Western USA		Mascarenes
	Africa		Remaining USA		Seychelles
	Northern	X	Mexico	N	Madagascar
	Tropical				
	Southern		*Southern America*		*Pacific Islands*
		X	Tropical		Micronesia
	Asia	X	Caribbean		Galapagos
	Temperate		Chile, Argentina		Hawaii
	Tropical				Melanesia, Polynesia

Invaded Habitats
Floodplains, riparian habitats and forests, woodland, coastal beaches.

Description
A woody climber >15 m tall or a many-stemmed shrub of 1-3 m height, with greyish brown and slender stems. Stems are either branched, leaf-bearing and up to 2 m long, or unbranched, leafless, 3-8 m long and climbing over any supports. The dark green, simple leaves are opposite, glossy, elliptic, 6-10 cm long and 2-4 cm wide, entire and with short petioles. Flowers are white inside and pale purple outside, 4-5 cm long and *c.* 5 cm in diameter. Fruits are greenish pods of 10-12 cm length and 3-4 cm width, borne in pairs on common stalks. Each fruit contains 200-250 brown, flat and ovate seeds that have long fine hairs of 10-15 mm length. The root system reaches a depth of up to 12 m.

Ecology and Control
This fast growing vine grows naturally in the western coastal plain of Madagascar. Where invasive, it covers trees with a dense carpet, causing damage to them during floods due to debris accumulation. Seedling growth rate is slow at first but increases rapidly thereafter, and plants may reach 4-5 m height in the first year. The climber forms dense impenetrable thickets and smothers all native vegetation up to 40 m height. Seeds are effectively dispersed by wind and water, and can withstand periods of drought up to 8 months.

Scattered plants may be treated by basal-bark application of picloram, triclopyr, or 2,4-D esters. Foliar treatments with 2,4-D, dicamba or imazapyr are also effective. Fire is used for the control of dense infestations. If vines are cut, regrowth and emerging seedlings must be controlled in follow-up programmes.

References
89, 90, 91, 92, 493, 494, 495, 498, 541, 607, 608, 845, 986, 1297, 1345.

Cupaniopsis anacardioides (Rich.) Redkf.   Sapindaceae

LF: Evergreen tree
SN: *Cupania anacardioides* Rich.
CU: Ornamental

Geographic Distribution

Europe	Australasia	Atlantic Islands
Northern	N Australia	Cape Verde
British Isles	New Zealand	Canary, Madeira
Central, France		Azores
Southern	*Northern America*	South Atlantic Isl.
Eastern	Canada, Alaska	
Mediterranean Isl.	• Southeastern USA	*Indian Ocean Islands*
	Western USA	Mascarenes
Africa	Remaining USA	Seychelles
Northern	Mexico	N Madagascar
Tropical		
Southern	*Southern America*	*Pacific Islands*
	Tropical	N Micronesia
Asia	Caribbean	Galapagos
Temperate	Chile, Argentina	Hawaii
N Tropical		Melanesia, Polynesia

Invaded Habitats
Mangrove forests, cypress swamps, tropical hammocks, coastal dunes and forests.

Description
A slender tree reaching 10 m height or more, with a broad, irregular crown, a dark grey outer bark and an orange inner bark. Leaves are alternate, pinnately compound with 4-12 leaflets, and have petioles that are swollen at the base. The oblong and leathery leaflets are up to 20 cm long and 7.5 cm wide, shiny yellowish green, and have short stalks and entire margins. Numerous white to greenish yellow flowers of *c.* 8 mm diameter are borne in branched, axillary clusters of up to 35 cm length. Flowers have five petals and 6-8 stamens. Fruits are yellow-orange capsules of *c.* 2 cm diameter with three segments that open to expose three seeds. These are oval, shiny black, 5-14 mm long, 4-9 mm wide and covered by orange arils.

Ecology and Control
Native habitats include coastal areas, lowland to upland rain forests, woodland and riverine forests. The tree is fast growing, salt-tolerant, grows in full sun and shade, and is adapted to poor soils. Where invasive, it forms dense thickets that crowd out native vegetation. Seedlings and saplings can reach high densities and alter the understorey habitat. The brightly coloured fruits are attractive to birds that disperse seeds to new places. The tree has a prolific seed production.

Seedlings can be hand pulled and mature trees controlled with a triclopyr herbicide mixed with a oil diluent and applied to the basal bark. Trees are also cut and the cut stumps treated with glyphosate or triclopyr.

References
544, 545, 715, 753, 959.

Cylindropuntia imbricata (Haw.) Knuth   Cactaceae

LF: Succulent perennial   CU: Ornamental
SN: *Opuntia arborescens* Engelm., *Opuntia imbricata* (Haw.) DC.

Geographic Distribution

	Europe		Australasia		Atlantic Islands
	Northern	X	Australia		Cape Verde
	British Isles		New Zealand		Canary, Madeira
	Central, France				Azores
X	Southern		*Northern America*		South Atlantic Isl.
X	Eastern		Canada, Alaska		
	Mediterranean Isl.		Southeastern USA		*Indian Ocean Islands*
		N	Western USA		Mascarenes
	Africa		Remaining USA		Seychelles
X	Northern	N	Mexico		Madagascar
	Tropical				
●	Southern		*Southern America*		*Pacific Islands*
			Tropical		Micronesia
	Asia		Caribbean		Galapagos
	Temperate	X	Chile, Argentina		Hawaii
	Tropical				Melanesia, Polynesia

Invaded Habitats
Grass- and heathland, rocky places.

Description
A jointed upright growing cylindrical cactus of 2-4 m height, shrubby or tree-like with a short trunk. Stem segments are cylindrical, grey-green, straight or curved, 8-25 cm long and 1.5-4 cm in diameter, with stout, silver, yellow or reddish-brown spines. Flowers are dark-pink to magenta. Fruits are obovoid, fleshy, yellow, spineless, 2.5-4.5 cm long and 2-4 cm wide.

Ecology and Control
In the native range, several varieties have been described. The plant is invasive because it forms large and dense, impenetrable thickets that affect wildlife and impede the growth and regeneration of native trees and shrubs.

    Small plants may be hand pulled or dug out. Chemical control is done by spot spraying 2,4,5-T, or by injecting MSMA into the stems. Large plants can be killed if they are felled immediately after MSMA application and if the cut stumps are also treated. Triclopyr is also effective.

References
17, 549, 809, 846, 1089.

Cynara cardunculus L.   Asteraceae

LF: Perennial herb
SN: -
CU: Ornamental, food

Geographic Distribution

Europe
    Northern
X British Isles
    Central, France
N Southern
    Eastern
N Mediterranean Isl.

Africa
N Northern
    Tropical
    Southern

Asia
N Temperate
    Tropical

Australasia
● Australia
X New Zealand

Northern America
    Canada, Alaska
    Southeastern USA
● Western USA
    Remaining USA
    Mexico

Southern America
X Tropical
    Caribbean
● Chile, Argentina

Atlantic Islands
    Cape Verde
N Canary, Madeira
    Azores
    South Atlantic Isl.

Indian Ocean Islands
    Mascarenes
    Seychelles
    Madagascar

Pacific Islands
    Micronesia
    Galapagos
    Hawaii
    Melanesia, Polynesia

Invaded Habitats
Grassland, rangeland, riparian habitats, disturbed sites.

Description
A large erect perennial herb of 80-180 cm height, with a taproot up to 2 m long and strongly ribbed stems branching near the top. Leaves are greyish green above, whitish pubescent below, deeply divided almost to the midrib, each lobe having secondary lobes and yellow-orange spines along margins. Petioles are ribbed and with spines. Rosette leaves are up to 90 cm long and 30 cm wide, stem leaves are smaller. Flowerheads are globular, 7-13 cm in diameter including the subtending stout bracts. Each bract ends in a rigid spine. Flowerheads are borne singly at the ends of branches. Seeds are brown to black, 6-8 mm long, smooth, and have a pappus of feathery hairs up to 4 cm length.

Ecology and Control
Where invasive, the species grows in dense populations with up to 22,000 plants per acre, displacing native vegetation, reducing species diversity and making an impassable barrier for wildlife. Seeds are usually not dispersed over long distances and fall mainly within a few metres of the plants, although they may be dispersed by animals and water. Establishment depends on disturbance but once established, the plant persists and dominates the vegetation, competes for soil moisture and nutrients, and shades out native species.

    Scattered plants may be grubbed but much of the taproot must be removed to prevent regrowth. Seedlings can be pulled by hand. Several herbicides are used, e.g. glyphosate, picloram, or dicamba. MCPA is used for selective control of seedlings in grassland.

References
133, 215, 924, 986.

Cynodon dactylon (L.) Pers.   Poaceae

LF: Perennial herb   CU: Erosion control, forage
SN: -

Geographic Distribution

Europe		*Australasia*		*Atlantic Islands*	
N	Northern	●	Australia	X	Cape Verde
X	British Isles	X	New Zealand	N	Canary, Madeira
N	Central, France			N	Azores
N	Southern	*Northern America*			South Atlantic Isl.
N	Eastern	X	Canada, Alaska		
N	Mediterranean Isl.	X	Southeastern USA	*Indian Ocean Islands*	
		X	Western USA		Mascarenes
Africa		X	Remaining USA		Seychelles
N	Northern	X	Mexico		Madagascar
N	Tropical				
X	Southern	*Southern America*		*Pacific Islands*	
		X	Tropical	X	Micronesia
Asia		X	Caribbean	X	Galapagos
X	Temperate	X	Chile, Argentina	●	Hawaii
N	Tropical			X	Melanesia, Polynesia

Invaded Habitats
Grass- and heathland, sclerophyll forests, riparian habitats, freshwater wetlands, rocky sites.

Description
A variable grass, creeping by means of stolons and rhizomes, with culms of 5-50 cm height or more, rooting at the lower nodes. Leaf blades are flat, 3-6 cm long and 1.5-4 mm wide, and have smooth sheaths. The ligules are short, fringed membranes of 0.2-0.5 mm length. Each flowering culm bears a panicle at its end composed of 3-6 or more digitately arranged spikes. These are 3-6 cm long and have sessile spikelets of 2-3 mm length with one floret each. Glumes are small and compressed. The lemma is broadly ovate and 1.8-2.5 mm long. Fruits are ellipsoid caryopses of 2.2-2.7 mm length.

Ecology and Control
In the native range, the grass frequently grows in grassland, disturbed places and cultivated areas. Its natural distribution is obscure due to its widespread use, and many cultivars have been bred. It is invasive because it spreads quickly by rhizomes and stolons, forming solid mats that crowd out native species and prevent their natural regeneration. Rhizomes grow in soil depths ranging from a few centimetres to a metre or more. Seeds are sparsely produced. The grass is drought tolerant and tolerates salinity well but salt slows down growth. It tolerates heavy grazing and resprouts easily after fires.

Small patches can be dug out but all rhizomes and stolons must be removed. Solarization by plastic sheeting is used in sunny locations. Young plants are effectively controlled by applying paraquat, or glyphosate applied in spring or autumn when rhizomes are growing.

References
215, 351, 580, 1190.

Cyperus eragrostis Lam.
Cyperaceae

LF: Perennial herb
SN: -
CU: -

Geographic Distribution

Europe	*Australasia*	*Atlantic Islands*
Northern	● Australia	Cape Verde
X British Isles	X New Zealand	X Canary, Madeira
X Central, France		X Azores
● Southern	*Northern America*	South Atlantic Isl.
Eastern	N Canada, Alaska	
Mediterranean Isl.	Southeastern USA	*Indian Ocean Islands*
	N Western USA	Mascarenes
Africa	Remaining USA	Seychelles
Northern	Mexico	Madagascar
Tropical		
X Southern	*Southern America*	*Pacific Islands*
	N Tropical	Micronesia
Asia	Caribbean	Galapagos
Temperate	N Chile, Argentina	Hawaii
X Tropical		Melanesia, Polynesia

Invaded Habitats
Forests, riparian habitats and freshwater wetlands.

Description
A tufted erect grass-like herb up to 80 cm tall, with a short thick rhizome. Stems are slender, slightly swollen at the base, three-angled, smooth and striated. Leaves are shorter than stems, smooth, arising from or near the base, 4-10 mm wide, and slightly toothed on margins. Inflorescences are simple or compound umbels of clustered, flattened spikelets. These are yellow to golden brown, and 7-15 mm long. Inflorescences are on stalks of unequal length and subtended by leaf-like bracts that exceed the inflorescences in length. Spikelets have several florets and compressed glumes. Fruits are brown achenes of *c.* 1 mm length.

Ecology and Control
A sedge that is confined to damp situations. Once established, it is very persistent and forms dense infestations that block channels and drains, reduce native species diversity and impede growth of desired species. It colonizes small streams with quiet water and alters the vegetation structure of riparian plant communities. The plant has an extensive root system of coarse fibrous roots. It is an important agricultural weed.

    In agricultural areas, the species is controlled by cultivation. Chemical control is done by spraying 2,4-D. Smaller infestations can be removed manually, but all rhizomes must be removed to prevent regrowth.

References
215, 986, 1211.

Cytisus multiflorus (L'Hér.) Sweet — Fabaceae

LF: Evergreen shrub
CU: Ornamental
SN: *Cytisus albus* (Lam.) Link, *Cytisus lusitanicus* Willk.

Geographic Distribution

Europe	*Australasia*	*Atlantic Islands*
Northern	● Australia	Cape Verde
X British Isles	X New Zealand	Canary, Madeira
Central, France		Azores
N Southern	*Northern America*	South Atlantic Isl.
Eastern	Canada, Alaska	
Mediterranean Isl.	Southeastern USA	*Indian Ocean Islands*
	X Western USA	Mascarenes
Africa	Remaining USA	Seychelles
Northern	Mexico	Madagascar
Tropical		
Southern	*Southern America*	*Pacific Islands*
	Tropical	Micronesia
Asia	Caribbean	Galapagos
Temperate	Chile, Argentina	Hawaii
Tropical		Melanesia, Polynesia

Invaded Habitats
Dry sclerophyll forests, woodland, scrubland.

Description
A much-branched and erect shrub of 1-3 m height, with five-angled and flexible branches, and twigs becoming glabrous with age. Leaves of lower branches are compound with three leaflets, leaves on upper branches are simple. Leaflets are linear-lanceolate to oblong, silvery and up to 10 mm long. Flowers are borne in clusters of 1-3 and have pedicels of *c.* 10 mm length. The white corollas are 9-12 mm long. Fruits are strongly compressed, oblong pods of 15-25 mm length and 5-9 mm width. Seeds are greyish brown, compressed and 2.5-3 mm long.

Ecology and Control
In the native range, this shrub grows in heath- and woodland, on river banks, and is generally calcifuge. Where invasive, the shrub forms extensive thickets that crowd out native species and prevent natural forest regeneration. It is nitrogen-fixing and increases soil fertility levels. It can hybridize with *Cytisus scoparius* where both species are present.

    Specific control methods for this species are not available. The same methods as for the following species may apply.

References
215, 830, 1097.

Cytisus scoparius (L.) Link
Fabaceae

LF: Deciduous shrub
CU: Ornamental, soil stabilization
SN: *Sarothamnus scoparius* (L.) Wimm., *Spartium scoparium* L.

Geographic Distribution

Europe
- N   Northern
- N   British Isles
- N   Central, France
- N   Southern
- N   Eastern
- N   Mediterranean Isl.

Africa
-    Northern
-    Tropical
- X   Southern

Asia
- X   Temperate
- X   Tropical

Australasia
- •   Australia
- •   New Zealand

Northern America
- X   Canada, Alaska
-    Southeastern USA
- •   Western USA
-    Remaining USA
-    Mexico

Southern America
-    Tropical
-    Caribbean
-    Chile, Argentina

Atlantic Islands
-    Cape Verde
- X   Canary, Madeira
-    Azores
-    South Atlantic Isl.

Indian Ocean Islands
-    Mascarenes
-    Seychelles
-    Madagascar

Pacific Islands
-    Micronesia
-    Galapagos
- X   Hawaii
-    Melanesia, Polynesia

Invaded Habitats
Grass- and heathland, sclerophyll forests, riparian habitats, coastal scrub.

Description
An erect shrub of 1-3 m height, with green to brownish green and ridged stems, and numerous branchlets. Leaves grow singly or in clusters, each leaf consisting of three softly hairy, entire, ovate to lanceolate leaflets of 1-2 cm length. Bright yellow flowers of 2-2.5 cm length are borne singly or in pairs in the axils of the upper leaves. Fruits are brown to black, flattened pods of *c.* 5 cm length and 1 cm width, hairy on the margins, each having 6-22 seeds. These are yellowish brown, shiny, rounded, 3-4 mm long and *c.* 2 mm wide. The plant has a stout taproot.

Ecology and Control
Where native, this nitrogen-fixing shrub grows commonly on heaths and wasteland. Establishment depends on disturbance, the plant then forms extensive thickets that crowd out native species and affect wildlife. It smothers large shrubs and prevents establishment of native species. It produces large amounts of long-lived seeds and resprouts from the root crown if damaged. The shrub is flammable and may increase fire intensities in fire prone areas.

Cutting close to the ground may prevent resprouting. Mature plants can be removed with a weed wrench. Chemical control includes applying triclopyr mixed with an oil surfactant to the basal bark just after flowering. Repeated prescribed burning is used to deplete the soil seed bank. Emerging seedlings can be hand pulled, sprayed or burned.

References
102, 130, 131, 132, 133, 215, 218, 358, 590, 607, 621, 765, 861, 937, 982, 983, 986, 996, 997, 1012, 1144, 1173, 1202, 1252, 1272, 1394, 1413.

Dactylis glomerata L.   Poaceae

LF: Perennial herb
SN: *Dactylis glaucescens* Willd.
CU: Ornamental, forage

Geographic Distribution

Europe
- N Northern
- N British Isles
- N Central, France
- N Southern
- N Eastern
- N Mediterranean Isl.

Africa
- N Northern
-     Tropical
- X Southern

Asia
- N Temperate
- N Tropical

Australasia
- ● Australia
- X New Zealand

Northern America
-     Canada, Alaska
-     Southeastern USA
- X Western USA
- X Remaining USA
-     Mexico

Southern America
- X Tropical
-     Caribbean
- X Chile, Argentina

Atlantic Islands
-     Cape Verde
- N Canary, Madeira
- N Azores
-     South Atlantic Isl.

Indian Ocean Islands
- X Mascarenes
-     Seychelles
-     Madagascar

Pacific Islands
-     Micronesia
-     Galapagos
- ● Hawaii
-     Melanesia, Polynesia

Invaded Habitats
Grass- and heathland, forests, riparian habitats, freshwater wetlands, coastal areas.

Description
A tussock forming perennial grass with erect or spreading culms of 30-150 cm height. Ligules are 2-10 mm long, sometimes pubescent. The glaucous leaf blades are strongly keeled and have minutely toothed margins. Inflorescences are erect and stiff panicles of 1-20 cm length, comprising close branches with almost sessile spikelets of 5-9 mm length, each containing 2-5 florets. The glumes are lanceolate to ovate, and long-acute. Fruits are *c.* 2.5 mm long caryopses.

Ecology and Control
In the native range, this grass is found in meadows, pastures, sand dunes and disturbed sites. It is a highly variable species with several varieties in Europe, and several cultivars have been developed differing in phenology. The grass establishes in disturbed sites and forms dense swards that suppress native grasses and forbs. Seed production is usually high, and seeds are dispersed by wind, water and by adhering to animals.

    Plants can be dug out, the crown must be removed to prevent regrowth. Larger stands are cut before flowering commences. In Britain, heavy grazing during autumn may lead to complete elimination of this grass. Follow-up programmes are necessary to treat seedlings.

References
49, 93, 121, 215, 924, 1117.

Datura stramonium L.   Solanaceae

LF: Annual herb   CU: -
SN: *Datura tatula* L., *Datura inermis* Juss. Ex Jacq.

Geographic Distribution

Europe		*Australasia*		*Atlantic Islands*	
X	Northern	X	Australia	X	Cape Verde
X	British Isles	X	New Zealand	X	Canary, Madeira
X	Central, France			X	Azores
X	Southern	*Northern America*			South Atlantic Isl.
X	Eastern	X	Canada, Alaska		
X	Mediterranean Isl.	X	Southeastern USA	*Indian Ocean Islands*	
		X	Western USA		Mascarenes
Africa		X	Remaining USA		Seychelles
X	Northern	N	Mexico		Madagascar
X	Tropical				
●	Southern	*Southern America*		*Pacific Islands*	
		N	Tropical		Micronesia
Asia		X	Caribbean	●	Galapagos
X	Temperate	X	Chile, Argentina	X	Hawaii
X	Tropical			X	Melanesia, Polynesia

Invaded Habitats
Sandy places, disturbed sites.

Description
A large erect and smooth annual of 50-200 cm height with glabrous to pubescent stems. The ovate to elliptical leaves are acute, 5-20 cm long and 4-15 cm wide. The calyx is angled and 3-5 cm long. The large flowers are white or pale purple, the corolla is 5-10 cm long. Fruits are ovoid and erect, dehiscent capsules of 3-7 cm length and 2-3.5 cm width, densely covered with rather slender spines of up to 15 mm length. Seeds are dark brown to black and 3-3.5 mm long.

Ecology and Control
A widespread weed with a prolific seed production. In tropical America, several varieties have been described. Plants are toxic to humans and animals. Seedlings establish quickly, and large thickets are built up that shade out all surrounding vegetation due to the large leaves.

    Isolated plants can be hand pulled or hoed before fruits are formed. Effective pre-emergence herbicides are atrazine or dicamba. Seedlings and larger plants can be sprayed with dicamba, 2,4-D or bromoxyril.

References
581, 985, 986, 1089, 1313, 1364.

Delairea odorata Lem.   Asteraceae

LF: Evergeen vine   CU: Ornamental
SN: *Senecio mikanioides* Otto ex Walp.

Geographic Distribution

	Europe		*Australasia*		*Atlantic Islands*
	Northern	●	Australia		Cape Verde
X	British Isles	X	New Zealand	X	Canary, Madeira
	Central, France			X	Azores
X	Southern		*Northern America*		South Atlantic Isl.
	Eastern		Canada, Alaska		
	Mediterranean Isl.		Southeastern USA		*Indian Ocean Islands*
		●	Western USA		Mascarenes
	Africa		Remaining USA		Seychelles
X	Northern		Mexico		Madagascar
	Tropical				
N	Southern		*Southern America*		*Pacific Islands*
			Tropical		Micronesia
	Asia		Caribbean		Galapagos
	Temperate		Chile, Argentina	X	Hawaii
	Tropical				Melanesia, Polynesia

Invaded Habitats
Heathland, forests, riparian habitats, coastal grassland and scrub, shady disturbed sites.

Description
A perennial vine climbing 6 m tall or more, with much-branched and slender stems that are woody at least below. Each node bears one green to yellowish-green and shiny leaf. The leave blades are 3-10 cm long, usually with petioles longer than the blades, and have 5-9 lobes ending in sharp points. The plant produces waxy above ground stolons and purplish rhizomes. Flowerheads with up to 40 bright yellow disc florets are borne in dense axillary and terminal panicles. Flowerheads are 5-7 mm in diameter. Fruits are glabrous achenes.

Ecology and Control
This fast growing vine spreads extensively by vegetative growth, and its rhizomes grow to a depth of 90 cm. The plant smothers trees and shrubs by forming a dense cover, reducing light and impeding growth. Native species richness is strongly reduced under stands of the vine. It spreads most vigorously in moist, semi-shaded sites. In drier areas, it dies back during the dry season but regrows rapidly during the wet season. The stolons easily fragment and the fragments can quickly produce new plants. Plants in California do not produce seeds.

Control is difficult because of rapid regrowth. Manual removal requires clearing the infested area and pulling the roots and stems out of the soil. An effective chemical control is a foliar application of glyphosate plus triclopyr, or by spraying with a clopyralid herbicide. Follow-up treatments are necessary to control regrowth and emerging seedlings.

References
121, 133, 215, 396, 607, 608, 1415.

Dioscorea alata L.   Dioscoreaceae

LF: Evergeen vine
SN: *Dioscorea rubella* Roxb.
CU: Ornamental, food

Geographic Distribution

Europe	*Australasia*	*Atlantic Islands*
Northern	X Australia	Cape Verde
British Isles	New Zealand	Canary, Madeira
Central, France		Azores
Southern	*Northern America*	South Atlantic Isl.
Eastern	Canada, Alaska	
Mediterranean Isl.	• Southeastern USA	*Indian Ocean Islands*
	Western USA	Mascarenes
Africa	Remaining USA	Seychelles
Northern	Mexico	Madagascar
• Tropical		
Southern	*Southern America*	*Pacific Islands*
	X Tropical	Micronesia
Asia	Caribbean	Galapagos
Temperate	Chile, Argentina	Hawaii
N Tropical		Melanesia, Polynesia

Invaded Habitats
Forests and forest edges.

Description
A glabrous herb with winged and twining stems reaching many metres in length, with mostly opposite leaves. Globose bulbils are frequently produced on the stems. Leaves are ovate and have petioles of 3-18 cm length. Leaf blades are 5-25 cm long and 3-15 cm wide. Male flowers are yellow and borne in axillary spikes of 1-3 cm length. Female flowers appear in racemes of 10-60 cm length. Fruits are broadly ovate smooth capsules of 17-20 mm length, releasing winged seeds. The plant has a tuberous root system, with globose to pear-shaped, often lobed tubers. The plant is dioecious.

Ecology and Control
This plant is widely cultivated for its edible tubers, and many varieties exist. It establishes well in disturbed sites and grows rapidly. The vine smothers native plants and shades out understorey species. The plant spreads mainly vegetatively, and stem fragments can become rooted. It does not set seeds in Australia.

Smaller plants can be dug out or hand pulled. The roots and tubers must be removed to prevent regrowth. Larger patches are cut or treated with herbicide.

References
449, 715, 727, 1168.

Dipogon lignosus (L.) Verdc.   Fabaceae

LF: Evergreen climber   CU: Ornamental
SN: *Dolichos gibbosus* Thunb., *Verdcourtia lignosa* (L.) Wilczek

Geographic Distribution

Europe	*Australasia*	*Atlantic Islands*
Northern	● Australia	Cape Verde
British Isles	X New Zealand	Canary, Madeira
Central, France		Azores
Southern	*Northern America*	South Atlantic Isl.
Eastern	Canada, Alaska	
Mediterranean Isl.	Southeastern USA	*Indian Ocean Islands*
	Western USA	Mascarenes
Africa	Remaining USA	Seychelles
Northern	Mexico	Madagascar
Tropical		
N Southern	*Southern America*	*Pacific Islands*
	Tropical	Micronesia
Asia	Caribbean	Galapagos
Temperate	Chile, Argentina	Hawaii
Tropical		Melanesia, Polynesia

Invaded Habitats
Grass- and heathland, forests, riparian habitats, coastal beaches.

Description
A woody, scrambling and climbing vine with many stems up to 6 m long and slender, long branches. Leaves are alternate and composed of three rhomboid leaflets. These are 3-6 cm long and 15-40 mm wide, and have petioles up to 6 cm long. Red purple flowers of 10-15 mm length are borne in solitary and axillary long-stalked racemes containing 3-6 flowers each. Fruits are brownish and stalked, glabrous pods of 2-5 cm length and 5-10 mm width, each with 3-6 black seeds of *c.* 4 mm length.

Ecology and Control
A native of scrub and open forests that grows best in sunny locations, but seedlings can establish in shade. It is a prolific seed producer and seeds can remain dormant for several years. They are dispersed mainly by birds. The species is nitrogen-fixing and increases soil fertility levels. The plant forms dense infestations that smother all other vegetation and prevent regeneration of native shrubs and trees.

    Seedlings and small plants can be hand pulled or dug out, roots must be removed. Larger vines can be cut and the roots dug out, or the cut stumps treated with herbicide. Cutting before fruits develop prevents seed dispersal. Follow-up treatments are necessary to control regrowth and seedlings.

References
121, 215, 607, 924.

Dipsacus fullonum L.   Dipsacaceae

LF: Biennial herb
SN: *Dipsacus sylvestris* Huds.
CU: Ornamental

Geographic Distribution

Europe
-     Northern
- N British Isles
- N Central, France
- N Southern
- N Eastern
- N Mediterranean Isl.

Africa
- N Northern
-     Tropical
- X Southern

Asia
- N Temperate
-     Tropical

Australasia
- X Australia
- X New Zealand

Northern America
- X Canada, Alaska
-     Southeastern USA
- X Western USA
- ● Remaining USA
-     Mexico

Southern America
-     Tropical
-     Caribbean
- X Chile, Argentina

Atlantic Islands
-     Cape Verde
-     Canary, Madeira
-     Azores
-     South Atlantic Isl.

Indian Ocean Islands
-     Mascarenes
-     Seychelles
-     Madagascar

Pacific Islands
-     Micronesia
-     Galapagos
-     Hawaii
-     Melanesia, Polynesia

Invaded Habitats
Grassland, savannas, forest edges, riparian habitats, wet meadows.

Description
A few-branched herb of 50-250 cm height, with ridged stems and curved prickles on the ridges, and rosette leaves up to 50 cm long. Basal leaves are lanceolate and sessile, the surface being rough with prickles. Stem leaves are smaller, opposite, united at the base and with prickles along the lower side of the midvein. Solitary and cylindrical flowerheads of 5-12 cm length appear at the ends of branches and are surrounded by long bracts. They contain numerous pink, purple or lilac florets. Bracts are stiff and longer than the florets. Fruits are yellow to greyish brown achenes of 3-5 mm length and 1-1.5 mm width. The plant has a deep and thick taproot reaching 75 cm depth.

Ecology and Control
In the native range, the plant is found along streams, in forest gaps and other mesic sites. Seeds may be dispersed by streams. It is invasive because it forms large monospecific colonies that displace native vegetation. The large leaves of the rosettes form a continuous cover shading out all other species. Areas covered by this plant impede movement of wildlife. The plant produces large amounts of seed that germinate readily and at high rates.

Repeated cutting before flowering starts can eliminate a population. The weed is effectively controlled by 2,4-D applied to plants in the rosette stage.

References
459, 571, 985, 986, 1386.

Dipsacus laciniatus L. Dipsacaceae

LF: Biennial herb
SN: -
CU: Ornamental

Geographic Distribution

Europe	*Australasia*	*Atlantic Islands*
Northern	Australia	Cape Verde
British Isles	New Zealand	Canary, Madeira
N Central, France		Azores
N Southern	*Northern America*	South Atlantic Isl.
N Eastern	Canada, Alaska	
Mediterranean Isl.	Southeastern USA	*Indian Ocean Islands*
	Western USA	Mascarenes
Africa	● Remaining USA	Seychelles
Northern	Mexico	Madagascar
Tropical		
Southern	*Southern America*	*Pacific Islands*
	Tropical	Micronesia
Asia	Caribbean	Galapagos
N Temperate	Chile, Argentina	Hawaii
Tropical		Melanesia, Polynesia

Invaded Habitats
Grass- and woodland, wet meadows.

Description
A stout herb of 50-300 cm height and with prickly spines on leaves and stems. Stem leaves are deeply lobed and ciliate at the margins. Flowerheads are ovoid to cylindrical, with lanceolate-subulate bracts that are curved upwards. Florets are pale pink or white, and have corollas with four lobes. Fruits are blackish-brown achenes of *c.* 5 mm length, and are covered with appressed hairs.

Ecology and Control
Where native, the species grows in forest edges, riparian forests, meadows, and disturbed places. The plant forms monospecific colonies where invasive and crowds out native vegetation. Seeds are abundantly produced and germinate readily. Seeds are not adapted for long-distance dispersal but may be carried by streams.

    Individual rosettes can be grubbed, but as much of the root as possible must be removed to prevent resprouting. Chemical control is done by foliar spraying of glyphosate or 2,4-D amine during early spring. Flowering stalks are cut once flowering has initiated and removed from the site. Emerging seedlings require a follow-up treatment.

References
459, 571, 1208, 1209.

Echinochloa polystachya (Kunth) Hitchc.   Poaceae

LF: Perennial herb   CU: Forage
SN: *Echinochloa spectabilis* (Nees) Link., *Oplismenus polystachyus* HBK.

Geographic Distribution

Europe		*Australasia*		*Atlantic Islands*	
	Northern	●	Australia		Cape Verde
	British Isles		New Zealand		Canary, Madeira
	Central, France				Azores
	Southern	*Northern America*			South Atlantic Isl.
	Eastern		Canada, Alaska		
	Mediterranean Isl.		Southeastern USA	*Indian Ocean Islands*	
			Western USA		Mascarenes
Africa		X	Remaining USA		Seychelles
	Northern	N	Mexico		Madagascar
X	Tropical				
X	Southern	*Southern America*		*Pacific Islands*	
		N	Tropical		Micronesia
Asia		N	Caribbean		Galapagos
	Temperate	X	Chile, Argentina	X	Hawaii
X	Tropical				Melanesia, Polynesia

Invaded Habitats
Freshwater wetlands, riparian habitats.

Description
A grass with coarse and glabrous culms of 1-2 m height, and with a long creeping root base. Leaf blades are 2-5 cm wide and have ligules consisting of stiff, yellowish hairs. Inflorescences are dense panicles of 10-30 cm length and ascending racemes, with almost sessile spikelets of *c.* 5 mm length. Awns are 2-10 mm long. Fruits are rather soft caryopses of *c.* 4 mm length.

Ecology and Control
A C_4 aquatic grass of floodplains and estuaries with a very high productivity. The grass does not tolerate drought and frosts, and grows well in standing water. It is adapted to wet and very wet soils, and to varying water levels. The fast growing grass forms extensive monotypic stands on the lower levels of floodplains, crowding out native water plants and altering the vegetation structure. It grows rapidly during floods and decays rapidly during water retreat, resulting in a high release of nutrients during the low water period. This nutrient flush may affect water quality and reduce the oxygen content.

     Specific control methods for this species are not available. Scattered plants may be removed manually, larger stands treated with herbicide.

References
43, 291, 431, 607, 902, 1019, 1034, 1039, 1190.

Echium plantagineum L.   Boraginaceae

LF: Annual, biennial herb
SN: *Echium lycopsis* L.
CU: Ornamental, bee plant

Geographic Distribution

Europe		*Australasia*		*Atlantic Islands*	
	Northern	●	Australia		Cape Verde
N	British Isles	X	New Zealand	N	Canary, Madeira
N	Central, France			N	Azores
N	Southern		*Northern America*		South Atlantic Isl.
N	Eastern		Canada, Alaska		
N	Mediterranean Isl.		Southeastern USA		*Indian Ocean Islands*
		X	Western USA		Mascarenes
Africa		X	Remaining USA		Seychelles
N	Northern		Mexico		Madagascar
	Tropical				
X	Southern		*Southern America*		*Pacific Islands*
			Tropical		Micronesia
Asia			Caribbean		Galapagos
N	Temperate	X	Chile, Argentina		Hawaii
	Tropical				Melanesia, Polynesia

Invaded Habitats
Grass- and shrubland, dry forests, riparian habitats, disturbed sites.

Description
A much branched herb of 30-80 cm height, occasionally up to 150 cm tall, with several stems arising from the base. Stems and leaves are covered with stout white hairs or bristles. Rosette leaves are up to 25 cm long, oval to oblong and stalked. Stem leaves are narrower and smaller, clasping the stem or sessile. Large, purple flowers of 2-3 cm length are arranged in curved spikes that are crowded along one side of the stem. Fruits consist each of four nutlets surrounded by the persistent calyx. Nutlets are brown to grey, 2-3 mm long and strongly wrinkled. The plant has a stout taproot with many lateral roots.

Ecology and Control
A common species of grassy and sandy places, and disturbed ground in the native range. It forms extensive swards where invasive that crowd out native species. The plant is a prolific seed producer, leading to a seed rain of more than 30,000 m^{-2} in dense infestations. Seeds are dispersed by attaching to animal furs and by streams. The plant accumulates a soil seed bank, and seeds may remain viable for several years. A high early growth rate enables the plant to outcompete seedlings of native plant species.

Single plants and small patches are cut or plants pulled by hand. Control on pastures include combinations of cultivation, grazing, and burning with the aim to stimulate seeds in the soil to germinate, and then killing the seedlings. Herbicides such as 2,4-D or bromoxyril are most effective when applied to rosettes.

References
215, 419, 506, 680, 985, 986, 1029, 1031, 1318, 1435.

Egeria densa Planch.   Hydrocharitaceae

LF: Submerged aquatic   CU: Ornamental
SN: *Anacharis densa* (Planch.) Vict., *Elodea densa* (Planch.) Casp.

Geographic Distribution

Europe
- X   Northern
- X   British Isles
- X   Central, France
- X   Southern
- X   Eastern
-      Mediterranean Isl.

Africa
-      Northern
-      Tropical
- X   Southern

Asia
- X   Temperate
- X   Tropical

Australasia
- ●   Australia
- ●   New Zealand

Northern America
-      Canada, Alaska
- X   Southeastern USA
- ●   Western USA
-      Remaining USA
-      Mexico

Southern America
- N   Tropical
-      Caribbean
- N   Chile, Argentina

Atlantic Islands
-      Cape Verde
-      Canary, Madeira
-      Azores
-      South Atlantic Isl.

Indian Ocean Islands
-      Mascarenes
-      Seychelles
-      Madagascar

Pacific Islands
-      Micronesia
-      Galapagos
- X   Hawaii
-      Melanesia, Polynesia

Invaded Habitats
Freshwater wetlands, lakes, ditches, ponds, slow streams.

Description
A submerged and much-branched aquatic herb with green to brownish stems of 3-5 m length and 2-2.5 mm width. Nodes are crowded in the upper part of the stems. The minutely toothed leaves are sessile, 15-30 mm long and 3-5 mm wide. Upper leaves are in whorls of 3-7. Flowers are white and 10-20 mm in diameter. The plant is dioecious, e.g. male and female flowers appear on different individuals.

Ecology and Control
This fast growing water plant is a native of mild to warm freshwater ponds and lakes, and slowly moving streams. Where invasive, it has replaced native aquatic vegetation due to dense infestations, resulting mostly from eutrophication of lakes. It forms extensive thick mats of intertwining stems between 2 and 5 m depth, outcompeting native macrophytes, retarding water flow and reducing light intensity. It tolerates a wide range of nutrient levels and grows well under low light intensities. Plants are usually firmly rooted in the substrate but also occur as free-floating mats. The plant spreads vegetatively; in Australia, only male plants are present.

     Mechanical removal such as cutting or hand pulling may encourage fragmentation of the plant and spread of the fragments. Herbicidal control methods include treating infestations of flowing waters with acrolein or diquat.

References
133, 215, 332, 404, 523, 986, 1274, 1275, 1385, 1415.

Ehrharta calycina Sm. Poaceae

LF: Perennial herb
SN: -
CU: Forage, revegetation

Geographic Distribution

Europe	*Australasia*	*Atlantic Islands*
Northern	● Australia	Cape Verde
British Isles	New Zealand	Canary, Madeira
Central, France		Azores
Southern	*Northern America*	South Atlantic Isl.
Eastern	Canada, Alaska	
Mediterranean Isl.	Southeastern USA	*Indian Ocean Islands*
	● Western USA	Mascarenes
Africa	Remaining USA	Seychelles
Northern	Mexico	Madagascar
Tropical		
N Southern	*Southern America*	*Pacific Islands*
	Tropical	Micronesia
Asia	Caribbean	Galapagos
Temperate	Chile, Argentina	Hawaii
Tropical		Melanesia, Polynesia

Invaded Habitats
Grass- and heathland, woodland, coastal dunes.

Description
A tufted grass with 30-120 cm tall stems, often with short rhizomes, and relatively few culms. Leaf blades are 7-25 cm long and 2-7 mm wide. The inflorescence is a more or less open panicle of 10-25 cm length, with the secondary branches mostly hanging downwards, and with spikelets almost sessile or on short pedicels. Spikelets are 5-8 mm long, glumes 5-7 mm long, becoming purplish. Sterile lemmas have an awn or are pointed. Glumes are membranous and purple to brown.

Ecology and Control
In the native range, this grass grows mostly on disturbed sandy soils. It is adapted to nutrient-poor soils and establishes in open sites. It forms dense and tall patches that eliminate all native herbaceous species. Seed production is abundant, and seeds are dispersed by wind, water and animals. The grass accumulates large quantitites of dried plants thus increasing fire intensity and frequency. Fires themselves stimulate germination and regrowth, favouring the spread of this grass.

Roots and rhizomes are shallow and plants are easy to uproot. All rhizomes must be removed to prevent regrowth. Herbicides are best applied before flowering stems elongate. Seedlings need to be treated in a follow-up programme.

References
133, 215, 876, 1205.

Ehrharta erecta Lam.   Poaceae

LF: Perennial herb
SN: *Ehrharta panicea* Sm.
CU: -

Geographic Distribution

Europe	Australasia	Atlantic Islands
Northern	● Australia	Cape Verde
British Isles	● New Zealand	Canary, Madeira
Central, France		Azores
Southern	*Northern America*	South Atlantic Isl.
Eastern	Canada, Alaska	
Mediterranean Isl.	Southeastern USA	*Indian Ocean Islands*
	● Western USA	Mascarenes
Africa	Remaining USA	Seychelles
Northern	Mexico	Madagascar
Tropical		
N Southern	*Southern America*	*Pacific Islands*
	Tropical	Micronesia
Asia	Caribbean	Galapagos
Temperate	Chile, Argentina	Hawaii
Tropical		Melanesia, Polynesia

Invaded Habitats
Grass- and woodland, dry forests, riparian habitats, rock outcrops.

Description
A sparsely tufted grass with loose culms of 30-90 cm height and leaf blades of 5-20 cm length and 4-9 mm width. Inflorescences are panicles compact and narrow at first, becoming open and spreading, and are 10-40 cm long. Spikelets are sessile or almost so, stiff and 3-6 mm long. Glumes are 1.5-3 mm long. The lemmas are glabrous and without awns. Rhizomes are short and slender. Seeds are *c.* 3 mm long.

Ecology and Control
In the native range, this grass grows in tropical and subtropical forests, coastal dune forests, grassland and disturbed sites, thriving best in shaded habitats. Three varieties occur in South Africa: var. *erecta*, var. *natalensis* and var. *abyssinica*. It is capable of entering undisturbed native vegetation and invades *Leptospermum laevigatum* scrub. It forms dense swards and produces seedlings in high densities. At drier sites, tillers die back at the end of the growing season, leading to accumulation of large amounts of grass litter that impede the growth of native plants. Seeds are shed in a dormant state and germinate readily after *c.* 12 months. They are dispersed by water, wind and birds.

Plants are easy to pull or dig out, rhizomes must be removed to prevent regrowth. Hot fires are used to kill mature plants and to stimulate germination of seeds. Seedlings and regrowth are then treated with herbicide 4-6 weeks later.

References
133, 215, 847, 1415.

Eichhornia crassipes (Mart.) Solms   Pontederiaceae

LF: Perennial aquatic   CU: Ornamental
SN: *Pontederia crassipes* Mart. & Zucc.

Geographic Distribution

Europe		Australasia		Atlantic Islands	
	Northern	●	Australia		Cape Verde
	British Isles	X	New Zealand		Canary, Madeira
	Central, France			X	Azores
●	Southern		*Northern America*	X	South Atlantic Isl.
	Eastern		Canada, Alaska		
	Mediterranean Isl.	●	Southeastern USA		*Indian Ocean Islands*
		●	Western USA		Mascarenes
Africa			Remaining USA	X	Seychelles
X	Northern	N	Mexico	X	Madagascar
●	Tropical				
●	Southern		*Southern America*		*Pacific Islands*
		N	Tropical		Micronesia
Asia		X	Caribbean		Galapagos
	Temperate	X	Chile, Argentina	X	Hawaii
●	Tropical			X	Melanesia, Polynesia

Invaded Habitats
Freshwater wetlands, ponds.

Description
A free-floating aquatic herb with leaves in a rosette and a stem-like rhizome consisting of numerous nodes and short internodes. Leaves have spongy and inflated petioles up to 30 cm length. Leaf blades are circular to kidney-shaped, smooth and glossy green, 4-15 cm long and wide. The inflorescence is a spike of up to 30 cm length and has 5-8 flowers. The showy flowers are lilac to bluish-purple and have a narrow tube. The rhizomes vary in length from 1-30 cm and are 1-2.5 cm in diameter.

Ecology and Control
A widespread aquatic spreading by rhizomes and rapidly colonizing large areas. It forms solid mats on the water surface that may completely cover the water body. These mats crowd out native species and their shade kills submerged plants and affect water temperatures. They are colonized by other plant species and greatly accelerate fresh water succession. Plants can root and survive on land if sufficient moisture is available. Mats may grow at a rate of *c.* 60 cm lateral expansion per month.

     Plants from small infestations may be removed by hand. Harvester machines are used for clearing larger areas. Chemical control is done by using glyphosate or 2,4-D dimethylamine approved for use in aquatic environments.

References
7, 133, 227, 229, 232, 284, 307, 474, 508, 510, 511, 580, 607, 715, 1005, 1089, 1122, 1284, 1285, 1319.

Elaeagnus angustifolia L. — Elaeagnaceae

LF: Deciduous shrub, tree
SN: *Elaeagnus hortensis* Bieb.
CU: Erosion control, ornamental

Geographic Distribution

Europe	*Australasia*	*Atlantic Islands*
Northern	Australia	Cape Verde
British Isles	New Zealand	Canary, Madeira
X Central, France		Azores
X Southern	*Northern America*	South Atlantic Isl.
N Eastern	X Canada, Alaska	
Mediterranean Isl.	Southeastern USA	*Indian Ocean Islands*
	● Western USA	Mascarenes
Africa	● Remaining USA	Seychelles
X Northern	Mexico	Madagascar
Tropical		
Southern	*Southern America*	*Pacific Islands*
	Tropical	Micronesia
Asia	Caribbean	Galapagos
N Temperate	Chile, Argentina	Hawaii
N Tropical		Melanesia, Polynesia

Invaded Habitats
Forests, riparian habitats.

Description
A large shrub or small tree up to 8 m tall, with distinctive silver-grey leaves and young branches, a reddish brown bark and woody thorns. Leaves are oblong-lanceolate to linear-lanceolate, 4-8 cm long and 1-2.5 cm wide, dull green above and silvery beneath, and with petioles of 5-8 mm length. Flowers are borne on short stalks in clusters of 1-3 on the lower parts of branches. The fragrant flowers are *c.* 1 cm long, silvery outside and yellow inside, and borne in small axillary clusters. Fruits are hard and resemble olives in shape. They are 10-20 mm long, ellipsoid, yellow, fleshy and covered with silvery scales. Each fruit contains one seed of *c.* 9 mm length.

Ecology and Control
The shrub grows commonly in wet places but also tolerates poor soils and drought. The species quickly spreads and forms dense thickets that crowd out native vegetation and prevent the establishment of native trees. It replaces native riparian forests in North America and causes loss of wildlife habitat. Seeds are dispersed by birds and small mammals. The species resprouts from its root system.

Smaller trees may be removed with a weed wrench, larger plants should be cut at ground level. If stumps cannot be completely buried, they should be treated with a herbicide. Effective herbicides are 2,4-D or 2,4,5-T.

References
133, 141, 159, 250, 593, 688, 737, 738, 961, 1114, 1160.

Elaeagnus umbellata Thunb.   Elaeagnaceae

LF: Deciduous shrub, tree   CU: Ornamental
SN: *Elaeagnus crispa* Thunb.

Geographic Distribution

Europe	Australasia	Atlantic Islands
Northern	Australia	Cape Verde
British Isles	New Zealand	Canary, Madeira
Central, France		Azores
Southern	*Northern America*	South Atlantic Isl.
Eastern	● Canada, Alaska	
Mediterranean Isl.	Southeastern USA	*Indian Ocean Islands*
	Western USA	Mascarenes
Africa	● Remaining USA	Seychelles
Northern	Mexico	Madagascar
Tropical		
Southern	*Southern America*	*Pacific Islands*
	Tropical	Micronesia
Asia	Caribbean	Galapagos
N Temperate	Chile, Argentina	● Hawaii
Tropical		Melanesia, Polynesia

Invaded Habitats
Grass- and woodland, disturbed sites.

Description
A more or less spiny shrub of 2-4 m height, with slender and spreading branches. The elliptic to ovate-oblong leaves are 4-8 cm long and 1-2.5 cm wide, the lower surface being densely covered with white scales, as are the 5-10 mm long petioles. Flowers are borne in axillary umbels densely covered with white scales. Pedicels are 3-6 mm long. Fruits are red almost globose to broadly ellipsoid berries of 6-8 mm length.

Ecology and Control
This fast growing shrub spreads rapidly in mesic to wet habitats and disturbed areas. It forms dense thickets displacing native vegetation and preventing the growth and regeneration of native plants. Fruit production is prolific and seeds are dispersed by birds. The shrub resprouts quickly after burning or cutting. The plant is nitrogen-fixing and grows well in soils of low fertility.

Repeated cutting or burning may prevent spread. Small plants and seedlings can be removed manually. Larger individuals are cut and the cut stumps treated with glyphosate. Basal applications of triclopyr applied in early spring are also effective.

References
222, 367, 368, 690, 1210, 1267.

Elodea canadensis Michx.   Hydrocharitaceae

LF: Aquatic perennial   CU: Ornamental
SN: *Anacharis canadensis* (Michx.) Planch.

Geographic Distribution

Europe
X   Northern
X   British Isles
X   Central, France
●   Southern
X   Eastern
X   Mediterranean Isl.

Africa
X   Northern
     Tropical
X   Southern

Asia
X   Temperate
X   Tropical

Australasia
●   Australia
●   New Zealand

Northern America
N   Canada, Alaska
N   Southeastern USA
N   Western USA
N   Remaining USA
X   Mexico

Southern America
X   Tropical
X   Caribbean
X   Chile, Argentina

Atlantic Islands
     Cape Verde
     Canary, Madeira
     Azores
     South Atlantic Isl.

Indian Ocean Islands
X   Mascarenes
     Seychelles
     Madagascar

Pacific Islands
     Micronesia
     Galapagos
X   Hawaii
     Melanesia, Polynesia

Invaded Habitats
Freshwater wetlands, clear and cool waterbodies, standing or slowly moving streams.

Description
A submersed aquatic herb with branching stems, mostly 20-100 cm long, easily fragmenting and rooting at the nodes. Leaves are up to 17 mm long, 2-4 mm wide, oblong-linear and in whorls of three. Flower spathes are borne in the upper leaf axils. The hypanthium elongates up to 30 cm length to bring the white or pale purple flowers to the water surface. The plant is dioecious. Male flowers have peduncles of 10-20 cm length, female flowers shorter ones. Fruits are capsules of 6-9 mm length. The cylindric seeds are 4-5 mm long.

Ecology and Control
The species grows between 0.5 m and 7 m depth in standing or slowly moving, mainly base-rich waters that are clear and cool. It is invasive because the rapid vegetative growth leads to dense closed stands that reduce water flow up to 80%, alter light conditions and impede growth of native aquatics. Species richness in stands of this plant is greatly reduced. The stems fragment easily and fragments can quickly regenerate new plants. Plants in Europe are exclusively female, plants in Australia exclusively male.

     Mechanical control such as cutting or hand pulling may encourage fragmentation and spread of the fragments. Shading the water surface of infested water bodies substantially reduces the growth rate. Effective herbicides are diquat, dichlobenil, or terbutryne. The decomposition of killed infestations may lead to a mass release of nutrients.

References
148, 215, 313, 332, 581, 628, 986, 1181, 1182, 1220.

Elytrigia juncea (L.) Nevski   Poaceae

LF: Perennial herb   CU: -
SN: *Thinopyrum junceum* (L.) Löve, *Agropyron junceum* (L.) Beauv.

Geographic Distribution

Europe	*Australasia*	*Atlantic Islands*
Northern	● Australia	Cape Verde
British Isles	New Zealand	Canary, Madeira
Central, France		Azores
N Southern	*Northern America*	South Atlantic Isl.
N Eastern	Canada, Alaska	
N Mediterranean Isl.	Southeastern USA	*Indian Ocean Islands*
	Western USA	Mascarenes
Africa	Remaining USA	Seychelles
N Northern	Mexico	Madagascar
Tropical		
Southern	*Southern America*	*Pacific Islands*
	Tropical	Micronesia
Asia	Caribbean	Galapagos
N Temperate	Chile, Argentina	Hawaii
Tropical		Melanesia, Polynesia

Invaded Habitats
Grassland.

Description
A grass with culms of 30-80 cm height and long rhizomes. Leaves are 2-6 mm wide and have prominent and minutely pubescent ribs on the upperside. Inflorescences are spikes up to 20 cm length, with the spikelets having 5-8 florets each. Spikelets are 15-30 mm long. The glumes are without awns and have 7-11 veins.

Ecology and Control
In the native range, this grass occurs mostly on coastal dunes and in coastal grassland. Where invasive it forms dense swards that crowd out native species and prevent their regeneration. Plants in Australia are referred to subsp. *juncea*.

Specific control methods for this species are not available. Small patches can be hand pulled or dug out, all rhizomes must be removed to prevent recolonization. Larger populations can be sprayed with herbicides.

References
215.

Elytrigia repens (L.) Desv. ex Nevski   Poaceae

LF: Perennial herb   CU: -
SN: *Agropyron repens* (L.) Beauv., *Elymus repens* L.

Geographic Distribution

Europe
N   Northern
N   British Isles
N   Central, France
N   Southern
N   Eastern
N   Mediterranean Isl.

Africa
N   Northern
     Tropical
X   Southern

Asia
N   Temperate
N   Tropical

Australasia
X   Australia
     New Zealand

Northern America
●   Canada, Alaska
     Southeastern USA
X   Western USA
     Remaining USA
     Mexico

Southern America
     Tropical
     Caribbean
     Chile, Argentina

Atlantic Islands
     Cape Verde
X   Canary, Madeira
     Azores
     South Atlantic Isl.

Indian Ocean Islands
●   Mascarenes
     Seychelles
     Madagascar

Pacific Islands
     Micronesia
     Galapagos
     Hawaii
     Melanesia, Polynesia

Invaded Habitats
Grassland.

Description
A 30-150 cm tall grass with bluish-green stems and long, branching rhizomes. Leaf blades are flat, linear, 6-30 cm long and 2-10 mm wide, sometimes pubescent above. The membranous ligule is *c.* 1 mm long. Inflorescences are contracted and narrowly oblong spikes of 5-20 cm length, with *c.* 20 spikelets. Spikelets are 10-20 mm long, have 3-8 florets and fall entirely at maturity. Glumes are 5-12 mm long, lemmas 6-12 mm long.

Ecology and Control
This grass is common in arable fields, forest edges and is a significant agricultural weed. It is a very variable species. Where invasive, it eliminates native vegetation by forming dense and pure swards. Once established, the grass persists and spreads extensively by creeping rhizomes.

     Eradication is difficult due to the perennation by its rhizomes. Scattered plants can be dug out, all rhizomes must be removed to prevent regrowth. Larger infestations are treated with herbicide.

References
230, 825, 971, 1139, 1387.

Eragrostis curvula (Schrad.) Nees   Poaceae

LF: Perennial herb
CU: Erosion control, forage
SN: *Eragrostis robusta* Stent, *Eragrostis subulata* Nees

Geographic Distribution

Europe	*Australasia*	*Atlantic Islands*
Northern	● Australia	Cape Verde
X British Isles	X New Zealand	X Canary, Madeira
Central, France		Azores
Southern	*Northern America*	South Atlantic Isl.
Eastern	Canada, Alaska	
Mediterranean Isl.	Southeastern USA	*Indian Ocean Islands*
	X Western USA	Mascarenes
Africa	Remaining USA	Seychelles
X Northern	Mexico	Madagascar
N Tropical		
N Southern	*Southern America*	*Pacific Islands*
	X Tropical	Micronesia
Asia	Caribbean	Galapagos
X Temperate	Chile, Argentina	Hawaii
X Tropical		Melanesia, Polynesia

Invaded Habitats
Heath- and shrubland, grass- and woodland, seasonal freshwater wetlands.

Description
A robust, densely tufted perennial grass of 60-120 cm height, with many long and hanging, rigid, narrow leaves. Leaf blades are blue-green to dark-green, 20-40 cm long and 3-5 mm wide. Inflorescences are narrow panicles up to 30 cm length, with green or dark grey spikelets of 4-10 mm length, each having six florets. The yellow to brown seeds are 0.5-1 mm long.

Ecology and Control
Where native, this grass is found in moist sandy soils and in disturbed woodlands. Several cultivars have been developed. It is drought tolerant and very tolerant to soil salinity; germination can take place under high soil salinity levels. It does not grow on wet, seepy soils and does not tolerate standing water. The grass establishes easily and persists well under grazing. It survives fire and is a prolific seed producer. Seeds are dispersed by water, wind and animals. Plants in Britain are annual. Where invasive, it becomes dominant on low-fertility soils and the dense tussocks displace native vegetation.

    Smaller plants can be dug out, the crowns must be removed to prevent regrowth. Burning before flowering starts is used to remove topgrowth. Regrowth and seedlings are sprayed with grass-selective herbicides. Re-establishment of desirable overstorey species can shade out the grass.

References
215, 876, 986, 1190.

Eragrostis lehmanniana Nees   Poaceae

LF: Perennial herb   CU: Erosion control, forage
SN: -

Geographic Distribution

Europe	*Australasia*	*Atlantic Islands*
Northern	Australia	Cape Verde
British Isles	New Zealand	Canary, Madeira
Central, France		Azores
Southern	*Northern America*	South Atlantic Isl.
Eastern	Canada, Alaska	
Mediterranean Isl.	Southeastern USA	*Indian Ocean Islands*
	● Western USA	Mascarenes
Africa	Remaining USA	Seychelles
Northern	X Mexico	Madagascar
N Tropical		
N Southern	*Southern America*	*Pacific Islands*
	X Tropical	Micronesia
Asia	Caribbean	Galapagos
Temperate	Chile, Argentina	Hawaii
Tropical		Melanesia, Polynesia

Invaded Habitats
Desert scrub, arid grass- and rangeland, alluvial plains.

Description
A tufted perennial grass with often branched culms of 20-90 cm height and narrow leaves up to 10 cm long and 1-3 mm wide. Sometimes rooting at the lower nodes. Inflorescences are open panicles of 10-20 cm length, with the lower branches single or in pairs. The lower leaf-sheaths are membranous. Seeds are *c.* 0.6 mm long.

Ecology and Control
The grass grows naturally in semi-arid summer rainfall areas of Africa, and two varieties exist: var. *lehmanniana* and var. *chaunantha*. Strains introduced to the USA have prostrate stems that root at the nodes. The species is highly drought tolerant. It is invasive because it spreads quickly and covers native desert vegetation, displacing native grasses and forbs. Dense stands have a reduced faunal diversity. The grass invades desert grassland seemingly undisturbed by livestock or humans. Seed production is high, and seeds are dispersed by water and wind. A large seed bank is accumulated in the soil. The grass increases fire frequency due to accumulation of a large amount of litter and dead stems.
    Controlling infested areas include a spring burning followed by spraying regrowth and seedlings with herbicides. Seeding native grasses may suppress re-establishment of this grass. Seedling density can be reduced by mowing or mulching.

References
31, 112, 211, 280, 671, 842, 1116, 1190, 1244.

Erica arborea L. — Ericaceae

LF: Evergreen shrub, tree
SN: -
CU: Ornamental

Geographic Distribution

Europe
- Northern
- X British Isles
- Central, France
- N Southern
- N Eastern
- N Mediterranean Isl.

Africa
- N Northern
- N Tropical
- Southern

Asia
- N Temperate
- Tropical

Australasia
- ● Australia
- X New Zealand

Northern America
- Canada, Alaska
- Southeastern USA
- Western USA
- Remaining USA
- Mexico

Southern America
- Tropical
- Caribbean
- Chile, Argentina

Atlantic Islands
- Cape Verde
- N Canary, Madeira
- Azores
- South Atlantic Isl.

Indian Ocean Islands
- Mascarenes
- Seychelles
- Madagascar

Pacific Islands
- Micronesia
- Galapagos
- Hawaii
- Melanesia, Polynesia

Invaded Habitats
Riparian habitats, forests, shrubland.

Description
A shrub or small tree of 1-8 m height, sometimes more, with the young branches being densely pubescent. The dark green, glabrous and linear leaves are 3-6 mm long and borne in whorls of 3-5. White flowers appear in lateral racemes, usually clustered in panicles. The flowers are bell-shaped, have four lobes and are 2.5-4 mm across. Pedicels are glabrous. Fruits are broad-oblong, red capsules of 2-3 mm length, containing numerous minute seeds.

Ecology and Control
Where native, this shrub grows mainly on acid soils of woods, evergreen scrub, and stream banks. The dense growth habit leads to extensive thickets that crowd out native vegetation. The shrub vigorously resprouts after damage by cutting or fire. The short-lived seeds are dispersed by wind.

Repeated cutting may decrease the resprouting vigour. Seedlings can be hand pulled or dug out, larger stems cut and the cut stumps treated with herbicide to prevent regrowth.

References
215, 866, 867, 1078, 1079.

Erica lusitanica Rudolphi — Ericaceae

LF: Evergreen shrub
SN: -
CU: Ornamental

Geographic Distribution

Europe	*Australasia*	*Atlantic Islands*
Northern	● Australia	Cape Verde
X British Isles	● New Zealand	Canary, Madeira
Central, France		Azores
N Southern	*Northern America*	South Atlantic Isl.
Eastern	Canada, Alaska	
Mediterranean Isl.	Southeastern USA	*Indian Ocean Islands*
	X Western USA	Mascarenes
Africa	Remaining USA	Seychelles
Northern	Mexico	Madagascar
Tropical		
Southern	*Southern America*	*Pacific Islands*
	Tropical	Micronesia
Asia	Caribbean	Galapagos
Temperate	Chile, Argentina	Hawaii
Tropical		Melanesia, Polynesia

Invaded Habitats
Grass- and heathland, dry forests, riparian habitats.

Description
An erect and heathy shrub of 1-3 m height. Leaves are crowded and arranged in whorls of 3-4, glabrous and 3-7 mm long. Pink to white flowers of 4-5 mm diameter are borne in small racemes that are usually densely crowded. Fruits are capsules of *c.* 3 mm length, glabrous, broad-ellipsoid, containing numerous minute seeds of *c.* 0.4 mm length.

Ecology and Control
This shrub may completely dominate the shrub canopy, thereby eliminating all native vegetation and preventing the regeneration of native plants. Seed production is prolific, and seeds are dispersed by wind and water. They accumulate in a soil seed bank. The shrub vigorously resprouts after damage such as fire. Seedling establishment is often enhanced after fires.

    Seedlings and small plants are easy to hand pull, roots must be removed to prevent regrowth. If cut, the stumps should be treated with herbicide.

References
121, 215, 607, 608, 832, 924.

Erigeron karvinskianus DC.   Asteraceae

LF: Perennial herb
SN: *Erigeron mucronatus* DC.
CU: Ornamental

Geographic Distribution

Europe		*Australasia*		*Atlantic Islands*	
	Northern		Australia		Cape Verde
X	British Isles	●	New Zealand		Canary, Madeira
X	Central, France			●	Azores
X	Southern	*Northern America*			South Atlantic Isl.
	Eastern		Canada, Alaska		
X	Mediterranean Isl.		Southeastern USA	*Indian Ocean Islands*	
		X	Western USA	X	Mascarenes
Africa			Remaining USA		Seychelles
	Northern	N	Mexico		Madagascar
X	Tropical				
	Southern	*Southern America*		*Pacific Islands*	
		N	Tropical		Micronesia
Asia			Caribbean		Galapagos
	Temperate	X	Chile, Argentina	X	Hawaii
	Tropical				Melanesia, Polynesia

Invaded Habitats
Ravines, rock outcrops, moist disturbed sites.

Description
A sprawling to erect 50-100 cm tall herb with much-branched erect to ascending stems that are woody at the base and sparsely hairy above. Leaves are simple, entire, elliptic, 1-5 cm long and *c.* 1 cm wide. Flowerheads of 7-10 mm diameter are borne on long peduncles in lax corymbs. The numerous ray florets are pinkish-white to purple, the disk florets yellow. Fruits are achenes of *c.* 1 mm length, with a pappus.

Ecology and Control
This mat forming plant spreads rapidly in disturbed sites and displaces native species by competing for space, water and nutrients. Little is known on the ecology of this species.

Specific control methods for this species are not available. Scattered plants can be hand pulled or dug out, larger populations treated with herbicide.

References
153, 316, 1415.

Erodium botrys (Cav.) Bertol.   Geraniaceae

LF: Annual herb   CU: -
SN: *Geranium botrys* Cav.

Geographic Distribution

Europe		*Australasia*		*Atlantic Islands*	
	Northern	●	Australia		Cape Verde
	British Isles	X	New Zealand	N	Canary, Madeira
	Central, France				Azores
N	Southern	*Northern America*			South Atlantic Isl.
N	Eastern		Canada, Alaska		
N	Mediterranean Isl.		Southeastern USA	*Indian Ocean Islands*	
		●	Western USA		Mascarenes
Africa			Remaining USA		Seychelles
N	Northern		Mexico		Madagascar
	Tropical				
	Southern	*Southern America*		*Pacific Islands*	
			Tropical		Micronesia
Asia			Caribbean		Galapagos
X	Temperate	X	Chile, Argentina		Hawaii
	Tropical				Melanesia, Polynesia

Invaded Habitats
Grassland, woodland.

Description
A short-hairy herb with prostrate to ascending stems of 10-90 cm height, and with lobed or dissected leaves. The lower leaves have petioles and blades of 3-15 cm length, the blade being ovate to oblong, glabrous or sparsely pubescent. Inflorescences are umbels with 1-4 flowers. The lavender flowers have sepals of 10-13 mm length and slightly larger petals. Fruits are indehiscent, 8-11 mm long, and have a beak of 5-11 cm length that becomes twisted.

Ecology and Control
Where native, this plant is commonly found in dry and disturbed places. The plant forms dense swards that may cover large areas and crowd out native grasses and forbs where invasive. The dense foliage shades out native plants and prevents their regeneration.

Scattered plants can be hand pulled. Larger patches are best sprayed with herbicide during active growth.

References
215, 1193.

Erodium cicutarium (L.) L'Hér.   Geraniaceae

LF: Annual herb   CU: -
SN: *Geranium cicutarium* L.

Geographic Distribution

Europe		*Australasia*		*Atlantic Islands*	
N	Northern	●	Australia		Cape Verde
N	British Isles	X	New Zealand	N	Canary, Madeira
N	Central, France			N	Azores
N	Southern	*Northern America*			South Atlantic Isl.
N	Eastern	X	Canada, Alaska		
N	Mediterranean Isl.		Southeastern USA	*Indian Ocean Islands*	
		●	Western USA		Mascarenes
Africa		X	Remaining USA		Seychelles
N	Northern	X	Mexico		Madagascar
	Tropical				
	Southern	*Southern America*		*Pacific Islands*	
		X	Tropical		Micronesia
Asia		X	Caribbean		Galapagos
X	Temperate	X	Chile, Argentina	X	Hawaii
X	Tropical				Melanesia, Polynesia

Invaded Habitats
Grass- and woodland, dry open forests, shrubland, disturbed sites.

Description
A 5-50 cm tall herb with decumbent to ascending, glandular-hairy stems and compound leaves. The lower leaves are 3-10 cm long, ovate to oblanceolate, sparsely hairy, and have 9-13 deeply dissected leaflets. Flowers are red-lavender with a purple base and petals of 3-5 mm length. The fruit is an indehiscent mericarp with a fruit body length of 4-7 mm and a beak of 2-5 cm length, becoming twisted.

Ecology and Control
A native of warm, dry and ruderal places whose establishment is facilitated by disturbances. Plants are extremely variable in size and shape. Once established, it forms dense stands that eliminate native vegetation and successfully compete with native grasses and forbs.

Individual plants can be hand pulled or cut. Larger patches are cut or treated with herbicide.

References
180, 215, 862, 1138.

Eucalyptus cladocalyx Muell.   Myrtaceae

LF: Evergreen tree   CU: Ornamental, wood
SN: -

Geographic Distribution

Europe	*Australasia*	*Atlantic Islands*
Northern	● N Australia	Cape Verde
British Isles	New Zealand	Canary, Madeira
Central, France		Azores
Southern	*Northern America*	South Atlantic Isl.
Eastern	Canada, Alaska	
Mediterranean Isl.	Southeastern USA	*Indian Ocean Islands*
	X Western USA	Mascarenes
Africa	Remaining USA	Seychelles
Northern	Mexico	Madagascar
Tropical		
● Southern	*Southern America*	*Pacific Islands*
	Tropical	Micronesia
Asia	Caribbean	Galapagos
Temperate	Chile, Argentina	X Hawaii
Tropical		Melanesia, Polynesia

Invaded Habitats
Grass- and woodland, dry forests, riparian habitats, rock outcrops, coastal areas.

Description
A 8-35 m tall tree, generally straight, with a smooth, pale bark being shed in large irregular patches. Adult leaves are broad-lanceolate and have petioles of 12-20 mm length, the leaf blade is 11-15 cm long and 2-2.5 cm wide. Creamy-white flowers are borne in umbels of 4-16 and have white stamens. The hypanthium is less than 10 mm long. Fruits are barrel-shaped and ribbed, 9-16 mm long and 6-10 mm wide.

Ecology and Control
In the native range, this fast growing tree is a small to tall woodland or forest tree. Where invasive, the tree recruits dense cohorts of seedlings following fires, threatening native plants by shading them out. It coppices freely after damage. Over time, extensive stands dry out the soils due to a high water consumption and change the species composition of the associated flora.

    Seedlings and saplings can be dug out, roots must be removed to prevent regrowth. Larger trees are cut and the cut stumps treated with herbicide. Regrowth and emerging seedlings can be sprayed. The drill-fill application of herbicide is also practicable for large trees.

References
215, 549, 924, 1062, 1089.

Eucalyptus diversicolor Muell.   Myrtaceae

LF: Evergreen tree   CU: Bee plant, wood
SN: -

Geographic Distribution

Europe	*Australasia*	*Atlantic Islands*
Northern	**N** Australia	Cape Verde
British Isles	New Zealand	Canary, Madeira
Central, France		Azores
Southern	*Northern America*	South Atlantic Isl.
Eastern	Canada, Alaska	
Mediterranean Isl.	Southeastern USA	*Indian Ocean Islands*
	Western USA	Mascarenes
Africa	Remaining USA	Seychelles
Northern	Mexico	Madagascar
Tropical		
● Southern	*Southern America*	*Pacific Islands*
	Tropical	Micronesia
Asia	Caribbean	Galapagos
Temperate	Chile, Argentina	Hawaii
Tropical		Melanesia, Polynesia

Invaded Habitats
Grass- and woodland, scrub.

Description
A tree reaching 90 m height with a smooth bark. Adult leaves are broadly-lanceolate, with blades of 9-12 cm length and 2-3 cm width, conspicuous lateral veins and channelled petioles of 10-20 mm length. Flowers are borne in umbels of 6-8. Peduncles are 18-28 mm long, pedicels 3-6 mm long. The cylindrical to obconical hypanthium is 7-8 mm long and 5-7 mm wide. Fruits are 8-12 mm long and 7-10 mm wide.

Ecology and Control
A native of hills and tall, open forest. Where invasive, it forms extensive stands that compete for water and light. Species richness is strongly reduced in stands of this tree.

Specific control methods for this species are not available. The same methods as for the previous species may apply.

References
156, 549, 841, 950, 951, 1089.

Eucalyptus globulus Labill.   Myrtaceae

LF: Evergreen tree   CU: Bee plant, ornamental, wood
SN: -

Geographic Distribution

Europe	Australasia	Atlantic Islands
Northern	N Australia	Cape Verde
British Isles	X New Zealand	X Canary, Madeira
Central, France		X Azores
● Southern	*Northern America*	South Atlantic Isl.
Eastern	Canada, Alaska	
X Mediterranean Isl.	Southeastern USA	*Indian Ocean Islands*
	● Western USA	Mascarenes
Africa	Remaining USA	Seychelles
Northern	Mexico	Madagascar
Tropical		
● Southern	*Southern America*	*Pacific Islands*
	Tropical	Micronesia
Asia	Caribbean	Galapagos
Temperate	Chile, Argentina	● Hawaii
Tropical		Melanesia, Polynesia

Invaded Habitats
Grass- and heathland, forests, riparian habitats.

Description
A large tree reaching 45 m height or more, with a straight, bluish grey and smooth trunk, and a bark shedding in irregular strips. Young stems are more or less square and bear soft and oval leaves, while older leaves are sickle-shaped, 10-20 cm long and 2.5-4 cm wide, narrow-lanceolate, hanging and bluish due to a waxy cover. They are very aromatic. Flowers appear solitary in the axils of leaves and are creamy white to yellow. The bud cup is flat-hemispheric, conspicuously knobbed and bluish white. Fruits are woody, 20-30 mm in diameter, glaucous, ribbed and blue-grey.

Ecology and Control
This fast growing tree grows in a wide range of soils. Four subspecies are recognized in Australia. The tree competes for water and light, and produces large quantitities of litter preventing the establishment of native species and posing a fire hazard. Forests dominated by this tree are species poor. The tree easily resprouts from stumps. Clearing can result in mass establishment of seedlings. The tree's high water consumption leads to decreased soil moisture contents.

Since the tree easily resprouts, cut stumps must be treated immediately with herbicides such as glyphosate, or sprouts must be regularly removed over several years. Stumps may also be ground to a depth of 20-30 cm below the soil surface to prevent resprouting.

References
133, 318, 520, 664, 1089, 1191.

Eugenia uniflora L.   Myrtaceae

LF: Evergreen shrub, tree   CU: Ornamental, food
SN: *Eugenia michelii* Lam., *Eugenia brasiliana* (L.) Aubl.

Geographic Distribution

Europe
   Northern
   British Isles
   Central, France
   Southern
   Eastern
   Mediterranean Isl.

Africa
   Northern
   Tropical
● Southern

Asia
   Temperate
   Tropical

Australasia
   Australia
   New Zealand

Northern America
   Canada, Alaska
● Southeastern USA
   Western USA
   Remaining USA
   Mexico

Southern America
N Tropical
● Caribbean
   Chile, Argentina

Atlantic Islands
   Cape Verde
   Canary, Madeira
   Azores
   South Atlantic Isl.

Indian Ocean Islands
   Mascarenes
   Seychelles
   Madagascar

Pacific Islands
   Micronesia
   Galapagos
   Hawaii
   Melanesia, Polynesia

Invaded Habitats
Coastal scrub, forests and forest edges, riverbanks.

Description
A spreading shrub or small tree up to 6 m height with opposite leaves. These are ovate to ovate-lanceolate, 2.5-6.5 cm long, up to 3.5 cm wide, shiny dark green above and paler beneath. Petioles are 3-4 mm long. White flowers of *c.* 15 mm diameter are borne in axillary clusters. Pedicels are 5-20 mm long. Fruits are globose-ovoid, red when ripe, and have eight longitudinal ribs.

Ecology and Control
A fast growing species, growing best in rich, well-drained soils. Little is known on the ecology of this species. It forms dense thickets that displace native plants and prevents their regeneration.
    Specific control methods for this species are not available. Seedlings and saplings may be hand pulled or dug out, larger trees cut and the cut stumps treated with herbicide.

References
549, 715, 816.

Euphorbia esula L. Euphorbiaceae

LF: Perennial herb
SN: -
CU: -

Geographic Distribution

Europe
- Northern
- X British Isles
- N Central, France
- N Southern
- N Eastern
- Mediterranean Isl.

Africa
- Northern
- Tropical
- Southern

Asia
- N Temperate
- Tropical

Australasia
- Australia
- New Zealand

Northern America
- ● Canada, Alaska
- Southeastern USA
- ● Western USA
- ● Remaining USA
- Mexico

Southern America
- Tropical
- Caribbean
- X Chile, Argentina

Atlantic Islands
- Cape Verde
- Canary, Madeira
- Azores
- South Atlantic Isl.

Indian Ocean Islands
- Mascarenes
- Seychelles
- Madagascar

Pacific Islands
- Micronesia
- Galapagos
- X Hawaii
- Melanesia, Polynesia

Invaded Habitats
Grassland, rangeland.

Description
A 30-100 cm tall perennial herb with stems usually unbranched at the base but with axillary non-flowering branches. All plant parts contain a white latex. Leaves are linear to lanceolate, acute, sessile, 15-90 mm long and 0.5-15 mm wide. Flowers are inconspicuous and borne in cup-shaped cyathia that have four yellowish-green nectar glands. Cyathia are grouped in an umbel-like cluster at the end of the stem. Fruits are glabrous capsules of 2.5-3 mm length and *c.* 3.5 mm width, containing grey to brownish, ovoid and smooth seeds of *c.* 2 mm length.

Ecology and Control
Where native, this forb grows in grassland, hedges, and ruderal places. It is a highly variable species in Europe, with different ecotypes introduced to northern America. The plant spreads by seeds and by vegetative growth from the rootstock. It rapidly expands and forms large and dense patches that displace native grasses and forbs. The plant accumulates a thick litter layer impeding the growth of other species.

Continous grazing by sheep can reduce density, but recovery is rapid after grazing ceases. 2,4-D applied during flowering reduces production of viable seeds. Other effective herbicides are glyphosate plus 2,4-D or picloram.

References
15, 40, 100, 107, 133, 571, 707, 789, 1139, 1362, 1396.

Euphorbia paralias L.   Euphorbiaceae

LF: Perennial herb   CU: -
SN: -

Geographic Distribution

Europe	*Australasia*	*Atlantic Islands*
Northern	● Australia	Cape Verde
N British Isles	New Zealand	N Canary, Madeira
N Central, France		Azores
N Southern	*Northern America*	South Atlantic Isl.
N Eastern	Canada, Alaska	
N Mediterranean Isl.	Southeastern USA	*Indian Ocean Islands*
	Western USA	Mascarenes
Africa	Remaining USA	Seychelles
N Northern	Mexico	Madagascar
Tropical		
Southern	*Southern America*	*Pacific Islands*
	Tropical	Micronesia
Asia	Caribbean	Galapagos
N Temperate	Chile, Argentina	Hawaii
Tropical		Melanesia, Polynesia

Invaded Habitats
Coastal dunes and beaches.

Description
A glabrous and glaucous herb up to 70 cm tall, with blue-green stems that are branched from the base. The entire leaves are obovate to lanceolate-elliptic, 3-30 mm long and 2-15 mm wide. Fruits are deeply sulcate capsules of 3-5 mm length and 4.5-6 mm width, containing broadly ovoid, smooth and pale grey seeds of 2.5-3.5 mm length.

Ecology and Control
A native of coastal beaches, sand dunes, and shingle. As a strandline pioneer it tolerates sand accretion and dry conditions. Seeds can float for several years without losing viability. The plant is invasive because it forms dense stands that eliminate native vegetation and prevents establishment of native species.

Specific control methods for this species are not available. Scattered plants may be hand pulled or dug out. Larger patches can be sprayed with herbicide.

References
121, 215, 377, 559, 656.

Fallopia japonica (Houtt.) Ronse Decr.   Polygonaceae

LF: Perennial herb   CU: Ornamental
SN: *Polygonum cuspidatum* Sieb. & Zucc., *Reynoutria japonica* Houtt.

Geographic Distribution

Europe
     Northern
     • British Isles
     • Central, France
     • Southern
     • Eastern
     Mediterranean Isl.

Africa
     Northern
     Tropical
     Southern

Asia
N   Temperate
     Tropical

Australasia
X   Australia
   • New Zealand

Northern America
   • Canada, Alaska
     Southeastern USA
   • Western USA
   • Remaining USA
     Mexico

Southern America
     Tropical
     Caribbean
     Chile, Argentina

Atlantic Islands
     Cape Verde
     Canary, Madeira
     Azores
     South Atlantic Isl.

Indian Ocean Islands
     Mascarenes
     Seychelles
     Madagascar

Pacific Islands
     Micronesia
     Galapagos
     Hawaii
     Melanesia, Polynesia

Invaded Habitats
Woodland, forest edges, riparian habitats, wetlands, disturbed sites.

Description
A large and erect herbaceous perennial with annual stems of 1-4 m height and up to 4 cm diameter, and an extensive below-ground rhizome system. Stems are dark red to purple in early spring, hollow and have distinct nodes. Leaves are heart-shaped and up to 12 cm long. Flowers are cream to white and borne in axillary clusters of 8-12 cm length. The plant is dioecious. Fruits are winged achenes, each containing a dark brown, shiny seed of *c.* 3 mm length.

Ecology and Control
Where native, this plant grows along streams and as a pioneer species up to 2,500 m elevation. The dense foliage and the vegetative growth leads to dense and tall thickets that completely shade out all other vegetation, degrading native plant and wildlife habitats and eliminating native species. Extensive stands reduce the water carrying capacity of streams and promote soil erosion. The stout rhizomes form a deep mat and can grow more than one metre in depth and become 15-20 m long. Fragments of rhizomes easily resprout and are carried by streams. During winter, the species' standing biomass may become a fire hazard in fire prone regions.
     Control is difficult due to the vigorous rhizomes. Frequent cutting, e.g. at least every 4 weeks, reduces rhizome growth but cutting alone does not eliminate the plant. Chemical control includes spraying with glyphosate or 2,4-D amine.

References
58, 59, 60, 95, 96, 98, 99, 111, 150, 160, 243, 244, 245, 246, 267, 284, 540, 563, 579, 589, 831, 1152, 1153, 1243, 1253, 1415.

Festuca arundinacea Schreb.   Poaceae

LF: Perennial herb
SN: *Festuca elatior* L.
CU: Erosion control, forage, ornamental

Geographic Distribution

Europe			*Australasia*		*Atlantic Islands*	
N	Northern		●	Australia		Cape Verde
N	British Isles		●	New Zealand	X	Canary, Madeira
N	Central, France				X	Azores
N	Southern		*Northern America*			South Atlantic Isl.
N	Eastern		X	Canada, Alaska		
N	Mediterranean Isl.		X	Southeastern USA	*Indian Ocean Islands*	
			X	Western USA		Mascarenes
Africa			●	Remaining USA		Seychelles
N	Northern			Mexico		Madagascar
	Tropical					
X	Southern		*Southern America*		*Pacific Islands*	
				Tropical		Micronesia
Asia				Caribbean		Galapagos
N	Temperate			Chile, Argentina		Hawaii
N	Tropical					Melanesia, Polynesia

Invaded Habitats
Grass- and heathland, woodland, riparian habitats, freshwater and saline wetlands.

Description
A coarse densely caespitose to short-rhizomatous grass with hollow culms of 0.5-2 m height. Leaves are stiff and rigid, 10-70 cm long and 2-10 mm wide. Older leaves are dark green. Ligules are up to 2 mm long. Inflorescences are erect or pendent, open to narrowly branched panicles up to 40 cm length. Spikelets are 9-14 mm long and 3 mm wide, and have 3-8 florets. Lemmas are 4-9 mm long and have an awn up to 4 mm length.

Ecology and Control
A grass native in moist forests and grassland, reed swamps, riparian habitats, and seashores. It is a highly variable species in Europe with five subspecies and many varieties. In addition, numerous cultivars have been developed. It is a persistent grass that strongly competes with native species. It forms dense and species poor stands where invasive that displace native herbaceous vegetation and reduce species richness. The grass is mostly infected with an endophytic fungus making it more drought tolerant and increasing its nitrogen utilization efficiency.

    Small infestations are controlled by applying glyphosate, metsulfuron, or imazapic before flowers appear. Heavy infestations are controlled by prescribed burning at the flowering stage.

References
215, 327, 450, 1226, 1415.

Ficus carica L.   Moraceae

LF: Deciduous shrub, tree
SN: -
CU: Food, ornamental

Geographic Distribution

Europe		Australasia		Atlantic Islands	
	Northern	●	Australia	X	Cape Verde
X	British Isles	X	New Zealand	X	Canary, Madeira
	Central, France			X	Azores
X	Southern	*Northern America*			South Atlantic Isl.
	Eastern		Canada, Alaska		
X	Mediterranean Isl.	X	Southeastern USA	*Indian Ocean Islands*	
		●	Western USA		Mascarenes
Africa			Remaining USA		Seychelles
N	Northern		Mexico		Madagascar
	Tropical				
X	Southern	*Southern America*		*Pacific Islands*	
			Tropical		Micronesia
Asia			Caribbean	X	Galapagos
N	Temperate		Chile, Argentina		Hawaii
N	Tropical				Melanesia, Polynesia

Invaded Habitats
Riparian habitats, forests, disturbed sites.

Description
A deciduous shrub or small tree up to 10 m tall, with often multiple trunks and with a smooth, light grey and flaky bark. The bright green leaves are usually deeply lobed with 3-7 lobes, are 10-20 cm long and about as wide, usually heart-shaped at the base, with the lobes being irregularly toothed. Petioles are 2-5 cm long. Flowers are borne inside a hollow, greenish to brownish violet receptacle that develops at maturity into a green to purplish syncarp of 5-8 cm length, containing achenes of *c.* 2 mm length.

Ecology and Control
This is a rapidly growing tree that spreads mainly vegetatively. It forms dense thickets that displace native trees and understorey shrubs. The dense foliage casts heavy shade and prevents establishment of native plants. The tree resprouts vigorously after cutting or other damage. Seeds are dispersed by birds.

Young plants can easily be pulled by hand and larger saplings removed with a weed wrench. Repeated and frequent cutting may eventually exhaust the root system. Chemical control includes applying a triclopyr herbicide to cut stumps over a period of several years; treating the basal areas of uncut stems is also effective in the case of smaller trees.

References
133, 215, 751.

Ficus microcarpa L.fil. Moraceae

LF: Evergreen tree
SN: *Ficus aggregata* Vahl
CU: Ornamental

Geographic Distribution

Europe	*Australasia*	*Atlantic Islands*
Northern	N Australia	Cape Verde
British Isles	New Zealand	Canary, Madeira
Central, France		Azores
Southern	*Northern America*	South Atlantic Isl.
Eastern	Canada, Alaska	
Mediterranean Isl.	● Southeastern USA	*Indian Ocean Islands*
	Western USA	Mascarenes
Africa	Remaining USA	Seychelles
Northern	Mexico	Madagascar
Tropical		
Southern	*Southern America*	*Pacific Islands*
	X Tropical	Micronesia
Asia	Caribbean	Galapagos
N Temperate	Chile, Argentina	● Hawaii
N Tropical		Melanesia, Polynesia

Invaded Habitats
Pine rockland, hardwood forests, disturbed places.

Description
A tree or shrub variable in growth habit, growing epiphytic or as a tree, with a greyish-white and glabrous bark. The tree reaches 25 m height and has large branches, a dense canopy, and numerous aerial roots hanging from the trunk and branches, sometimes growing to the ground and supporting the limbs. Leaves are oblong, elliptic or broadly elliptic, usually 5-8 cm long and 3-5 cm wide, glabrous, entire, and have petioles of 0.5-2 cm length. Flowers are borne inside a sessile receptacle of 6-10 mm diameter. Fruits are red, occasionally yellow syncarps, becoming dark when ripe, globose, and 9-12 mm in diameter.

Ecology and Control
Where native, the tree grows in lowland rain forest, riverbanks, tidal floodplains, exposed rocky coasts, and in swamps. It grows fast and is very variable in its growth form, ranging from dwarfed to creeping bushes on windy places to epiphytes on host trees. Large individuals form impenetrable thickets due to the hanging aerial roots, shading out all other plants. Fruits are eaten and distributed by birds, and seeds often germinate in the forks of trees, growing as epiphytes and sending aerial roots to the ground. Host trees may be killed due to the constricting roots and competition for light and nutrients.

Fig trees are very sensitive to triclopyr herbicides which are best applied as a basal or cut stump treatment. Caution must be taken if young fig trees growing as epiphytes are treated with herbicides to ensure that the host tree is not getting damaged.

References
275, 411, 715.

Foeniculum vulgare Mill.   Apiaceae

LF: Perennial herb   CU: Food
SN: -

Geographic Distribution

Europe		*Australasia*		*Atlantic Islands*	
	Northern	●	Australia	X	Cape Verde
X	British Isles	X	New Zealand	N	Canary, Madeira
N	Central, France			N	Azores
N	Southern	*Northern America*			South Atlantic Isl.
N	Eastern		Canada, Alaska		
N	Mediterranean Isl.	X	Southeastern USA	*Indian Ocean Islands*	
		●	Western USA		Mascarenes
Africa			Remaining USA		Seychelles
N	Northern	X	Mexico		Madagascar
	Tropical				
X	Southern	*Southern America*		*Pacific Islands*	
		X	Tropical	X	Micronesia
Asia			Caribbean		Galapagos
N	Temperate	X	Chile, Argentina	X	Hawaii
N	Tropical				Melanesia, Polynesia

Invaded Habitats
Grass- and woodland, coastal scrub, riparian habitats, rock outcrops, disturbed sites.

Description
A stout, glabrous and glaucous, erect herb with branched stems of 90-150 cm height, sometimes exceeding 2 m. Leaves are glaucous to dark green, *c.* 30 cm long and 40 cm wide, dissected into linear or filiform segments, and have petioles of 3-10 cm length that are broadly sheathing. Leaf segments are 1-3 cm long and narrow. Yellow flowers are borne in large, compound umbels of 4-10 cm diameter on stout terminal and axillary peduncles. Fruits are ovoid-oblong, 4-10 mm long, somewhat flattened, and with prominent ribs. The plant has a stout taproot and widespreading shallow roots.

Ecology and Control
Within the native range, this plant is found in coastal scrub, grassland, rocky places, and in disturbed places. The plant forms dense mats that reduce species richness and alter the composition and structure of native grassland communities. It successfully competes with native perennials in coastal sage communities. Its success may relate to its ability to exploit resources during the summer when most species are not active.

    Seedlings and smaller plants can be dug out, the crowns and most of the taproots must be removed to prevent regrowth. Effective herbicides are glyphosate, triclopyr, or 2,4-D ester.

References
88, 133, 157, 215, 924, 985, 986.

Frangula alnus Mill.   Rhamnaceae

LF: Deciduous shrub, tree
CU: Ornamental
SN: *Rhamnus frangula* L.

Geographic Distribution

Europe
- N   Northern
- N   British Isles
- N   Central, France
- N   Southern
- N   Eastern
-      Mediterranean Isl.

Africa
- N   Northern
-      Tropical
-      Southern

Asia
- N   Temperate
-      Tropical

Australasia
- Australia
- New Zealand

Northern America
- ● Canada, Alaska
-     Southeastern USA
-     Western USA
- ● Remaining USA
-     Mexico

Southern America
- Tropical
- Caribbean
- Chile, Argentina

Atlantic Islands
- Cape Verde
- Canary, Madeira
- Azores
- South Atlantic Isl.

Indian Ocean Islands
- Mascarenes
- Seychelles
- Madagascar

Pacific Islands
- Micronesia
- Galapagos
- Hawaii
- Melanesia, Polynesia

Invaded Habitats
Open forests and forest margins, riparian habitats.

Description
A many-stemmed shrub or small tree up to 6 m tall, with a grey-brown bark and the young branches pubescent. Leaves are opposite, obovate, 3-7 cm long, entire, dark green above and light green and sometimes pubescent beneath. Petioles are 6-12 mm long. Each leaf has 7-9 pairs of veins. Small and greenish flowers are borne solitary or in axillary clusters of 2-10 on the current season's twigs. Pedicels are 8-12 mm long. Fruits are globose drupes of 6-10 mm diameter, red at first, becoming black. Each fruit contains two seeds of 4-5 mm diameter.

Ecology and Control
Native habitats of this shrub include moist open forests, heathland, riparian forests, and bogs. Where invasive, the shrub spreads rapidly and crowds out native woody and herbaceous plants. It is intolerant of heavy shade and becomes dominant at forest edges and in cleared areas. It is a prolific seed producer and seeds are dispersed by birds and small mammals. Seeds can remain viable in the soil for several years. Cut or otherwise damaged plants resprout vigorously.

Prescribed burning may kill seedlings but not necessarily larger plants. A foliar application of glyphosate during spring growth proved to be effective. Glyphosate can also be used to treat freshly cut stumps.

References
221, 467, 546, 571, 594, 1042, 1396.

Fraxinus rotundifolia Mill.   Oleaceae

LF: Deciduous shrub, tree   CU: Ornamental, shelter
SN: *Fraxinus angustifolia* Vahl

Geographic Distribution

Europe
-     Northern
-     British Isles
- N Central, France
- N Southern
- N Eastern
- N Mediterranean Isl.

Africa
- N Northern
-     Tropical
- X Southern

Asia
- N Temperate
-     Tropical

Australasia
- ● Australia
-    New Zealand

Northern America
- Canada, Alaska
- Southeastern USA
- Western USA
- Remaining USA
- Mexico

Southern America
- Tropical
- Caribbean
- Chile, Argentina

Atlantic Islands
- Cape Verde
- Canary, Madeira
- Azores
- South Atlantic Isl.

Indian Ocean Islands
- Mascarenes
- Seychelles
- Madagascar

Pacific Islands
- Micronesia
- Galapagos
- Hawaii
- Melanesia, Polynesia

Invaded Habitats
Grass- and woodland, riparian habitats.

Description
A shrub or small tree up to 5 m tall, with slender and often purplish branches and compound leaves of 12-20 cm length, each having 5-13 leaflets. Leaflets are broad-elliptic to orbicular-obovate, serrated, sessile, 3-7 cm long and 1-2 cm wide. Fruits are samaras of 3-5 cm length containing one seed each.

Ecology and Control
Naturalized plants in Australia belong to *F. angustifolia* subsp. *angustifolia*. The tree tolerates periodic inundation and forms dense monospecific stands that shade out native vegetation and prevent any regeneration of native shrubs and trees. The tree coppices freely after damage. It produces large amounts of wind and water dispersed seeds.

Seedlings and small plants can be pulled or dug out. Larger plants are cut and the cut stumps treated with herbicide. Foliar sprays are best applied before leaf senescence. Regrowth and emerging seedlings can be controlled in a follow-up programme.

References
121, 215, 488, 924.

Fraxinus uhdei (Wenz.) Lingelsh.   Oleaceae

LF: Deciduous tree   CU: Ornamental
SN: *Fraxinus americana* var. *uhdei* Wenzig

Geographic Distribution

Europe	Australasia	Atlantic Islands
Northern	Australia	Cape Verde
British Isles	New Zealand	Canary, Madeira
Central, France		Azores
Southern	*Northern America*	South Atlantic Isl.
Eastern	Canada, Alaska	
Mediterranean Isl.	Southeastern USA	*Indian Ocean Islands*
	Western USA	Mascarenes
Africa	Remaining USA	Seychelles
Northern	N Mexico	Madagascar
Tropical		
Southern	*Southern America*	*Pacific Islands*
	N Tropical	Micronesia
Asia	Caribbean	Galapagos
Temperate	Chile, Argentina	● Hawaii
Tropical		Melanesia, Polynesia

Invaded Habitats
Forests and forest edges.

Description
A dioecious tree up to 26 m tall, with a grey or brown, furrowed bark, young branches pubescent, and pinnately compound leaves. These are opposite, dull green above and pale green below, 15-28 cm long, and have 5-9 leaflets of 6-11 cm length and 2-5 cm width. Margins are irregularly toothed, the petioles are 6-10 cm long. Flowers are borne in panicles of 13-20 cm length, Fruits are samaras of *c.* 3 cm length and 4 mm width, each with one terminal wing and one seed.

Ecology and Control
A native of alluvial soils, growing along streams. It is fast growing and has a juvenile period of *c.* 15 years. Where invasive, it colonizes disturbed areas in forests and precludes the establishment of native plants. The large canopies shade out most understorey species and prevent forest succession. The tree regenerates vigorously from root and shoot sprouts. Fruits are dispersed by wind and water. Seeds may remain viable in the soil for 6 years or longer.
    Specific control methods for this species are not available. The same methods may apply as for the previous species.

References
48, 228, 1063.

Freesia leichtlinii Klatt   Iridaceae

LF: Perennial herb   CU: Ornamental
SN: -

Geographic Distribution

Europe	Australasia	Atlantic Islands
Northern	● Australia	Cape Verde
British Isles	New Zealand	Canary, Madeira
Central, France		Azores
Southern	Northern America	South Atlantic Isl.
Eastern	Canada, Alaska	
Mediterranean Isl.	Southeastern USA	Indian Ocean Islands
	Western USA	Mascarenes
Africa	Remaining USA	Seychelles
Northern	Mexico	Madagascar
Tropical		
N Southern	Southern America	Pacific Islands
	Tropical	Micronesia
Asia	Caribbean	Galapagos
Temperate	Chile, Argentina	Hawaii
Tropical		Melanesia, Polynesia

Invaded Habitats
Coastal dunes, heath- and woodland, dry sclerophyll forests.

Description
A 10-40 cm tall herb with a corm of up to 15 mm diameter. Leaves are all basal, linear, 8-27 cm long and 5-10 mm wide. Stems are unbranched and bear flower spikes with 3-7 flowers. Flowers are white to cream, have bright yellow markings, and are purplish outside the tube. The perianth tube is 2-3.5 cm long, filaments are *c.* 2 cm long. Fruits are green capsules of 10-15 mm length, containing seeds of 3-4 mm diameter.

Ecology and Control
Where native, this herb grows in sandy soils near the coast. Naturalized plants are probably of hybrid origin. The plant is able to reproduce vegetatively by small bulbils produced in the axils of leaves. The plant forms extensive and dense stands that impede the growth and regeneration of native species. The species establishes quickly after disturbance. Seeds and bulbils are spread by water and wind.

   Plants can carefully dug out, all corms must be removed to prevent re-establishment. Herbicides are best applied before the flower stems elongate.

References
215.

Fuchsia magellanica Lam.   Onagraceae

LF: Deciduous shrub   CU: Ornamental
SN: *Fuchsia gracilis* Lindl., *Fuchsia macrostemma* Ruiz. & Pav.

Geographic Distribution

	Europe		*Australasia*		*Atlantic Islands*
	Northern	●	Australia		Cape Verde
X	British Isles	X	New Zealand	X	Canary, Madeira
	Central, France			X	Azores
	Southern		*Northern America*		South Atlantic Isl.
	Eastern		Canada, Alaska		
	Mediterranean Isl.		Southeastern USA		*Indian Ocean Islands*
			Western USA	●	Mascarenes
	Africa		Remaining USA		Seychelles
	Northern		Mexico		Madagascar
	Tropical				
	Southern		*Southern America*		*Pacific Islands*
			Tropical		Micronesia
	Asia		Caribbean		Galapagos
	Temperate	N	Chile, Argentina	X	Hawaii
	Tropical				Melanesia, Polynesia

Invaded Habitats
Forest margins, scrub, woodland, riparian habitats.

Description
A spreading shrub up to 3 m tall, with mostly opposite leaves and a peeling papery bark on the main stem. Leaves are ovate to elliptic and have petioles up to 15 mm length. Leaf blades are 2-9 cm long and 0.5-4 cm wide, and acuminate. Bright red flowers are borne solitary on pedicels of 2-6 cm length. The petals are violet and 6-16 mm long, the sepals red, 12-24 mm long and 4-10 mm wide. Fruits are black berries of 1-2 cm length.

Ecology and Control
A variable shrub with a number of varieties and cultivars. It often forms dense thickets and scrambles over other shrubs and small trees. Extensive stands develop in disturbed sites, impeding the growth and regeneration of native trees and shrubs. Seeds are dispersed by birds and water.

Seedlings and small plants can be hand pulled or dug out. Larger shrubs are cut and the cut stumps treated with herbicide. Larger stands are also slashed to near ground level, and regrowth treated with herbicide.

References
924, 1278, 1299.

Galium aparine L. — Rubiaceae

LF: Annual herb
SN: -
CU: -

Geographic Distribution

Europe
- N Northern
- N British Isles
- N Central, France
- N Southern
- N Eastern
- N Mediterranean Isl.

Africa
- X Northern
-    Tropical
-    Southern

Asia
- X Temperate
-    Tropical

Australasia
- ● Australia
- X New Zealand

Northern America
- X Canada, Alaska
-    Southeastern USA
- X Western USA
- X Remaining USA
-    Mexico

Southern America
-    Tropical
-    Caribbean
- X Chile, Argentina

Atlantic Islands
-    Cape Verde
- N Canary, Madeira
- N Azores
-    South Atlantic Isl.

Indian Ocean Islands
-    Mascarenes
-    Seychelles
-    Madagascar

Pacific Islands
-    Micronesia
-    Galapagos
-    Hawaii
-    Melanesia, Polynesia

Invaded Habitats
Coastal dunes, grass- and woodland, forests, riparian habitats, rock outcrops.

Description
A herb with 80-180 cm long, ascending four-angled stems that have strongly recurved prickles. Leaves are in whorls of 6-9, simple, 2-7 cm long and 3-8 mm wide, narrowly to widely oblanceolate. Flowers are borne in clusters of 1-3 on a peduncle. Corollas are white, 1-1.7 mm in diameter and have a very short tube. Fruits are 3-5 mm in diameter, densely covered with hooked bristles, and composed of two carpels. Seeds are 2-3 mm in diameter.

Ecology and Control
A native of arable land, hedgerows, scrub, and woods. The plant establishes well after disturbance and forms dense patches that crowd out native species and prevent their regeneration. Seeds are dispersed by wind, water and animals. They are rather short-lived and remain viable in the soil for less than 2 years.

Control should aim at preventing seed formation. Plants can be killed with herbicides based on chlorsulfuron or fluoroxypyr. In arable fields, growth is suppressed by covering the soil with a mulch at germination time.

References
121, 215, 580, 812.

Genista linifolia L.   Fabaceae

LF: Evergreen shrub   CU: Ornamental
SN: *Teline linifolia* (L.) Webb. & Berthel., *Cytisus linifolius* (L.) Lam.

Geographic Distribution

Europe	*Australasia*	*Atlantic Islands*
Northern	● Australia	Cape Verde
British Isles	New Zealand	N Canary, Madeira
Central, France		Azores
N Southern	*Northern America*	South Atlantic Isl.
Eastern	Canada, Alaska	
N Mediterranean Isl.	Southeastern USA	*Indian Ocean Islands*
	Western USA	Mascarenes
Africa	Remaining USA	Seychelles
N Northern	Mexico	Madagascar
Tropical		
Southern	*Southern America*	*Pacific Islands*
	Tropical	Micronesia
Asia	Caribbean	Galapagos
Temperate	Chile, Argentina	Hawaii
Tropical		Melanesia, Polynesia

Invaded Habitats
Grassland, heath- and shrubland, dry forests, riparian habitats, coastal areas.

Description
An unarmed erect to spreading shrub up to 3 m tall, with densely silky-hairy twigs. Leaves are sessile or have short petioles, and are compound with three leaflets. These are narrowly oblanceolate to narrowly elliptic, 10-60 mm long, and are densely white-hairy on the lower surface. Yellow flowers of 10-15 mm length are formed in terminal racemes with 5-20 flowers each. Pedicels are short, the calyx is 5-15 mm long and densely silky-hairy. Fruits are dehiscent and silky-hairy pods of 15-35 mm length, each containing 2-4 globose seeds of *c.* 3 mm length.

Ecology and Control
A variable shrub that grows in woodland and scrub within the native range. It is nitrogen-fixing and increases soil fertility. It is usually calcifuge and prefers dry soils. Where invasive, it forms dense thickets that displace native vegetation and prevent regeneration of native herbs and shrubs. Seeds are abundantly produced.

    Seedlings and smaller plants can be pulled or dug out. Larger shrubs are dug out or burned before seeding. Smaller bushes can be slashed. In any case, seedlings and regrowth must be controlled in a follow-up programme. An effective herbicide is 2,4,5-T ester applied to the foliage. Glyphosate plus picloram are effective when applied to plants in full leaf before flowering commences.

References
215, 607, 924, 985, 986.

Genista monspessulana (L.) Johnson Fabaceae

LF: Evergreen shrub
CU: Ornamental
SN: *Cytisus monspessulanus* L., *Teline monspessulana* (L.) Koch

Geographic Distribution

Europe	*Australasia*	*Atlantic Islands*
Northern	● Australia	Cape Verde
X British Isles	X New Zealand	Canary, Madeira
Central, France		X Azores
N Southern	*Northern America*	South Atlantic Isl.
Eastern	Canada, Alaska	
N Mediterranean Isl.	Southeastern USA	*Indian Ocean Islands*
	● Western USA	Mascarenes
Africa	Remaining USA	Seychelles
N Northern	Mexico	Madagascar
Tropical		
X Southern	*Southern America*	*Pacific Islands*
	Tropical	Micronesia
Asia	Caribbean	Galapagos
N Temperate	X Chile, Argentina	X Hawaii
Tropical		Melanesia, Polynesia

Invaded Habitats
Grass- and heathland, sclerophyll forests, riparian habitats, rock outcrops, coastal areas.

Description
An unarmed erect shrub with 1-3 m tall stems and silvery-hairy twigs. Leaves are compound with three obovate to oblanceolate leaflets of 10-20 mm length and 2-10 mm width, and short petioles. The upper surface is generally glabrous, the lower surface appressed-hairy. Yellow flowers are borne in clusters on axillary short-shoots. The inflorescences are 15-60 mm long and have each 4-10 flowers. Fruits are dehiscent, narrow-oblong and densely silky-hairy pods of 15-25 mm length. Each pod contains 3-8 brown to black, round to oval and shiny seeds of 2-3 mm length.

Ecology and Control
In the native range, this shrub grows in scrub and in rocky places. It forms extensive and dense thickets where invasive that displace native plant and forage species. It is a fast-growing shrub that becomes reproductive within 2-3 years. Seeds are copiously produced and long-lived, and the soil seed bank may contain >6,000 seeds m^{-2}. Seedlings are fairly shade tolerant. The shrub easily resprouts from the root crown after damage.

    Manual control methods include hand pulling of seedlings and young plants and removing larger individuals with a weed wrench. Cutting at or below ground of shrubs after seed set, and mowing the following summer to kill seedlings is another method. It needs to be repeated over several years to deplete the soil seed bank. Chemical control is done by foliar spraying of glyphosate, 2,4-D, or picloram plus triclopyr applied to plants in full leaf.

References
133, 215, 607, 765, 924, 985, 986.

Gladiolus undulatus L.   Iridaceae

LF: Perennial herb   CU: Ornamental
SN: -

Geographic Distribution

Europe	*Australasia*	*Atlantic Islands*
Northern	● Australia	Cape Verde
British Isles	X New Zealand	Canary, Madeira
Central, France		Azores
Southern	*Northern America*	South Atlantic Isl.
Eastern	Canada, Alaska	
Mediterranean Isl.	Southeastern USA	*Indian Ocean Islands*
	Western USA	Mascarenes
Africa	Remaining USA	Seychelles
Northern	Mexico	Madagascar
Tropical		
N Southern	*Southern America*	*Pacific Islands*
	Tropical	Micronesia
Asia	X Caribbean	Galapagos
Temperate	Chile, Argentina	Hawaii
X Tropical		Melanesia, Polynesia

Invaded Habitats
Grassland, forest edges, riparian habitats, seasonal freshwater wetlands.

Description
A cormous perennial of 0.7-1.4 m height, with a corm of 2-3 cm diameter and numerous subterranean bulbils. Basal leaves are linear, 25-75 cm long and 5-15 mm wide, flat, and have prominent veins. Stems are unbranched and bear flower spikes with 3-8 flowers. Flowers are funnel-shaped, white to cream, with the perianth tube being 5-7 cm long.

Ecology and Control
This plant does not set fruits in Australia and spreads solely vegetatively by bulbils and corms. These are carried by streams and in soil. The plant is invasive because it can form extensive and dense populations that crowd out native vegetation and reduce species richness. Little is known on the ecology of this species.

    Specific control methods for this species are not available. The same methods may apply as for other monocots, e.g. *Moraea flaccida* or *Freesia leichtlinii*.

References
215.

Gleditsia triacanthos L.   Fabaceae

LF: Deciduous tree, shrub   CU: Erosion control, fodder, ornamental
SN: -

Geographic Distribution

Europe		*Australasia*		*Atlantic Islands*	
	Northern	●	Australia		Cape Verde
	British Isles		New Zealand		Canary, Madeira
X	Central, France				Azores
X	Southern	*Northern America*			South Atlantic Isl.
	Eastern	N	Canada, Alaska		
	Mediterranean Isl.	N	Southeastern USA	*Indian Ocean Islands*	
			Western USA		Mascarenes
Africa		N	Remaining USA		Seychelles
	Northern		Mexico		Madagascar
	Tropical				
●	Southern	*Southern America*		*Pacific Islands*	
			Tropical		Micronesia
Asia			Caribbean		Galapagos
	Temperate	●	Chile, Argentina		Hawaii
	Tropical				Melanesia, Polynesia

Invaded Habitats
Forests, grassland, riparian habitats and freshwater wetlands.

Description
A tree reaching 25 m height or more, with the trunk and branches being protected by numerous stout and sharp spines of 8-20 cm length. Leaves are pinnately compound with 12-28 ovate-lanceolate leaflets of 1-2.5 cm length. Most plants of this species are dioecious. Female trees have racemes with 10-40 small yellow-green flowers. Male trees produce dense racemes of 3-12 cm length with usually more than 100 greenish flowers. Fruits are flattened pods of 20-30 cm length with up to 30 seeds.

Ecology and Control
Where native, this tree occurs in woodlands, rocky slopes, on floodplains and abandoned pastures. It is a vigorously growing species that forms dense and spiny thickets, displacing native vegetation and affecting wildlife by impeding movement and restricting access to water. A thornless form is widely cultivated as an ornamental and shade tree. The plant is tolerant of poor soils and develops an extensive lateral root system. Seeds require damage or passage through the digestive systems of mammals to break dormancy. Secondary dispersers include mammals and floodwater. The tree resprouts after damage and forms multiple stems.

    Cutting trees requires treating the cut stumps with herbicides to prevent regrowth. Seedlings and saplings can be hand pulled or dug out.

References
116, 290, 549, 1217, 1219, 1238, 1392.

Glyceria fluitans (L.) R.Br.   Poaceae

LF: Perennial herb   CU: Food
SN: *Panicularia fluitans* (L.) Kuntze

Geographic Distribution

Europe		*Australasia*		*Atlantic Islands*	
N	Northern		Australia		Cape Verde
N	British Isles	●	New Zealand		Canary, Madeira
N	Central, France			N	Azores
N	Southern	*Northern America*			South Atlantic Isl.
N	Eastern	N	Canada, Alaska		
N	Mediterranean Isl.		Southeastern USA	*Indian Ocean Islands*	
			Western USA		Mascarenes
Africa			Remaining USA		Seychelles
N	Northern		Mexico		Madagascar
	Tropical				
	Southern	*Southern America*		*Pacific Islands*	
			Tropical		Micronesia
Asia			Caribbean		Galapagos
N	Temperate		Chile, Argentina		Hawaii
	Tropical				Melanesia, Polynesia

Invaded Habitats
Margins of freshwater wetlands, ditches, swamps, riparian habitats.

Description
A 40-140 cm tall rhizomatous grass with decumbent to ascending stems that root at the nodes. Plants are spreading to form loose clumps or they are sprawling across the water surface, with the leaves floating. Inflorescences are sparsely branched and narrow panicles of 10-40 cm length. Lemmas are entire and 5.5-7 mm long, and have seven prominent veins. The anthers are usully purple.

Ecology and Control
Where native, this grass is found in standing and slowly moving waters, ditches, in shallow water by ponds, marshes and wet meadows. It invades similar habitats in New Zealand and forms dense swards. It grows sometimes as an aquatic in shallow water. Little is known on the ecology of this species.

    Scattered plants may be dug out or pulled by hand. Larger patches can be sprayed with grass-selective or non-selective herbicides approved for use in aquatic environments.

References
1415.

Glyceria maxima (Hartm.) Holmb.   Poaceae

LF: Perennial herb   CU: Fodder, ornamental
SN: *Panicularia aquatica* (L.) Kuntze, *Poa aquatica* L.

Geographic Distribution

Europe		*Australasia*		*Atlantic Islands*	
N	Northern	●	Australia		Cape Verde
N	British Isles	●	New Zealand		Canary, Madeira
N	Central, France				Azores
N	Southern		*Northern America*		South Atlantic Isl.
N	Eastern	X	Canada, Alaska		
N	Mediterranean Isl.	X	Southeastern USA		*Indian Ocean Islands*
			Western USA		Mascarenes
Africa			Remaining USA		Seychelles
	Northern		Mexico		Madagascar
	Tropical				
	Southern		*Southern America*		*Pacific Islands*
			Tropical		Micronesia
Asia			Caribbean		Galapagos
N	Temperate		Chile, Argentina		Hawaii
	Tropical				Melanesia, Polynesia

Invaded Habitats
Freshwater wetlands, riparian habitats.

Description
A stout, erect rhizomatous grass of 80-200 cm height, with numerous stems up to 12 mm in diameter. Leaves are bright green, 3-6 cm long and 1-2 cm wide, and smooth above. The ligules are up to 4 mm long. Inflorescences are much-branched and rather dense panicles of 10-40 cm length. The spikelets are ovate, 5-10 mm long, green or tinged with purple, each having 5-8 florets. The lemmas are entire and 3-4 mm long. Seeds are dark brown and 1-2 mm long.

Ecology and Control
This plant grows in reed swamps, lakes, ditches, riparian habitats, and ponds within the native range. It usually occurs in deeper water than the previous species and grows also in brackish water. Vegetative stems are easily broken and become rooted. It spreads rapidly by rhizomes and a single rootstock may cover 25 m^2 in 3 years. Trailing stems growing in water produce roots and new lateral shoots abundantly. The grass is thus able to rapidly cover the water surface, excluding native aquatic plants.
    Control is difficult because the extensive rhizome system quickly re-establishes the population. Scattered plants may be dug out, all rhizomes must be removed. Glyphosate applied in late summer and autumn, when the plant is in full flower, has proved to be effective.

References
32, 215, 607, 712, 986, 1246, 1247.

Gunnera tinctoria (Molina) Mirb.   Gunneraceae

LF: Perennial herb   CU: Ornamental
SN: *Gunnera chilensis* Lam.

Geographic Distribution

Europe		*Australasia*		*Atlantic Islands*	
	Northern		Australia		Cape Verde
●	British Isles	●	New Zealand	X	Canary, Madeira
	Central, France			●	Azores
X	Southern		*Northern America*		South Atlantic Isl.
	Eastern		Canada, Alaska		
	Mediterranean Isl.		Southeastern USA		*Indian Ocean Islands*
		X	Western USA		Mascarenes
Africa			Remaining USA		Seychelles
	Northern		Mexico		Madagascar
	Tropical				
	Southern		*Southern America*		*Pacific Islands*
			Tropical		Micronesia
Asia			Caribbean		Galapagos
	Temperate	N	Chile, Argentina		Hawaii
	Tropical				Melanesia, Polynesia

Invaded Habitats
Grassland, riparian habitats, coastal scrub and cliffs, shaded damp areas.

Description
A large perennial herb with simple and large leaves emerging from a rhizome. The leaf blades are 1-2 m long, thick, round in shape, lobed and irregularly toothed. The petioles are 1-1.5 m long. Leaves are covered with stiff prickles. The inflorescence is a spike of 50-75 cm length and up to 10 cm width, consisting of many lateral branches of 2-5 cm length, bearing very small flowers. These have two petals. Fruits are conspicuous, red drupes of 1.5-2 mm length. Seeds are 1-1.3 mm in diameter.

Ecology and Control
In the native range, this plant occurs in frost-free habitats with moderate to heavy rainfall, generally on leached soils. The plant is nitrogen-fixing due to a symbiotic association with a cyanobacterium. Seedling growth rate is slow, and rhizomes take several years to develop. The rapid shoot growth then coincides with the formation of a large rhizome. The large size of the plant and the large leaves shade out native vegetation and form tall thickets along streams. The plant spreads by both seeds and rhizomes. Discarded rhizomes may rapidly regenerate new plants.

    Specific control methods for this species are not available. Small plants may be hand pulled or dug out, rhizomes must be removed. Larger plants may be treated with herbicide applied to the foliage.

References
560, 561, 963.

Gymnocoronis spilanthoides (Don) DC.   Asteraceae

LF: Aquatic perennial herb   CU: Ornamental
SN: -

Geographic Distribution

Europe	*Australasia*	*Atlantic Islands*
Northern	● Australia	Cape Verde
British Isles	● New Zealand	Canary, Madeira
Central, France		Azores
Southern	*Northern America*	South Atlantic Isl.
Eastern	Canada, Alaska	
Mediterranean Isl.	Southeastern USA	*Indian Ocean Islands*
	Western USA	Mascarenes
Africa	Remaining USA	Seychelles
Northern	N Mexico	Madagascar
Tropical		
Southern	*Southern America*	*Pacific Islands*
	N Tropical	Micronesia
Asia	Caribbean	Galapagos
Temperate	N Chile, Argentina	Hawaii
Tropical		Melanesia, Polynesia

Invaded Habitats
Wet marshes, still and slowly flowing water bodies.

Description
A perennial herb that forms rounded bushes or mats of tangled stems. Stems are pale green, erect at first, becoming prostrate and scrambling, 1-1.5 m long and up to 2 cm in diameter. Internodes are hollow, inflated and buoyant. The stems root at the nodes. The opposite leaves are dark green, ovate to lanceolate, 5-20 cm long and 2.5-5 cm wide, with short stalks and serrate margins. Flowerheads are 1.5-2 cm in diameter, borne in clusters at the ends of branches, and contain numerous whitish florets. Fruits are yellow-brown and ribbed achenes of *c.* 5 mm diameter, without a pappus.

Ecology and Control
In the native range, this plant grows in wet soils and in still to very slowly moving waters. It spreads by both seeds and vegetative growth. The plant is fast-growing, and the long branching stems form a dense mat that, growing out from the bank, covers the water surface. The plant grows also when completely submersed. Dense infestations impede water flow and growth of native plant species. Seeds are dispersed by water or in mud sticking to animals. The plant is also spread by stem fragments which quickly develop adventitious roots.
    Control is achieved by mechanical harvesting such as raking the plant material from the water surface, or treating with herbicides approved for use in water.

References
986, 1415.

Hakea drupacea (Gaertn.) Roem. & Schult.   Proteaceae

LF: Evergreen tree, shrub   CU: Ornamental
SN: *Hakea suaveolens* R. Br., *Conchium drupaceum* Gaertn.

Geographic Distribution

Europe	*Australasia*	*Atlantic Islands*
Northern	● N Australia	Cape Verde
British Isles	X New Zealand	Canary, Madeira
Central, France		Azores
Southern	*Northern America*	South Atlantic Isl.
Eastern	Canada, Alaska	
Mediterranean Isl.	Southeastern USA	*Indian Ocean Islands*
	Western USA	Mascarenes
Africa	Remaining USA	Seychelles
Northern	Mexico	Madagascar
Tropical		
● Southern	*Southern America*	*Pacific Islands*
	Tropical	Micronesia
Asia	Caribbean	Galapagos
Temperate	Chile, Argentina	Hawaii
Tropical		Melanesia, Polynesia

Invaded Habitats
Scrub- and woodland.

Description
A dense, spreading or erect shrub or small tree, 1-4 m tall, with pubescent branchlets. Leaves are simple or compound, 3-13 cm long and 1-1.6 mm wide, and glabrous to pubescent. Inflorescences are axillary racemes with 45-85 flowers each. The perianth is pale pink to white and 3-4 mm long. Fruits are pale brown, ovate, 2-2.5 cm long and 1.5-2 cm wide, and covered with blackish horns of 2-4 mm length. The winged seeds are 16-20 mm long and *c.* 6 mm wide.

Ecology and Control
This fire adapted shrub forms extensive stands where invasive, displacing native shrubs and trees with species poor stands. Seeds are not released unless fire or other damage kills the branch or tree, and fire usually leads to large-scale release of seeds and subsequent seedling emergence.

Mechanical control is effective if all plants are removed. Seedlings and young plants can be pulled or dug out, or treated with herbicides. Larger trees are cut or girdled.

References
215, 284, 549, 1089.

Hakea gibbosa (Sm.) Cav.   Proteaceae

LF: Evergreen shrub, tree
SN: *Banksia gibbosa* Sm.
CU: Ornamental

Geographic Distribution

Europe
   Northern
   British Isles
   Central, France
   Southern
   Eastern
   Mediterranean Isl.

Africa
   Northern
   Tropical
● Southern

Asia
   Temperate
   Tropical

Australasia
   N Australia
   ● New Zealand

Northern America
   Canada, Alaska
   Southeastern USA
   Western USA
   Remaining USA
   Mexico

Southern America
   Tropical
   Caribbean
   Chile, Argentina

Atlantic Islands
   Cape Verde
   Canary, Madeira
   Azores
   South Atlantic Isl.

Indian Ocean Islands
   Mascarenes
   Seychelles
   Madagascar

Pacific Islands
   Micronesia
   Galapagos
   Hawaii
   Melanesia, Polynesia

Invaded Habitats
Scrub- and woodland, disturbed places.

Description
A 1-4 m tall shrub or small tree, with densely brown-villous branchlets. The densely villous leaves are narrowly divergent to widely spreading, 2.5-8.5 cm long and 9-15 mm wide. Inflorescences are axillary umbels with 2-4 flowers each. Pedicels are 1.8-4.5 mm long and densely villous. Flowers are cream-yellow and 4-5.5 mm long. Fruits are 2.5-4.5 cm long and 2-3 cm wide, and have a small beak. The black and winged seeds are 30-33 mm long and 10-14 mm wide.

Ecology and Control
Where native, this shrub grows in dry forests, heathland and coastal scrub, usually on sandy soils. Where invasive, the shrub forms extensive and dense thickets in which the native vegetation is suppressed and its regeneration prevented. The plant is shade tolerant.
    Individual shrubs can be killed by cutting at ground level. An effective chemical control of seedlings is spraying triclopyr or tebuthiuron.

References
284, 349, 549, 1089, 1415.

Hakea salicifolia (Vent.) Burtt   Proteaceae

LF: Evergreen shrub, tree   CU: Ornamental, shade tree
SN: *Hakea saligna* (Andr.) Knight

Geographic Distribution

Europe	*Australasia*	*Atlantic Islands*
Northern	● N Australia	Cape Verde
British Isles	● New Zealand	Canary, Madeira
Central, France		Azores
X Southern	*Northern America*	South Atlantic Isl.
Eastern	Canada, Alaska	
Mediterranean Isl.	Southeastern USA	*Indian Ocean Islands*
	Western USA	Mascarenes
Africa	Remaining USA	Seychelles
Northern	Mexico	Madagascar
Tropical		
X Southern	*Southern America*	*Pacific Islands*
	Tropical	Micronesia
Asia	Caribbean	Galapagos
Temperate	Chile, Argentina	Hawaii
Tropical		Melanesia, Polynesia

Invaded Habitats
Dry forests, heathland, scrub.

Description
A multi-stemmed erect shrub up to 4 m tall with a grey and smooth bark. Leaves are narrowly elliptic to linear-oblong, 5-10 cm long and 5-20 mm wide. Flowers are white, numerous, very small and borne in axillary clusters. The perianth is 2-3.5 mm long. Fruits are tardily dehiscent, woody follicles of *c.* 3 cm length opening with two valves. Each fruit contains two winged seeds of 15-20 mm length and 5-7 mm width.

Ecology and Control
This shrub is adapted to nutrient-poor soils and forms dense pure stands that reduce species richness and eliminate the original vegetation. As a light dependent species, it cannot regenerate under a closed canopy, and establishment depends on forest openings created by disturbances. Seeds remain viable on trees for a long period of time. Once released, however, seeds germinate rapidly under suitable conditions.
    Seedlings and saplings can be hand pulled or dug out. Larger shrubs can be cut and the cut stumps treated with herbicide. Fire may lead to mass release of seeds.

References
215, 1414.

Hakea sericea Schrad.
Proteaceae

LF: Evergreen shrub, tree
CU: Erosion control, shelter
SN: *Hakea tenuifolia* (Salisb.) Domin., *Hakea acicularis* (Vent.) R. Br.

Geographic Distribution

Europe
- Northern
- British Isles
- Central, France
- X Southern
- Eastern
- Mediterranean Isl.

Africa
- Northern
- Tropical
- ● Southern

Asia
- Temperate
- Tropical

Australasia
- ● N Australia
- X New Zealand

Northern America
- Canada, Alaska
- Southeastern USA
- Western USA
- Remaining USA
- Mexico

Southern America
- Tropical
- Caribbean
- Chile, Argentina

Atlantic Islands
- Cape Verde
- Canary, Madeira
- Azores
- South Atlantic Isl.

Indian Ocean Islands
- Mascarenes
- Seychelles
- Madagascar

Pacific Islands
- Micronesia
- Galapagos
- Hawaii
- Melanesia, Polynesia

Invaded Habitats
Forests, coastal grassland, mountain scrub.

Description
A much-branched and very prickly shrub of 2-5 m height. Young branches have short hairs. The stiff and needle-like leaves are dark-green to grey-green, glabrous and 2-6 cm long, and have a sharp point at the end. Flowers are white or pink and borne in axillary clusters. The perianth is 4-5 mm long. Fruits are woody, purplish brown to greyish capsules of 2.5-4 cm length, with two beaks at the end, and persisting for years. Each fruit contains two winged seeds.

Ecology and Control
The shrub grows naturally in heaths, scrubs, and in the understorey of open forests. Fruits accumulate on the tree throughout their lifetime. Seeds are released from the fruits when the parent tree dies, e.g. after fire. Fires lead to mass release of seeds and stimulate germination, leading to dense seedling populations. The shrub forms dense and impenetrable thickets, reducing native species richness, affecting wildlife, reducing water surface resources and increasing fire hazards.

In fire-adapted communities, mechanical control includes felling trees and leaving them for 12-18 months until seeds have been released. Subsequent burning then kills seeds and seedlings. Chemical control is done by treating shrubs with tebuthiuron, seedlings may also controlled by spraying triclopyr.

References
137, 164, 215, 284, 349, 366, 429, 549, 683, 795, 1085, 1089, 1333, 1395.

Harungana madagascariensis Lam. ex Poir.   Clusiaceae

LF: Evergreen shrub, tree   CU: -
SN: *Harongana madagascariensis* (Lam. ex Poir.) Choisy

Geographic Distribution

Europe	*Australasia*	*Atlantic Islands*
Northern	● Australia	Cape Verde
British Isles	New Zealand	Canary, Madeira
Central, France		Azores
Southern	*Northern America*	South Atlantic Isl.
Eastern	Canada, Alaska	
Mediterranean Isl.	Southeastern USA	*Indian Ocean Islands*
	Western USA	Mascarenes
Africa	Remaining USA	Seychelles
Northern	Mexico	N Madagascar
N Tropical		
Southern	*Southern America*	*Pacific Islands*
	Tropical	Micronesia
Asia	Caribbean	Galapagos
Temperate	Chile, Argentina	Hawaii
Tropical		Melanesia, Polynesia

Invaded Habitats
Forests and forest edges, disturbed sites.

Description
A shrub or small tree of 3-18 m height, with a greyish bark and broadly ovate and opposite leaves. The plant releases a bright orange juice when torn. Young shoots are covered with a short rusty felt or with scurfs. Leaves are 10-20 cm long and 5-10 cm wide, ovate to elliptic, acute, dark green and glossy above, sometimes scurfy beneath. Each leaf has 12-20 pairs of prominent lateral veins. Petioles are stout and 12-18 mm long. White and fragrant flowers of *c.* 2 mm diameter are borne in many-branched inflorescences of 15-30 cm diameter. Inflorescences are terminal and loose flattened heads. Fruits are yellow at first, becoming brown, globose, *c.* 2 mm in diameter and crowned by the persistent styles.

Ecology and Control
The plant grows naturally in forest clearings, in savanna and secondary forests. It can form dense populations that exclude native vegetation and prevents the regeneration of native shrubs and trees. Primarily it colonizes disturbed sites. Little is known on the ecology of this species.
    Specific control methods for this species are not available. Seedlings and saplings may be hand pulled or dug out. Larger trees can be cut and the cut stumps treated with herbicide.

References
607, 608, 1346.

Hedera helix L.   Araliaceae

LF: Evergreen climber
SN: -
CU: Ornamental

Geographic Distribution

Europe
- N Northern
- N British Isles
- N Central, France
- N Southern
- N Eastern
- N Mediterranean Isl.

Africa
- Northern
- Tropical
- Southern

Asia
- Temperate
- Tropical

Australasia
- ● Australia
- ● New Zealand

Northern America
- X Canada, Alaska
-    Southeastern USA
- ● Western USA
-    Remaining USA
-    Mexico

Southern America
- Tropical
- Caribbean
- Chile, Argentina

Atlantic Islands
-    Cape Verde
- N Canary, Madeira
- N Azores
-    South Atlantic Isl.

Indian Ocean Islands
- Mascarenes
- Seychelles
- Madagascar

Pacific Islands
-    Micronesia
-    Galapagos
- X Hawaii
-    Melanesia, Polynesia

Invaded Habitats
Forests, forest edges, rocky places.

Description
A woody vine creeping or climbing up to 30 m by adventitious roots, and with simple, shiny, dark green leaves. Young shoots and inflorescences are densely covered with stellate hairs. Leaves of sterile branches are broadly ovate, 4-10 cm long, and have 3-5 lobes. Those of flowering branches are ovate to rhombic and entire. Yellowish-green flowers of *c.* 5 mm diameter are borne in terminal umbels. Fruits are dark blue to purplish, globose drupes of 8-10 mm diameter, each containing 2-5 black seeds.

Ecology and Control
Native habitats include forest floors and trees, and rocky and shady places. In New Zealand, two subspecies have become naturalized: subsp. *canariensis* and subsp. *helix*. The vine is shade tolerant and climbs along tree trunks but grows also on the forest floor. It forms dense populations that inhibit the regeneration of native herbaceous species, trees and shrubs. If growing as a climber, it may smother the host tree. The vine is not very useful to native wildlife in areas where it is introduced. Fruits are eaten by birds, frequently by exotic birds such as starlings. Stem fragments root easily.

Mechanical control measures include cutting the vines and pulling them from trees and from the floor. Herbicides are not very effective due to the waxy leaves of the vine. Repeated burning individual plants with a blow torch proved to be partially successful.

References
121, 133, 215, 216, 924, 1141, 1300, 1415.

Hedychium coronarium König   Zingiberaceae

LF: Perennial herb   CU: Ornamental
SN: -

Geographic Distribution

Europe		*Australasia*		*Atlantic Islands*	
	Northern	X	Australia		Cape Verde
	British Isles		New Zealand		Canary, Madeira
	Central, France			X	Azores
	Southern	*Northern America*			South Atlantic Isl.
	Eastern		Canada, Alaska		
	Mediterranean Isl.		Southeastern USA	*Indian Ocean Islands*	
			Western USA	X	Mascarenes
Africa			Remaining USA		Seychelles
	Northern		Mexico		Madagascar
	Tropical				
●	Southern	*Southern America*		*Pacific Islands*	
		X	Tropical	X	Micronesia
Asia		X	Caribbean		Galapagos
N	Temperate		Chile, Argentina	●	Hawaii
N	Tropical				Melanesia, Polynesia

Invaded Habitats
Wet forests, swamp edges, riparian habitats.

Description
A perennial herb with tuberous rhizomes and erect, leafy pseudostems of 1-3 m height. The sessile leaves have blades that are oblong-lanceolate, 20-60 cm long and 4-10 cm wide, and pubescent beneath. A 2-3 cm long and membranous ligule is present. Inflorescences are terminal spikes of 10-20 cm length, the bracts being imbricate, ovate, 4-5 cm long and 2.5-4 cm wide. Each bract is with 2-3 flowers inside. Flowers are white, fragrant, zygomorphic, and have a *c.* 8 cm long and slender corolla tube. The lobes are lanceolate and up to 5 cm long; the lip is *c.* 6 cm long and wide. Fruits are globose to ovate-triangular capsules containing numerous, bright red seeds of *c.* 5 mm length. Each seed has an orange-red aril attached.

Ecology and Control
Within the native range, this plant grows in forests, in Yunnan, China, up to 2,500 m elevation. Where invasive, it forms large and dense thickets along creeks and in wet forests, crowding out native vegetation and preventing the establishment of native shrubs and trees. The plant does not produce seeds in large numbers and spreads mainly by vegetative growth of the rhizomes.
    Seedlings and small plants can be hand pulled or dug out, rhizomes must be removed to prevent regrowth.

References
549, 1063.

Hedychium flavescens Carey ex Roscoe   Zingiberaceae

LF: Perennial herb   CU: Ornamental
SN: -

Geographic Distribution

Europe	Australasia	Atlantic Islands
Northern	Australia	Cape Verde
British Isles	● New Zealand	Canary, Madeira
Central, France		X Azores
Southern	*Northern America*	South Atlantic Isl.
Eastern	Canada, Alaska	
Mediterranean Isl.	Southeastern USA	*Indian Ocean Islands*
	Western USA	● Mascarenes
Africa	Remaining USA	Seychelles
Northern	Mexico	Madagascar
Tropical		
● Southern	*Southern America*	*Pacific Islands*
	Tropical	Micronesia
Asia	Caribbean	Galapagos
Temperate	Chile, Argentina	● Hawaii
N Tropical		Melanesia, Polynesia

Invaded Habitats
Forests and forest edges, riverbanks.

Description
A perennial herb with tuberous rhizomes and erect, leafy pseudostems of 1-3 m height. The sessile leaves have slightly pubescent sheaths; the ligule is 3-5 cm long and membranous. Leaf blades are elliptic-lanceolate to lanceolate, 20-50 cm long and 4-10 cm wide, and pubescent beneath. Inflorescences are oblong spikes of 15-20 cm length, the bracts being imbricate, oblong to ovate, 3-4 cm long and 2-4 cm wide. Flowers are yellow or yellowish white, fragrant, with a *c.* 8 cm long and slender corolla tube. The lobes of the zygomorphic flowers are lanceolate and 3-3.5 cm long. Fruits are globose capsules with three valves, containing numerous seeds.

Ecology and Control
As the previous species, this plant forms dense thickets in forest openings and along creeks, impeding growth and regeneration of native plants. It spreads mainly by vegetative growth.

Specific control methods for this species are not available. The same methods may apply as for *H. gardnerianum*.

References
1278, 1415.

Hedychium gardnerianum Shepp. ex Gawl.   Zingiberaceae

LF: Perennial herb   CU: Ornamental
SN: -

Geographic Distribution

Europe	*Australasia*	*Atlantic Islands*
Northern	● Australia	Cape Verde
British Isles	● New Zealand	X Canary, Madeira
Central, France		● Azores
Southern	*Northern America*	South Atlantic Isl.
Eastern	Canada, Alaska	
Mediterranean Isl.	Southeastern USA	*Indian Ocean Islands*
	Western USA	● Mascarenes
Africa	Remaining USA	Seychelles
Northern	Mexico	Madagascar
Tropical		
● Southern	*Southern America*	*Pacific Islands*
	Tropical	Micronesia
Asia	● Caribbean	Galapagos
Temperate	Chile, Argentina	● Hawaii
N Tropical		Melanesia, Polynesia

Invaded Habitats
Forests and forest edges, moist places.

Description
A large herb with tuberous rhizomes and erect, leafy and simple stems of 1-3 m height. Leaves are elliptic-oblong, 24-40 cm long and *c.* 15 cm wide, acuminate and with a petiole appressed to the stem. The very fragrant, yellow flowers are borne in spikes of 20-35 cm length. The green and slender corolla tube is 4-5 cm long. Fruits are subglobose capsules containing seeds that are red at first, becoming grey, and have an aril.

Ecology and Control
This tall herb spreads rapidly and may become dominant over large areas, especially under canopy openings or in cleared areas. It forms dense thickets that penetrate into undisturbed forests and impede the growth and regeneration of native plants. Seedlings of shrubs and trees are unable to establish in stands of this plant. Once established, the plant is shade-tolerant and very persistent. Seeds are dispersed by birds. The species spreads both by seeds and by rhizome growth. The latter allows a rapid expansion of established plants.

Small plants can be hand pulled or dug out, rhizomes must be removed to prevent regrowth. Larger plants can be slashed at ground level and regrowth treated chemically. An effective herbicide is metsulfuron-methyl.

References
37, 39, 121, 1063, 1415.

Helianthus tuberosus L.   Asteraceae

LF: Perennial herb
SN: *Helianthus tomentosus* Michx.
CU: Fodder, food

Geographic Distribution

Europe		*Australasia*		*Atlantic Islands*	
	Northern		Australia		Cape Verde
X	British Isles	X	New Zealand		Canary, Madeira
●	Central, France			X	Azores
	Southern	*Northern America*			South Atlantic Isl.
●	Eastern	N	Canada, Alaska		
	Mediterranean Isl.	N	Southeastern USA	*Indian Ocean Islands*	
			Western USA		Mascarenes
Africa		N	Remaining USA		Seychelles
	Northern		Mexico		Madagascar
	Tropical				
	Southern	*Southern America*		*Pacific Islands*	
		X	Tropical		Micronesia
Asia			Caribbean		Galapagos
X	Temperate		Chile, Argentina		Hawaii
	Tropical				Melanesia, Polynesia

Invaded Habitats
Forest edges, woodland, riparian habitats.

Description
A stout perennial of 1-2.8 m height, usually branched above and with tuber-bearing rhizomes. Leaves are opposite except the upper most ones, broad-elliptic to elliptic-lanceolate, 10-25 cm long and 7-15 cm wide, coarsely toothed and white-pubescent beneath. Flowerheads are 4-8 cm in diameter, erect and usually borne in clusters. The ligules are 3-4 cm long, the inner florets are yellow. Fruits are glabrous or hairy achenes of 5-6 mm length.

Ecology and Control
This tall herb forms extensive and dense populations due to vegetative growth, crowding out native vegetation and preventing natural regeneration by shrubs and trees. It is a variable species with several cultivars including tall, unbranched varieties and plants with a rather bushy growth habit. Naturalized plants strongly compete for space, water and nutrients. The species spreads both by seeds and by tubers. Seeds are dispersed by wind, tubers carried by streams.

Chemical control is most effective at the pre-flowering stage. Herbicides include glyphosate, dicamba plus 2,4-D, or 2,4-D ester. Mechanical control is difficult because the numerous underground rhizomes and tubers allow a rapid recolonization.

References
540, 1258.

Heracleum mantegazzianum Sommier & Levier — Apiaceae

LF: Perennial herb
SN: -
CU: Ornamental

Geographic Distribution

Europe	Australasia	Atlantic Islands
● Northern	X Australia	Cape Verde
● British Isles	X New Zealand	Canary, Madeira
X Central, France		Azores
Southern	*Northern America*	South Atlantic Isl.
● Eastern	X Canada, Alaska	
Mediterranean Isl.	X Southeastern USA	*Indian Ocean Islands*
	X Western USA	Mascarenes
Africa	Remaining USA	Seychelles
Northern	Mexico	Madagascar
Tropical		
Southern	*Southern America*	*Pacific Islands*
	Tropical	Micronesia
Asia	Caribbean	Galapagos
N Temperate	Chile, Argentina	Hawaii
Tropical		Melanesia, Polynesia

Invaded Habitats
Riparian habitats, grassland, forest edges, disturbed sites.

Description
A large and stout herb with hollow stems of 2-5 m height and up to 10 cm diameter, usually softly pubescent and with conspicuous purple blotches. Leaves are up to 3 m long, pinnately divided or lobed, with lateral segments being up to 1.3 m long. Flowers appear in large umbels up to 50 cm in diameter that have 50-150 rays. The white or pinkish petals are 10-12 mm long. Fruits are glabrous or pubescent, 8-14 mm long and 6-8 mm wide. The plant has a thick taproot.

Ecology and Control
Naturalized plants are very variable in perennation, size and shape. The species forms extensive populations whose large rosettes crowd out native species and reduce species richness. The plant causes health problems by leading to photosensitization of the skin, which may result in strong dermatitis. Seeds are abundantly produced and dispersed by water and wind. On riverbanks, large stands may destabilize the soil and increase soil erosion.

Control must aim at depleting the soil seed bank. Repeated cutting before flowering may reduce the plant's vigour. Plants can be killed if the taproot is cut 8-12 cm below ground level. Seedlings and young plants can be hand pulled. Effective herbicides are glyphosate, triclopyr, or imazapyr applied in spring.

References
35, 181, 191, 192, 345, 540, 549, 786, 965, 1054, 1055, 1057, 1058, 1128, 1293, 1294, 1295, 1353, 1421.

Hesperis matronalis L.   Brassicaceae

LF: Perennial herb
SN: -
CU: Ornamental

Geographic Distribution

Europe
      Northern
X   British Isles
N   Central, France
N   Southern
N   Eastern
      Mediterranean Isl.

Africa
      Northern
      Tropical
      Southern

Asia
N   Temperate
      Tropical

Australasia
      Australia
X   New Zealand

Northern America
X   Canada, Alaska
      Southeastern USA
X   Western USA
●   Remaining USA
      Mexico

Southern America
      Tropical
      Caribbean
      Chile, Argentina

Atlantic Islands
      Cape Verde
      Canary, Madeira
      Azores
      South Atlantic Isl.

Indian Ocean Islands
      Mascarenes
      Seychelles
      Madagascar

Pacific Islands
      Micronesia
      Galapagos
      Hawaii
      Melanesia, Polynesia

Invaded Habitats
Forests and forest gaps, grassland, hedges.

Description
A short-lived perennial herb of 40-150 cm height, with erect, glabrous to sparsely pubescent and often branching stems. Stem leaves have petioles, are lanceolate to ovate and have sharply toothed margins. The lower leaves are up to 50 cm long and 10 cm wide, upper leaves are much smaller. Flowers appear in loose clusters and have purplish, white or lilac corollas with four petals of 14-25 mm length. Fruits are stalked siliques of 3-10 cm length and 1.5-3 mm width.

Ecology and Control
In the native range, this species comprises a group of closely related taxa, growing in damp or shaded places, riparian habitats, and disturbed places. The plant spreads rapidly from seed and forms dense infestations that exclude native species and prevent their regeneration. It may become the dominant understorey species in forests, excluding native shrubs.

    Smaller infestations can be contained by hand pulling the plants, larger populations may be treated with a glyphosate herbicide. In any case, monitoring the area is necessary to detect and eliminate seedlings emerging from the seed bank.

References
571, 887, 1111.

Hexaglottis lewisiae Goldblatt   Iridaceae

LF: Perennial herb   CU: Ornamental
SN: -

Geographic Distribution

Europe	*Australasia*	*Atlantic Islands*
Northern	● Australia	Cape Verde
British Isles	New Zealand	Canary, Madeira
Central, France		Azores
Southern	*Northern America*	South Atlantic Isl.
Eastern	Canada, Alaska	
Mediterranean Isl.	Southeastern USA	*Indian Ocean Islands*
	Western USA	Mascarenes
Africa	Remaining USA	Seychelles
Northern	Mexico	Madagascar
Tropical		
N Southern	*Southern America*	*Pacific Islands*
	Tropical	Micronesia
Asia	Caribbean	Galapagos
Temperate	Chile, Argentina	Hawaii
Tropical		Melanesia, Polynesia

Invaded Habitats
Wetlands, grass- and woodland.

Description
A cormous herb with flexuous and rigid stems of 20-60 cm height. Corms are 15-20 mm in diameter. The lower leaves are 30-80 cm long and 2-6 mm wide, long-linear and grooved. Spathes are 20-40 mm long. Flowers are yellow and have perianth segments of 16-20 mm length. Fruits are cylindric capsules of 10-15 mm length.

Ecology and Control
The species spreads rapidly by seeds and corms, and its dense patches crowd out native vegetation. In Australia, it invades native *Themeda* grassland. Regeneration of native plants is impeded by the shading effect of the large leaves of this herb.

Specific control methods for this species are not available. The same methods as for *Moraea flaccida* or *Freesia leichtlinii* may apply.

References
215.

Hieracium pilosella Asteraceae

LF: Perennial herb CU: -
SN: *Pilosella officinarum* Schultz & Bip.

Geographic Distribution

Europe
- N Northern
- N British Isles
- N Central, France
- N Southern
- N Eastern
- N Mediterranean Isl.

Africa
- Northern
- Tropical
- Southern

Asia
- N Temperate
- Tropical

Australasia
- Australia
- ● New Zealand

Northern America
- X Canada, Alaska
- Southeastern USA
- X Western USA
- Remaining USA
- Mexico

Southern America
- Tropical
- Caribbean
- Chile, Argentina

Atlantic Islands
- Cape Verde
- Canary, Madeira
- Azores
- South Atlantic Isl.

Indian Ocean Islands
- Mascarenes
- Seychelles
- Madagascar

Pacific Islands
- Micronesia
- Galapagos
- Hawaii
- Melanesia, Polynesia

Invaded Habitats
Grassland, pastures, rock outcrops, disturbed sites.

Description
A densely hairy herb with slender stolons and erect flowering stems of 5-15 cm height. Rosette leaves are dull green above, white-hairy beneath, oblanceolate, 2-7 cm long and 1-2 cm wide. Flowerheads are borne solitary, the involucre is 8-13 mm long. Florets are yellow. Fruits are dark brown achenes of *c.* 2 mm length, having a pappus of 5-6 mm length.

Ecology and Control
In the native range, this species is a highly variable complex. It invades native tussock grasslands in New Zealand and forms dense patches that displace native grasses and forbs. It spreads mainly by vegetative growth of the stolons, and individual clones quickly cover the soil. Seeds are wind dispersed. The rapidly growing plants outcompete native plants and transform invaded areas into pure stands. Fertilization enhances spread of this herb.

Long term restoration of invaded grasslands may require reducing fertilizer input and exclusion from grazing. Scattered plants may be dug out. Chemical control is done by spraying 2,4-D esters.

References
113, 114, 134, 265, 400, 686, 806, 1147, 1148, 1427.

Hiptage benghalensis (L.) Kurz   Malpighiaceae

LF: Evergreen liana, shrub   CU: Ornamental
SN: *Banisteria benghalensis* L., *Hiptage madablota* Gaertn.

Geographic Distribution

Europe	Australasia	Atlantic Islands
Northern	Australia	Cape Verde
British Isles	New Zealand	Canary, Madeira
Central, France		Azores
Southern	*Northern America*	South Atlantic Isl.
Eastern	Canada, Alaska	
Mediterranean Isl.	Southeastern USA	*Indian Ocean Islands*
	Western USA	● Mascarenes
Africa	Remaining USA	Seychelles
Northern	Mexico	Madagascar
Tropical		
Southern	*Southern America*	*Pacific Islands*
	Tropical	Micronesia
Asia	Caribbean	Galapagos
N Temperate	Chile, Argentina	Hawaii
N Tropical		Melanesia, Polynesia

Invaded Habitats
Forests and forest edges, disturbed sites.

Description
A large woody, straggling or climbing shrub with young branches being grey tomentose. The opposite and entire leaves are oblong to ovate-lanceolate, 9-21 cm long and 4-9 cm wide, acute or acuminate, glabrous, and have petioles of *c.* 1 cm length. White and fragrant flowers of 2-3 cm diameter are borne in erect, pubescent racemes of 10-20 cm length, the pedicels being 15-20 mm long. Flowers have a yellow centre and orbicular to elliptic petals that are hairy outside. Fruits are samaras with three wings each, the middle wing being 4-6 cm long and the lateral wings 2-3 cm long.

Ecology and Control
A variable species with three varieties distinguished in India: var. *longifolia*, var. *benghalensis*, and var. *rothinii*. Where invasive, the plant climbs over native shrubs and trees, smothering them and preventing their growth and regeneration.

Specific control methods for this species are not available. Stems may be cut and pulled from trees, and the cut stumps treated with herbicide.

References
1227, 1239, 1278.

Holcus lanatus L.
Poaceae

LF: Perennial herb
CU: Erosion control
SN: *Holcus argenteus* Agardh ex Roem. & Schult.

Geographic Distribution

Europe
- N   Northern
- N   British Isles
- N   Central, France
- N   Southern
- N   Eastern
- N   Mediterranean Isl.

Africa
- N   Northern
-     Tropical
- X   Southern

Asia
- N   Temperate
- X   Tropical

Australasia
- ●   Australia
- X   New Zealand

Northern America
- X   Canada, Alaska
-     Southeastern USA
- ●   Western USA
- X   Remaining USA
-     Mexico

Southern America
- X   Tropical
-     Caribbean
- X   Chile, Argentina

Atlantic Islands
-     Cape Verde
- N   Canary, Madeira
- N   Azores
- X   South Atlantic Isl.

Indian Ocean Islands
- ●   Mascarenes
-     Seychelles
-     Madagascar

Pacific Islands
-     Micronesia
-     Galapagos
- ●   Hawaii
-     Melanesia, Polynesia

Invaded Habitats
Coastal grassland and scrub, disturbed sites.

Description
A 30-100 cm tall grass with densely tufted and softly hairy, erect or ascending stems. Leaves are narrowly lanceolate, 3-10 mm wide, and pubescent on both sides; the ligules are 1-4 mm long. The inflorescence is an open to very dense, lanceolate, whitish to dark purple panicle of 3-20 cm length and 1-8 cm width. Spikelets are 4-6 mm long and have lanceolate glumes that are ciliate on keels and veins. Each spikelet has 2-3 florets. The lemma of the upper floret has a strongly hooked awn.

Ecology and Control
Where native, this grass is found in mesic to wet meadows, open forests, reed swamps, grassland, and open woods. It is tolerant of a wide range of soils. It is invasive because it forms dense swards that reduce native species richness and eliminate native grasses and forbs. The grass is a prolific seed producer, and seeds are dispersed by wind.

Intensive mowing or grazing suppresses the establishment and spread of this grass. Effective herbicides are atrazine or diuron.

References
94, 121, 215, 382, 1278, 1290.

Hordeum marinum Huds. Poaceae

LF: Annual herb
SN: *Critesion marinum* (Huds.) Löve
CU:

Geographic Distribution

Europe	*Australasia*	*Atlantic Islands*
Northern	● Australia	Cape Verde
N British Isles	X New Zealand	N Canary, Madeira
N Central, France		Azores
N Southern	*Northern America*	South Atlantic Isl.
N Eastern	Canada, Alaska	
N Mediterranean Isl.	Southeastern USA	*Indian Ocean Islands*
	Western USA	Mascarenes
Africa	Remaining USA	Seychelles
N Northern	Mexico	Madagascar
Tropical		
Southern	*Southern America*	*Pacific Islands*
	Tropical	Micronesia
Asia	Caribbean	Galapagos
N Temperate	Chile, Argentina	Hawaii
Tropical		Melanesia, Polynesia

Invaded Habitats
Seasonal freshwater wetlands, saline wetlands, coastal beaches.

Description
An erect or ascending grass with culms of 15-80 cm height. Leaf blades are flat, linear, 25-80 mm long and 1.3-2.3 mm wide. The ligule is *c.* 0.5 mm long. The inflorescence is a single spike, contracted, oblong to narrowly ovate, 5-12 cm long, and has 25-60 spikelets. These are in groups of three. Seeds are oblong, compressed, and 3-4 mm long.

Ecology and Control
In the native range, this grass is found in saline meadows, and is a variable species with at least two subspecies. Where invasive, it spreads quickly and forms dense stands that eliminate native plant species and reduce species richness.

Specific control methods for this species are not available. Scattered plants can be hand pulled. Larger infestations can be treated with herbicides or cut to prevent seed dispersal.

References
85, 215, 1350.

Hordeum murinum L.   Poaceae

LF: Annual herb
SN: -
CU: -

Geographic Distribution

Europe	*Australasia*	*Atlantic Islands*
Northern	X Australia	Cape Verde
N British Isles	X New Zealand	N Canary, Madeira
N Central, France		N Azores
N Southern	*Northern America*	South Atlantic Isl.
N Eastern	Canada, Alaska	
N Mediterranean Isl.	Southeastern USA	*Indian Ocean Islands*
	● Western USA	Mascarenes
Africa	X Remaining USA	Seychelles
N Northern	Mexico	X Madagascar
Tropical		
X Southern	*Southern America*	*Pacific Islands*
	Tropical	Micronesia
Asia	Caribbean	Galapagos
X Temperate	X Chile, Argentina	X Hawaii
N Tropical		Melanesia, Polynesia

Invaded Habitats
Deserts and desert scrub, arid grassland.

Description
A tufted grass with erect or ascending, smooth stems of 5-50 cm height and flat leaves. Leaves are glabrous or slightly pubescent, 4-10 cm long and 2-8 mm wide. The membranous ligule is *c.* 1 mm long. Inflorescences are oblong and strongly compressed spikes of 2-10 cm length with three spikelets at each node. These have usually one floret each. The lemmas have awns of 1-5 cm length and the glumes have awns of 1-3 cm length. Fruits are yellow caryopses of 5-7 mm length.

Ecology and Control
A native of dry grassland and disturbed places with three subspecies within the native range. Seeds are dispersed by adhering to animals and by wind. The grass is invasive because it forms dense swards that displace native grasses and forbs by competing for water, nutrients and space. The grass is a highly competitive to native annuals in arid areas.

    Specific control methods for this species are not available. Scattered plants may be hand pulled or cut. Larger populations can be sprayed with grass-selective herbicides.

References
85, 110, 180, 581, 626.

Hydrilla verticillata (L.f.) Royle — Hydrocharitaceae

LF: Aquatic perennial herb
SN: *Hottonia serrata* Willd.
CU: Ornamental

Geographic Distribution

Europe		*Australasia*		*Atlantic Islands*	
	Northern	N	Australia		Cape Verde
X	British Isles	●	New Zealand	X	Canary, Madeira
X	Central, France				Azores
X	Southern	*Northern America*			South Atlantic Isl.
	Eastern		Canada, Alaska		
	Mediterranean Isl.	●	Southeastern USA	*Indian Ocean Islands*	
		●	Western USA	X	Mascarenes
Africa		X	Remaining USA	X	Seychelles
X	Northern	X	Mexico	X	Madagascar
X	Tropical				
	Southern	*Southern America*		*Pacific Islands*	
		X	Tropical	X	Micronesia
Asia			Caribbean		Galapagos
N	Temperate		Chile, Argentina		Hawaii
N	Tropical			X	Melanesia, Polynesia

Invaded Habitats
Freshwater lakes and ponds, coastal estuaries.

Description
A wholly submerged aquatic plant usually firmly rooted in the substrate, with pale green and slender, branching stems up to 8 m long, rooting at the lower nodes. Vegetative reproductive organs include tubers and turions. Leaves are pale to bright green, linear-lanceolate, 1-3 cm long, and borne in whorls of 3-8. Flowers are solitary, inconspicuous, *c.* 3 mm in diameter, and exserted from leaf axils. The plant is dioecious. Sepals of female flowers are 1.5-3 mm long. Petals are transparent with a few red streaks. Fruits are 5-15 mm long and 3-6 mm wide, containing 1-6 seeds of 2-2.5 mm length.

Ecology and Control
The plant grows in water up to 12 m deep and spreads mainly by vegetative reproduction, e.g. by stolons, tubers, turions and stem fragments. It forms dense and large mats that fill the water, restrict water flow, crowd out native plants, and reduce habitat for wildlife. Growth is strongly enhanced with increasing nutrient content of the water. The plant accumulates a bank of tubers and turions in the soil that may persist for several years. The plant usually dies back above-ground at the end of the growing season.

Smaller infestations may be removed by hand. Larger infestations in flowing waters are treated with herbicides based on acrolein or fluridone, or plants are killed by draining. For control in still waters, herbicides used include 2,4-D, diquat, or paraquat.

References
133, 270, 332, 387, 506, 581, 710, 715, 986, 1233, 1262, 1322, 1323, 1415.

Hygrophila polysperma (Roxb.) T. Anders.   Acanthaceae

LF: Aquatic perennial herb   CU: Ornamental
SN: *Justicia polysperma* Roxb.

Geographic Distribution

Europe
    Northern
    British Isles
    Central, France
    Southern
    Eastern
    Mediterranean Isl.

Africa
    Northern
    Tropical
    Southern

Asia
    Temperate
N   Tropical

Australasia
    Australia
    New Zealand

Northern America
    Canada, Alaska
●   Southeastern USA
    Western USA
    Remaining USA
    Mexico

Southern America
    Tropical
    Caribbean
    Chile, Argentina

Atlantic Islands
    Cape Verde
    Canary, Madeira
    Azores
    South Atlantic Isl.

Indian Ocean Islands
    Mascarenes
    Seychelles
    Madagascar

Pacific Islands
    Micronesia
    Galapagos
    Hawaii
    Melanesia, Polynesia

Invaded Habitats
Freshwater lakes and ponds, riparian habitats.

Description
A mostly submersed aquatic plant with ascending to creeping squared stems, rooted in the soil and rooting at the nodes. It also produces shoots and leaves above the water surface. The opposite leaves are up to 8 cm long and 2 cm wide, with the bases joined at nodes. Flowers are small, bluish-white and borne solitary in the uppermost leaf axils. Fruits are narrow capsules of 6-8 mm length, releasing numerous tiny round seeds.

Ecology and Control
Stems easily fragment and develop new plants by rooting. The plant forms dense and pure stands in waters up to 3 m depth or more, impeding water flow and displacing native aquatic species. It is adapted to low light conditions and expands rapidly, e.g. from 0.1 acre to 10 acres in one year. The plant tends to grow more vigorously in flowing than in standing waters.

    The plant is difficult to control due to stem fragmentation. Small infestations may be removed by hand. Larger infestations can be treated with herbicides approved for use in aquatic environments.

References
41, 715, 1330, 1335.

Hymenachne amplexicaulis (Rudge) Nees Poaceae

LF: Perennial herb
SN: *Panicum amplexicaulis* Rudge
CU: Forage

Geographic Distribution

Europe		*Australasia*		*Atlantic Islands*	
	Northern	X	Australia		Cape Verde
	British Isles		New Zealand		Canary, Madeira
	Central, France				Azores
	Southern	*Northern America*			South Atlantic Isl.
	Eastern		Canada, Alaska		
	Mediterranean Isl.	●	Southeastern USA	*Indian Ocean Islands*	
			Western USA		Mascarenes
Africa			Remaining USA		Seychelles
	Northern	N	Mexico		Madagascar
	Tropical				
	Southern	*Southern America*		*Pacific Islands*	
		N	Tropical		Micronesia
Asia		N	Caribbean		Galapagos
	Temperate		Chile, Argentina		Hawaii
X	Tropical				Melanesia, Polynesia

Invaded Habitats
Marshes, ditches, riparian habitats, freshwater wetlands.

Description
A robust stoloniferous grass with stout glabrous culms of 1-2.5 m height, rooting at the lower nodes and growing as a semi-aquatic plant. Stems are floating, creeping or ascending, and sparingly branched. Leaves are lanceolate and cordate, the blades are flat, up to 35 cm long and 4 cm wide, clasping the stem, and have long hairs on their lower margins. Ligules are membranous. The inflorescence is a terminal, dense and spike-like panicle of *c.* 8 mm width and 20-50 cm length. Spikelets are 3.3-4.3 mm long.

Ecology and Control
In the native range, this grass occurs in shallow waters of swamp margins and river banks. It is adapted to fluctuating water levels and massive regeneration by seed may occur after drought. It tolerates extended periods of flooding and water depths up to 1.2 m. Seeds are dispersed by water. Where invasive, it spreads rapidly and forms extensive pure colonies that displace native marsh communities.

Established plants are difficult to control. Scattered plants may be removed manually, all stolons and roots must be removed to prevent regrowth. Larger patches can be treated with herbicides approved for use in aquatic environments.

References
607, 666, 715, 1190.

Hyparrhenia rufa (Nees) Stapf

Poaceae

LF: Perennial herb
CU: Fodder, forage
SN: *Andropogon rufus* (Nees) Kunth, *Cymbopogon rufus* (Nees) Rendle

Geographic Distribution

Europe
- Northern
- British Isles
- Central, France
- Southern
- Eastern
- Mediterranean Isl.

Africa
- Northern
- N Tropical
- N Southern

Asia
- Temperate
- X Tropical

Australasia
- Australia
- New Zealand

Northern America
- Canada, Alaska
- Southeastern USA
- Western USA
- Remaining USA
- Mexico

Southern America
- ● Tropical
- X Caribbean
- Chile, Argentina

Atlantic Islands
- Cape Verde
- Canary, Madeira
- Azores
- South Atlantic Isl.

Indian Ocean Islands
- X Mascarenes
- Seychelles
- N Madagascar

Pacific Islands
- Micronesia
- Galapagos
- X Hawaii
- Melanesia, Polynesia

Invaded Habitats
Grassland, savanna, disturbed sites.

Description
A variable perennial grass of 60-240 cm height. Leaf blades are flat, elongate, and 2-8 mm wide. Inflorescences are loose and narrow panicles of 20-60 cm length, with slightly spreading or contiguous racemes. The shortly hairy or glabrous spikelets are 3.5-5 mm long and borne in pairs. One is sessile and has two florets, the other pedicelled and usually longer than the sessile spikelet. Inflorescences are panicles of 20-60 cm length, consisting of pairs of racemes bearing paired spikelets. Fertile spikelets are pubescent with dark reddish brown hairs. Awns are 15-20 mm long.

Ecology and Control
This grass occurs as a native in seasonally flooded grassland and open woodland. It is a drought tolerant species and withstands dry seasons of several months, seasonal burning but also temporary flooding. The grass produces seeds abundantly and spreads rapidly. It forms dense swards where invasive that displace native grasses and forbs, preventing the establishment of other species and transforming native savannas into species pure stands.

Specific control methods for this grass are not available. Cutting before seeds ripen prevents seed dispersal.

References
6, 74, 984, 1033, 1190.

Hypericum androsaemum L.   Clusiaceae

LF: Deciduous shrub   CU: Ornamental
SN: *Androsaemum officinale* All.

Geographic Distribution

Europe		*Australasia*	*Atlantic Islands*
	Northern	● Australia	Cape Verde
N	British Isles	● New Zealand	Canary, Madeira
N	Central, France		Azores
N	Southern	*Northern America*	South Atlantic Isl.
N	Eastern	Canada, Alaska	
N	Mediterranean Isl.	Southeastern USA	*Indian Ocean Islands*
		Western USA	Mascarenes
Africa		Remaining USA	Seychelles
N	Northern	Mexico	Madagascar
	Tropical		
X	Southern	*Southern America*	*Pacific Islands*
		Tropical	Micronesia
Asia		Caribbean	Galapagos
N	Temperate	Chile, Argentina	Hawaii
	Tropical		Melanesia, Polynesia

Invaded Habitats
Bushland, grassland, forest edges, riparian habitats.

Description
A glabrous shrub with spreading stems of 30-150 cm height. Leaves are glabrous and opposite, entire, broadly ovate to ovate-oblong, sessile, and 4-15 cm long. There are minute oil glands on the lower surface. Pale yellow flowers of 2-3 cm diameter are borne in clusters at the ends of branches. Stamens are as long as the petals. Petals sometimes have black dots along the margins. Fruits are fleshy berries, reddish at first and becoming black, broadly ellipsoid to globose, and 7-12 mm long. They contain numerous brown seeds of *c.* 1 mm length.

Ecology and Control
Where native, this plant grows in damp woods, shady hedges and other moist places. It is invasive because its rather large leaves shade out native species and the shrub forms dense thickets. It covers extensive areas and displaces native vegetation. The plant grows both in shade and full sun. Seeds are presumably dispersed by birds.

Isolated plants can be hand pulled or dug out. Roots must be removed to prevent regrowth. An effective herbicide is picloram, best applied before fruits develop.

References
215, 607, 924, 985, 986, 1415.

Hypericum calycinum L.   Clusiaceae

LF: Evergreen shrub   CU: Ornamental
SN: -

Geographic Distribution

Europe		*Australasia*		*Atlantic Islands*	
	Northern	●	Australia		Cape Verde
X	British Isles	X	New Zealand		Canary, Madeira
X	Central, France				Azores
N	Southern		*Northern America*		South Atlantic Isl.
N	Eastern		Canada, Alaska		
	Mediterranean Isl.		Southeastern USA		*Indian Ocean Islands*
			Western USA		Mascarenes
Africa			Remaining USA		Seychelles
	Northern		Mexico		Madagascar
	Tropical				
	Southern		*Southern America*		*Pacific Islands*
			Tropical		Micronesia
Asia			Caribbean		Galapagos
N	Temperate		Chile, Argentina		Hawaii
	Tropical				Melanesia, Polynesia

Invaded Habitats
Grassland, forest edges, disturbed sites.

Description
A glabrous small shrub with stems up to 60 cm tall and stout rhizomes. Leaves are sessile, ovate, 3-10 cm long and 1.5-3 cm wide, the lower surface being greyish-green. The bright yellow flowers are up to 8 cm in diameter and borne solitary or in cymes of 2-3. The usually asymmetric petals are 2.5-4 cm long. Fruits are 1-2 cm long capsules, releasing broad cylindric seeds of 1.5-2 mm length.

Ecology and Control
A drought-resistant and fire-tolerant species that forms dense colonies due to the extensively creeping rhizomes. It impedes the growth and regeneration of native shrubs and trees by competing for nutrients and space. The plant establishes readily after disturbances.
    Seedlings and small plants can be pulled or dug out. Rhizomes must be removed to prevent regrowth. Larger plants can be dug out or cut and the cut stumps treated with herbicide.

References
215, 607.

Hypericum perforatum L. Clusiaceae

LF: Perennial herb
SN: -
CU: Essential oils

Geographic Distribution

	Europe			Australasia			Atlantic Islands
N	Northern		●	Australia			Cape Verde
N	British Isles		X	New Zealand		N	Canary, Madeira
N	Central, France					N	Azores
N	Southern			*Northern America*			South Atlantic Isl.
N	Eastern		X	Canada, Alaska			
N	Mediterranean Isl.			Southeastern USA			*Indian Ocean Islands*
			X	Western USA			Mascarenes
	Africa		X	Remaining USA			Seychelles
N	Northern			Mexico			Madagascar
	Tropical						
X	Southern			*Southern America*			*Pacific Islands*
				Tropical			Micronesia
	Asia		X	Caribbean			Galapagos
N	Temperate		X	Chile, Argentina		X	Hawaii
N	Tropical						Melanesia, Polynesia

Invaded Habitats
Grass- and woodland, riverbanks, disturbed sites.

Description
A rhizomatous perennial with glabrous and often reddish, erect stems of 30-80 cm height, with dark glands on two opposite ridges along the stems. Several stems arise from a woody crown or rootstock. Rhizomes are long and slender. Leaves are opposite, ovate to linear, sessile, 1.5-3 cm long and have numerous small oil glands. Numerous golden yellow flowers of *c.* 2 cm diameter are borne in terminal clusters. Petals often have black dots along the margins. Fruits are sticky capsules of 5-10 mm length with three persistent styles. Seeds are dark brown to black, densely pitted and *c.* 1 mm long. The plant has a taproot reaching 1 m depth.

Ecology and Control
This plant forms extensive and dense colonies where invasive, eliminating native vegetation. Seed production is prolific and seeds are dispersed by water, soil and agricultural activities. Fire stimulates germination, leading to dense seedling growth. Seedlings are sensitive to competition but established plants compete successfully with native forbs and grasses. The weed cannot survive in densely shaded areas.

Seedlings and smaller plants can be hand pulled or dug out, crowns and rhizomes must be removed. Repeated defoliation can reduce the plant's density. Effective herbicides are 2,4-D, glyphosate, or picloram. Bare ground resulting from control measures is likely to be colonized by emerging seedlings.

References
121, 200, 201, 215, 283, 292, 536, 689, 924, 986.

Hypochaeris radicata L. — Asteraceae

LF: Perennial herb
SN: -
CU: -

Geographic Distribution

Europe
- Northern
- N British Isles
- N Central, France
- N Southern
- N Eastern
- Mediterranean Isl.

Africa
- N Northern
- Tropical
- X Southern

Asia
- N Temperate
- Tropical

Australasia
- ● Australia
- X New Zealand

Northern America
- X Canada, Alaska
- Southeastern USA
- X Western USA
- X Remaining USA
- Mexico

Southern America
- X Tropical
- Caribbean
- X Chile, Argentina

Atlantic Islands
- Cape Verde
- N Canary, Madeira
- N Azores
- South Atlantic Isl.

Indian Ocean Islands
- Mascarenes
- Seychelles
- Madagascar

Pacific Islands
- Micronesia
- Galapagos
- ● Hawaii
- Melanesia, Polynesia

Invaded Habitats
Grassland, riparian habitats, seasonal freshwater wetlands, coastal beaches.

Description
An erect or ascending, usually branched herb with 20-80 cm tall stems, more or less glabrous. Leaves are oblong to elliptical or oblanceolate, 5-25 cm long, and mostly pubescent. One to few flowerheads of 2-3 cm diameter appear on the stems, and have involucral bracts in several rows. Ligules are bright yellow. Fruits are long-beaked achenes of 8-17 mm length; the marginal achenes are sometimes smaller. The pappus consists of two rows of hairs and is 10-13 mm long.

Ecology and Control
A native of grassy places and meadows, this plant forms extensive colonies where invasive that shade out native grasses and forbs, and prevent the establishment of shrubs and trees. It tolerates a wide range of soils and strongly competes with native species. Establishment occurs mainly in disturbed sites. The plant propagates by seeds and vegetatively by perennating buds. Seeds are wind dispersed.

Scattered plants can be dug out, crowns must be removed to prevent regrowth. Effective herbicides are MCPA salts, or 2,4-D amine and esters.

References
1, 197, 215, 1304.

Ilex aquifolium L.   Aquifoliaceae

LF: Evergreen shrub, tree   CU: Ornamental
SN: -

Geographic Distribution

Europe	*Australasia*	*Atlantic Islands*
Northern	● Australia	Cape Verde
N British Isles	X New Zealand	Canary, Madeira
N Central, France		Azores
N Southern	*Northern America*	South Atlantic Isl.
N Eastern	Canada, Alaska	
N Mediterranean Isl.	Southeastern USA	*Indian Ocean Islands*
	X Western USA	Mascarenes
Africa	Remaining USA	Seychelles
N Northern	Mexico	Madagascar
Tropical		
Southern	*Southern America*	*Pacific Islands*
	Tropical	Micronesia
Asia	Caribbean	Galapagos
N Temperate	Chile, Argentina	X Hawaii
Tropical		Melanesia, Polynesia

Invaded Habitats
Forests, forest edges, scrub- and woodland.

Description
A much-branched shrub or small tree of 2-20 m height, with a conical shape and with a pale grey bark. The thick, glabrous and glossy leaves are ovate to elliptic, 5-12 cm long and 2.5-5.5 cm wide, dark green above and paler beneath. The lower leaves are usually undulate and strongly spinose at the margins. Small and greenish-white, fragrant flowers of *c.* 8 mm diameter appear in crowded axillary cymes. The drupe-like fruits are bright red, globose, 8-10 mm in diameter, and contain 2-4 seeds of *c.* 8 mm length.

Ecology and Control
In the native range, the plant occurs in woods, hedges and scrub. Numerous cultivars and hybrids have been developed. The plant forms dense thickets on the floor and on trees that change the structure of invaded forests by adding a tall and species poor shrub layer. Native plants are impeded in their growth and regeneration. Seeds are dispersed by frugivorous birds. The species responds to cutting by vigorously resprouting, and it suckers from lateral roots. The juvenile period lasts 5-12 years.

    Small plants and seedlings are pulled or dug out, the roots must be removed. Freshly cut stumps should be treated with a glyphosate herbicide, otherwise the species will resprout. Fruit-bearing plants should be removed first to prevent seed dispersal.

References
121, 215, 607, 924, 946, 947, 948, 949, 1009.

Impatiens glandulifera Royle — Balsaminaceae

LF: Annual herb
SN: *Impatiens roylei* Walp.
CU: Ornamental

Geographic Distribution

Europe		*Australasia*		*Atlantic Islands*
●	Northern		Australia	Cape Verde
●	British Isles	X	New Zealand	Canary, Madeira
●	Central, France			Azores
	Southern	*Northern America*		South Atlantic Isl.
X	Eastern		Canada, Alaska	
	Mediterranean Isl.		Southeastern USA	*Indian Ocean Islands*
		X	Western USA	Mascarenes
Africa		X	Remaining USA	Seychelles
	Northern		Mexico	Madagascar
	Tropical			
	Southern	*Southern America*		*Pacific Islands*
			Tropical	Micronesia
Asia			Caribbean	Galapagos
X	Temperate		Chile, Argentina	Hawaii
N	Tropical			Melanesia, Polynesia

Invaded Habitats
Riparian habitats, moist woods, forest edges.

Description
A large, glabrous annual of 50-200 cm height, with stout and purplish stems, simple or sometimes branched. Leaves are 5-18 cm long and 2.5-7 cm wide, lanceolate to elliptical, and have serrated margins. Inflorescences are axillary racemes with 5-12 deep to pale purplish-pink, rarely white flowers. These are 2.5-4 cm long, zygomorphic, with a sepal-sac of 12-20 mm length that is abruptly contracted to a spur of 2-8 mm length. Fruits are capsules of 15-30 mm length, dehiscing elastically and coiling. They release black seeds of 2-3 mm length.

Ecology and Control
Native habitats include moist sites in forest openings, shrubland, and hedges from 1,800-3,200 m elevation. Where invasive, this large and fast-growing herb forms dense infestations that support only few native species and prevent the establishment of shrubs and trees. Seeds are ejected explosively and easily dispersed by streams over long distances. A dense population can produce up to 30,000 seeds m^{-2}, but the plant does not have a persistent seed bank.

Single plants and smaller infestations can be hand pulled. Larger infestations are mown close to the ground, best shortly before flowering commences. Plants damaged early in the season are able to resprout and still produce new seeds.

References
96, 97, 298, 361, 540, 721, 744, 1008, 1048, 1056, 1230, 1292, 1422.

Imperata cylindrica (L.) P. Beauv.   Poaceae

LF: Perennial herb   CU: Erosion control, ornamental
SN: *Imperata arundinacea* Cirillo

Geographic Distribution

Europe
    Northern
    British Isles
    Central, France
N   Southern
    Eastern
    Mediterranean Isl.

Africa
X   Northern
N   Tropical
N   Southern

Asia
X   Temperate
●   Tropical

Australasia
X   Australia
X   New Zealand

Northern America
    Canada, Alaska
●   Southeastern USA
    Western USA
    Remaining USA
    Mexico

Southern America
X   Tropical
X   Caribbean
X   Chile, Argentina

Atlantic Islands
X   Cape Verde
    Canary, Madeira
    Azores
    South Atlantic Isl.

Indian Ocean Islands
X   Mascarenes
X   Seychelles
X   Madagascar

Pacific Islands
    Micronesia
    Galapagos
●   Hawaii
X   Melanesia, Polynesia

Invaded Habitats
Grass- and woodland, open forests, disturbed sites.

Description
A tufted grass up to 120 cm tall, with unbranched culms and narrow, rigid and glabrous leaves arising from the base and with silky-white plumes. Leaves are hard and stiff, and have prominent midribs. Leaf blades are up to 1 m long, often less, and 3-10 mm wide. Inflorescences are dense and cylindrical panicles of 5-15 cm length, with paired spikelets being surrounded by white silky hairs of 10-15 mm length. Spikelets are lanceolate, 4-5 mm long, contain two florets and have membranous glumes of about the same length as the spikelet.

Ecology and Control
Where native, this grass is found in poorly drained, damp soils and on riverbanks. It is a highly variable species with at least five varieties. It spreads readily by long rhizomes and by seeds. Pieces of rhizomes easily regenerate new plants. It is invasive because it outcompetes native plant species, forms dense stands, provides poor wildlife habitat and is highly flammable. The grass is encouraged by burning, which does little damage to the underground rhizomes. It is, however, not shade tolerant and is outcompeted by shading plants.
    Control measures include spraying glyphosate, dalapon or TCA in autumn. Burning several months before spraying apparently enhances uptake of the herbicide. Planting trees can shade out the grass.

References
161, 169, 240, 241, 580, 669, 670, 715, 750, 791, 999, 1190, 1276, 1279, 1316, 1405, 1406, 1407, 1408.

Ipomoea aquatica Forssk.   Convolvulaceae

LF: Perennial vine
SN: *Ipomoea reptans* Poir.
CU: Food, fodder

Geographic Distribution

Europe	*Australasia*	*Atlantic Islands*
Northern	N Australia	Cape Verde
British Isles	New Zealand	Canary, Madeira
Central, France		Azores
Southern	*Northern America*	South Atlantic Isl.
Eastern	Canada, Alaska	
Mediterranean Isl.	● Southeastern USA	*Indian Ocean Islands*
	X Western USA	Mascarenes
Africa	X Remaining USA	Seychelles
Northern	X Mexico	Madagascar
N Tropical		
Southern	*Southern America*	*Pacific Islands*
	N Tropical	X Micronesia
Asia	N Caribbean	Galapagos
N Temperate	Chile, Argentina	X Hawaii
N Tropical		X Melanesia, Polynesia

Invaded Habitats
Freshwater wetlands, lakes, ponds.

Description
A glabrous aquatic herb with thick, hollow and trailing stems up to 3 m long that root at the nodes. Leaves are variable in shape and size, but usually lanceolate to oblong with basal lobes pointing outwards. Each node has one leaf. Petioles are 3-14 cm long. Leaf blades are 3.5-17 cm long and 1-9 cm wide. Flowers have a corolla of 4-7 cm diameter and are purple, pink, or white with a purple centre. Inflorescences are axillary cymoses with 1-5 flowers and peduncles of 1.5-9 cm length. The fruit is an ovoid to globose capsule of *c.* 1 cm diameter, containing 4-6 densely pubescent seeds.

Ecology and Control
In the native range, this plant grows in shallow water of lakes, seasonal ponds and swamps. The plant grows either as a terrestrial or as a floating aquatic plant. It is a variable species with several distinct races and many cultivars. Where invasive, this profusely branching plant forms dense floating mats of intertwined stems, shading out native aquatic plants and impeding water flow. Stems easily root at nodes and produce new plants when fragmented. Stem fragments are carried by streams.

    Control methods include mechanical removal or chemical treatment with 2,4-D sodium salt, 2,4,5-T or MCPA.

References
581, 715, 989.

Ipomoea cairica (L.) Sweet Convolvulaceae

LF: Perennial herb
SN: *Ipomoea palmata* Forssk.
CU: Ornamental

Geographic Distribution

Europe	*Australasia*	*Atlantic Islands*
Northern	● Australia	X Cape Verde
British Isles	X New Zealand	Canary, Madeira
Central, France		Azores
Southern	*Northern America*	South Atlantic Isl.
Eastern	Canada, Alaska	
Mediterranean Isl.	X Southeastern USA	*Indian Ocean Islands*
	X Western USA	Mascarenes
Africa	Remaining USA	Seychelles
Northern	Mexico	Madagascar
N Tropical		
Southern	*Southern America*	*Pacific Islands*
	X Tropical	Micronesia
Asia	Caribbean	Galapagos
Temperate	X Chile, Argentina	N Hawaii
Tropical		Melanesia, Polynesia

Invaded Habitats
Coastal dunes and bluffs, forest edges, riparian habitats.

Description
A glabrous twining or trailing herbaceous vine with slender stems climbing up to 5 m height. Petioles are up to 5 cm long, the leaf blades are 2.5-8 cm long, palmately lobed with five to seven elliptic-lanceolate lobes. Flowers are purplish or white with a dark centre, and borne solitary or in clusters of a few. The funnel-shaped corolla is 4.5-6 cm wide and 5-8 cm long. Fruits are ovoid capsules of *c.* 1 cm length, containing 2-4 finely pubescent seeds.

Ecology and Control
The exact native range of this plant is obscure. The fast growing plant is invasive because its trailing and climbing stems can completely smother native shrubs and trees, impeding their growth and preventing their regeneration. Stems easily root at nodes. Seeds are dispersed by wind or water.
    Small infestations can be removed manually, but all roots and all stems touching the ground must be removed. Chemical control is done by cutting vines at breast height, laying the lower portions on the ground and spraying them with herbicide.

References
607, 608, 924.

Ipomoea indica (Burm.) Merr.   Convolvulaceae

LF: Perennial vine   CU: Ornamental
SN: *I. acuminata* (Vahl) Roem., *I. congesta* R. Br., *Convolvulus acuminatus* Vahl

Geographic Distribution

Europe	*Australasia*	*Atlantic Islands*
Northern	• Australia	Cape Verde
British Isles	• New Zealand	X Canary, Madeira
Central, France		X Azores
• Southern	*Northern America*	South Atlantic Isl.
Eastern	Canada, Alaska	
Mediterranean Isl.	Southeastern USA	*Indian Ocean Islands*
	Western USA	Mascarenes
Africa	Remaining USA	Seychelles
Northern	Mexico	Madagascar
Tropical		
• Southern	*Southern America*	*Pacific Islands*
	N Tropical	Micronesia
Asia	N Caribbean	Galapagos
Temperate	Chile, Argentina	N Hawaii
N Tropical		X Melanesia, Polynesia

Invaded Habitats
Forest edges, riparian habitats, woodland, seasonal freshwater wetlands.

Description
A twining or prostrate herb with stems of 3-6 m length. The leaves have petioles of 2-18 cm length and ovate to orbicular blades of 5-15 cm length and 3.5-14 cm width. They are densely and soft pubescent beneath and sparsely pubescent above. Leaf margins are entire or lobed. Inflorescences are dense umbel-like cymes with peduncles of 4-20 cm length and linear to lanceolate bracts. The corollas are bright blue or purplish, with a paler centre, and 5-8 cm in diameter. Fruits are globose capsules of 10-13 mm diameter, containing 4-6 seeds of *c.* 5 mm length.

Ecology and Control
In the native range, this herb grows in coastal sites, moist forests and disturbed places. It originates probably from South America, but the exact native range is obscure. The trailing stems lead to dense infestations that crowd out native vegetation by smothering shrubs and trees. Plants in New Zealand and Australia do not form viable seeds. Stems easily root at nodes and the plant spreads by stem fragments. Cut stems resprout vigorously.
    Specific control methods for this species are not available. The same methods may apply as for the previous species.

References
121, 215, 549, 607, 924, 1415.

Iris pseudacorus L.   Iridaceae

LF: Perennial herb   CU: Ornamental
SN: -

Geographic Distribution

Europe
- N   Northern
- N   British Isles
- N   Central, France
- N   Southern
- N   Eastern
- N   Mediterranean Isl.

Africa
- N   Northern
-     Tropical
-     Southern

Asia
- N   Temperate
-     Tropical

Australasia
- ●   Australia
- ●   New Zealand

Northern America
- X   Canada, Alaska
- X   Southeastern USA
- X   Western USA
-     Remaining USA
-     Mexico

Southern America
-     Tropical
-     Caribbean
- X   Chile, Argentina

Atlantic Islands
-     Cape Verde
- N   Canary, Madeira
-     Azores
-     South Atlantic Isl.

Indian Ocean Islands
-     Mascarenes
-     Seychelles
-     Madagascar

Pacific Islands
-     Micronesia
-     Galapagos
-     Hawaii
-     Melanesia, Polynesia

Invaded Habitats
Freshwater wetlands, riparian habitats.

Description
A usually branched perennial herb with a strong rhizome system and slightly compressed stems of 60-150 cm height, bearing several leaves. Rhizomes are 1-4 cm in diameter. The basal leaves are 50-90 cm long and 10-30 mm wide, and have a conspicuous midrib. The large yellow zygomorphic flowers are 7-10 cm in diameter and have pedicels of 2-5 cm length. They are borne in clusters of 5-10. Fruits are cylindrical capsules of 4-8 cm length and with a short beak. They contain numerous dark brown and smooth seeds.

Ecology and Control
This herb grows in sites with a continuously high soil-water content. It occurs in freshwater habitats but also in salt marshes. The plant forms radially spreading clones which become fragmented as they develop. The thick rhizomes grow over the soil or may form firm mats floating in water. Pieces of rhizomes are carried by streams. Seeds are mainly dispersed by water and are able to float for a long time. Established stands of this herb completely eliminate the native vegetation.

Scattered plants can be dug out, all rhizomes must be removed to prevent regrowth. Larger infestations may be removed manually or sprayed with herbicide applied during active growth.

References
215, 1250, 1251.

Jacaranda mimosifolia D. Don
Bignoniaceae

LF: Deciduous tree
SN: *Jacaranda ovalifolia* R. Br.
CU: Ornamental, wood

Geographic Distribution

Europe
- Northern
- British Isles
- Central, France
- Southern
- Eastern
- Mediterranean Isl.

Africa
- Northern
- Tropical
- • Southern

Asia
- Temperate
- Tropical

Australasia
- Australia
- New Zealand

Northern America
- Canada, Alaska
- Southeastern USA
- Western USA
- Remaining USA
- Mexico

Southern America
- N Tropical
- Caribbean
- X Chile, Argentina

Atlantic Islands
- Cape Verde
- Canary, Madeira
- Azores
- South Atlantic Isl.

Indian Ocean Islands
- Mascarenes
- Seychelles
- Madagascar

Pacific Islands
- Micronesia
- Galapagos
- Hawaii
- Melanesia, Polynesia

Invaded Habitats
Bushland, grassland, wooded ravines and riverbanks.

Description
A tree up to 15 m tall with a rounded, spreading crown. Leaves are opposite, 20-80 cm long, and bipinnately compound with 20-40 pairs of pinnae, each pinna having 19-45 leaflets. The dark green leaflets are sharply pointed. Large, bluish to lilac, rarely white flowers of 4-5 cm length are borne in loose and hanging panicles of 20-30 cm length at the ends of branches. They are bell-shaped and usually appear before the leaves. Fruits are woody, broadly oval and flat capsules of 5-7 cm diameter, containing numerous winged seeds of *c.* 10 mm length.

Ecology and Control
A fast growing tree that resprouts easily if damaged. The spreading growth habit and the dense foliage shade out native plants and prevent their regeneration. Little is known on the ecology of this species.

Specific control methods for this species are not available. Seedlings and saplings may be pulled or dug out, roots must be removed to prevent resprouting. Larger trees can be cut and the cut stumps treated with herbicide.

References
549, 792, 1089, 1218.

Jasminum fluminense Vell.   Oleaceae

LF: Evergreen climber   CU: Ornamental
SN: -

Geographic Distribution

Europe	Australasia	Atlantic Islands
Northern	Australia	Cape Verde
British Isles	New Zealand	Canary, Madeira
Central, France		Azores
Southern	*Northern America*	South Atlantic Isl.
Eastern	Canada, Alaska	
Mediterranean Isl.	● Southeastern USA	*Indian Ocean Islands*
	Western USA	N Mascarenes
Africa	Remaining USA	N Seychelles
Northern	Mexico	Madagascar
N Tropical		
Southern	*Southern America*	*Pacific Islands*
	Tropical	Micronesia
Asia	X Caribbean	Galapagos
N Temperate	Chile, Argentina	● Hawaii
Tropical		Melanesia, Polynesia

Invaded Habitats
Forests and forest gaps, woodland, hedges, disturbed sites.

Description
A scrub or woody climber up to 6 m tall, with young stems being densely hairy. The opposite leaves are compound with three leathery and stalked leaflets, the terminal leaflet being up to 7 cm long. Leaflets are broadly ovate, pubescent on both sides, and have pointed tips. White or yellowish, fragrant flowers of 5-7 mm diameter are borne in broad, branched clusters in the axils of leaves. The petals are fused into a narrow tube of 2-2.5 cm length, with the 5-7 lobes being shorter. Fruits are small, round, black and fleshy berries.

Ecology and Control
This fast-growing plant climbs high into the canopy of forests, forms dense infestations that completely enshroud native vegetation and reduce native diversity. Regeneration of native shrubs and trees is suppressed under stands of this vine. Seeds are dispersed by birds and small mammals.
    Specific control methods for this species are not available. Seedlings can be hand pulled, large vines may be controlled by cutting and treating regrowth with herbicides, or cutting the vines at breast height, laying the lower portions on the floor and applying a foliar spray.

References
715.

Jasminum humile L.

Oleaceae

LF: Evergreen shrub
SN: *Jasminum revolutum* Sims
CU: Ornamental

Geographic Distribution

Europe	*Australasia*	*Atlantic Islands*
Northern	Australia	Cape Verde
British Isles	● New Zealand	Canary, Madeira
Central, France		Azores
X Southern	*Northern America*	South Atlantic Isl.
Eastern	Canada, Alaska	
Mediterranean Isl.	Southeastern USA	*Indian Ocean Islands*
	Western USA	Mascarenes
Africa	Remaining USA	Seychelles
Northern	Mexico	Madagascar
Tropical		
X Southern	*Southern America*	*Pacific Islands*
	Tropical	Micronesia
Asia	Caribbean	Galapagos
N Temperate	Chile, Argentina	Hawaii
N Tropical		Melanesia, Polynesia

Invaded Habitats
Forest edges and gaps, scrub- and woodland.

Description
A glabrous shrub of 0.3-3 m height, with spreading branches and alternate leaves. Leaves are simple or compound with 3-9 but mostly five leaflets. These are ovate to elliptic or oblong, dark green above and paler beneath, 6-60 mm long and 2-20 mm wide. Yellow flowers of *c.* 10 mm diameter are borne in axillary cymes with 1-15 flowers. The corolla tube is 8-16 mm long. Fruits are black and glossy berries of 6-8 mm diameter.

Ecology and Control
In the native range, this shrub grows in woods and thickets from 1,100-3,800 m elevation. Where invasive, it forms dense thickets that shade out native vegetation and prevent natural regeneration of forest understorey species. The dense foliage shades out almost all species under its canopies. Little is known on the ecology of this plant as an invasive species.

    Specific control methods for this species are not available. Seedlings and young plants may be hand pulled or dug out. Larger shrubs can be cut and the cut stumps treated with herbicide.

References
549, 1415.

Jasminum polyanthum Franch.   Oleaceae

LF: Evergreen shrub, climber   CU: Ornamental
SN: -

Geographic Distribution

Europe	*Australasia*	*Atlantic Islands*
Northern	Australia	Cape Verde
British Isles	● New Zealand	Canary, Madeira
Central, France		Azores
Southern	*Northern America*	South Atlantic Isl.
Eastern	Canada, Alaska	
Mediterranean Isl.	Southeastern USA	*Indian Ocean Islands*
	Western USA	Mascarenes
Africa	Remaining USA	Seychelles
Northern	Mexico	Madagascar
Tropical		
Southern	*Southern America*	*Pacific Islands*
	Tropical	Micronesia
Asia	Caribbean	Galapagos
N Temperate	Chile, Argentina	Hawaii
Tropical		Melanesia, Polynesia

Invaded Habitats
Forests and forest edges, scrubland.

Description
A woody and twining climber of 1-10 m height, with glabrous branchlets and with opposite, pinnately compound leaves. These have usually 5-7 ovate-lanceolate leaflets of 2-5 cm length and 1-2.5 cm width, and are thin leathery. Petioles are 5-30 mm long. Flowers are very fragrant and borne in terminal and axillary racemes or panicles with 5-50 flowers each. Pedicels are 5-25 mm long. The corolla is white inside and reddish outside, the corolla tube is 15-25 mm long and has five lobes. Fruits are black, somewhat globose berries of 6-11 mm diameter.

Ecology and Control
In the native range, this plant is found from 1,400-3,000 m elevation in ravines, riparian habitats, woods and thickets. Little is known on the ecology of this shrub. It forms dense thickets and smothers native shrubs and trees, impeding their growth and regeneration. Fruits are rarely produced in New Zealand.

    Specific control methods for this species are not available. Seedlings and young plants may be hand pulled or dug out. Larger shrubs can be cut and the cut stumps treated with herbicide.

References
1415.

Juncus acutus L. — Juncaceae

LF: Perennial herb
SN: -
CU: -

Geographic Distribution

Europe
- Northern
- N British Isles
- N Central, France
- N Southern
- N Eastern
- N Mediterranean Isl.

Africa
- N Northern
- Tropical
- N Southern

Asia
- Temperate
- Tropical

Australasia
- ● Australia
- X New Zealand

Northern America
- Canada, Alaska
- Southeastern USA
- X Western USA
- Remaining USA
- Mexico

Southern America
- Tropical
- Caribbean
- X Chile, Argentina

Atlantic Islands
- Cape Verde
- N Canary, Madeira
- N Azores
- South Atlantic Isl.

Indian Ocean Islands
- Mascarenes
- Seychelles
- Madagascar

Pacific Islands
- Micronesia
- Galapagos
- Hawaii
- Melanesia, Polynesia

Invaded Habitats
Coastal flats and saline areas, grassland, riparian habitats, disturbed sites.

Description
A very densely tufted grassoid with numerous erect, stiff and unbranched stems of 20-200 cm height and 2-4 mm diameter, arising from rhizomes, and each having 2-5 leaves. Leaves are dark green, similar to the stems, 25-150 cm long and 2.5-5 mm wide, and tapering to a sharp spine. Flowers are green to reddish brown, sessile and very small, and borne in globose, dense flowerheads. The perianth-segments have a wide margin. Fruits are obovoid to ovoid capsules of 4-6 mm length containing numerous seeds of *c.* 1 mm length.

Ecology and Control
Native habitats of this plant include sandy seashores, drier parts of salt marshes, and other sandy places. It is a highly variable species with numerous varieties. Seeds are dispersed mainly by water. The large and dense tussocks lead to dense patches that support little native vegetation. In infested watercourses, the plant restricts water flow.

Single tussocks may be removed mechanically. Cutting at ground level is used for larger infestations, and a follow-up programme is necessary to control regrowth and emerging seedlings. Effective herbicides are 2,4-D ester or hexazinone.

References
215, 639, 985, 986.

Juncus articulatus L. Juncaceae

LF: Perennial herb   CU: -
SN: *Juncus compressus* Relhan, *Juncus lampocarpus* Ehrh. ex Hoffm.

Geographic Distribution

Europe	*Australasia*	*Atlantic Islands*
N Northern	● Australia	Cape Verde
N British Isles	● New Zealand	N Canary, Madeira
N Central, France		N Azores
N Southern	*Northern America*	South Atlantic Isl.
N Eastern	N Canada, Alaska	
N Mediterranean Isl.	N Southeastern USA	*Indian Ocean Islands*
	N Western USA	Mascarenes
Africa	N Remaining USA	Seychelles
N Northern	Mexico	Madagascar
Tropical		
Southern	*Southern America*	*Pacific Islands*
	Tropical	Micronesia
Asia	Caribbean	Galapagos
N Temperate	Chile, Argentina	Hawaii
N Tropical		Melanesia, Polynesia

Invaded Habitats
Freshwater wetlands, lagoons.

Description
A glabrous grassoid with erect or ascending stems of 15-80 cm height, each having 3-6 leaves. The dark green leaves are 3-30 mm long and 0.5-3 mm wide, long-sheathing at the base. The inflorescence is usually wide, much-branched, with 5-20 flowerheads or more, each containing 5-20 florets. The perianth segments are 2.5-3.5 mm long, ovate, and dark brown to blackish. Fruits are capsules of 2.5-4 mm length. Seeds are ovoid and 0.5-0.6 mm long.

Ecology and Control
In the native range, this plant occurs in damp grass- and heathland, marshes, coastal meadows and dune-slacks, and riparian habitats. It is a highly variable species with several subspecies in the native range, differing in size and growth habit. It is well adapted to fluctuating water levels and competes successfully with native grasses due to the dense clumps. Scattered clumps collate into a continuous matrix, completely eliminating the original vegetation. The plant accumulates a long-lived seed bank. It spreads both by seeds and rhizomes.

Specific control methods for this species are not available. The same methods may apply as for the previous species.

References
215, 382, 1206, 1231, 1415.

Juncus effusus L.   Juncaceae

LF: Perennial herb   CU: Ornamental
SN: -

Geographic Distribution

Europe
- N Northern
- N British Isles
- N Central, France
- N Southern
- N Eastern
- N Mediterranean Isl.

Africa
-     Northern
-     Tropical
-     Southern

Asia
- X Temperate
-     Tropical

Australasia
- ● Australia
- X New Zealand

Northern America
-     Canada, Alaska
- N Southeastern USA
- N Western USA
-     Remaining USA
- X Mexico

Southern America
-     Tropical
-     Caribbean
-     Chile, Argentina

Atlantic Islands
-     Cape Verde
- N Canary, Madeira
- N Azores
-     South Atlantic Isl.

Indian Ocean Islands
-     Mascarenes
-     Seychelles
-     Madagascar

Pacific Islands
-     Micronesia
-     Galapagos
- X Hawaii
-     Melanesia, Polynesia

Invaded Habitats
Freshwater wetlands, wet grassland.

Description
A densely tufted, pale green herb with erect, smooth and glossy stems of 30-150 cm height and 3-5 mm width, forming large tussocks. Sheaths reach 15 cm length or more and are often reddish-brown. The involucral leaf is *c.* 10 cm long and tapering to a sharp apex. The inflorescence is compact or diffuse, with many greenish or pale brown flowers of 2-3 mm length. Fruits are ovoid to globose capsules, usually rounded at the end. Seeds are *c.* 0.5 mm long, minutely ribbed and obliquely ovoid.

Ecology and Control
Where native, this plant is found in wet ground, cypress swamps, shores of ponds, marshes, bogs, riparian habitats, and damp woods. It grows mostly on acid soils. The species is a troublesome weed of pastures and also invades native communities. The plant forms dense patches that suppress native species by shading them out, and becomes dominant over large areas. Single tussocks collate into a continuous cover over time. Seeds are persistent and allow a rapid colonization of disturbed sites.

Cutting at ground level reduces the vigour but a follow-up programme is necessary to treat regrowth and seedlings. Continuous grazing can eliminate the plant from pastures.

References
121, 215, 239, 383, 384, 385, 519, 703, 864, 865, 1083.

Lagarosiphon major (Ridl.) Moss   Hydrocharitaceae

LF: Aquatic perennial herb   CU: Ornamental
SN: -

Geographic Distribution

Europe	*Australasia*	*Atlantic Islands*
X Northern	Australia	Cape Verde
X British Isles	● New Zealand	Canary, Madeira
X Central, France		Azores
X Southern	*Northern America*	South Atlantic Isl.
X Eastern	Canada, Alaska	
Mediterranean Isl.	Southeastern USA	*Indian Ocean Islands*
	Western USA	● Mascarenes
Africa	Remaining USA	Seychelles
Northern	Mexico	Madagascar
Tropical		
N Southern	*Southern America*	*Pacific Islands*
	Tropical	Micronesia
Asia	Caribbean	Galapagos
Temperate	Chile, Argentina	Hawaii
Tropical		Melanesia, Polynesia

Invaded Habitats
Freshwater wetlands, lakes, ponds, slow rivers.

Description
A submerged aquatic plant with branched stems up to 3 m long, rooted in the substrate, and with alternate leaves. These are dark green, linear, 6-30 mm long and 2-4 mm wide, and densely crowded at the ends of the branches. The plant is dioecious. Inflorescences appear in the axils of leaves and have two united bracts. Male inflorescences have numerous flowers, female ones contain only one flower. Male flowers are very small, becoming detached in bud and opening on the water surface. The pinkish female flowers have long slender hypanthia which bring them to the water surface. Fruits are ovoid capsules of 4-5 mm length. Seeds are *c.* 2 mm long.

Ecology and Control
This plant grows best in nutrient rich waters. Where invasive, the plant forms dense weed beds in sheltered sites in water of 2-6.5 m depth. The tall, closely packed stands exclude almost all other species, reducing light levels and affecting water chemistry. The stems reach the water surface and dense mats may attract large herbivorous birds which may adversely affect the native flora. The plant spreads by stem fragmentation. In New Zealand, all plants of this species are female.

     Small infestations may be removed mechanically, but stem fragmentation must be avoided. Chemical control includes spraying with diquat approved for use in aquatic environments.

References
284, 592, 628, 834, 986, 1354, 1415.

Lagurus ovatus L. — Poaceae

LF: Annual herb
SN: -
CU: Ornamental

Geographic Distribution

Europe
- Northern
- X British Isles
- Central, France
- N Southern
- N Eastern
- N Mediterranean Isl.

Africa
- N Northern
- Tropical
- X Southern

Asia
- N Temperate
- Tropical

Australasia
- ● Australia
- X New Zealand

Northern America
- Canada, Alaska
- Southeastern USA
- X Western USA
- Remaining USA
- Mexico

Southern America
- Tropical
- Caribbean
- X Chile, Argentina

Atlantic Islands
- Cape Verde
- N Canary, Madeira
- N Azores
- South Atlantic Isl.

Indian Ocean Islands
- Mascarenes
- Seychelles
- Madagascar

Pacific Islands
- Micronesia
- Galapagos
- Hawaii
- Melanesia, Polynesia

Invaded Habitats
Grassland, coastal marshes, sand dunes.

Description
A grass of 8-60 cm height with erect or ascending, greyish-pubescent, simple or branched stems. Leaves are softly pubescent, long-acute, 4-17 cm long and 5-10 mm wide. The papery ligules are 1.5-2.5 mm long. Inflorescences are very compact ovoid panicles of 1-7 cm length and 5-20 mm width, and very softly and densely hairy. Spikelets are 7-9 mm long (excluding awns), the lemmas are lanceolate and each have an awn of 8-20 mm length.

Ecology and Control
Where native, this grass grows mainly in maritime sands and dry places near the sea. Seeds are dispersed by wind and animals. The grass is invasive because it spreads rapidly and its dense populations compete for water and space with native grasses and forbs, thereby displacing the native vegetation. Little is known on the ecology of this plant.

Specific control methods for this species are not available. Single plants and small patches can be removed manually. Larger infestations may be sprayed with herbicides. Cutting before seeds ripen prevents seed dispersal.

References
121, 215.

Lantana camara L.   Verbenaceae

LF: Evergreen shrub   CU: Ornamental
SN: *Lantana aculeata* L., *Lantana nivea* Vent.

Geographic Distribution

Europe		*Australasia*		*Atlantic Islands*	
	Northern	●	Australia	X	Cape Verde
	British Isles	●	New Zealand	X	Canary, Madeira
	Central, France			●	Azores
●	Southern	*Northern America*		X	South Atlantic Isl.
	Eastern		Canada, Alaska		
	Mediterranean Isl.	●	Southeastern USA	*Indian Ocean Islands*	
		X	Western USA	●	Mascarenes
Africa			Remaining USA	●	Seychelles
	Northern	N	Mexico	X	Madagascar
●	Tropical				
●	Southern	*Southern America*		*Pacific Islands*	
		N	Tropical	X	Micronesia
Asia		N	Caribbean	●	Galapagos
X	Temperate		Chile, Argentina	●	Hawaii
●	Tropical			X	Melanesia, Polynesia

Invaded Habitats
Savanna, bushland, riparian habitats, forests, disturbed sites.

Description
A much-branched, scrambling shrub of 2-5 m height, with scattered recurved prickles on the stems and pubescent young twigs. Leaves are ovate, 5-8 cm long, serrated and usually covered with rough hairs. Flowers are borne in compact and flat flowerheads of 15-25 mm diameter, changing the colour as they fade. The corolla tube is 7-8 mm long. Colours range from pink, yellow to orange, and red. The globose fruits are fleshy drupes, purplish black, and *c.* 5 mm in diameter.

Ecology and Control
A highly variable species with a large number of cultivars differing mainly in flower colours. The shrub grows well on poor soils and regenerates easily from the base after damage. Seeds are dispersed by birds. The dense thickets eliminate native vegetation and transform natural forests into shrubland. The shrub rapidly colonizes disturbed sites. On the Galapagos Islands, dense thickets of this shrub decrease the number of certain bird species by impeding their flight.

Physical removal or burning is often combined with planting desirable trees to shade out shrubs of this plant. Individual plants can be pulled out but the root system must be removed. Cleared areas are rapidly reinfested by seedlings and coppice growth. Effective herbicides are 2,4-D, MCPA, dicamba, triclopyr, glyphosate, or picloram.

References
108, 252, 284, 289, 408, 443, 580, 607, 608, 692, 715, 725, 833, 906, 986, 1089, 1129, 1168, 1189, 1199, 1239, 1259, 1264, 1280, 1313, 1415.

Lavandula stoechas L.   Lamiaceae

LF: Evergreen shrub   CU: Ornamental
SN: -

Geographic Distribution

Europe		*Australasia*		*Atlantic Islands*	
	Northern	●	Australia		Cape Verde
	British Isles	X	New Zealand	N	Canary, Madeira
	Central, France				Azores
N	Southern		*Northern America*		South Atlantic Isl.
	Eastern		Canada, Alaska		
N	Mediterranean Isl.		Southeastern USA		*Indian Ocean Islands*
			Western USA		Mascarenes
Africa			Remaining USA		Seychelles
N	Northern		Mexico		Madagascar
	Tropical				
	Southern		*Southern America*		*Pacific Islands*
			Tropical		Micronesia
Asia			Caribbean		Galapagos
N	Temperate		Chile, Argentina		Hawaii
	Tropical				Melanesia, Polynesia

Invaded Habitats
Grass- and woodland, riparian habitats, rocky places.

Description
A small, densely branched, tomentose shrub up to 1 m tall. The linear to oblong-lanceolate leaves are entire, grey-tomentose, strongly aromatic and 1-4 cm long. Inflorescences are spikes of 2-3 cm length with the upper bracts usually being purple and 1-5 cm long. Flowers are dark purple, zygomorphic and have corollas of 6-8 mm length. Fruits are broad-oblong nutlets of *c.* 2 mm length.

Ecology and Control
This shrub is common in rocky dry places and scrubland within the native range, and several subspecies have been recognized. The shrub fruits prolifically and the long-lived seeds are dispersed mainly by water and wind. It forms dense thickets which eliminate most other vegetation. It establishes well on bare soil and in disturbed sites, persisting and becoming dominant once established. The shrub resprouts vigorously from the base after damage.

    Individual plants can be grubbed and the main roots removed. Slashing or burning in autumn is used to kill topgrowth of large infestations. Regrowth and seedlings can then be sprayed with herbicides such as 2,4-D ester.

References
215, 607, 924, 985, 986.

Leersia oryzoides (L.) Sw.   Poaceae

LF: Perennial herb   CU: -
SN: *Homalocenchrus oryzoides* (L.) H., *Oryza oryzoides* (L.) Brand, *Phalaris oryzoides* L.

Geographic Distribution

Europe		*Australasia*		*Atlantic Islands*	
N	Northern	•	Australia		Cape Verde
N	British Isles		New Zealand		Canary, Madeira
N	Central, France			N	Azores
N	Southern	*Northern America*			South Atlantic Isl.
N	Eastern	N	Canada, Alaska		
N	Mediterranean Isl.	N	Southeastern USA	*Indian Ocean Islands*	
		N	Western USA		Mascarenes
Africa		N	Remaining USA		Seychelles
	Northern	N	Mexico		Madagascar
	Tropical				
	Southern	*Southern America*		*Pacific Islands*	
			Tropical		Micronesia
Asia			Caribbean		Galapagos
N	Temperate		Chile, Argentina		Hawaii
	Tropical				Melanesia, Polynesia

Invaded Habitats
Riparian habitats, freshwater wetlands.

Description
A rhizomatous perennial grass with decumbent culms of 30-120 cm height, rooting at the lower nodes, and with sheaths and leaves having rough downwardly pointed bristles. Ligules are 0.5-1.5 mm long. Leaves are 8-30 cm long and 5-10 mm wide, yellowish-green, flat and long-acute. Inflorescences are panicles of 5-17 cm length, with flexuous branches. The pale green spikelets have short pedicels and are 5-6 mm long. Rhizomes are long and slender.

Ecology and Control
Where native, this grass is found in wet meadows, ditches, riparian habitats, and wet places. It displaces native plants in invaded areas and spreads rapidly by rhizome growth. It builds up dense swards that expand laterally and reduce species richness. The tall size prevents the establishment of many native grasses and forbs.

    Specific control methods for this species are not available. Single plants and small patches may be removed manually, all rhizomes must be removed to prevent regrowth. Larger infestations can be treated with herbicide, or mown and regrowth treated chemically.

References
215, 1107.

Lepidium draba L.   Brassicaceae

LF: Perennial herb
CU: -
SN: *Cardaria draba* (L.) Desv., *Cochlearia draba* L.

Geographic Distribution

Europe		*Australasia*		*Atlantic Islands*	
X	Northern	X	Australia		Cape Verde
X	British Isles	X	New Zealand		Canary, Madeira
X	Central, France				Azores
N	Southern		*Northern America*		South Atlantic Isl.
N	Eastern	●	Canada, Alaska		
N	Mediterranean Isl.		Southeastern USA		*Indian Ocean Islands*
		●	Western USA		Mascarenes
Africa		●	Remaining USA		Seychelles
N	Northern		Mexico		Madagascar
	Tropical				
X	Southern		*Southern America*		*Pacific Islands*
		X	Tropical		Micronesia
Asia			Caribbean		Galapagos
N	Temperate	X	Chile, Argentina		Hawaii
	Tropical				Melanesia, Polynesia

Invaded Habitats
Grass- and rangeland, riparian habitats, disturbed sites.

Description
A stout herb of 20-50 cm height with greyish stems branching above. Plants are glabrous or almost so at the top and densely hairy below. Leaves are bluish green, 1.5-8 cm long and broadly ovate to obovate. The lower leaves are longer and more slender, tapering to a short petiole. Upper leaves are stem clasping. The leaf surfaces are weakly to densely hairy. Numerous white flowers are borne in a somewhat flattened corymb of racemes. Flowers have four petals and are *c.* 2 mm wide. The heart shaped fruits are 3-4 mm long and have distinct beaks on the upper ends. They contain 2-4 dark red-brown seeds of 2-3 mm length. The plant has a deep taproot.

Ecology and Control
A significant agricultural weed that also invades natural plant communities such as prairies and the edges of riparian habitats. It is a variable species and plants in USA are referred to subsp. *draba*. The plant spreads rapidly in disturbed sites and excludes the establishment of native grasses and forbs. The species easily regenerates from the root system if damaged. Seeds are numerous and dispersed by water and soil movement.

Control is difficult once established. Repeated cutting of plants in full flower reduces the plant's vigour and is often combined with spraying 2,4-D, or grazing. Scattered plants can be removed manually but most of the root system must be removed. Further herbicides include MCPA, amitrole, or sulfometuron.

References
133, 898, 917, 918, 986, 1150, 1237.

Lepidium latifolium L.   Brassicaceae

LF: Perennial herb
SN: -
CU: Food

Geographic Distribution

Europe		*Australasia*		*Atlantic Islands*	
N	Northern	X	Australia		Cape Verde
N	British Isles		New Zealand		Canary, Madeira
N	Central, France			N	Azores
N	Southern	*Northern America*			South Atlantic Isl.
N	Eastern	X	Canada, Alaska		
N	Mediterranean Isl.		Southeastern USA	*Indian Ocean Islands*	
		●	Western USA		Mascarenes
Africa		X	Remaining USA		Seychelles
N	Northern	X	Mexico		Madagascar
	Tropical				
	Southern	*Southern America*		*Pacific Islands*	
			Tropical		Micronesia
Asia			Caribbean		Galapagos
N	Temperate		Chile, Argentina	X	Hawaii
N	Tropical				Melanesia, Polynesia

Invaded Habitats
Grassland, riparian habitats, freshwater wetlands, coastal estuaries and marshes.

Description
A 50-150 cm tall, glabrous and erect herb that is much-branched above. Basal and lower stem leaves are ovate, up to 30 cm long, toothed at margins, and have long petioles. Upper leaves are sessile, ovate to lanceolate, and entire. Flowers appear in large, dense and pyramidal panicles. Flowers are whitish and have four petals of 1.8-2.5 mm length. The central root can become woody and enlarged to 8-12 cm diameter.

Ecology and Control
A flood tolerant herb that survives extended periods of flooding. It adapts to flooding by developing a shallow root system and by adventitious root formation on the stem base. The species is highly variable in the native range with regard to flower colour and degree of leaf pubescence. It spreads by seeds and root fragments carried by streams. It is invasive because it forms dense thickets that exclude almost all other species and affect wildlife habitats. Dense infestations on riverbanks may destabilize the soil and promote soil erosion.

Scattered plants can be pulled or dug out, roots must be removed to prevent regrowth. Effective herbicides are 2,4-D or chlorsulfuron.

References
118, 133, 237, 350, 877, 1448, 1449.

Leptospermum laevigatum (Gaertn.) Muell.   Myrtaceae

LF: Evergreen shrub, tree   CU: Erosion control, ornamental, shelter
SN: -

Geographic Distribution

Europe
    Northern
    British Isles
    Central, France
    Southern
    Eastern
    Mediterranean Isl.

Africa
    Northern
    Tropical
● Southern

Asia
    Temperate
    Tropical

Australasia
  ● N Australia
  X New Zealand

Northern America
    Canada, Alaska
    Southeastern USA
  X Western USA
    Remaining USA
    Mexico

Southern America
    Tropical
    Caribbean
    Chile, Argentina

Atlantic Islands
    Cape Verde
    Canary, Madeira
    Azores
    South Atlantic Isl.

Indian Ocean Islands
    Mascarenes
    Seychelles
    Madagascar

Pacific Islands
    Micronesia
    Galapagos
● Hawaii
    Melanesia, Polynesia

Invaded Habitats
Heath- and scrubland, grassland, coastal dunes.

Description
A grey-green shrub or small tree of 2-10 m height, with spreading, rigid and fissured stems, and a more or less smooth trunk. The bark is flaking in long thin strips. Leaves are alternate, obovate-oblong, dull grey-green, 15-30 mm long and 6-12 mm wide, and have three veins. The white flowers are sessile and solitary in the axils of leaves, 15-22 mm in diameter, and have numerous stamens. Fruits are woody, flat capsules of 7-8 mm length and with 8-10 valves, releasing numerous small seeds.

Ecology and Control
A native of coastal heath communities, this shrub is tolerant of salt spray and invades coastal vegetation. It can form extensive and dense thickets that displace the native vegetation. The masses of fine roots in the top soil allow an efficient exploitation of soil moisture and competition with native plants. Fruits remain on the tree until fire or other damage kills it, leading to mass release of seeds. These are dispersed by wind and water.
    Control includes cutting trees and burning the area about 4 years later to kill emerging seedlings.

References
121, 137, 183, 215, 284, 549, 1089.

Leucaena leucocephala (Lam.) de Wit Fabaceae

LF: Evergreen shrub, tree
SN: *Leucaena glauca* Benth.
CU: Fodder, forage

Geographic Distribution

Europe	Australasia	Atlantic Islands
Northern	● Australia	X Cape Verde
British Isles	New Zealand	X Canary, Madeira
Central, France		Azores
● Southern	*Northern America*	South Atlantic Isl.
Eastern	Canada, Alaska	
Mediterranean Isl.	● Southeastern USA	*Indian Ocean Islands*
	Western USA	● Mascarenes
Africa	Remaining USA	● Seychelles
Northern	N Mexico	Madagascar
● Tropical		
● Southern	*Southern America*	*Pacific Islands*
	●N Tropical	X Micronesia
Asia	N Caribbean	● Galapagos
● Temperate	Chile, Argentina	● Hawaii
● Tropical		X Melanesia, Polynesia

Invaded Habitats
Coastal heath- and scrubland.

Description
A shrub or tree of 5-20 m height with a trunk up to 10 cm in diameter. Leaves are alternate and bipinnately compound, have 2-6 pairs of pinnae and a rachis of 6-14 cm length. Each pinna bears 5-20 pairs of oblong to lanceolate leaflets of 7-15 mm length and 1.5-4 mm width. Petioles are 2-5 cm long. Greenish to creamy-white flowers are borne in globular flowerheads of *c.* 2 cm diameter, with peduncles of 2-4 cm length. Fruits are thin and flat pods of 12-18 cm length and 1.2-2 cm width. Pods appear in clusters of 15-20. Each pod contains 15-25 elliptical, shiny brown seeds of 7-8 mm length and *c.* 4 mm width.

Ecology and Control
This plant is evergreen unless moisture is limited seasonally. The shrub forms extensive and dense thickets displacing the original vegetation and reducing species richness. The tree is deep-rooted and the taproot may grow rapidly, e.g. more than 2 m in 1-year-old plants. The plant coppices freely and resprouts from cuttings, stumps and root collars. It is a nitrogen-fixing species that increases soil fertility levels. Seed production is prolific.

Seedlings and small plants can be pulled or dug out. Larger stands can be slashed and the regrowth treated with herbicide. If cut, the cut stumps need to be treated with herbicide to prevent regrowth.

References
284, 486, 549, 607, 767, 987, 1171, 1239, 1313.

Leycesteria formosa Wall.   Caprifoliaceae

LF: Deciduous shrub
SN: -
CU: Ornamental

Geographic Distribution

Europe	*Australasia*	*Atlantic Islands*
Northern	● Australia	Cape Verde
X British Isles	X New Zealand	X Canary, Madeira
Central, France		● Azores
X Southern	*Northern America*	South Atlantic Isl.
Eastern	Canada, Alaska	
Mediterranean Isl.	Southeastern USA	*Indian Ocean Islands*
	Western USA	Mascarenes
Africa	Remaining USA	Seychelles
Northern	Mexico	Madagascar
Tropical		
Southern	*Southern America*	*Pacific Islands*
	Tropical	Micronesia
Asia	Caribbean	Galapagos
N Temperate	Chile, Argentina	Hawaii
Tropical		Melanesia, Polynesia

Invaded Habitats
Riparian habitats, scrub- and woodland.

Description
A semi-woody glabrous shrub of 1-3 m height, with hollow stems, glabrous or with short purple glandular hairs. Leaves are opposite, broad-ovate to ovate-lanceolate, 4-17 cm long and 2-6 cm wide, and have lateral veins in four pairs. The petiole is 7-15 mm long. Flowers are borne in drooping spikes of 3-10 cm length, the flowers being sessile in the axils of purplish bracts. Spikes are terminal and axillary. The white to purplish corolla is funnel-shaped and 1.5-2 cm long. Fruits are ovoid, red to purple berries of 7-10 mm diameter, covered with short glandular hairs, and containing numerous small seeds. These are 1.1-1.5 mm long.

Ecology and Control
Where native, this shrub grows in shady forests and riparian thickets, and is found up to 3,500 m elevation. Fruits are abundantly produced and seeds dispersed by birds, water and mammals. The shrub is invasive because it forms dense thickets that shade out all other vegetation, and thickets rapidly expand laterally. Stems touching the ground become rooted. Native shrubs and trees are impeded in their growth and regeneration.

Seedlings and smaller plants can be hand pulled or dug out, all crowns and layered stems must be removed. Dense infestations are slashed in winter and regrowth controlled in spring. Larger shrubs are effectively killed by applying herbicide with the drill-fill method, best before flowering starts.

References
121, 215, 607, 924.

Ligustrum lucidum Aiton   Oleaceae

LF: Semi-deciduous shrub   CU: Ornamental
SN: -

Geographic Distribution

Europe		*Australasia*		*Atlantic Islands*
	Northern	●	Australia	Cape Verde
	British Isles	●	New Zealand	Canary, Madeira
	Central, France			Azores
	Southern	*Northern America*		South Atlantic Isl.
	Eastern		Canada, Alaska	
	Mediterranean Isl.		Southeastern USA	*Indian Ocean Islands*
			Western USA	Mascarenes
Africa			Remaining USA	Seychelles
	Northern		Mexico	Madagascar
	Tropical			
X	Southern	*Southern America*		*Pacific Islands*
			Tropical	Micronesia
Asia			Caribbean	Galapagos
N	Temperate	X	Chile, Argentina	Hawaii
	Tropical			Melanesia, Polynesia

Invaded Habitats
Forests and forest edges, coastal cliffs.

Description
A shrub or small tree of 3-15 m height or more, with spreading branches and an evergreen to deciduous foliage. Leaves are ovate to elliptic or ovate-lanceolate, simple and glabrous above, 6-17 cm long and 3-8 cm wide, with 6-8 pairs of conspicuous veins on each surface. Petioles are 1-3 cm long. Flowers are borne in panicles of 8-20 cm length and about the same width at the ends of branches. Flowers are sessile or almost so and have corollas of 4-5 mm diameter. Fruits are oblong and dark blue to black berries of 7-10 mm length and 4-6 mm width. Each berry contains 1-3 seeds of *c.* 7 mm length.

Ecology and Control
Evergreen plants of this species have been described as var. *lucidum*, whereas deciduous plants as var. *latifolium*. It is a shade tolerant species that can invade closed forests wherever gaps occur. Once established, it persists and forms dense understorey thickets that smother the native ground flora, shrubs and trees. It germinates in the shade and seedlings rapidly grow to shrubs once the canopy is opened. Damaged plants resprout from the root crown. Seeds are dispersed by birds.

Seedlings and young plants can be pulled or dug out. Control should aim at removing fruit bearing plants first. Burning infested areas in dry winters proved to be successful in Australia. Effective herbicides are 2,4,5-T ester, glyphosate, triclopyr, or dicamba.

References
121, 607, 924, 1263, 1415.

Ligustrum robustum (Rox.) Blume — Oleaceae

LF: Evergreen shrub
SN: -
CU: Ornamental

Geographic Distribution

Europe	*Australasia*	*Atlantic Islands*
Northern	Australia	Cape Verde
British Isles	New Zealand	Canary, Madeira
Central, France		Azores
Southern	*Northern America*	South Atlantic Isl.
Eastern	Canada, Alaska	
Mediterranean Isl.	● Southeastern USA	*Indian Ocean Islands*
	Western USA	● Mascarenes
Africa	Remaining USA	Seychelles
Northern	Mexico	Madagascar
Tropical		
Southern	*Southern America*	*Pacific Islands*
	Tropical	Micronesia
Asia	Caribbean	Galapagos
N Temperate	Chile, Argentina	Hawaii
N Tropical		Melanesia, Polynesia

Invaded Habitats
Forests and forest edges.

Description
A shrub up to 5 m tall, with arched stems and branchlets covered with small white dots. The opposite leaves are entire, nearly glabrous, and ovate to ovate-lanceolate. The white flowers are almost sessile, and borne in panicles at the ends of branches. The bracts are linear and deciduous.

Ecology and Control
A shrub with two subspecies, subsp. *robustum* and subsp. *walkeri*. The invasive plants are referred to subsp. *walkeri*. The rapidly growing shrub is highly shade tolerant and forms a dominant shrub layer in invaded forests, threatening the regeneration of native species. Dense stands change the structure and light conditions and affect wildlife. Fruits are abundantly produced and seeds dispersed by birds. Seedlings may form a dense and carpet-like ground cover. Cut plants resprout vigorously from the base.

Specific control methods for this species are not available. The same methods may apply as for other *Ligustrum* species.

References
284, 723, 724, 769, 770, 1239.

Ligustrum sinense Lour.   Oleaceae

LF: Evergreen shrub   CU: Ornamental
SN: *Ligustrum indicum* (Lour.) Merr., *Ligustrum microcarpum* Kaneh. & Sasaki

Geographic Distribution

Europe	*Australasia*	*Atlantic Islands*
Northern	● Australia	Cape Verde
British Isles	● New Zealand	Canary, Madeira
Central, France		Azores
Southern	*Northern America*	South Atlantic Isl.
Eastern	Canada, Alaska	
Mediterranean Isl.	● Southeastern USA	*Indian Ocean Islands*
	Western USA	Mascarenes
Africa	Remaining USA	Seychelles
Northern	Mexico	Madagascar
Tropical		
X Southern	*Southern America*	*Pacific Islands*
	Tropical	Micronesia
Asia	Caribbean	Galapagos
N Temperate	Chile, Argentina	Hawaii
N Tropical		Melanesia, Polynesia

Invaded Habitats
Forests and woodland, coastal cliffs.

Description
A stout and much-branched shrub up to 4 m tall, with spreading branches and usually pubescent branchlets. Leaves have short petioles and are elliptic to elliptic-oblong, 3-7 cm long and 1-3 cm wide, densely villous or sparsely pubescent, and dull green above. Flowers appear in loose and pubescent panicles of 5-11 cm length and 3-8 cm width in the axils of leaves or at the ends of branches. Flowers have short pedicels and corollas of 3.5-5.5 mm length. Fruits are subglobose berries of 5-8 mm diameter.

Ecology and Control
In the native range, this shrub grows in valleys, along streams, in mixed forests and ravines from 200-2,700 m elevation. Several varieties are widely used as ornamentals. The shrub is a short-lived forest pioneer species establishing in disturbed sites. Where invasive, it forms impenetrable thickets and thus crowds out native vegetation. It displaces the native shrub layer of invaded forests and prevents regeneration of native species. It is a prolific fruit producer, and seeds are dispersed by birds. The shrub suckers from roots after damage.

Seedlings and small plants can be pulled or dug out. Large plants are cut and the cut stumps treated with herbicide. Effective herbicides are glyphosate, triclopyr, or 2,4-D plus picloram. Control should aim at removing fruit-bearing plants first.

References
607, 715, 765, 924, 1263.

Ligustrum vulgare L.   Oleaceae

LF: Evergreen shrub
SN: *Ligustrum insulare* Decne.
CU: Ornamental

Geographic Distribution

Europe		*Australasia*		*Atlantic Islands*	
N	Northern	●	Australia		Cape Verde
N	British Isles	X	New Zealand		Canary, Madeira
N	Central, France			X	Azores
N	Southern	*Northern America*			South Atlantic Isl.
N	Eastern		Canada, Alaska		
N	Mediterranean Isl.	●	Southeastern USA	*Indian Ocean Islands*	
		X	Western USA		Mascarenes
Africa			Remaining USA		Seychelles
N	Northern		Mexico		Madagascar
	Tropical				
X	Southern	*Southern America*		*Pacific Islands*	
			Tropical		Micronesia
Asia			Caribbean		Galapagos
N	Temperate		Chile, Argentina		Hawaii
	Tropical				Melanesia, Polynesia

Invaded Habitats
Riparian habitats, forest edges.

Description
A stout and much-branched shrub up to 5 m tall, with slender and spreading branches and a smooth, grey bark. Young branches are pubescent and brownish. Leaves are oblong-ovate to lanceolate, 3-6 cm long and 1-2 cm wide, glabrous, obtuse or acute, and have petioles of 3-10 mm length. White to creamish and fragrant flowers are borne in dense panicles of 3-6 cm length. The anthers are exceeding the tube, and the flowers have four petals. Fruits are sub-globose to ovoid black berries of 6-8 mm length. Seeds are obovoid to ellipsoid and 4-5 mm long.

Ecology and Control
A frequent shrub of open forests, thickets, riparian forests, wood margins and scrub in the native range. Where invasive, it forms dense and impenetrable thickets that crowd out native species. The shrub produces numerous fruits that are eaten by birds, dispersing the seeds to new places. It also spreads by suckers. The species is somewhat calcicole and grows best on base-rich soils.

Once established, the plant is difficult to control. Smaller plants can be dug out, larger plants can be cut and the cut stumps treated with glyphosate.

References
215, 691.

Litsea glutinosa (Lour.) Rob.   Lauraceae

LF: Evergreen tree   CU: -
SN: *Litsea laurifolia* (Jacq.) Bailey, *Tetranthera laurifolia* Jacq.

Geographic Distribution

Europe	*Australasia*	*Atlantic Islands*
Northern	N Australia	Cape Verde
British Isles	New Zealand	Canary, Madeira
Central, France		Azores
Southern	*Northern America*	South Atlantic Isl.
Eastern	Canada, Alaska	
Mediterranean Isl.	Southeastern USA	*Indian Ocean Islands*
	Western USA	● Mascarenes
Africa	Remaining USA	Seychelles
Northern	Mexico	Madagascar
Tropical		
● Southern	*Southern America*	*Pacific Islands*
	Tropical	Micronesia
Asia	Caribbean	Galapagos
N Temperate	Chile, Argentina	Hawaii
N Tropical		Melanesia, Polynesia

Invaded Habitats
Forests and forest edges, moist gullies.

Description
Usually a small tree with simple and alternate leaves. These have petioles, are elliptic to obovate, entire, the leaf blades being 7-28 cm long and 3-19 cm wide, and are often white pubescent on the upper surface. The small flowers are white, cream, green or yellowish, and borne in racemes or umbels in the axils of leaves or at the ends of branches. The plant has male and female flowers. Fruits are black and fleshy, indehiscent, globular and 8-11 mm in diameter.

Ecology and Control
This tree grows in monsoon forests, riparian habitats and open woodland within the native range. It is invasive because it spreads rapidly and builds up dense thickets that displace native vegetation. Little is known on the ecology of this species.

    Specific control methods for this species are not available. Seedlings and saplings may be pulled or dug out, larger plants cut and the cut stumps treated with herbicide.

References
549, 1089, 1239.

Lolium perenne L.

Poaceae

LF: Perennial herb
SN: *Lolium boucheanum* Kunth
CU: Erosion control, ornamental

Geographic Distribution

Europe	*Australasia*	*Atlantic Islands*
N Northern	● Australia	Cape Verde
N British Isles	● New Zealand	N Canary, Madeira
N Central, France		N Azores
N Southern	*Northern America*	South Atlantic Isl.
N Eastern	X Canada, Alaska	
N Mediterranean Isl.	X Southeastern USA	*Indian Ocean Islands*
	X Western USA	X Mascarenes
Africa	Remaining USA	Seychelles
N Northern	Mexico	Madagascar
Tropical		
X Southern	*Southern America*	*Pacific Islands*
	Tropical	Micronesia
Asia	Caribbean	Galapagos
N Temperate	X Chile, Argentina	X Hawaii
N Tropical		Melanesia, Polynesia

Invaded Habitats
Heath- and shrubland, riparian habitats, freshwater wetlands, coastal beaches.

Description
A grass with 8-90 cm tall stems that have 2-4 nodes. Leafblades are 5-30 cm long and 2-4 mm wide, glabrous and shiny beneath. Ligules are 1.5-2.5 mm long. Inflorescences are spikes of 3-30 cm length with a slender and flexuose rachis. Spikelets have 3-14 florets, and are 5-23 mm long and 1-7 mm wide. The glumes are lanceolate or narrowly oblong. Lemmas are oblong to oblong-lanceolate and mostly without an awn.

Ecology and Control
In Europe, this grass is frequently found in grassland, pastures, meadows and disturbed places. Where invasive, it spreads quickly and forms dense swards that displace native grass and forb species and reduce species richness. The grass is highly competitive.

Specific control methods for this species are not available. Single plants can be hand pulled or dug out. Larger populations may be sprayed with herbicides.

References
215, 642, 1415.

Lonicera japonica Thunb.   Caprifoliaceae

LF: Semi-evergreen climber   CU: Ornamental
SN: -

Geographic Distribution

	Europe		*Australasia*		*Atlantic Islands*
	Northern		● Australia		Cape Verde
X	British Isles		● New Zealand	X	Canary, Madeira
●	Central, France			X	Azores
	Southern		*Northern America*		South Atlantic Isl.
	Eastern		Canada, Alaska		
	Mediterranean Isl.	●	Southeastern USA		*Indian Ocean Islands*
		X	Western USA		Mascarenes
Africa		●	Remaining USA		Seychelles
	Northern		Mexico		Madagascar
X	Tropical				
X	Southern		*Southern America*		*Pacific Islands*
			Tropical		Micronesia
Asia			Caribbean		Galapagos
N	Temperate	X	Chile, Argentina	●	Hawaii
	Tropical				Melanesia, Polynesia

Invaded Habitats
Riparian forests, grass- and heathland, forest openings, edges of wetlands.

Description
A woody vine or twining shrub reaching 10 m height in the presence of support, with pubescent branchlets. Leaves are ovate to oblong-ovate, sometimes lobed, 3-8 cm long and 2-4 cm wide, pubescent when young and with ciliate margins. Fragrant white and zygomorphic flowers grow in pairs, turning yellow with age. The peduncles are 5-10 mm long, the corollas 3-5 cm long. The corolla is white tinged with purple and has a narrow tube. Fruits are black to purple berries of 5-7 mm diameter, containing a few seeds of *c.* 2 mm diameter.

Ecology and Control
Several cultivars of this plant have become naturalized. The vine spreads rapidly and overtops and smothers small trees and shrubs. It grows as a ground layer on forest floors and quickly climbs into canopies in tree gaps where light is increased. It often forms a dense curtain of vines on forest edges and displaces the understorey shrub layer. Stems easily root at nodes. Seeds are dispersed by fruit eating birds.

    Whereas small plants can be hand-pulled, established vines are effectively controlled by a foliar application of glyphosate shortly after the first killing frost. Other effective herbicides are bromacil, hexazinone, or picloram. Pulling, cutting or mowing may stimulate dense resprouting. If plants are dug out, all crowns and rooting stems must be removed.

References
57, 68, 215, 216, 338, 607, 715, 765, 1134, 1137, 1146, 1240, 1415, 1419, 1420.

Lonicera maackii (Rupr.) Maxim. Caprifoliaceae

LF: Deciduous shrub
SN: -
CU: Ornamental

Geographic Distribution

Europe	*Australasia*	*Atlantic Islands*
Northern	Australia	Cape Verde
British Isles	New Zealand	Canary, Madeira
Central, France		Azores
Southern	*Northern America*	South Atlantic Isl.
Eastern	Canada, Alaska	
Mediterranean Isl.	Southeastern USA	*Indian Ocean Islands*
	Western USA	Mascarenes
Africa	● Remaining USA	Seychelles
Northern	Mexico	Madagascar
Tropical		
Southern	*Southern America*	*Pacific Islands*
	Tropical	Micronesia
Asia	Caribbean	Galapagos
N Temperate	Chile, Argentina	Hawaii
Tropical		Melanesia, Polynesia

Invaded Habitats
Forests and forest edges, grassland, disturbed sites.

Description
An upright and many-stemmed shrub growing either tall and tree-like or short and densely branched, up to 5 m tall, with pubescent branchlets and small and ovoid winter-buds. Leaves are ovate-elliptic to ovate-lanceolate, 5-8 cm long, dark green above and light green beneath, pubescent on the veins, and have petioles of 3-5 mm length. The fragrant flowers are white, becoming yellowish, and have corollas of 15-20 mm length. The peduncles are shorter than petioles. Fruits are dark red berries with a few seeds each.

Ecology and Control
The growth form of this shrub depends on the habitat: in forests it grows as a tall tree-like shrub whereas in open and sunny habitats, it is shorter and compact. The plant establishes well in sunny places and transforms native prairies into scrub. In forests, it reduces native plant diversity, prevents the growth of tree seedlings, and becomes dominant in the shrub layer. The shrub easily resprouts after fire or other damage. Seeds are dispersed by birds.

Young plants can be hand pulled. Removal of dense infestations will stimulate seed germination. Adult plants can be cut and the stumps treated with glyphosate. In shady habitats, the plant can be killed by repeated clipping. Seeds in the soil are not long-lived and removal of emerging seedlings over a few years should eradicate the population.

References
315, 477, 615, 780, 781, 782, 783, 784, 785, 945, 1409.

Lonicera morrowii Gray   Caprifoliaceae

LF: Deciduous shrub   CU: Ornamental
SN: *Lonicera insularis* Nakai

Geographic Distribution

Europe	*Australasia*	*Atlantic Islands*
Northern	Australia	Cape Verde
British Isles	New Zealand	Canary, Madeira
Central, France		Azores
Southern	*Northern America*	South Atlantic Isl.
Eastern	Canada, Alaska	
Mediterranean Isl.	● Southeastern USA	*Indian Ocean Islands*
	Western USA	Mascarenes
Africa	● Remaining USA	Seychelles
Northern	Mexico	Madagascar
Tropical		
Southern	*Southern America*	*Pacific Islands*
	Tropical	Micronesia
Asia	Caribbean	Galapagos
N Temperate	Chile, Argentina	Hawaii
Tropical		Melanesia, Polynesia

Invaded Habitats
Riparian habitats, forests and forest edges, grassland.

Description
A dense and wide-spreading shrub reaching 2 m height. Young branches are pubescent. The elliptic to ovate-oblong leaves are 3-5 cm long, rounded at the base, almost glabrous above and pubescent beneath, and ending in a sharp point. Petioles are shorter than the peduncles. Flowers are white becoming yellow with age, *c.* 15 mm long, and borne in pairs on pubescent peduncles of 5-15 mm length. The upper lip of the corolla is divided to the base. Fruits are dark red berries, each having a few seeds.

Ecology and Control
This vigorously growing shrub establishes well in areas of sparse vegetation. Once established, it persists and becomes the dominant species, building up dense thickets that crowd out native plants. The dense foliage reduces light levels and shades out the ground flora. Seeds are dispersed by birds. Damaged shrubs resprout vigorously.

    In fire adapted communities, repeated prescribed burning in spring is used to kill seedlings. Seedlings can also hand pulled when the soil is moist. Large plants are cut and the cut stumps treated with glyphosate. Follow-up programmes are necessary to control seedlings and regrowth.

References
571, 945.

Lonicera tatarica L. Caprifoliaceae

LF: Deciduous shrub
SN: -

CU: Ornamental

Geographic Distribution

Europe	*Australasia*	*Atlantic Islands*
Northern	Australia	Cape Verde
British Isles	New Zealand	Canary, Madeira
Central, France		Azores
N Southern	*Northern America*	South Atlantic Isl.
N Eastern	● Canada, Alaska	
Mediterranean Isl.	Southeastern USA	*Indian Ocean Islands*
	X Western USA	Mascarenes
Africa	● Remaining USA	Seychelles
Northern	Mexico	Madagascar
Tropical		
Southern	*Southern America*	*Pacific Islands*
	Tropical	Micronesia
Asia	Caribbean	Galapagos
N Temperate	Chile, Argentina	Hawaii
Tropical		Melanesia, Polynesia

Invaded Habitats
Riparian habitats, forests, grassland.

Description
An upright and compact, glabrous shrub growing to 4 m height, with older branches being grey. The ovate to ovate-lanceolate leaves are dark green above and light green beneath, 3-8 cm long and 1.5-4.5 cm wide, ending in a sharp point, and have petioles of 2-6 mm length. Flowers are borne on slender peduncles of 15-30 mm length, are pink to white with two-lipped corollas of 15-20 mm length. Fruits are globose yellow, orange or red berries.

Ecology and Control
This vigorously growing shrub builds up dense stands that shade out other species, displacing native shrubs and trees and impeding forest regeneration. Seeds are dispersed by birds. The shrub may hybridize with *L. morrowii* if both are present in the same area. Little is known on the ecology of this shrub, although it may be similar to *L. maackii* and *L. morrowii*.

Specific control methods for this species are not available. The same methods as for other *Lonicera* species may apply.

References
571, 945, 1396, 1437.

Lotus corniculatus L. Fabaceae

LF: Perennial herb CU: Fodder
SN: *Lotus ambiguus* Besser ex Spreng., *Lotus balticus* Miniaer

Geographic Distribution

Europe			*Australasia*		*Atlantic Islands*
N	Northern		● Australia		Cape Verde
N	British Isles		X New Zealand		Canary, Madeira
N	Central, France				N Azores
N	Southern		*Northern America*		South Atlantic Isl.
N	Eastern		X Canada, Alaska		
N	Mediterranean Isl.		Southeastern USA		*Indian Ocean Islands*
			X Western USA		Mascarenes
Africa			● Remaining USA		Seychelles
N	Northern		Mexico		Madagascar
N	Tropical				
	Southern		*Southern America*		*Pacific Islands*
			Tropical		Micronesia
Asia			Caribbean		Galapagos
N	Temperate		X Chile, Argentina		Hawaii
N	Tropical				Melanesia, Polynesia

Invaded Habitats
Grassland, riparian habitats, freshwater wetlands, coastal beaches.

Description
A perennial herb with procumbent to ascending, glabrous or pubescent stems up to 50 cm length, either sprawling or growing erect. Leaves are compound with five leaflets. These are lanceolate to suborbicular, 4-18 mm long and 1-10 mm wide, and have minute stipules. Flowers are bright yellow to orange, often with red strikes, and appear in umbel-like flowerheads of 2-8 flowers. The corolla is 8-18 mm long. Fruits are cylindrical and straight pods of 15-30 mm length and 2-3 mm width, containing many rounded seeds of 1.3-1.5 mm length. The plant has a taproot reaching 1 m depth.

Ecology and Control
Where native, this plant grows mainly in meadows, grass- and heathland, and disturbed sites. It is a widespread variable complex in Europe with many closely related taxa. The plant can form thick and dense mats, crowding out native vegetation. It also forms dense fibrous roots in the topsoil. Burning increases seed germination and facilitates the establishment of new seedlings. It is nitrogen-fixing and increases soil fertility levels, thereby affecting the floristic composition of the vegetation.

An effective manual control is mowing more than once every 3 weeks. Effective chemical control is done by spraying 2,4-D-mecoprop, dicamba, MCPA, or clopyralid. Dalapon applied as a pre-emergence herbicide reduces seed germination. Seedlings are killed by atrazine or bromacil.

References
215, 633, 1065, 1306.

Lotus uliginosus Schkuhr   Fabaceae

LF: Perennial herb   CU: Forage
SN: *Lotus decumbens* Poir.

Geographic Distribution

Europe
- N   Northern
- N   British Isles
- N   Central, France
- N   Southern
- N   Eastern
- N   Mediterranean Isl.

Africa
- N   Northern
-     Tropical
-     Southern

Asia
-     Temperate
-     Tropical

Australasia
- ●   Australia
-     New Zealand

Northern America
-     Canada, Alaska
-     Southeastern USA
- X   Western USA
-     Remaining USA
-     Mexico

Southern America
-     Tropical
-     Caribbean
-     Chile, Argentina

Atlantic Islands
-     Cape Verde
- N   Canary, Madeira
- N   Azores
-     South Atlantic Isl.

Indian Ocean Islands
-     Mascarenes
-     Seychelles
-     Madagascar

Pacific Islands
-     Micronesia
-     Galapagos
- X   Hawaii
-     Melanesia, Polynesia

Invaded Habitats
Forests, riparian habitats, freshwater wetlands, coastal beaches.

Description
A 30-100 cm tall herb with erect or ascending and hollow stems. Leaves are glaucous beneath, compound and have obovate leaflets of 8-25 mm length and 3-15 mm width. Flowerheads have 5-12 flowers on short pedicels. The corolla is 10-18 mm long. Fruits are pods of 15-35 mm length and 2-3 mm width.

Ecology and Control
Where native, this plant grows commonly in marshes and wet grasslands. The plant is nitrogen-fixing and thus increasing soil fertility levels, which may change the floristic composition of the invaded vegetation. The dense growth habit crowds out native plants and leads to pure stands that prevent forest regeneration.

Specific control methods for this plant are not available. The same methods as for the previous species may apply.

References
123, 124, 215, 629, 1377.

Lupinus arboreus Sims   Fabaceae

LF: Evergreen shrub   CU: Erosion control, ornamental
SN: -

Geographic Distribution

Europe	*Australasia*	*Atlantic Islands*
Northern	● Australia	Cape Verde
X British Isles	● New Zealand	Canary, Madeira
Central, France		Azores
Southern	*Northern America*	South Atlantic Isl.
Eastern	Canada, Alaska	
Mediterranean Isl.	Southeastern USA	*Indian Ocean Islands*
	● N Western USA	Mascarenes
Africa	Remaining USA	Seychelles
Northern	Mexico	Madagascar
Tropical		
Southern	*Southern America*	*Pacific Islands*
	Tropical	Micronesia
Asia	Caribbean	Galapagos
X Temperate	X Chile, Argentina	Hawaii
Tropical		Melanesia, Polynesia

Invaded Habitats
Coastal grass- and heathland, coastal scrub and dunes.

Description
A much-branched, glabrous or silvery-hairy shrub of 1-3 m height. The compound leaves have 5-12 leaflets, each 2-6 cm long and 5-10 mm wide, and with petioles of 2-4 cm length. Inflorescences are racemes of 10-30 cm length, with peduncles of 4-10 cm length. The flowers are 14-18 mm long, yellow, sometimes lilac to purple. The upper lip is 5-9 mm long and has two teeth, the lower lip is 5-7 mm long. Fruits are brown to blackish and hairy pods of 4-8 cm length, each containing 8-12 seeds. These are dark brown, ellipsoid, and 4-5 mm long.

Ecology and Control
This fast growing and short-lived shrub grows on coastal bluffs and dunes in the native range. It is a nitrogen-fixing species and forms dense thickets that alter the community structure and fauna richness, and reduces plant species richness. The enriched soil promotes invasion by exotic weeds under dead bushes of this plant. In New Zealand, decline of this shrub has occurred in pine plantations as a result of infection by the fungus *Colletotrichum gloeosporioides*.

Mechanical removal and re-establishing native plants after removal is a current control measure. In cleared areas, all nonnative plants and litter should be removed in order to discourage germination by seeds of this shrub. Treatment must be repeated for at least 3 years in order to deplete the seed bank.

References
133, 215, 308, 336, 607, 821, 822, 893, 970, 1017, 1018, 1415.

Lupinus polyphyllus Lindl. — Fabaceae

LF: Perennial herb
SN: -
CU: Ornamental

Geographic Distribution

Europe		Australasia		Atlantic Islands	
X	Northern		Australia		Cape Verde
X	British Isles	●	New Zealand		Canary, Madeira
●	Central, France				Azores
X	Southern	*Northern America*			South Atlantic Isl.
●	Eastern	N	Canada, Alaska		
	Mediterranean Isl.		Southeastern USA	*Indian Ocean Islands*	
		N	Western USA		Mascarenes
Africa		N	Remaining USA		Seychelles
	Northern		Mexico		Madagascar
	Tropical				
	Southern	*Southern America*		*Pacific Islands*	
			Tropical		Micronesia
Asia			Caribbean		Galapagos
	Temperate		Chile, Argentina		Hawaii
	Tropical				Melanesia, Polynesia

Invaded Habitats
Grassland, forest edges, heath- and woodland.

Description
A stout herb of 50-150 cm height, with usually unbranched and minutely pubescent stems. Leaves are compound and have 9-17 obovate-lanceolate leaflets of 7-15 cm length and 1.5-3 cm width. Inflorescences are rather dense racemes of 15-60 cm length, each having up to 80 flowers, and with the flowers borne in whorls. Peduncles are 3-8 cm long. The flowers are blue, purple, pink or white, and 12-14 mm long. Fruits are brown and sparsely hairy pods of 25-40 mm length, each containing 5-9 seeds of *c.* 4 mm length.

Ecology and Control
A large herb forming dense patches that can rapidly expand and crowd out almost all other species. In Europe, extensive stands degrade species rich dry grasslands. The tall size of the plant makes it highly competitive to native grasses and forbs. The plant is nitrogen-fixing and increases soil fertility levels, which may change the floristic composition of the invaded vegetation. It is a prolific seed producer and spreads by both seeds and rhizomes.

Seedlings and small plants may be pulled or dug out. If larger plants are dug out, rhizomes must be removed to prevent regrowth. Cutting prevents seed formation. Regrowth can be treated with herbicide.

References
34, 399, 1415.

Lycium ferocissimum Miers   Solanaceae

LF: Evergreen shrub   CU: Ornamental, shelter
SN: *Lycium macrocalyx*

Geographic Distribution

Europe	*Australasia*	*Atlantic Islands*
Northern	● Australia	Cape Verde
British Isles	● New Zealand	Canary, Madeira
Central, France		Azores
Southern	*Northern America*	South Atlantic Isl.
Eastern	Canada, Alaska	
Mediterranean Isl.	● Southeastern USA	*Indian Ocean Islands*
	Western USA	Mascarenes
Africa	Remaining USA	Seychelles
Northern	Mexico	Madagascar
Tropical		
N Southern	*Southern America*	*Pacific Islands*
	Tropical	Micronesia
Asia	Caribbean	Galapagos
Temperate	Chile, Argentina	Hawaii
Tropical		Melanesia, Polynesia

Invaded Habitats
Grassland, riparian habitats, coastal beaches.

Description
A tall erect, much-branched, stiff shrub of up to 5 m height and 3 m diameter, with brown to grey stems, spines to 15 cm length on the main stems, and smaller spines on the branchlets. Leaves are glabrous, more or less succulent, ovate to elliptical, up to 3.5 cm long and 2 cm wide. White and fragrant flowers of *c.* 1 cm diameter are borne singly or in pairs in the axils of leaves. The bell-shaped flowers haver five petals and lilac or purple markings in the centre. Fruits are orange-red, globular to ovoid, smooth berries of *c.* 1 cm diameter. Seeds are light brown or yellow, flattened, *c.* 2.5 mm long and 1.5 mm wide. The plant has a deep taproot.

Ecology and Control
This plant forms extensive and spiny thickets that impede wildlife and crowd out native vegetation. The dense stands prevent any regeneration by native plants. Along watercourses, it denies animal access to watering points. The shrub easily resprouts from the root crown after damage and forms root suckers. Seeds are dispersed by birds and mammals.

    Physical removal and burning proved to be effective. Cutting plants at ground level is followed by treating all exposed surfaces with amine 2,4-D or triclopyr to prevent regrowth. Herbicides applied as foliar sprays include glyphosate, picloram, or a mixture of picloram and triclopyr. A follow-up programme is necessary to control seedlings.

References
121, 215, 607, 924, 985, 986, 1415.

Lygodium japonicum (Thunb.) Sw.   Schizaeaceae

LF: Climbing fern   CU: Ornamental
SN: *Ophioglossum japonicum* Thunb.

Geographic Distribution

Europe	Australasia	Atlantic Islands
Northern	N Australia	Cape Verde
British Isles	New Zealand	Canary, Madeira
Central, France		Azores
Southern	*Northern America*	South Atlantic Isl.
Eastern	Canada, Alaska	
Mediterranean Isl.	● Southeastern USA	*Indian Ocean Islands*
	Western USA	Mascarenes
Africa	Remaining USA	Seychelles
Northern	Mexico	Madagascar
Tropical		
Southern	*Southern America*	*Pacific Islands*
	Tropical	Micronesia
Asia	Caribbean	Galapagos
N Temperate	Chile, Argentina	Hawaii
N Tropical		Melanesia, Polynesia

Invaded Habitats
Tropical hammocks, riparian habitats, disturbed places.

Description
A large, bright green and twining fern with short-creeping rhizomes. Leaves are up to 5 m long, bipinnately compound with widely spaced pinnae along the rachis. Pinnae are fertile or sterile, 10-18 cm long and 7-10 cm wide, and consist of 2-3 leaflets of up to 7 cm length and 2.5 cm width. Fertile leaflets are flabellate and have eight or more narrow lobes. Sterile pinnae are triangular, with the sterile leaflets being palmate to deeply lobed. Sporangia open upwardly and are attached on free veinlets.

Ecology and Control
This fast growing fern twines around small stems and branches of trees and shrubs. It quickly climbs to the canopy and forms dense mats there, shading out the host trees and any other supporting vegetation. It can weaken or even kill smothered trees. The dry dead fronds are flammable and in fire-prone regions the fern carries fires from the ground to the forest canopies, thus intensifying wild fires.

    Small infestations can be controlled by repeated hand pulling or cutting. In Florida, the vines are pulled from the trees and treated with a foliar spray of a mixture of triclopyr and an oil surfactant.

References
715, 928.

Lygodium microphyllum (Cav.) R. Br. — Schizaeaceae

LF: Climbing fern
CU: Ornamental
SN: *Lygodium scandens* (L.) Swartz, *Ugena microphylla* Cav.

Geographic Distribution

Europe	*Australasia*	*Atlantic Islands*
Northern	N Australia	Cape Verde
British Isles	New Zealand	Canary, Madeira
Central, France		Azores
Southern	*Northern America*	South Atlantic Isl.
Eastern	Canada, Alaska	
Mediterranean Isl.	• Southeastern USA	*Indian Ocean Islands*
	Western USA	N Mascarenes
Africa	Remaining USA	Seychelles
Northern	Mexico	Madagascar
N Tropical		
Southern	*Southern America*	*Pacific Islands*
	Tropical	N Micronesia
Asia	X Caribbean	Galapagos
N Temperate	Chile, Argentina	Hawaii
N Tropical		N Melanesia, Polynesia

Invaded Habitats
Freshwater wetlands, cypress swamps, forests, mangrove swamps, wet grassland.

Description
A large fern with a short-creeping rhizome and fronds reaching 10 m length. Petioles are 7-25 cm long. Sterile pinnae are oblong, 5-12 cm long and 3-6 cm wide, fertile pinnae are 3-14 cm long and 2.5-6 cm wide.

Ecology and Control
A fast growing fern that is native in wet tropical and subtropical regions. Where invasive, it smothers the native vegetation by climbing high into and over the crowns of shrubs and trees. It also grows horizontally on the floor and the numerous fronds build thick mats that completely smother whole plant communities and cover the ground. Tall infestations by this fern can be a fire hazard because the dry dead fronds are flammable and carry fires into the canopies of trees.

Manual removal is labour intensive. Herbicide may applied to the foliage, follow-up applications are necessary to treat regrowth.

References
155, 715, 928, 1002, 1004.

Lythrum salicaria L.   Lythraceae

LF: Perennial herb
SN: -
CU: Ornamental

Geographic Distribution

Europe	*Australasia*	*Atlantic Islands*
N Northern	X Australia	Cape Verde
N British Isles	X New Zealand	Canary, Madeira
N Central, France		Azores
N Southern	*Northern America*	South Atlantic Isl.
N Eastern	● Canada, Alaska	
N Mediterranean Isl.	Southeastern USA	*Indian Ocean Islands*
	● Western USA	Mascarenes
Africa	● Remaining USA	Seychelles
N Northern	Mexico	Madagascar
Tropical		
X Southern	*Southern America*	*Pacific Islands*
	Tropical	Micronesia
Asia	Caribbean	Galapagos
N Temperate	Chile, Argentina	Hawaii
N Tropical		Melanesia, Polynesia

Invaded Habitats
Riparian habitats, lake shores, marshes, freshwater wetlands.

Description
An erect, glabrous to pubescent sparingly branched herb of 50-150 cm height with 30-50 stems, and with leaves opposite or in whorls of three. The leaves are sessile, lanceolate-oblong to ovate, and 3-10 cm long. The trimorphic and purple flowers are clustered in whorls, forming long terminal spikes. The petals are 8-10 mm long. Fruits are ovoid capsules of 3-4 mm length, containing numerous seeds of *c.* 0.4 mm length. The plant has a taproot and a spreading rootstock.

Ecology and Control
In the native range, this forb grows in margins of lakes and ponds, along streams, marshes, riverbanks, damp places, and swamps. Where invasive, the plant forms extensive and persistent stands in wetland habitats, replacing native vegetation and threatening rare and endangered plant species. Large stands reduce the availability of shelter and food for wildlife, and restrict access to open water. The plant can resprout from fragments of the root system and stems.

Pulling individual plants by hand should be done before seed set. Larger populations are difficult to control; frequent cutting at ground level is effective if done over a period of several years. Chemical control is achieved by spot treatments of a glyphosate herbicide.

References
36, 83, 122, 133, 371, 432, 562, 571, 807, 810, 811, 1015, 1288, 1396.

Macfadyena unguis-cati (L.) Gentry Bignoniaceae

LF: Woody climber CU: Ornamental
SN: *Bignonia tweedieana* Lindl., *B. unguis-cati* L., *Doxantha unguis-cati* (L.) Miers

Geographic Distribution

Europe		*Australasia*		*Atlantic Islands*	
	Northern	●	Australia	X	Cape Verde
	British Isles		New Zealand		Canary, Madeira
	Central, France				Azores
	Southern	*Northern America*		X	South Atlantic Isl.
	Eastern		Canada, Alaska		
	Mediterranean Isl.	●	Southeastern USA	*Indian Ocean Islands*	
			Western USA	X	Mascarenes
Africa			Remaining USA		Seychelles
	Northern	N	Mexico		Madagascar
	Tropical				
●	Southern	*Southern America*		*Pacific Islands*	
		N	Tropical		Micronesia
Asia		N	Caribbean		Galapagos
	Temperate		Chile, Argentina	X	Hawaii
X	Tropical				Melanesia, Polynesia

Invaded Habitats
Forests and forest gaps, woodland, tropical hammocks, floodplains.

Description
A glabrous trailing and climbing liana adhering to host trees by recurved tendrils and adventitious roots. Leaves have petioles of 10-20 cm length, and are compound with two elliptic and entire leaflets of 2-8 cm length between which lies a tendril ending in three claws. Leaflets of juvenile plants are much smaller. The large and yellow trumpet-shaped flowers are up to 10 cm long and wide, and borne in axillary panicles. Fruits are long and narrow, linear and flat capsules of 15-50 cm length and *c.* 1 cm width. They contain very flat and winged seeds of 2-4 cm length. The plant has underground tubers of 20-40 cm length.

Ecology and Control
A native of closed forests, this fast growing climber rapidly invades disturbed rainforests and other communities. It climbs on walls, tree trunks and other vegetation, forming a thick carpet on the forest floor and climbing over tree canopies. Host trees may be killed by the vine's weight and shading effect, and the vine outcompetes understorey plants. The plant reproduces by seeds and vegetatively by tuberous roots. Seeds are dispersed by wind and water. Stems root at nodes upon contact with the soil.

Mechanical control is difficult as the tuber-like roots easily break off and regenerate new plants. Seedlings and small plants can be dug out, roots must be removed carefully. Larger vines are cut and the cut stumps treated with herbicide.

References
549, 607, 715, 924, 1411.

Mahonia aquifolium (Pursh) Nutt. Berberidaceae

LF: Evergreen shrub CU: Ornamental
SN: *Berberis aquifolium* Pursh, *Berberis diversifolia* (Sweet) Steud.

Geographic Distribution

Europe	*Australasia*	*Atlantic Islands*
Northern	Australia	Cape Verde
X British Isles	X New Zealand	Canary, Madeira
● Central, France		Azores
X Southern	*Northern America*	South Atlantic Isl.
● Eastern	N Canada, Alaska	
Mediterranean Isl.	Southeastern USA	*Indian Ocean Islands*
	N Western USA	Mascarenes
Africa	N Remaining USA	Seychelles
Northern	Mexico	Madagascar
Tropical		
Southern	*Southern America*	*Pacific Islands*
	Tropical	Micronesia
Asia	Caribbean	Galapagos
Temperate	Chile, Argentina	Hawaii
Tropical		Melanesia, Polynesia

Invaded Habitats

Forests, heath- and woodland.

Description

An unarmed shrub of 1-2 m height, sparingly branched, and with ascending stems. Leaves are pinnately compound, each having 5-9 sessile leaflets. These are ovate to oblong-ovate, 4-8 cm long and 2-4 cm wide, dark green and shining, and spinose-dentate. Yellow flowers with six petals each are borne in clustered erect racemes of 5-8 cm length. Fruits are dark blue to black and globose berries of *c.* 8 mm diameter, having a few seeds each.

Ecology and Control

A variable species with several varieties and cultivars differing mainly in growth habit. Plants naturalized in Europe are often hybrids of closely related taxa of this genus. The plant is somewhat stoloniferous and suckering from roots. Stems touching the ground may become rooted. Vegetative growth leads to expanding patches that reduce native species richness and prevent regeneration of native shrubs and trees. The shrub may completely dominate the understorey of invaded forests and prevent forest regeneration. The plant produces fruits abundantly and seeds are dispersed by birds. Seedlings can reach high densities.

Specific control methods for this species are not available. Seedlings and small plants can be hand pulled or dug out. If larger shrubs are dug out, roots must be removed to prevent regrowth. Shrubs may also be cut at ground level and the cut stumps treated with herbicide.

References

53, 1207.

Marrubium vulgare L.   Lamiaceae

LF: Perennial herb   CU: Ornamental, flavouring
SN: -

Geographic Distribution

Europe		*Australasia*		*Atlantic Islands*	
	Northern	●	Australia	X	Cape Verde
N	British Isles	X	New Zealand	N	Canary, Madeira
N	Central, France			N	Azores
N	Southern	*Northern America*			South Atlantic Isl.
N	Eastern		Canada, Alaska		
N	Mediterranean Isl.		Southeastern USA	*Indian Ocean Islands*	
		X	Western USA		Mascarenes
Africa			Remaining USA		Seychelles
N	Northern		Mexico		Madagascar
	Tropical				
X	Southern	*Southern America*		*Pacific Islands*	
		X	Tropical		Micronesia
Asia			Caribbean		Galapagos
N	Temperate	X	Chile, Argentina	X	Hawaii
N	Tropical				Melanesia, Polynesia

Invaded Habitats
Dry forests, scrub- and woodland, arid rangelands, disturbed sites.

Description
A herb or sometimes subshrub with ascending to erect stems of 10-60 cm height, whitish-tomentose or densely pubescent, with many short flowerless branches. The leaves have petioles of 7-15 mm length, the blades are broadly ovate with the base being rounded to lobed, 1.5-5.5 cm long, and tomentose beneath. Numerous white flowers of *c.* 9 mm diameter are borne in dense whorls. The calyx has ten teeth and is soft-hairy. Fruits are ovoid nutlets.

Ecology and Control
A native of dry grassland and open places, this drought tolerant plant forms dense and pure stands where invasive that may extend over large areas. These stands reduce native species richness and alter the community structure. Establishment and growth of tree and shrub seedlings is strongly reduced in invaded areas. Seeds are dispersed by animals and streams. They germinate whenever sufficient rains occur, and a soil seed bank may be accumulated in large stands.
    In fire adapted communities, prescribed burning is used to kill plants and to reduce the soil seed bank. Follow-up programmes are necessary to treat seedlings. An effective herbicide is 2,4-D ester.

References
121, 215, 749, 985, 986, 1379, 1380, 1447.

Melaleuca quinquenervia (Cav.) Blake — Myrtaceae

LF: Evergreen tree
SN: -
CU: Ornamental, essential oils

Geographic Distribution

Europe	*Australasia*	*Atlantic Islands*
Northern	N Australia	Cape Verde
British Isles	New Zealand	Canary, Madeira
Central, France		Azores
Southern	*Northern America*	South Atlantic Isl.
Eastern	Canada, Alaska	
Mediterranean Isl.	● Southeastern USA	*Indian Ocean Islands*
	Western USA	Mascarenes
Africa	Remaining USA	Seychelles
Northern	Mexico	Madagascar
Tropical		
X Southern	*Southern America*	*Pacific Islands*
	Tropical	Micronesia
Asia	● Caribbean	Galapagos
N Temperate	Chile, Argentina	X Hawaii
Tropical		N Melanesia, Polynesia

Invaded Habitats
Wet forests, cypress swamps, marshes, wet grassland.

Description
A slender tree reaching 15-20 m in height, with often drooping and irregular branches, and a thick, whitish and soft bark. Leaves are alternate, elliptic to lanceolate-elliptic, 5-12 cm long, bright green, and smell like camphor when crushed. White and sessile flowers are borne in many-flowered spikes along the branches. Petals are 3-4 mm long. Fruits are short-cylindrical woody capsules of 3-5 mm length, aggregated in tightly packed clusters around the branches. Each capsule contains 200-350 tiny seeds.

Ecology and Control
In the native range, this tree often grows in pure stands on creek banks and in swamps. Where invasive, it rapidly colonizes freshwater wetlands and almost completely displaces the native vegetation, thereby degrading prime wildlife habitat. The tree grows well on poor soils, is tolerant of flooding and is a prolific seed producer. Seeds may remain on the trees for more than 10 years and are released upon fire or other damage. Fallen seeds are dispersed by water. The tree resprouts easily after damage. Stands of this tree can carry fires into the canopies, thereby killing other tree species.

    Seedlings and saplings can be hand pulled. Cut stumps are treated with imazapyr. Trees can also be girdled and the cut area treated with herbicide. In fire tolerant communities, prescribed fires during the dry season are used to control seedlings, but mature trees will release huge numbers of seeds after fire.

References
56, 284, 394, 441, 490, 715, 716, 925, 1436.

Melia azedarach L. — Meliaceae

LF: Deciduous tree
SN: *Melia toosendan* Siebold & Zucc.
CU: Ornamental, wood

Geographic Distribution

Europe	*Australasia*	*Atlantic Islands*
Northern	N Australia	Cape Verde
British Isles	New Zealand	Canary, Madeira
Central, France		Azores
X Southern	*Northern America*	South Atlantic Isl.
Eastern	Canada, Alaska	
Mediterranean Isl.	● Southeastern USA	*Indian Ocean Islands*
	X Western USA	Mascarenes
Africa	Remaining USA	Seychelles
X Northern	X Mexico	Madagascar
X Tropical		
● Southern	*Southern America*	*Pacific Islands*
	X Tropical	Micronesia
Asia	X Caribbean	X Galapagos
N Temperate	Chile, Argentina	● Hawaii
N Tropical		N Melanesia, Polynesia

Invaded Habitats
Riparian habitats, forests, grassland, disturbed sites.

Description
A tree up to 15 m tall, with a grey-brown and smooth bark. Leaves are unpleasantly scented when crushed, up to 50 cm long, deep green and glossy, usually bipinnately compound with 4-6 pairs of pinnae. Each pinna bears 3-6 lanceolate, almost glabrous leaflets up to 5 cm length and 2.5 cm wide, with deeply toothed margins. Fragrant pale lilac flowers of 15-20 mm diameter are borne in large, axillary panicles. Fruits are small, pale-yellow and globose drupes of *c.* 12 mm length. Each fruit contains 1-6 seeds of 8-10 mm length and 6-7 mm width.

Ecology and Control
In the native range, this rapidly growing tree occurs in lowland to highland rainforests. Where invasive, it builds up dense stands that outcompete native plant species and prevent regeneration of shrubs and trees. The tree flowers throughout the year and produces huge numbers of seeds that are widely dispersed by birds. The juvenile period is very short and plants may begin to flower even at seedling stage.

Seedlings can be hand pulled and larger trees can be treated with a triclopyr herbicide. Seed producing trees especially should be removed first to prevent seed dispersal.

References
284, 549, 550, 715, 765, 793, 1089, 1298.

Melianthus comosus Vahl Melianthaceae

LF: Evergreen shrub
SN: *Melianthus minor* L.
CU: Ornamental

Geographic Distribution

Europe	*Australasia*	*Atlantic Islands*
Northern	• Australia	Cape Verde
British Isles	New Zealand	Canary, Madeira
Central, France		Azores
Southern	*Northern America*	South Atlantic Isl.
Eastern	Canada, Alaska	
Mediterranean Isl.	Southeastern USA	*Indian Ocean Islands*
	Western USA	Mascarenes
Africa	Remaining USA	Seychelles
Northern	Mexico	Madagascar
Tropical		
N Southern	*Southern America*	*Pacific Islands*
	Tropical	Micronesia
Asia	Caribbean	Galapagos
Temperate	Chile, Argentina	Hawaii
Tropical		Melanesia, Polynesia

Invaded Habitats
Riparian habitats, forest edges.

Description
An erect shrub reaching 2.5 m height, often smaller, with several branched stems arising from a crown. Leaves are up to 20 cm long, pinnately compound with 6-7 pairs of sessile leaflets. These are coarsely toothed and downy white on the lower surface. Flowers are red and borne in loose clusters. Fruits are papery inflated capsules of 2.5-4 cm length, containing 2-8 seeds. These are dark brown to black, smooth and shiny, and *c.* 10 mm in diameter.

Ecology and Control
The plant has an extensive shallow root system and forms dense clumps that shade out all other vegetation. It competes successfully for water, space and nutrients. It grows in margins of water courses and the dense stands may impede water flow and restrict animal movement. Fruits are carried by streams and are able to float for long periods of time.

Isolated plants can be pulled or dug out, roots must be removed to prevent regrowth. Effective herbicides for foliar spray are 2,4,5-T ester or picloram.

References
607, 608, 985, 986.

Melianthus major L.   Melianthaceae

LF: Semi-deciduous shrub   CU: Ornamental
SN: -

Geographic Distribution

Europe	*Australasia*	*Atlantic Islands*
Northern	Australia	Cape Verde
British Isles	● New Zealand	Canary, Madeira
Central, France		Azores
Southern	*Northern America*	South Atlantic Isl.
Eastern	Canada, Alaska	
Mediterranean Isl.	Southeastern USA	*Indian Ocean Islands*
	Western USA	Mascarenes
Africa	Remaining USA	Seychelles
Northern	Mexico	Madagascar
Tropical		
N Southern	*Southern America*	*Pacific Islands*
	Tropical	Micronesia
Asia	Caribbean	Galapagos
Temperate	Chile, Argentina	Hawaii
Tropical		Melanesia, Polynesia

Invaded Habitats
Sand dunes, scrubland, disturbed sites.

Description
A stout shrub up to 2 m tall, with hollow and soft-wooded stems, often dark purplish. Leaves are up to 50 cm long, greyish green and compound with 11-21 toothed leaflets of 8-15 cm length and 3.5-7.5 cm width. Flowers are dark reddish brown and borne in terminal racemes of 30-40 cm length. Fruits are papery capsules of 2.5-5 cm length, releasing shining black seeds of 5-6 mm length.

Ecology and Control
This fast growing drought and frost tolerant shrub occurs on sandstone slopes and along streams in the native range. Where invasive, it forms dense thickets that expand rapidly and crowd out native plants. The shrub suckers from roots and resprouts after damage. Seed production is prolific and seeds are dispersed by water.
     Specific control methods for this species are not available. The same methods may apply as for the previous species.

References
121, 1415.

Melilotus albus Medik.

Fabaceae

LF: Annual, biennial herb
SN: -
CU: Forage, soil improvement

Geographic Distribution

Europe		Australasia		Atlantic Islands	
	Northern	X	Australia		Cape Verde
X	British Isles	X	New Zealand	X	Canary, Madeira
N	Central, France				Azores
N	Southern	*Northern America*			South Atlantic Isl.
N	Eastern	●	Canada, Alaska		
N	Mediterranean Isl.		Southeastern USA	*Indian Ocean Islands*	
		X	Western USA	X	Mascarenes
Africa		●	Remaining USA		Seychelles
N	Northern		Mexico		Madagascar
	Tropical				
X	Southern	*Southern America*		*Pacific Islands*	
		X	Tropical		Micronesia
Asia		X	Caribbean		Galapagos
N	Temperate	X	Chile, Argentina	X	Hawaii
N	Tropical				Melanesia, Polynesia

Invaded Habitats
Riparian habitats, grassland, disturbed places.

Description
A much-branched, erect herb of 50-250 cm height with a strong taproot. Leaves are compound and have three narrowly oblong-obovate to lanceolate leaflets of 15-50 mm length. White flowers of 4-5 mm length are borne in elongated and slender racemes of 8-15 cm length. Fruits are greyish-brown and glabrous pods of 3-5 mm length and 2-2.5 mm width, usually with one seed. Seeds are ovate, yellowish and 2-2.5 mm long.

Ecology and Control
This herb grows on a wide range of soils and forms dense stands where invasive. It colonizes disturbed sites and may become dominant, outcompeting native grasses and forbs. It is nitrogen-fixing and increases soil fertility levels, thereby changing the floristic composition of invaded areas. Seeds are dispersed by water and may remain viable in the soil for more than 20 years. Burning facilitates the establishment of this plant in grassland communities.

Scattered plants can be hand pulled, most of the root must be removed. Cutting close to the ground is effective if done before flowering occurs. Prescribed burning is used to control this plant in North American prairies. Effective herbicides are 2,4-D amine, dicamba, or MCPA.

References
257, 395, 571, 675, 1307.

Melilotus indicus (L.) All.   Fabaceae

LF: Annual herb
SN: -

CU: Erosion control, forage

Geographic Distribution

Europe
- Northern
- X British Isles
- Central, France
- N Southern
- N Eastern
- N Mediterranean Isl.

Africa
- N Northern
- X Tropical
- X Southern

Asia
- N Temperate
- N Tropical

Australasia
- ● Australia
- X New Zealand

Northern America
- Canada, Alaska
- Southeastern USA
- X Western USA
- Remaining USA
- X Mexico

Southern America
- Tropical
- X Caribbean
- X Chile, Argentina

Atlantic Islands
- Cape Verde
- N Canary, Madeira
- N Azores
- South Atlantic Isl.

Indian Ocean Islands
- X Mascarenes
- Seychelles
- Madagascar

Pacific Islands
- Micronesia
- Galapagos
- X Hawaii
- Melanesia, Polynesia

Invaded Habitats
Grassland, freshwater wetlands, riparian habitats, coastal beaches.

Description
An erect or ascending glabrous or sparsely hairy herb of 20-60 cm height. Leaves have petioles of 5-25 mm length and are compound with three elliptic to obovate leaflets of 6-25 mm length. Flowers are yellow, 2-3 mm long and borne in dense and slender racemes. Fruits are glabrous pods of 2-4 mm length, containing 1-2 light brown seeds of 1.5-2 mm length.

Ecology and Control
As the previous species, this plant is invasive because it rapidly colonizes suitable sites and builds up dense stands that reduce native species richness. The plant is nitrogen-fixing and increases soil fertility levels. Little is known on the ecology of this plant where invasive.

Specific control methods for this species are not available. The same methods as for the previous species may apply.

References
215.

Melilotus officinalis (L.) Lam.   Fabaceae

LF: Annual, biennial herb   CU:
SN: *Melilotus arvensis* Wallr., *Melilotus officinale* L.

Geographic Distribution

Europe		*Australasia*		*Atlantic Islands*
	Northern		Australia	Cape Verde
X	British Isles	X	New Zealand	Canary, Madeira
N	Central, France			Azores
N	Southern		*Northern America*	South Atlantic Isl.
N	Eastern	●	Canada, Alaska	
N	Mediterranean Isl.		Southeastern USA	*Indian Ocean Islands*
		X	Western USA	Mascarenes
Africa		●	Remaining USA	Seychelles
	Northern		Mexico	Madagascar
	Tropical			
	Southern		*Southern America*	*Pacific Islands*
			Tropical	Micronesia
Asia			Caribbean	Galapagos
N	Temperate	X	Chile, Argentina	Hawaii
N	Tropical			Melanesia, Polynesia

Invaded Habitats
Grassland, riparian habitats, disturbed sites.

Description
An erect to decumbent and branched herb of 40-250 cm height. Leaves are compound with three leaflets of 15-25 mm length. The leaflets of the lower leaves are ovate to obovate, those of the upper leaves ovate-lanceolate. Yellow flowers of 4-7 mm length are borne in lax and slender racemes. The wings are longer than the keel. Fruits are glabrous and pale brown pods of 3-5 mm length, each containing usually one seed of 2-2.5 mm length.

Ecology and Control
A plant that grows in a wide range of soil conditions including saline soils. It forms dense stands that displace native grass and forbs. Seeds are dispersed by water and may remain viable in the soil for more than 20 years. It establishes well in grasslands after disturbances such as fires. The plant is nitrogen-fixing and increases soil fertility levels, which may change the species composition of invaded areas.

Cutting close to the ground reduces the plant's vitality. Scattered plants may be dug out, the roots must be removed. Effective herbicides are 2,4-D, dicamba, or MCPA.

References
257, 571, 1307, 1433.

Melinis minutiflora Beauv.　　　　　　　　　　　　　　　　　　Poaceae

LF: Perennial herb　　　　　　　CU: Forage, erosion control
SN: *Melinis tenuinervis* Stapf, *Panicum melinis* Trin.

Geographic Distribution

Europe		*Australasia*		*Atlantic Islands*	
	Northern	●	Australia	N	Cape Verde
	British Isles		New Zealand		Canary, Madeira
	Central, France				Azores
	Southern	*Northern America*			South Atlantic Isl.
	Eastern		Canada, Alaska		
	Mediterranean Isl.	X	Southeastern USA	*Indian Ocean Islands*	
			Western USA		Mascarenes
Africa		●	Remaining USA		Seychelles
	Northern		Mexico		Madagascar
N	Tropical				
N	Southern	*Southern America*		*Pacific Islands*	
		●	Tropical	X	Micronesia
Asia		X	Caribbean	●	Galapagos
	Temperate		Chile, Argentina	●	Hawaii
X	Tropical			X	Melanesia, Polynesia

Invaded Habitats
Grassland, pastures, disturbed sites.

Description
A tufted and stoloniferous grass of 1.5-2 m height. Leaf sheaths are pubescent or long-hairy and have ligules consisting of a ring of hairs. Leaf blades are flat, glabrous to pubescent and up to 13 mm wide. Inflorescences are contracted panicles of 8-20 cm long and contain small, glabrous spikelets of 1.5-2.5 mm length. Awns are 6-16 mm long.

Ecology and Control
This fast growing and fire adapted C_4 grass has a sprawling growth habit. It climbs over shrubs and forms dense and impenetrable mats up to 1.5 m deep on the floor, completely covering large areas and eliminating all native vegetation. The grass accumulates a large amount of dead biomass and increases fire hazards. It spreads by seeds and by vegetative growth. The grass establishes well after fires.
　　　The grass does not withstand grazing or cutting at soil level because the crowns are well above the ground. Thus, repeated cutting can be effective in killing the plants. An effective chemical control is spraying a mixture of 2,2-DPA and paraquat.

References
50, 284, 305, 306, 600, 984, 1033, 1041, 1190, 1199, 1200, 1313.

Memecylon floribundum Blume   Melastomataceae

LF: Evergreen shrub   CU: Ornamental
SN: *Memecylon caeruleum* L.

Geographic Distribution

Europe
   Northern
      British Isles
      Central, France
   Southern
      Eastern
      Mediterranean Isl.

Africa
   Northern
   Tropical
   Southern

Asia
   Temperate
 N Tropical

Australasia
   Australia
   New Zealand

Northern America
   Canada, Alaska
   Southeastern USA
   Western USA
   Remaining USA
   Mexico

Southern America
   Tropical
   Caribbean
   Chile, Argentina

Atlantic Islands
   Cape Verde
   Canary, Madeira
   Azores
   South Atlantic Isl.

Indian Ocean Islands
   Mascarenes
 ● Seychelles
   Madagascar

Pacific Islands
   Micronesia
   Galapagos
   Hawaii
   Melanesia, Polynesia

Invaded Habitats
Forests and forest gaps, mountain slopes, rock outcrops.

Description
A sparingly branched shrub or small tree of 1-4 m height with a glossy foliage. The sessile leaves are opposite, ovate, 8-14 cm long and 3.5-5 cm wide. Flowers are deep blue, *c.* 10 mm long, and borne in clusters in the axils of leaves. Fruits are globose berries of *c.* 18 mm length and 11 mm width, containing one seed.

Ecology and Control
Where native, this shrub is common on rocky and sandy shores. It is invasive because it becomes dominant in the shrub layer of forests and the dense canopies eliminate all vegetation by shading out. Forest gaps are quickly covered by a continuous cover of seedlings of this shrub. Establishment depends on light and occurs in openings and disturbed sites. Established stands have a very low plant and invertebrate species richness and prevent natural forest regeneration. Seeds are dispersed by birds.
    Specific control methods for this species are not available. Seedlings and saplings may be pulled or dug out. Larger stems can be cut and the cut stumps treated with herbicide.

References
446, 447.

Mentha pulegium L.   Lamiaceae

LF: Perennial herb
SN: *Pulegium vulgare* Mill.
CU: Ornamental, essential oils

Geographic Distribution

Europe
-     Northern
- N   British Isles
- N   Central, France
- N   Southern
- N   Eastern
- N   Mediterranean Isl.

Africa
- N   Northern
-     Tropical
-     Southern

Asia
- N   Temperate
-     Tropical

Australasia
- ●   Australia
- X   New Zealand

Northern America
-     Canada, Alaska
-     Southeastern USA
- ●   Western USA
-     Remaining USA
-     Mexico

Southern America
-     Tropical
-     Caribbean
- X   Chile, Argentina

Atlantic Islands
-     Cape Verde
- N   Canary, Madeira
- N   Azores
-     South Atlantic Isl.

Indian Ocean Islands
-     Mascarenes
-     Seychelles
-     Madagascar

Pacific Islands
-     Micronesia
-     Galapagos
- X   Hawaii
-     Melanesia, Polynesia

Invaded Habitats
Grassland, riparian habitats, freshwater wetlands, alluvial plains.

Description
A somewhat glabrous to tomentose herb with procumbent to ascending stems of 10-40 cm height, with a pungent scent and shortly petiolated leaves. These are narrowly elliptical, weakly toothed or almost entire, 8-30 mm long and 4-12 mm wide. The pale-mauve to lilac flowers are borne in axillary whorls and have corollas of 4-6 mm length. Fruits are small, pale brown nutlets of *c.* 0.8 mm length. The plant has rhizomes and creeping stolons. Stems root at the nodes.

Ecology and Control
A highly variable species with regard to growth habit and leaf-shape. Several varieties have been described in the native range where it hybridizes with several congeners. The plant forms dense stands that crowd out native vegetation and reduce species richness. It is a prolific seed producer, and seeds are dispersed by water and animals. Seedlings can emerge and establish under water, enabling the plant to invade flooded areas. It accumulates a soil seed bank. Damaged plants easily resprout.

Single plants and small patches are hand pulled or dug out. Roots and stolons must be removed to prevent regrowth. Follow-up programmes are necessary to treat regrowth and seedlings. Seedlings can be controlled with 2,4-D, established plants with glyphosate, metsulfuron-methyl, or triclopyr plus picloram.

References
133, 215, 976, 977, 986.

Mesembryanthemum crystallinum L.   Aizoaceae

LF: Annual herb
SN: -
CU: Ornamental, food

Geographic Distribution

Europe		*Australasia*		*Atlantic Islands*	
	Northern	●	Australia		Cape Verde
	British Isles		New Zealand	X	Canary, Madeira
	Central, France			X	Azores
X	Southern	*Northern America*			South Atlantic Isl.
	Eastern		Canada, Alaska		
X	Mediterranean Isl.		Southeastern USA	*Indian Ocean Islands*	
		●	Western USA		Mascarenes
Africa			Remaining USA		Seychelles
N	Northern	X	Mexico		Madagascar
	Tropical				
N	Southern	*Southern America*		*Pacific Islands*	
			Tropical		Micronesia
Asia			Caribbean		Galapagos
N	Temperate	X	Chile, Argentina		Hawaii
	Tropical				Melanesia, Polynesia

Invaded Habitats
Grassland, coastal estuaries and beaches, salt marshes.

Description
A succulent herb with trailing stems reaching 1 m in length. The whole plant is densely covered with crystal-like vesicles. The spathulate to broadly ovate leaves are flat, 2-20 cm long and up to 15 cm wide, and have petioles. Flowers are white, becoming pink with age, 10-20 mm in diameter, and borne in the axils of leaves or in cymes. They have five petals. Fruits are dehiscent capsules with five valves, releasing numerous round and compressed seeds.

Ecology and Control
A native of coastal sands and salt marshes, this plant establishes well in locally disturbed sites with bare soil. Individual stems then grow very large, and patches expand laterally. Scattered patches may collate into a single cover of this plant, eliminating native vegetation. Spread and increase in cover is promoted by intense grazing. The plants accumulate large amounts of salt, which is washed into the topsoil if plants dry out, thus raising soil salinity. Native grasses and forbs are inhibited in ther growth.

    Specific control methods for this species are not available. Dense patches may be cut close to the ground or sprayed with herbicide.

References
133, 215, 681, 765, 1349.

Miconia calvescens DC   Melastomataceae

LF: Evergreen shrub, tree   CU: Ornamental
SN: -

Geographic Distribution

Europe	*Australasia*	*Atlantic Islands*
Northern	Australia	Cape Verde
British Isles	New Zealand	Canary, Madeira
Central, France		Azores
Southern	*Northern America*	South Atlantic Isl.
Eastern	Canada, Alaska	
Mediterranean Isl.	Southeastern USA	*Indian Ocean Islands*
	Western USA	Mascarenes
Africa	Remaining USA	Seychelles
Northern	N Mexico	Madagascar
Tropical		
Southern	*Southern America*	*Pacific Islands*
	N Tropical	Micronesia
Asia	Caribbean	Galapagos
Temperate	N Chile, Argentina	● Hawaii
X Tropical		● Melanesia, Polynesia

Invaded Habitats
Forests and forest edges, grassland.

Description
A slender tree or shrub of 2-10 m height, sometimes more. The glabrous dark green leaves are oblong-elliptical to elliptical-ovate, 20-80 cm long and 8-30 cm wide, acuminate, and with an obtuse or rounded base. Margins are entire or slightly toothed. Inflorescences are panicles of 20-35 cm length. Flowers are sessile and have oblong-obovate petals of 2-3 mm length. Bracts are 2-4 mm long. Fruits are globose, purple black, 3.5-4.5 mm in diameter, containing ovoid to pyramidal seeds of *c.* 0.5 mm length.

Ecology and Control
In the native range, this plant is common in humid thickets and in riparian habitats from lowland to montane tropical forests. It is a fast growing, shade tolerant tree that completely transforms invaded communities into species poor stands. The tree creates dense shade that eliminates almost all other species under its canopies. Invaded slopes are prone to landslides as the weak root system does not hold the soil well and the soil lacks a herbaceous ground cover. Seeds are dispersed by wind, water and birds. Germination is stimulated by light.

    Seedlings and saplings can be hand pulled or dug out, larger plants cut down and the stumps treated with herbicide. Cleared areas must be checked in a follow-up programme for new seedlings.

References
75, 259, 856, 871, 872, 873.

Microstegium vimineum (Trin.) Camus   Poaceae

LF: Annual herb   CU: -
SN: *Eulalia viminea* (Trin.) Kuntze, *Pollinia imberbis* Nees ex Steudel.

Geographic Distribution

Europe	*Australasia*	*Atlantic Islands*
Northern	Australia	Cape Verde
British Isles	New Zealand	Canary, Madeira
Central, France		Azores
Southern	*Northern America*	
Eastern	Canada, Alaska	South Atlantic Isl.
Mediterranean Isl.	● Southeastern USA	
	Western USA	*Indian Ocean Islands*
Africa	● Remaining USA	Mascarenes
Northern	Mexico	Seychelles
Tropical		Madagascar
Southern	*Southern America*	
	Tropical	*Pacific Islands*
Asia	Caribbean	Micronesia
N Temperate	Chile, Argentina	Galapagos
N Tropical		Hawaii
		Melanesia, Polynesia

Invaded Habitats
Floodplains, riparian habitats, forests, swamps.

Description
A grass with slender culms of 50-120 cm height, rooting at the lower nodes. The narrowly elliptic leaf blades are 3.5-16 cm long and 5-15 mm wide, and glabrous to sparsely hairy. Ligules are 0.5-1 mm long, ciliate and hairy on the back. Inflorescences consist of clusters of terminal racemes of 5-9 cm length. Spikelets are 4.5-6 mm long, the lower glume is pale green, 4.5-5.5 mm long, lemmas are 0.8-1.3 mm long and the awns 3-8.5 mm long.

Ecology and Control
Although a C_4 grass, this plant is adapted to low light conditions. In the native range, it grows on shady river banks in broad-leaved forests and other damp stream sites. There are two varieties in Bhutan. The grass establishes on bare soil and spreads rapidly into disturbed areas. It produces sprawling colonies that expand laterally and form dense pure stands. Native plants are impeded in growth and regeneration. The native vegetation can be displaced within a few years. The amount of available nitrate in the soil has shown to increase under stands of this grass. Seeds are dispersed by water and animals, and remain viable in the soil for some few years.

    Specific control methods for this species are not available. Large patches can be sprayed with grass-selective herbicides.

References
67, 397, 451, 693, 694, 765, 1071.

Mikania micrantha Kunth   Asteraceae

LF: Perennial vine   CU: -
SN: -

Geographic Distribution

Europe	*Australasia*	*Atlantic Islands*
Northern	Australia	Cape Verde
British Isles	New Zealand	Canary, Madeira
Central, France		Azores
Southern	*Northern America*	South Atlantic Isl.
Eastern	Canada, Alaska	
Mediterranean Isl.	X Southeastern USA	*Indian Ocean Islands*
	Western USA	● Mascarenes
Africa	Remaining USA	Seychelles
Northern	N Mexico	Madagascar
Tropical		
Southern	*Southern America*	*Pacific Islands*
	N Tropical	Micronesia
Asia	X Caribbean	Galapagos
Temperate	Chile, Argentina	● Hawaii
● Tropical		X Melanesia, Polynesia

Invaded Habitats
Forests and forest edges.

Description
A slender vine with slightly pubescent stems and opposite leaves. Petioles are 3-7 cm long, leaf blades ovate-cordate, acuminate, 3.5-13 cm long and 2-9 cm wide, glabrous or sparsely pubescent. Margins are coarsely toothed. Flowerheads are 4-6 mm long and borne in axillary clusters with long peduncles. Each flowerhead has four phyllaries and four disk florets, ray florets are lacking. The disk florets are white above and green below. Fruits are oblong achenes of *c.* 2 mm length with a pappus consisting of numerous bristles.

Ecology and Control
In the native range, this shade tolerant plant grows in low vegetation near lakes and forest clearings, sometimes being very abundant locally. It is a vigorously creeping and climbing weed, smothering shrubs and small trees and building dense thickets by the numerous intermingled stems and stolons. It spreads rapidly after disturbances such as fires and populations expand by vegetative growth, preventing any natural forest regeneration. It flowers and fruits throughout the year and is ecologically a variable species. Seeds are wind dispersed.

Scattered vines may be hand pulled or dug out. Larger infestations can be sprayed with herbicides such as 2,4-D or MCPA.

References
72, 597, 620, 1254, 1255, 1256, 1257, 1261.

Mimosa pigra L.   Fabaceae

LF: Evergreen shrub   CU: -
SN: *Mimosa asperata* L.

Geographic Distribution

Europe	*Australasia*	*Atlantic Islands*
Northern	● Australia	Cape Verde
British Isles	New Zealand	Canary, Madeira
Central, France		Azores
Southern	*Northern America*	South Atlantic Isl.
Eastern	Canada, Alaska	
Mediterranean Isl.	● Southeastern USA	*Indian Ocean Islands*
	Western USA	X Mascarenes
Africa	Remaining USA	Seychelles
X Northern	N Mexico	X Madagascar
● Tropical		
● Southern	*Southern America*	*Pacific Islands*
	N Tropical	Micronesia
Asia	Caribbean	X Galapagos
Temperate	Chile, Argentina	X Hawaii
● Tropical		Melanesia, Polynesia

Invaded Habitats
Grassland, freshwater wetlands, riparian habitats, wet forests.

Description
A prickly much-branched shrub or liana, with purpled stems and a taproot of 1-2 m depth. Leaves are twice pinnately compound with 8-15 pairs of pinnae, each having 20-40 pairs of leaflets. Leaflets are pubescent below, 3-8 mm long and 0.5-1.3 mm wide. Flowerheads are axillary, solitary or clustered, globose and 8-10 mm in diameter and have peduncles up to 4 cm length. Flowers are mauve or pink. Fruits are straight or somewhat curved, brownish-yellow pods of 4-8 cm length and 1-1.3 cm width.

Ecology and Control
This nitrogen-fixing shrub establishes readily after disturbance and forms thorny, impenetrable thickets over large areas that exclude all other species. It shades out native tree seedlings in invaded swamp forests and transforms various wetland communities into pure stands with reduced bird and plant species richness. The shrub accumulates a large soil seed bank with more than 10,000 seeds m^{-2}. Seeds are dispersed mainly by water.

The shrub can be killed by cutting at a depth of *c.* 10 cm below ground level. Burning is used to kill topgrowth but a follow-up programme is necessary to treat emerging seedlings. Cut stumps are treated with glyphosate, hexazinone, imazapyr, or triclopyr. Other effective herbicides are dicamba or tebuthiuron. In any case, follow-up programmes over several years are necessary to control regrowth and seedlings.

References
152, 281, 284, 415, 526, 713, 715, 759, 760, 761, 762, 763, 764, 813, 986, 1136.

Montanoa bipinnatifida (Kunth) Koch   Asteraceae

LF: Evergreen shrub, tree   CU: Ornamental
SN: *Uhdea bipinnatifida* Kunth

Geographic Distribution

Europe
   Northern
   British Isles
   Central, France
   Southern
   Eastern
   Mediterranean Isl.

Africa
   Northern
   Tropical
● Southern

Asia
   Temperate
   Tropical

Australasia
   Australia
   New Zealand

Northern America
   Canada, Alaska
   Southeastern USA
   Western USA
   Remaining USA
N Mexico

Southern America
   Tropical
   Caribbean
   Chile, Argentina

Atlantic Islands
   Cape Verde
   Canary, Madeira
   Azores
   South Atlantic Isl.

Indian Ocean Islands
   Mascarenes
   Seychelles
   Madagascar

Pacific Islands
   Micronesia
   Galapagos
   Hawaii
   Melanesia, Polynesia

Invaded Habitats
Forests and forest edges, riparian habitats.

Description
A shrub or small tree with several stems of 3-6 m height and 3-5 cm diameter. Branchlets are terete with ridges and whitish at the nodes. The deeply lobed leaves have stiff and appressed hairs on the veins, and have petioles of 5-10 cm length. Leaf blades are broadly ovate, 20-60 cm long, about as wide as long, with 3-5 lanceolate to ovate lobes on each side. Inflorescences are widely branched panicles reaching 1 m length, with numerous flowerheads and branches of 30-50 cm length. Flowerheads are 5-8 cm in diameter and have peduncles of 2-6 cm length. Ray florets have white ligules and are 2-3.5 cm long. The numerous disk florets are yellow and 5-8 mm long. Fruits are greenish brown to black, smooth, compressed achenes of 3-4 mm length.

Ecology and Control
Where native, this plant grows in wooded slopes and ravines, in open forests and along streams. It is invasive because it forms dense thickets that crowd out native plants. Little is known on the ecology of this species.

    Specific control methods for this species are not available. Seedlings and saplings may be hand pulled or dug out, larger plants cut and the cut stumps treated with herbicide.

References
792.

Moraea flaccida (Sweet) Steud.   Iridaceae

LF: Perennial herb   CU: Ornamental
SN: *Homeria flaccida* Sweet

Geographic Distribution

Europe	*Australasia*	*Atlantic Islands*
Northern	● Australia	Cape Verde
British Isles	New Zealand	Canary, Madeira
Central, France		Azores
Southern	*Northern America*	South Atlantic Isl.
Eastern	Canada, Alaska	
Mediterranean Isl.	Southeastern USA	*Indian Ocean Islands*
	Western USA	Mascarenes
Africa	Remaining USA	Seychelles
Northern	Mexico	Madagascar
Tropical		
N Southern	*Southern America*	*Pacific Islands*
	Tropical	Micronesia
Asia	Caribbean	Galapagos
Temperate	Chile, Argentina	Hawaii
Tropical		Melanesia, Polynesia

Invaded Habitats
Grass- and heathland, woodland.

Description
A cormous herb of 30-60 cm height, with branched stems and basal leaves. The solitary leaf is 9-13 mm wide. Flowers are salmon with a yellow centre or entirely yellow, and have bright yellow markings. The flower spathe is 5.5-8 cm long. Fruits are capsules of 3-5 cm length, containing 100-150 small, brown seeds.

Ecology and Control
A plant that invades natural communities along disturbed edges. It spreads rapidly and forms dense stands that reduce growth and regeneration of native plants. It produces numerous corms that remain viable in the soil for a few years. They may accumulate in the soil to a large corm bank.

Plants can be dug out, all corms must be removed. Chemical control is done by spraying infestations with herbicide before flowering commences. Any control programme will have to be repeated over several years to exhaust the corm bank in the soil.

References
215, 981, 986.

Morella faya (Ait.) Wilbur   Myricaceae

LF: Evergreen shrub, tree   CU: Ornamental
SN: *Myrica faya* Ait.

Geographic Distribution

Europe	*Australasia*		*Atlantic Islands*
Northern	Australia		Cape Verde
British Isles	New Zealand	N	Canary, Madeira
Central, France		N	Azores
Southern	*Northern America*		South Atlantic Isl.
Eastern	Canada, Alaska		
Mediterranean Isl.	Southeastern USA		*Indian Ocean Islands*
	Western USA		Mascarenes
Africa	Remaining USA		Seychelles
Northern	Mexico		Madagascar
Tropical			
Southern	*Southern America*		*Pacific Islands*
	Tropical		Micronesia
Asia	Caribbean		Galapagos
Temperate	Chile, Argentina	●	Hawaii
Tropical			Melanesia, Polynesia

Invaded Habitats
Grassland, forests, volcanic soils.

Description
A shrub or small tree up to 8 m tall, sometimes reaching 20 m, the young twigs being covered with small and brownish glands. The alternate and simple leaves have short petioles, are oblanceolate, 4-11 cm long, 1.3-3 cm wide, entire and glabrous or shallowly toothed towards the apex. Flowers are borne in axillary catkins on the leafy part of the stems. Male catkins are branched and bear numerous yellow-green flowers. Fruits are slightly fleshy globose drupes of 4-8 mm diameter.

Ecology and Control
A nitrogen-fixing shrub or tree that increases soil fertility levels. In the native range, it is common in the understorey of evergreen forests and on mountain slopes. Where invasive, the shrub quickly develops a closed canopy forest and displaces native vegetation. It grows rapidly and forms dense pure stands. The large amount of litter inhibits germination of native trees and shrubs. The increased soil fertility promotes invasion by weedy forbs and grasses. Seeds are dispersed by birds and mammals. After dieback, replacement of this shrub with alien grasses increases fire hazard in Hawaii.

    Specific control methods for this plant are not available. Seedlings and saplings may be pulled or dug out. Larger shrubs can be cut and the cut stumps treated with herbicide.

References
5, 284, 720, 909, 1196, 1309, 1347, 1348, 1357, 1398, 1399.

Morus alba L.   Moraceae

LF: Deciduous tree   CU: Ornamental, food
SN: *Morus indica* L.

Geographic Distribution

	Europe		Australasia		Atlantic Islands
	Northern		Australia		Cape Verde
	British Isles		New Zealand		Canary, Madeira
X	Central, France				Azores
X	Southern		*Northern America*		South Atlantic Isl.
X	Eastern	X	Canada, Alaska		
X	Mediterranean Isl.		Southeastern USA		*Indian Ocean Islands*
		X	Western USA		Mascarenes
	Africa		Remaining USA		Seychelles
	Northern		Mexico		Madagascar
X	Tropical				
•	Southern		*Southern America*		*Pacific Islands*
			Tropical		Micronesia
	Asia		Caribbean		Galapagos
N	Temperate	X	Chile, Argentina	X	Hawaii
	Tropical				Melanesia, Polynesia

Invaded Habitats
Forests, forest edges, grassland, rocky places.

Description
A tree up to 15 m tall, with slender and smooth branches, a rounded crown and with a white latex. The ovate to broad-ovate leaves are 6-18 cm long, coarsely toothed and often irregularly lobed, glabrous or pubescent on the veins beneath, and with petioles of 10-25 mm length. The tiny flowers are greenish-white and borne in dense spikes of *c.* 10 mm length. Fruits are ovoid and compressed achenes covered by the fleshy calyx, and aggregating into a syncarp of 10-25 mm length of white to purplish-violet colour. The peduncles are about as along as the syncarp.

Ecology and Control
This fast growing tree is highly variable in size and shape, and many cultivars and varieties are in cultivation for silk production. It is invasive because it spreads rapidly and forms dense stands that crowd out native species and prevent forest regeneration.

Specific control methods for this plant are not available. Seedlings and saplings may be pulled or dug out. Larger shrubs can be cut and the cut stumps treated with herbicide.

References
317, 549, 792, 1089, 1186.

Myosotis silvatica Ehrh. ex Hoffm. — Boraginaceae

LF: Perennial herb
SN: *Myosotis oblongata* Link
CU: Ornamental

Geographic Distribution

Europe		*Australasia*		*Atlantic Islands*	
	Northern	●	Australia		Cape Verde
N	British Isles	X	New Zealand	X	Canary, Madeira
N	Central, France				Azores
N	Southern	*Northern America*			South Atlantic Isl.
N	Eastern		Canada, Alaska		
N	Mediterranean Isl.		Southeastern USA	*Indian Ocean Islands*	
			Western USA		Mascarenes
Africa			Remaining USA		Seychelles
	Northern		Mexico		Madagascar
	Tropical				
	Southern	*Southern America*		*Pacific Islands*	
			Tropical		Micronesia
Asia			Caribbean		Galapagos
N	Temperate		Chile, Argentina		Hawaii
N	Tropical				Melanesia, Polynesia

Invaded Habitats
Forests, riparian habitats, disturbed sites.

Description
An erect to ascending, much-branched herb up to 50 cm tall, with a very leafy inflorescence. The lower leaves are elliptic-spathulate or obovate, up to 10 cm long and 4 cm wide, stem leaves are smaller. Flowers appear in terminal and spiralled cymes. The corollas are bright blue. Fruits are ovoid and acute nutlets of 1.5-2 mm length and *c.* 1.2 mm width.

Ecology and Control
This plant grows in forests, woodland, and rocky places within the native range, and is a variable species with four subspecies in Europe. It is a prolific seed producer that establishes in locally disturbed sites within natural communities. The dense growth habit crowds out native plants and impedes their regeneration. It displaces invaded areas with monospecific stands. Seeds are dispersed by water and animals.

Single plants and small patches may be pulled or dug out. Dense stands can be sprayed with herbicide, best before flowering commences. Cutting before seed ripen prevents seed dispersal but does not eliminate the plant.

References
121, 215.

Myriophyllum aquaticum (Vell.) Verdc.   Haloragaceae

LF: Aquatic perennial   CU: Ornamental
SN: *Myriophyllum brasiliense* Cambess.

Geographic Distribution

Europe		*Australasia*		*Atlantic Islands*	
	Northern	●	Australia		Cape Verde
X	British Isles	●	New Zealand		Canary, Madeira
	Central, France				Azores
	Southern		*Northern America*		South Atlantic Isl.
	Eastern		Canada, Alaska		
	Mediterranean Isl.		Southeastern USA		*Indian Ocean Islands*
		X	Western USA		Mascarenes
Africa		●	Remaining USA		Seychelles
	Northern		Mexico		Madagascar
	Tropical				
●	Southern		*Southern America*		*Pacific Islands*
		N	Tropical		Micronesia
Asia			Caribbean		Galapagos
X	Temperate	N	Chile, Argentina	X	Hawaii
X	Tropical				Melanesia, Polynesia

Invaded Habitats
Freshwater wetlands, ponds, streams and lakes.

Description
A submerged aquatic plant with rhizomes and glaucous or grey-green stems up to 2 m long. Leaves are borne in whorls of 4-6, 15-35 mm long, and deeply dissected into 8-30 fine segments. Flowers are axillary and more or less emerging above the water surface. Fruits are groups of a few nutlets. The plant is dioecious.

Ecology and Control
This plant spreads vegetatively by stem fragmentation. Only female plants are present in Britain, California, and southern Africa. The plant has aerial stems floating on the water surface but grows also as a submerged plant. It rapidly colonizes wetlands and forms dense stands that exclude native water plants. Light is strongly reduced and water flow impeded.

    The plant can be removed by mechanical harvesters, although mechanical removal may enhance stem fragmentation. Effective herbicides are 2,4-D, diquat, or fluridone approved for use in aquatic environments.

References
133, 137, 284, 508, 549, 986, 1089, 1415.

Myriophyllum spicatum L.   Haloragaceae

LF: Aquatic perennial   CU: -
SN: -

Geographic Distribution

Europe		*Australasia*		*Atlantic Islands*	
N	Northern	X	Australia		Cape Verde
N	British Isles		New Zealand		Canary, Madeira
N	Central, France			N	Azores
N	Southern	*Northern America*			South Atlantic Isl.
N	Eastern	●	Canada, Alaska		
N	Mediterranean Isl.	●	Southeastern USA	*Indian Ocean Islands*	
		●	Western USA		Mascarenes
Africa		●	Remaining USA		Seychelles
N	Northern		Mexico		Madagascar
X	Tropical				
X	Southern	*Southern America*		*Pacific Islands*	
		X	Tropical		Micronesia
Asia			Caribbean		Galapagos
N	Temperate		Chile, Argentina		Hawaii
N	Tropical				Melanesia, Polynesia

Invaded Habitats
Freshwater wetlands, lakes, coastal estuaries.

Description
A submerged aquatic plant with branching stems of 0.5-7 m length, becoming reddish or olive-green when dry. Leaves appear in whorls of 3-5, are deeply dissected with 13-38 fine segments, and 25-45 mm long. Inflorescences are spikes of 7-25 cm length, with the flowers being in whorls of four. The lower flowers are female, the upper ones male. Fruits are sub-globose, 2-3 mm long and separate into four nutlets with one seed each.

Ecology and Control
This aquatic is most common in waters of 1-3 m depth, but can invade waters up to 10 m deep. Stems easily fragment and the plant spreads vegetatively. It forms dense stands that shade out other species and alter the temperature profile of the water body. It flourishes in lakes with nutrient rich water, and the spread of the plant is promoted by eutrophication.

Plants can be removed by mechanical harvesters, although this may enhance stem fragmentation. The plant is highly susceptible to 2,4-D. Regrowth is fast if plants are not killed, and repeated control is necessary.

References
8, 133, 256, 481, 571, 581, 668, 805, 1396.

Nandina domestica Thunb. Berberidaceae

LF: Evergreen shrub　　　CU: Ornamental
SN: -

Geographic Distribution

Europe	Australasia	Atlantic Islands
Northern	Australia	Cape Verde
British Isles	New Zealand	Canary, Madeira
Central, France		Azores
Southern	*Northern America*	South Atlantic Isl.
Eastern	Canada, Alaska	
Mediterranean Isl.	● Southeastern USA	*Indian Ocean Islands*
	Western USA	Mascarenes
Africa	Remaining USA	Seychelles
Northern	Mexico	Madagascar
Tropical		
Southern	*Southern America*	*Pacific Islands*
	Tropical	Micronesia
Asia	Caribbean	Galapagos
N　Temperate	Chile, Argentina	Hawaii
Tropical		Melanesia, Polynesia

Invaded Habitats
Woodland, floodplains, forest edges.

Description
An upright, glabrous shrub with erect unbranched, thin and bamboo-like stems up to 2.5 m tall, and with a glossy foliage. Leaves are 30-50 cm long, bi- or tripinnately compound, with oval to lanceolate leaflets of 3-25 mm length or more. Inflorescences are large erect panicles up to 30 cm long, growing from the uppermost leaf axil. Flowers are white, *c.* 6 mm in diameter, and have large yellow anthers. Fruits are bright red or purplish, globular berries of *c.* 8 mm diameter, containing two seeds. The plant has rhizomes.

Ecology and Control
The dense foliage of this shrub shades out native plants and prevents their regeneration. It forms extensive and dense stands displacing native vegetation. Populations of two rare plant species are threatened in Florida by this invader. Berries are abundantly produced and seeds dispersed by birds. The plant grows rather slowly and withstands cold weather.

　　Seedlings can easily be hand pulled or dug out. Since the plant develops a strong taproot, digging out large individuals is difficult. An effective control is cutting the stems close to the ground and treating the stumps with a glyphosate or triclopyr herbicide.

References
715.

Nassella neesiana (Trin. & Rupr.) Barkworth   Poaceae

LF: Perennial herb   CU: -
SN: *Stipa neesiana* Trin. & Rupr.

Geographic Distribution

Europe		*Australasia*		*Atlantic Islands*	
	Northern	●	Australia		Cape Verde
X	British Isles	X	New Zealand	X	Canary, Madeira
	Central, France				Azores
	Southern	*Northern America*			South Atlantic Isl.
	Eastern		Canada, Alaska		
	Mediterranean Isl.		Southeastern USA	*Indian Ocean Islands*	
			Western USA		Mascarenes
Africa			Remaining USA		Seychelles
	Northern		Mexico		Madagascar
	Tropical				
	Southern	*Southern America*		*Pacific Islands*	
		N	Tropical		Micronesia
Asia			Caribbean		Galapagos
	Temperate	N	Chile, Argentina		Hawaii
	Tropical				Melanesia, Polynesia

Invaded Habitats
Grass- and woodland, riparian habitats.

Description
A tussock-forming grass with erect culms of 60-200 cm height. Leaves are 10-30 cm long and 1-5 mm wide, the papery ligule is 1-3 mm long and has a fringe of hairs. Leaf sheaths are densely pubescent at the base. Inflorescences are open, with drooping branches, and up to 30 cm long. Glumes are *c.* 2 cm long and have a long and fine apex. Lemmas are *c.* 10 mm long and have a long terminal awn of 5-9 cm length. The lower part of the awn is stout and spirally twisted. Seeds are 6-10 mm long.

Ecology and Control
This grass invades mainly degraded and disturbed plant communities. It becomes dominant and forms dense swards that exclude native grasses and forbs, reduce species richness and accumulate large amounts of litter. It is a prolific seed producer, and seeds are dispersed by water and animals. The seed rain may exceed 20,000 seeds m^{-2}. Besides normal flowers, the grass produces axillary and basal cleistogenes, e.g. flowers concealed within the leaf sheath.

Once established, the grass is difficult to control. If plants are killed, rapid recruitment will take place from the soil seed bank. Control is thus best combined with sowing of desirable species. Plants can be dug out, the crown must be removed. Burning is used to clear large areas and regrowth is then treated with herbicide. Follow-up programmes are necessary to treat seedlings.

References
138, 139, 140, 215, 266, 437, 787.

Nassella tenuissima (Trin.) Barkworth   Poaceae

LF: Perennial herb   CU:
SN: *Stipa tenuissima* Trin.

Geographic Distribution

Europe	Australasia	Atlantic Islands
Northern	Australia	Cape Verde
British Isles	X New Zealand	Canary, Madeira
Central, France		Azores
Southern	*Northern America*	South Atlantic Isl.
Eastern	Canada, Alaska	
Mediterranean Isl.	Southeastern USA	*Indian Ocean Islands*
	N Western USA	Mascarenes
Africa	Remaining USA	Seychelles
Northern	N Mexico	Madagascar
Tropical		
● Southern	*Southern America*	*Pacific Islands*
	N Tropical	Micronesia
Asia	Caribbean	Galapagos
Temperate	N Chile, Argentina	Hawaii
Tropical		Melanesia, Polynesia

Invaded Habitats
Grass- and woodland, disturbed sites.

Description
A densely caespitose grass with numerous glabrous culms of 30-70 cm height. Ligules are 0.5-2.5 mm long and glabrous. Leaf blades are up to 60 cm long and 0.3-0.5 mm wide. Inflorescences are contracted and contain few spikelets, are green to purplish and 15-25 cm long. Spikelets are narrowly lanceolate and 5-10 mm long (excluding awn). Glumes are 5-10 mm long. Caryopses are ellipsoid, slightly compressed and 2-2.5 mm long.

Ecology and Control
In the native range, this grass occurs in grass- and shrubland, in arid woodland from near sea level to about 2,900 m elevation. The grass is almost indistinguishable from *N. trichotoma* unless flowers are available. Primarily it invades disturbed grassland and becomes dominant, crowding out native vegetation and preventing the establishment of native plants.
   Specific control methods for this species are not available. The same methods as for other *Nassella* species may apply.

References
549, 625, 1089.

Nassella trichotoma (Nees) Hack. ex Arechav.   Poaceae

LF: Perennial herb
SN: *Stipa trichotoma* Nees
CU: -

Geographic Distribution

Europe	*Australasia*	*Atlantic Islands*
Northern	● Australia	Cape Verde
British Isles	● New Zealand	X Canary, Madeira
Central, France		Azores
X Southern	*Northern America*	South Atlantic Isl.
Eastern	Canada, Alaska	
Mediterranean Isl.	X Southeastern USA	*Indian Ocean Islands*
	X Western USA	Mascarenes
Africa	X Remaining USA	Seychelles
Northern	Mexico	Madagascar
Tropical		
● Southern	*Southern America*	*Pacific Islands*
	N Tropical	Micronesia
Asia	Caribbean	Galapagos
Temperate	N Chile, Argentina	X Hawaii
Tropical		Melanesia, Polynesia

Invaded Habitats
Grass- and rangeland, alluvial plains, coastal sand dunes and beaches.

Description
A tussock-forming grass with a deep root system and culms of 20-60 cm height. Leaves are thin, tightly rolled, up to 50 cm long and *c.* 0.5 mm in diameter. Ligules are white and hairless, the leaf base is white. Inflorescences are much branched panicles up to 95 cm long. Seeds are *c.* 1.5 mm long and have a terminal awn of *c.* 25 mm length.

Ecology and Control
A drought resistant, cool season grass that grows well on poor soils. It invades mainly disturbed grassland and forms dense colonies that eliminate native vegetation. Species richness in areas colonized by this grass is strongly reduced. Seed production is prolific and seeds are dispersed by wind, water and animals. The long-lived seeds accumulate in a soil seed bank and may remain viable for more than 12 years. Fires often result in massive regeneration of seedlings.

    Once established, the grass is difficult to control. Burning is only useful if combined with subsequent control of regrowth and seedlings. Glyphosate applied before seedheads begin to emerge prevents seed formation. Other herbicides used are dalapon, tetrapion, or 2,2-DPA. Seedlings are weak competitors and the grass may be suppressed by planting native species after control.

References
55, 198, 199, 202, 203, 204, 215, 437, 991, 1089, 1343, 1389.

Nephrolepis cordifolia (L.) C. Presl   Nephrolepidaceae

LF: Climbing fern
CU: Ornamental
SN: *Nephrolepis auriculata* (L.) Trimen, *Polypodium cordifolium* L.

Geographic Distribution

Europe	Australasia	Atlantic Islands
Northern	● N Australia	N Cape Verde
British Isles	X New Zealand	X Canary, Madeira
Central, France		X Azores
Southern	Northern America	South Atlantic Isl.
Eastern	Canada, Alaska	
Mediterranean Isl.	● Southeastern USA	Indian Ocean Islands
	Western USA	Mascarenes
Africa	Remaining USA	Seychelles
Northern	Mexico	Madagascar
N Tropical		
Southern	Southern America	Pacific Islands
	N Tropical	Micronesia
Asia	Caribbean	Galapagos
X Temperate	Chile, Argentina	Hawaii
N Tropical		Melanesia, Polynesia

Invaded Habitats
Forests and forest edges, hammocks, cliffs, rocks, waste places.

Description
A stoloniferous fern with or without tubers, with spreading stems and a slender rootstock. Fronds are erect and stiff, 25-110 cm long, 3-7 cm wide, with a blade linear in outline, and with pinnae having overlapping auricles. Petioles are 3-20 cm long. Scales are pale brown and spreading. Numerous sori appear between the midvein and margin; they are terminal on the upper branch of a free veinlet. The indusia are attached along a broad sinus and are 1.1-1.7 mm wide.

Ecology and Control
This fern grows either as a terrestrial plant or as an epiphyte on trees. The exact native range is obscure, and several cultivars have been developed that are widely used as ornamentals. The fern grows in the understorey of forests or climbs high into the canopies of trees, smothering them and preventing regeneration of native shrubs and trees. The fern grows best in shady conditions. The large fronds reduce light levels and shade out native plants.

Specific control methods for this species are not available. Plants may be dug out, tubers and rhizomes must be removed to prevent regrowth.

References
715.

Nephrolepis multiflora (Rox.) Jarrett   Nephrolepidaceae

LF: Climbing fern   CU: Ornamental
SN: -

Geographic Distribution

Europe
    Northern
    British Isles
    Central, France
    Southern
    Eastern
    Mediterranean Isl.

Africa
    Northern
  N Tropical
    Southern

Asia
    Temperate
  N Tropical

Australasia
    Australia
    New Zealand

Northern America
    Canada, Alaska
  ● Southeastern USA
    Western USA
    Remaining USA
    Mexico

Southern America
  X Tropical
    Caribbean
    Chile, Argentina

Atlantic Islands
    Cape Verde
    Canary, Madeira
    Azores
    South Atlantic Isl.

Indian Ocean Islands
  ● Mascarenes
    Seychelles
    Madagascar

Pacific Islands
    Micronesia
    Galapagos
  X Hawaii
    Melanesia, Polynesia

Invaded Habitats
Forests and forest edges, hammocks.

Description
A medium to large fern with short and erect rhizomes. Leaf fronds are 30-250 cm long and 3-16 cm wide, and compound with pinnae of 3.5-12.5 cm length. Petioles are covered with appressed scales and are 5-45 cm long. The midribs of pinnae are densely hairy above. Sori are attached close to the margin. The circular to horseshoe-shaped indusia are 1.1-1.3 mm long.

Ecology and Control
Several cultivars of this fern are widely used as ornamentals. The fern builds dense populations and displaces native vegetation. It climbs high into the canopies of shrubs and trees, smothering them and preventing their regeneration. Fronds also form a thick mat on the ground, preventing any establishment of native plants. Little is known on the ecology of this species.
    Specific control methods for this species are not available. Plants may be dug out, rhizomes and stolons must be removed to prevent regrowth.

References
1093, 1261.

Neyraudia reynaudiana (Kunth) Keng ex Hitchc.   Poaceae

LF: Perennial herb   CU: -
SN: *Arundo reynaudiana* Kunth

Geographic Distribution

Europe	*Australasia*	*Atlantic Islands*
Northern	Australia	Cape Verde
British Isles	New Zealand	Canary, Madeira
Central, France		Azores
Southern	*Northern America*	South Atlantic Isl.
Eastern	Canada, Alaska	
Mediterranean Isl.	● Southeastern USA	*Indian Ocean Islands*
	Western USA	Mascarenes
Africa	Remaining USA	Seychelles
Northern	Mexico	Madagascar
Tropical		
Southern	*Southern America*	*Pacific Islands*
	Tropical	Micronesia
Asia	X Caribbean	Galapagos
N Temperate	Chile, Argentina	Hawaii
N Tropical		Melanesia, Polynesia

Invaded Habitats
Open forests and forest edges, disturbed sites.

Description
A tall and stout grass reaching 3 m height, with bluish nodes, glabrous-striate culms, and short, coarse rhizomes. Stems are often branched. Leaf sheaths are smooth, 10-25 cm long, and have ligules of hairs. Leaf blades are linear, 20-100 cm long and 8-25 mm wide, sparsely hairy above, and often deciduous. Inflorescences are large, silver-hairy panicles of 30-60 cm length, nodding and finely branched. Spikelets have 4-8 florets and are 6-8 mm long. Lemmas are long-hairy and have slender awns.

Ecology and Control
In the native range, this grass is found at bogs, in disturbed sites, often growing on infertile soils. The grass forms dense colonies that eliminate native vegetation. It invades pine rocklands in Florida and threatens populations of rare plant species. The grass accumulates dead biomass and is highly flammable, increasing fire frequency and intensity. Fires in turn promote the spread of this grass. It causes high mortality of *Pinus elliottii* in Florida due to increased fire frequency. It vigorously resprouts after damage. Seeds are dispersed by wind.

Specific control methods for this species are not available. Plants may be dug out or cut at ground level, rhizomes must be removed to prevent regrowth.

References
715, 1035.

Nicotiana glauca Graham — Solanaceae

LF: Evergreen shrub, tree
CU: -
SN: -

Geographic Distribution

Europe	*Australasia*	*Atlantic Islands*
Northern	● Australia	● Cape Verde
British Isles	X New Zealand	X Canary, Madeira
X Central, France		Azores
● Southern	*Northern America*	X South Atlantic Isl.
X Eastern	Canada, Alaska	
X Mediterranean Isl.	Southeastern USA	*Indian Ocean Islands*
	X Western USA	Mascarenes
Africa	Remaining USA	Seychelles
Northern	X Mexico	Madagascar
X Tropical		
● Southern	*Southern America*	*Pacific Islands*
	N Tropical	Micronesia
Asia	Caribbean	Galapagos
X Temperate	N Chile, Argentina	X Hawaii
Tropical		Melanesia, Polynesia

Invaded Habitats
Riverbanks, desert scrub, arid grassland, coastal beaches, rocky places.

Description
A slender and generally glabrous shrub or small tree of 2-8 m height, with a sparse crown and a soft wood. The ovate to elliptic or lanceolate leaves are 5-25 cm long and up to 12 cm wide, blue-green and glabrous, and have an asymmetric base. Yellow and narrowly tubular flowers of 30-50 mm length are borne in terminal drooping clusters. Corollas have short lobes and are 30-35 mm in diameter, with the filaments attached below the middle of the tube. Fruits are ellipsoid capsules of 7-15 mm length, containing numerous minute seeds.

Ecology and Control
A drought resistant plant that grows in a wide range of conditions, either as a stunted shrub or as a tree. The vigorous growth leads to dense pure stands that crowd out native species and prevent natural regeneration. The large leaves shade out all vegetation below its canopies. Little is known on the ecology of this shrub.

Seedlings and saplings can be hand pulled or dug out. Larger shrubs are cut and the cut stumps treated with herbicide.

References
215, 549, 554, 1089.

Olea europaea L. Oleaceae

LF: Evergreen tree
SN: -
CU: Food, ornamental

Geographic Distribution

Europe	*Australasia*	*Atlantic Islands*
Northern	● Australia	N Cape Verde
British Isles	X New Zealand	N Canary, Madeira
Central, France		Azores
N Southern	*Northern America*	South Atlantic Isl.
Eastern	Canada, Alaska	
N Mediterranean Isl.	Southeastern USA	*Indian Ocean Islands*
	X Western USA	N Mascarenes
Africa	Remaining USA	Seychelles
N Northern	Mexico	N Madagascar
N Tropical		
N Southern	*Southern America*	*Pacific Islands*
	Tropical	Micronesia
Asia	Caribbean	Galapagos
Temperate	Chile, Argentina	● Hawaii
Tropical		Melanesia, Polynesia

Invaded Habitats
Grass- and woodland, riparian habitats.

Description
A much-branched tree of 3-12 m height, sometimes more, with a broad crown, a thick trunk, and a grey and finely fissured bark. Leaves are mostly opposite, leathery, 2-8 cm long and 5-25 mm wide, almost sessile, dark greyish-green and glabrous above, silvery grey beneath. White flowers of 2.5-3 mm length, with four lobes and two stamens are borne in short racemes or panicles. Fruits are ellipsoid to subglobose brownish-green to black drupes of 10-35 mm length and 6-20 mm width.

Ecology and Control
Where native, this tree grows in woods, scrub, and dry rocky places up to 3000 m elevation. The growth form ranges from stunted shrubs to tall trees. The domestic olive is a group of more than 2500 cultivars. Naturalized plants belong to var. *africana*. Where invasive, the tree forms a dense and permanent canopy under which native shrubs and trees cannot grow, but seedlings of this tree grow well. Over time, it transforms native vegetation into a species poor shrubland. The tree propagates by seeds and vegetatively and it resprouts vigorously after damage. Seeds are dispersed by birds and mammals.

    Seedlings can be hand pulled. A dense seedling cover can be sprayed with glyphosate. Larger trees are best controlled by cutting and treating the cut stumps with herbicide or by basal bark treatments. Effective herbicides for these treatments are picloram plus triclopyr or 2,4-D ester.

References
121, 215, 607, 924, 986, 990, 1077, 1216, 1310.

Opuntia aurantiaca Lindl.   Cactaceae

LF: Succulent perennial   CU: -
SN: -

Geographic Distribution

Europe	*Australasia*	*Atlantic Islands*
Northern	● Australia	Cape Verde
British Isles	New Zealand	Canary, Madeira
Central, France		Azores
Southern	*Northern America*	South Atlantic Isl.
Eastern	Canada, Alaska	
Mediterranean Isl.	X Southeastern USA	*Indian Ocean Islands*
	Western USA	Mascarenes
Africa	Remaining USA	Seychelles
Northern	Mexico	Madagascar
Tropical		
● Southern	*Southern America*	*Pacific Islands*
	X Tropical	Micronesia
Asia	N Caribbean	Galapagos
Temperate	X Chile, Argentina	Hawaii
Tropical		Melanesia, Polynesia

Invaded Habitats
Grass- and rangeland, disturbed sites.

Description
A low-growing, much branched shrub up to 30 cm tall, with an underground tuber and many narrow joints. Stems are prostrate to erect, with dark green, linear to club shaped segments of 5-15 cm length and 10-15 cm width. Areoles are greyish white. Spines are brownish to yellowish, 1-3 cm long, and in clusters of 2-3 or more. Flowers are deep yellow to orange-yellow and up to 4 cm in diameter. The purplish red fruits are pear shaped, very spiny, and up to 3 cm long.

Ecology and Control
The exact native range of this cactus is obscure and it is probably of hybrid origin. The cactus reproduces entirely vegetatively, rooting from the joints that break off very easily. Such dislodged joints remain viable for many months and are carried by flood waters and animals. Seeds are dispersed by animals eating the fruits. The plant forms dense and tall thickets that crowd out native species and impede wildlife movement.

    Small plants can be pulled or dug out. Mechanical control of larger plants is difficult. Spraying with picloram iso-octylester proved to be an effective chemical control method.

References
215, 549, 899, 900, 986, 1089, 1329, 1460.

Opuntia dillenii (Ker Gawl.) Haw. Cactaceae

LF: Succulent perennial CU: Ornamental, shelter
SN: *Opuntia stricta* var. *dillenii* (Ker Gawl.) L. D. Benson

Geographic Distribution

Europe	*Australasia*	*Atlantic Islands*
Northern	Australia	Cape Verde
British Isles	New Zealand	Canary, Madeira
Central, France		Azores
● Southern	*Northern America*	South Atlantic Isl.
Eastern	Canada, Alaska	
Mediterranean Isl.	N Southeastern USA	*Indian Ocean Islands*
	N Western USA	Mascarenes
Africa	Remaining USA	Seychelles
Northern	N Mexico	● Madagascar
X Tropical		
X Southern	*Southern America*	*Pacific Islands*
	N Tropical	Micronesia
Asia	N Caribbean	Galapagos
● Temperate	Chile, Argentina	Hawaii
X Tropical		Melanesia, Polynesia

Invaded Habitats
Grass- and heathland, dry forests, shrubland, rock outcrops.

Description
An erect or sprawling shrubby succulent, up to 3 m tall, sometimes with a trunk, and with flattened fleshy, obovate, blue-green stem segments of 7-40 cm length and 6-9 cm width. Spines are extremely variable, borne in clusters of 1-5, sometimes absent, and up to 5 cm long. Flowers lemon yellow to yellowish orange, sometimes reddish, and 7-8 cm long. The purplish fruits are spineless, pear shaped to globose, and 5-7.5 cm long.

Ecology and Control
The joints of this cactus dislodge and root easily. They are carried by water streams and by animals. Seeds are dispersed by animals feeding on the fruits. The cactus forms dense and spiny thickets that crowd out native plants and affect wildlife. Scattered clumps expand laterally and may lead to thickets that cover large areas.
    Specific control methods for this species are not available. The same methods as for other *Opuntia* species may apply.

References
378, 826.

Opuntia ficus-indica (L.) Mill. — Cactaceae

LF: Succulent perennial
SN: *Opuntia gymnocarpa* Weber
CU: Ornamental, food

Geographic Distribution

Europe
- Northern
- British Isles
- Central, France
- ● Southern
- Eastern
- ● Mediterranean Isl.

Africa
- X Northern
- X Tropical
- ● Southern

Asia
- X Temperate
- Tropical

Australasia
- ● Australia
- New Zealand

Northern America
- Canada, Alaska
- Southeastern USA
- X Western USA
- Remaining USA
- N Mexico

Southern America
- Tropical
- X Caribbean
- Chile, Argentina

Atlantic Islands
- Cape Verde
- X Canary, Madeira
- X Azores
- South Atlantic Isl.

Indian Ocean Islands
- Mascarenes
- X Seychelles
- Madagascar

Pacific Islands
- Micronesia
- Galapagos
- X Hawaii
- Melanesia, Polynesia

Invaded Habitats
Arid bushland, grassland, coastal scrub, rocky places.

Description
A branched, succulent shrub or small tree of 3-5 m height, with some of the branches flattened to form leaf-like structures (cladodes). These are elliptic-obovate, greyish green, 25-60 cm long and 6-20 cm wide, almost spineless or with tufts of stout spines. The bright yellow to orange flowers are 7-10 cm in diameter and borne on the cladode margins. Filaments are pale green to pale pink. Fruits are oval, yellow-orange to purple and juicy berries of 5-9 cm length. Seeds are 4-5 mm long.

Ecology and Control
A variable cactus with several forms, including a thorny and a thornless one. In addition, numerous cultivars have been developed and are widely used. The plant is a large succulent that branches frequently and forms dense impenetrable thickets that crowd out native vegetation. It spreads by seeds and vegetatively by dislodged stem segments that easily root and regenerate new plants. A single stem segment is capable of building up a dense thicket. Seeds are dispersed by animals.

Specific control methods for this species are not available. The same methods as for other *Opuntia* species may apply.

References
42, 549, 734, 1089.

Opuntia stricta (Haw.) Haw.

Cactaceae

LF: Succulent perennial
SN: *Opuntia inermis* DC.
CU: Ornamental, shelter

Geographic Distribution

Europe
- Northern
  - British Isles
  - Central, France
- • Southern
  - Eastern
  - Mediterranean Isl.

Africa
- Northern
- Tropical
- • Southern

Asia
- Temperate
- Tropical

Australasia
- • Australia
- New Zealand

Northern America
- Canada, Alaska
- N Southeastern USA
- Western USA
- Remaining USA
- Mexico

Southern America
- Tropical
- N Caribbean
- Chile, Argentina

Atlantic Islands
- Cape Verde
- Canary, Madeira
- Azores
- South Atlantic Isl.

Indian Ocean Islands
- Mascarenes
- Seychelles
- Madagascar

Pacific Islands
- Micronesia
- Galapagos
- Hawaii
- Melanesia, Polynesia

Invaded Habitats
Grass- and shrubland, rangeland, disturbed sites.

Description
A succulent, much branched shrub up to 2 m tall, with compressed and flattened branches (stem segments) and leaves reduced to small deciduous scales. The green to bluish-green stem segments are narrowly obovate, 15-25 cm long and 8-13 cm wide. Areoles are distant, brownish, and with or without spines. Spines are yellow at maturity and 1-4 cm long. Light yellow to reddish flowers of 6-7 cm diameter are borne in few to many per stem segment. Filaments are yellow to greenish. The fruit is a purplish berry, obovoid, 4-6 cm long, containing 60-180 seeds.

Ecology and Control
In the native range, this cactus grows on coastal dunes, on shell mounds, and in coastal hammocks. There are two varieties in Florida: var. *stricta* and var. *dillenii*, the latter is often granted species status. The plant spreads rapidly and forms extensive thorny thickets that impede wildlife and replace native vegetation. Seeds are dispersed by birds and mammals, and remain viable for more than ten years. Dislodged stem segments are carried by streams and root easily, forming new infestations.

Smaller infestations may be removed manually. The best time for chemical control is before fruit swelling occurs, because seeds from unripe fruits are also viable. A very effective control is applying MSMA by stem injection. Frequent follow-up programmes are necessary to treat regrowth and seedlings.

References
215, 549, 575, 774, 775, 986, 1046, 1074, 1075, 1341.

Ornithogalum umbellatum L.   Liliaceae

LF: Perennial herb   CU: Ornamental
SN: -

Geographic Distribution

Europe	*Australasia*	*Atlantic Islands*
Northern	● Australia	Cape Verde
British Isles	New Zealand	Canary, Madeira
N Central, France		Azores
N Southern	*Northern America*	South Atlantic Isl.
N Eastern	Canada, Alaska	
N Mediterranean Isl.	Southeastern USA	*Indian Ocean Islands*
	Western USA	Mascarenes
Africa	● Remaining USA	Seychelles
N Northern	Mexico	Madagascar
Tropical		
Southern	*Southern America*	*Pacific Islands*
	Tropical	Micronesia
Asia	Caribbean	Galapagos
N Temperate	Chile, Argentina	Hawaii
Tropical		Melanesia, Polynesia

Invaded Habitats
Woodland, forests and forest edges, disturbed sites.

Description
A herb of 20-40 cm height with 5-9 basal leaves only. These are 2-8 mm wide, glabrous and have a white stripe on the upper surface. Flower stems are glabrous, and each flower has one bract shorter than or equalling the pedicels. Inflorescences have 8-20 flowers. The perianth-segments are 15-22 mm long and white with wide green stripes beneath. Fruits are oblong-ovoid capsules.

Ecology and Control
Where native, this herb grows in hedges, dry places and disturbed ground. It is a variable species, especially with regard to the shape and size of bulbs and leaves. The plant spreads both by seeds and vegetatively, and forms dense patches that displace native vegetation and reduce species richness. It establishes well in disturbed sites and serves as a vector for fungal rust transmission. It spreads probably exclusively by vegetative growth in invaded woodlands, since heavy shade prevents flowering.
    Plants can be pulled or dug out, bulbs and roots must be removed to prevent regrowth. Dense patches can be treated with herbicide.

References
320, 874, 901.

Oxalis latifolia Kunth   Oxalidaceae

LF: Perennial herb
SN: -
CU: Ornamental

Geographic Distribution

Europe	*Australasia*	*Atlantic Islands*
Northern	● Australia	Cape Verde
X British Isles	X New Zealand	X Canary, Madeira
Central, France		X Azores
X Southern	*Northern America*	South Atlantic Isl.
Eastern	Canada, Alaska	
Mediterranean Isl.	Southeastern USA	*Indian Ocean Islands*
	X Western USA	X Mascarenes
Africa	Remaining USA	X Seychelles
Northern	N Mexico	X Madagascar
X Tropical		
X Southern	*Southern America*	*Pacific Islands*
	N Tropical	Micronesia
Asia	X Caribbean	Galapagos
X Temperate	Chile, Argentina	Hawaii
X Tropical		X Melanesia, Polynesia

Invaded Habitats

Forests and forest edges, arable fields, waste places.

Description

A glabrous stemless herb with rhizomes of 2-30 cm length and bulblets often formed at the ends of rhizomes. The taproot is up to 6 cm long and bulbs of 1-2 cm diameter are formed at ground level. Leaves are compound and have three leaflets up to 6 cm long and 10 cm wide, with rounded to pointed lobes. Petioles are 10-25 cm long. Flowers are borne in umbel-like inflorescences on peduncles of 15-20 cm length, are pale to deep pink, occasionally white, and 8-13 mm in diameter. Fruits are capsules releasing numerous seeds of *c.* 0.7 mm length.

Ecology and Control

A variable species with regard to leaf shape. Numerous bulblets appear from the bulbs at the ends of rhizomes. These bulblets are easily dislodged and dispersed. The prolific production of bulbs and bulblets makes the plant a successful colonizer. Fruits are rarely produced among naturalized plants. Although mainly a weed of agroecosystems, it invades natural plant communities and crowds out native plants due to the dense stands.

Repeated clipping reduces the vigour of this plant. Covering plants with plastic or straw mulch can kill the plants, as does flooding. Effective herbicides are 2,4-D, diuron, or dalapon. Trifluralin is active against bulbs.

References

52, 215, 235, 236, 581, 627, 826, 827, 973, 986.

Oxalis pes-caprae L.   Oxalidaceae

LF: Perennial herb
SN: *Oxalis cernua* Thunb.
CU: Ornamental

Geographic Distribution

Europe	*Australasia*	*Atlantic Islands*
Northern	● Australia	Cape Verde
X British Isles	X New Zealand	X Canary, Madeira
X Central, France		X Azores
● Southern	*Northern America*	South Atlantic Isl.
Eastern	Canada, Alaska	
X Mediterranean Isl.	Southeastern USA	*Indian Ocean Islands*
	● Western USA	Mascarenes
Africa	Remaining USA	Seychelles
Northern	Mexico	Madagascar
Tropical		
N Southern	*Southern America*	*Pacific Islands*
	Tropical	Micronesia
Asia	Caribbean	Galapagos
Temperate	Chile, Argentina	Hawaii
X Tropical		Melanesia, Polynesia

Invaded Habitats
Forests, grassland, riparian habitats, coastal beaches, disturbed sites.

Description
A glabrous or sparsely pubescent, stemless herb with a rosette of leaves and flowers arising at soil level, and with a short annual underground stem emerging from a deeply buried bulb. The underground stem bears numerous bulblets. Leaves are compound and have three leaflets of 8-20 mm length and 12-30 mm width. Yellow flowers appear in umbellate cymes and are 20-25 mm in diameter. Both petioles and peduncles are up to 30 cm long. Fruits are short capsules, containing numerous small seeds.

Ecology and Control
A highly variable species with more than 30 naturalized variants found in Australia, differing mainly in their leaflet markings. The plant produces bulblets abundantly which easily break off and are the main propagule for spread. The plant forms extensive, almost pure stands and spreads rapidly. Although an important agricultural weed, it invades natural plant communities and displaces native plants by the dense stands.

    Individual scattered plants can be dug out, all bulbs and bulblets need to be removed to prevent recolonization. Constant weeding before bulblet formation may weaken the plant. Chemical control includes spraying 2,4-D, glyphosate, fenoprop, or chlorsulfuron.

References
154, 215, 235, 236, 714, 826, 875, 1001, 1049.

Oxalis purpurea L.   Oxalidaceae

LF: Perennial herb
SN: *Oxalis humilis* Thunb.
CU: Ornamental

Geographic Distribution

Europe	*Australasia*	*Atlantic Islands*
Northern	● Australia	Cape Verde
British Isles	X New Zealand	X Canary, Madeira
Central, France		X Azores
X Southern	*Northern America*	South Atlantic Isl.
Eastern	Canada, Alaska	
X Mediterranean Isl.	Southeastern USA	*Indian Ocean Islands*
	X Western USA	Mascarenes
Africa	Remaining USA	Seychelles
Northern	Mexico	Madagascar
Tropical		
N Southern	*Southern America*	*Pacific Islands*
	Tropical	Micronesia
Asia	X Caribbean	Galapagos
Temperate	Chile, Argentina	Hawaii
Tropical		Melanesia, Polynesia

Invaded Habitats
Grass- and heathland, shrubland, coastal beaches, disturbed sites.

Description
A pubescent or villous stemless herb with a bulb that emits an ascending underground stem. The stem forms a rosette of leaves and flower stalks at soil level. Petioles are 3-10 cm long, the leaves are compound with three rhombic leaflets of 8-23 mm length and 9-30 mm width. Leaflets are rhombic and often purplish beneath. Flowers are pale yellow, white or rose-purple and borne solitary on peduncles of 1-10 cm length. The corolla is 20-22 mm long. The sepals are lanceolate, the petals purplish-pink and white at the base, and 25-35 mm long. Fruits are capsules of *c.* 5 mm length.

Ecology and Control
This is a highly variable species within the native range with several varieties differing mainly in flower colour. The commonly naturalized form has rose to purple flowers. The plant spreads mainly vegetatively by bulbs which are easily dislodged and dispersed. The plant is invasive because it builds up dense colonies that displace native vegetation and reduce species richness. Established populations are persistent and highly competitive to native plants.

    Specific control methods for this species are not available. The same methods may apply as for other *Oxalis* species.

References
215.

Paederia foetida L.   Rubiaceae

LF: Deciduous climber   CU: Ornamental
SN: *Paederia scandens* (Lour.) Merr., *Paederia tomentosa* Blume

Geographic Distribution

Europe	*Australasia*	*Atlantic Islands*
Northern	Australia	Cape Verde
British Isles	New Zealand	Canary, Madeira
Central, France		Azores
Southern	*Northern America*	South Atlantic Isl.
Eastern	Canada, Alaska	
Mediterranean Isl.	● Southeastern USA	*Indian Ocean Islands*
	Western USA	X Mascarenes
Africa	Remaining USA	Seychelles
Northern	Mexico	Madagascar
Tropical		
Southern	*Southern America*	*Pacific Islands*
	X Tropical	Micronesia
Asia	Caribbean	Galapagos
N Temperate	Chile, Argentina	X Hawaii
N Tropical		X Melanesia, Polynesia

Invaded Habitats
Forests and forest edges, woodland, tropical hammocks.

Description
A climbing or twining shrub with slender stems of 1.5-7 m length and 2-5 mm diameter. Leaves are opposite, ovate to ovate-lanceolate, 5-12 cm long, acuminate or rounded at the end, dark green above, light green beneath and pubescent. Petioles are 1-5 cm long. Flowers are borne in axillary cymes that form long terminal panicles. The whitish to lilac flowers are pubescent outside and have a purple centre. The corolla is 10-15 mm long and 4-6 mm in diameter. Fruits are globose, orange, 5-6 mm in diameter, and each contain two nutlets of 3.5-5.5 mm length. The plant has a woody rootstock.

Ecology and Control
This highly variable and fast growing species grows naturally in openings of wet evergreen to dry deciduous forests and woodland. The plant has both climbing and creeping stems, the latter root at the nodes. The species forms dense curtains of intermingled stems, covering the floor, smothering all vegetation and altering the community structure. Trees may be killed by the weight of vines. The plant colonizes rapidly tree-fall gaps and persists once established, preventing natural forest regeneration.

    Prescribed burning is effective in slowing the spread of this vine in fire-adapted communities. Seedlings and small plants are easy to pull out. Cutting stems at ground level prevents host tree mortality but regrowth is rapid. An effective herbicide is glyphosate applied to the foliage and to basal as well as creeping stems.

References
334, 435, 715, 1044, 1050.

Panicum maximum Jacq. — Poaceae

LF: Perennial herb
SN: *Panicum hirsutissimum* Steud.
CU: Fodder, forage

Geographic Distribution

Europe
- Northern
- British Isles
- Central, France
- N Southern
- Eastern
- Mediterranean Isl.

Africa
- X Northern
- N Tropical
- N Southern

Asia
- N Temperate
- X Tropical

Australasia
- X Australia
- New Zealand

Northern America
- Canada, Alaska
- X Southeastern USA
- Western USA
- Remaining USA
- X Mexico

Southern America
- • Tropical
- X Caribbean
- Chile, Argentina

Atlantic Islands
- N Cape Verde
- X Canary, Madeira
- Azores
- South Atlantic Isl.

Indian Ocean Islands
- N Mascarenes
- N Seychelles
- N Madagascar

Pacific Islands
- Micronesia
- • Galapagos
- • Hawaii
- X Melanesia, Polynesia

Invaded Habitats
Grassland, pastures, rocky places, disturbed sites.

Description
A large, pale green tufted grass with stout and glabrous culms of 1-3.5 m height. Leaf sheaths are mostly shorter than the internodes, and short pubescent to almost glabrous. The ligule is 3-4 mm long and has long hairs. Leaf blades are linear, flat, up to 70 cm long and 1-3 cm wide, with a prominent white midvein. Inflorescences are terminal panicles of 20-50 cm length, with the primary branches in whorls and pendent. Spikelets have long pedicels, are 2-4 mm long, elliptical, and contain two florets each. The upper glume is as long as the spikelet. Fruits are greenish white and rugose caryopses of 2.5-3 mm length. The plant has a short creeping rhizome.

Ecology and Control
In the native range, this grass is found in grassland and open woodland, generally in shady places. The grass grows on a wide range of soil types but does not tolerate severe droughts and waterlogging. Several cultivars have been developed. The grass spreads slowly and forms dense tussocks, displacing native grasslands and other vegetation, thereby reducing native plant species richness. It is a prolific seed producer and the grass is resistant to fires.

Specific control methods for this species are not available. The grass regrows from its underground rhizomes if damaged, thus cutting alone does not kill the plant.

References
580, 984, 1033, 1041, 1190, 1313.

Panicum repens L.   Poaceae

LF: Perennial herb   CU: Erosion control
SN: *Panicum gouinii* Fourn., *Panicum littorale* Mohr ex Vasey

Geographic Distribution

	Europe		Australasia		Atlantic Islands
	Northern	X	Australia		Cape Verde
	British Isles		New Zealand	N	Canary, Madeira
X	Central, France			X	Azores
N	Southern		Northern America		South Atlantic Isl.
	Eastern		Canada, Alaska		
	Mediterranean Isl.	●	Southeastern USA		Indian Ocean Islands
		X	Western USA		Mascarenes
	Africa		Remaining USA	X	Seychelles
X	Northern		Mexico		Madagascar
N	Tropical				
N	Southern		Southern America		Pacific Islands
		X	Tropical		Micronesia
	Asia	X	Caribbean		Galapagos
	Temperate	X	Chile, Argentina	X	Hawaii
X	Tropical				Melanesia, Polynesia

Invaded Habitats
Grassland, riparian habitats, coastal beaches.

Description
A rhizomatous grass with rigid, decumbent culms of 30-90 cm length, branching from the rhizome. Leaf sheaths are fringed with long hairs. Leaves are 4-25 cm long and 2-15 mm wide, flat or folded. The inflorescence is an open panicle of 4-20 cm length with stiff and ascending branches. Spikelets are 2-3 mm long, elliptical, glabrous, and have two florets. The lower glume is shorter than the spikelet, the upper glume as long as it. Caryopses are lanceolate and straw-coloured.

Ecology and Control
Where native, this C_4 grass occurs in damp places, on lake shores and in swamps, generally on sandy soils and mostly near the coast. It is a very drought tolerant species due to persistent rhizomes and is highly salt tolerant. It spreads mainly by rhizomes and forms dense pure swards that replace native species. Rhizomes may reach 6 m length or more. It does not appear to produce viable seeds in Florida.
     Eradication is difficult due to the vigorous vegetative regrowth. Smaller patches can be dug out but rhizomes must be removed as far as possible. Larger infestations can be treated with grass-selective or non-selective herbicides.

References
580, 715, 1190, 1405, 1406.

Parapholis incurva (L.) Hubb.   Poaceae

LF: Annual herb   CU: -
SN: *Aegilops incurva* L.

Geographic Distribution

Europe
    Northern
  N British Isles
  N Central, France
  N Southern
  N Eastern
  N Mediterranean Isl.

Africa
  N Northern
    Tropical
  X Southern

Asia
  N Temperate
  N Tropical

Australasia
  ● Australia
  X New Zealand

Northern America
    Canada, Alaska
    Southeastern USA
  X Western USA
    Remaining USA
    Mexico

Southern America
  X Tropical
  X Caribbean
  X Chile, Argentina

Atlantic Islands
    Cape Verde
  N Canary, Madeira
    Azores
    South Atlantic Isl.

Indian Ocean Islands
    Mascarenes
    Seychelles
    Madagascar

Pacific Islands
    Micronesia
    Galapagos
    Hawaii
    Melanesia, Polynesia

Invaded Habitats
Coastal salt marshes and beaches.

Description
A smooth and glabrous grass with erect to ascending culms of 2-25 cm height. The leaves are up to 8 cm long and 1-2 mm wide, and have short membranuous ligules. Inflorescences are very slender cylindrical spikes of 1-10 cm length, with 10-20 alternate spikelets that are sunk in cavities of the spike axis. The spikelets have one floret each and are 4.5-7 mm long. Glumes are lanceolate and strongly veined.

Ecology and Control
This grass grows on coastal cliffs, in salt marshes and coastal sandy or rocky places. It is a cleistogamous species that spreads rapidly by seeds and builds up dense colonies that reduce species richness and displace native plants.

    Specific control methods for this species are not available. Scattered plants can be hand pulled, larger colonies may be cut to prevent seed formation or treated with herbicides.

References
126, 215.

Parapholis strigosa (Dumort.) Hubb.   Poaceae

LF: Annual herb   CU: -
SN: -

Geographic Distribution

Europe		*Australasia*		*Atlantic Islands*
N	Northern	●	Australia	Cape Verde
N	British Isles		New Zealand	Canary, Madeira
N	Central, France			Azores
	Southern		*Northern America*	South Atlantic Isl.
	Eastern		Canada, Alaska	
	Mediterranean Isl.		Southeastern USA	*Indian Ocean Islands*
		X	Western USA	Mascarenes
Africa			Remaining USA	Seychelles
	Northern		Mexico	Madagascar
	Tropical			
	Southern		*Southern America*	*Pacific Islands*
			Tropical	Micronesia
Asia			Caribbean	Galapagos
	Temperate		Chile, Argentina	Hawaii
	Tropical			Melanesia, Polynesia

Invaded Habitats
Riparian habitats, coastal estuaries and dunes, salt marshes.

Description
A tufted grass with erect or decumbent and slender stems up to 40 cm tall. The glabrous narrow leaves are up to 8 cm long and 0.5-1.5 mm wide, with ligules less than 0.5 mm. The inflorescence is a very slender, cylindrical spike of 2-20 cm length with 10-20 alternate spikelets. These are sunk in cavities of the spike's axis, 4-6 mm long and have one floret each. Glumes are 4-6 mm long, lemmas shorter and papery.

Ecology and Control
In the native range, this grass occurs mostly in salt marshes and on coastal cliffs. Where invasive, it spreads rapidly by seeds and forms dense swards, crowding out native plants. The dense swards prevent establishment of native species and reduce species richness.
     Specific control methods for this species are not available. Scattered plants can be hand pulled, larger populations cut to prevent seed formation or treated with herbicides.

References
125, 215.

Paraserianthes lophantha (Willd.) Nielsen   Fabaceae

LF: Evergreen shrub, tree   CU: Ornamental
SN: *Albizia lophantha* (Willd.) Benth.

Geographic Distribution

Europe	*Australasia*	*Atlantic Islands*
Northern	● N Australia	Cape Verde
British Isles	X New Zealand	Canary, Madeira
Central, France		X Azores
Southern	*Northern America*	South Atlantic Isl.
Eastern	Canada, Alaska	
Mediterranean Isl.	Southeastern USA	*Indian Ocean Islands*
	Western USA	Mascarenes
Africa	Remaining USA	Seychelles
Northern	Mexico	Madagascar
Tropical		
● Southern	*Southern America*	*Pacific Islands*
	Tropical	Micronesia
Asia	Caribbean	Galapagos
Temperate	Chile, Argentina	Hawaii
N Tropical		Melanesia, Polynesia

Invaded Habitats
Grass- and heathland, riparian habitats, forests, coastal beaches.

Description
A shrub or small tree with densely hairy and ribbed twigs. Leaves are alternate, twice pinnately compound with 8-15 pairs of pinnae, each having 20-40 pairs of leaflets. These are oblong, asymmetric with a prominent vein close to the upper margin, 5-10 mm long and 1-2.5 mm wide. Leaf petioles are 3-8 cm long. Numerous greenish yellow flowers are borne in axillary racemes. Fruits are glabrous straight pods of 8-15 cm length and 12-18 mm width, containing 8-10 dark brown to black, ellipsoid seeds of *c.* 7 mm length.

Ecology and Control
This tree grows commonly on moist soils of mountain slopes. The tree is nitrogen-fixing and increases soil fertility levels. It is fast growing with annual height increments exceeding 2 m under favourable conditions. It forms dense stands that shade out native species and impede overstorey regeneration. The long-lived seeds are dispersed by birds and ants. The tree does not coppice after fire, but fire may promote the spread by stimulating germination.

    Seedlings and saplings can be hand pulled or dug out, larger individuals cut. Fire can kill plants but emerging seedlings need to be controlled in a follow-up programme.

References
121, 215, 549, 844, 924, 1089.

Parkinsonia aculeata L.   Fabaceae

LF: Deciduous shrub, tree   CU: Erosion control, ornamental
SN: -

Geographic Distribution

Europe		*Australasia*		*Atlantic Islands*	
	Northern	●	Australia	X	Cape Verde
	British Isles		New Zealand		Canary, Madeira
	Central, France				Azores
	Southern	*Northern America*			South Atlantic Isl.
	Eastern		Canada, Alaska		
	Mediterranean Isl.	X	Southeastern USA	*Indian Ocean Islands*	
		X	Western USA		Mascarenes
Africa			Remaining USA		Seychelles
	Northern	N	Mexico		Madagascar
X	Tropical				
X	Southern	*Southern America*		*Pacific Islands*	
		N	Tropical		Micronesia
Asia		N	Caribbean		Galapagos
	Temperate	X	Chile, Argentina	X	Hawaii
	Tropical				Melanesia, Polynesia

Invaded Habitats
Savanna, woodland, riparian habitats, forest edges.

Description
A spiny shrub or small tree up to 10 m tall, with a characteristic green and usually short trunk, and sharp spines up to 3 cm long. The twice pinnately compound leaves are 15-40 cm long and have 2-3 pairs of pinnae. The leaf axis ends in a thorn. Pinnae are 20-30 cm long and have 20-30 pairs of leaflets each. These are oblong to obovate, 3-4 mm long and 1-1.5 mm wide. The slightly zygomorphic and fragrant flowers are yellow with red or brown dots on the largest petal, and have ten stamens. They are borne in axillary racemes. Fruits are light brown pods of 4-15 cm length and 6-8 mm width, constricted between the seeds, and have six or more dark brown seeds.

Ecology and Control
This fast growing tree is drought and salt tolerant but does not stand flooding. It grows as a tree or as a shrub, depending on the conditions. It produces seeds prolifically of two types: light brown ones that germinate readily and dark brown ones that have a hard seed coat and remain dormant. The tree is invasive because it spreads rapidly and forms dense stands that displace native vegetation and impede regeneration of native trees and shrubs. Once it has become dominant, species richness is reduced under stands of this tree.

Seedlings and saplings may be hand pulled or dug out, larger trees cut and the cut stumps treated with herbicide, or the basal bark treated. Effective herbicides are picloram plus 2,4-D or triclopyr.

References
608, 892, 986, 1432.

Paspalum conjugatum Bergius   Poaceae

LF: Perennial herb
SN: *Paspalum ciliatifolium* Trin.
CU: Forage

Geographic Distribution

Europe	Australasia	Atlantic Islands
Northern	X Australia	Cape Verde
British Isles	New Zealand	Canary, Madeira
Central, France		Azores
Southern	*Northern America*	South Atlantic Isl.
Eastern	Canada, Alaska	
Mediterranean Isl.	N Southeastern USA	*Indian Ocean Islands*
	Western USA	X Mascarenes
Africa	Remaining USA	X Seychelles
X Northern	Mexico	Madagascar
X Tropical		
Southern	*Southern America*	*Pacific Islands*
	N Tropical	X Micronesia
Asia	N Caribbean	X Galapagos
Temperate	Chile, Argentina	● Hawaii
X Tropical		X Melanesia, Polynesia

Invaded Habitats
Forest edges, grassland, moist places.

Description
A 30-100 cm tall grass with simple or sparingly branched culms and long creeping, leafy stolons up to 2 m long, rooting at the nodes. Leaf blades are 8-17 cm long and 5-15 mm wide, usually sparsely pubescent. Inflorescences are pairs of widely spreading racemes at the ends of culms, each raceme being 7-16 cm long. Spikelets grow solitary on short pedicels, are 1.4-2.2 mm long, ovate and long-ciliate. Fruits are ovate and flat caryopses of 1.4-2.2 mm length.

Ecology and Control
This grass spreads rapidly by means of stolons and forms dense swards that suppress or eliminate tree seedlings and herbs, reduce species richness and prevent regeneration of native plants. Fruits are dispersed by attaching to birds and animals. Little is known on the ecology of this grass.

Specific control methods for this species are not available. The same methods as for other *Paspalum* species may apply.

References
39, 580, 622, 735, 1121.

Paspalum dilatatum Poir. Poaceae

LF: Perennial herb
SN: -
CU: Erosion control, fodder, forage

Geographic Distribution

Europe		*Australasia*		*Atlantic Islands*	
	Northern	●	Australia		Cape Verde
	British Isles	X	New Zealand	X	Canary, Madeira
	Central, France			X	Azores
●	Southern	*Northern America*			South Atlantic Isl.
	Eastern		Canada, Alaska		
	Mediterranean Isl.		Southeastern USA	*Indian Ocean Islands*	
		X	Western USA	X	Mascarenes
Africa		X	Remaining USA		Seychelles
	Northern		Mexico		Madagascar
X	Tropical				
X	Southern	*Southern America*		*Pacific Islands*	
		N	Tropical		Micronesia
Asia		N	Caribbean		Galapagos
X	Temperate	N	Chile, Argentina	●	Hawaii
X	Tropical			X	Melanesia, Polynesia

Invaded Habitats
Grass- and heathland, forests, freshwater wetlands, riparian habitats.

Description
A densely tufted grass with ascending stems of 40-180 cm height, arising from shortly creeping rhizomes, and with membranous ligules. Leaves are up to 45 cm long and 3-15 mm wide. Inflorescences consist of 3-10 racemes on a common axis of 2-20 cm length. The racemes are 4-11 cm long and have spikelets arranged in two rows. These are yellowish-green, 2.8-4 mm long, ovate, and have two florets each. The upper glumes are sparsely hairy. Seeds are broadly elliptic and 2-3 mm long.

Ecology and Control
A native of moist grassland, this species grows best in moist soils of high fertility. The dense growth habit smothers all ground flora and prevents recruitment of native shrubs and trees. It produces large quantities of seeds which are dispersed by water and animals. Established plants are drought tolerant and tolerate heavy grazing.

Small plants can be dug out, rhizomes must be removed to prevent regrowth. Plants regenerate rapidly after cutting, so mowing is best combined with chemical control. An effective herbicide for young plants is paraquat, mature plants are best sprayed with glyphosate.

References
117, 215, 580, 609, 619, 771, 772, 1190, 1337.

Paspalum distichum L. Poaceae

LF: Perennial herb CU: Erosion control, fodder, forage
SN: *Paspalum paspalodes* (Michx.) Scribn., *Digitaria paspalodes* Michx.

Geographic Distribution

Europe		*Australasia*		*Atlantic Islands*	
	Northern	●	Australia		Cape Verde
X	British Isles	X	New Zealand	X	Canary, Madeira
	Central, France			X	Azores
●	Southern	*Northern America*			South Atlantic Isl.
X	Eastern		Canada, Alaska		
	Mediterranean Isl.	N	Southeastern USA	*Indian Ocean Islands*	
		X	Western USA	X	Mascarenes
Africa		X	Remaining USA		Seychelles
X	Northern	X	Mexico	X	Madagascar
X	Tropical				
N	Southern	*Southern America*		*Pacific Islands*	
		N	Tropical	X	Micronesia
Asia		N	Caribbean		Galapagos
X	Temperate	X	Chile, Argentina		Hawaii
X	Tropical			X	Melanesia, Polynesia

Invaded Habitats
Heath- and shrubland, riparian habitats, freshwater wetlands.

Description
An almost glabrous grass with long creeping rhizomes and extensively branched stolons, and erect culms of 15-60 cm height. The stiff and narrow leaves are 3-13 cm long and 2-6 mm wide. There are usually two racemes of 2-7 cm length at the ends of culms, with elliptical spikelets of 2.5-3.5 mm length that are appressed-pubescent on the upper glume. Lemmas are glabrous. Lower glumes are absent.

Ecology and Control
Within the native range, this C_4 grass is found in sands and muds near the seashore, in saline soils and swamps. It probably originates from tropical South America but the exact native range is obscure. The grass forms dense mats of stolons and shallow rhizomes, outcompeting and displacing native vegetation. The grass spreads mainly vegetatively. Pieces of stolons easily root and form new plants. Seeds remain viable in water for some while and are carried by streams long distances to start new infestations.
  Cutting close to the ground reduces stolon and rhizome growth but does not kill the plant. An effective herbicide is glyphosate.

References
215, 581, 595, 1190.

Passiflora edulis Sims Passifloraceae

LF: Evergreen climber
SN: -
CU: Food

Geographic Distribution

	Europe		Australasia		Atlantic Islands
	Northern		Australia		Cape Verde
	British Isles	X	New Zealand	X	Canary, Madeira
	Central, France			X	Azores
	Southern		*Northern America*		South Atlantic Isl.
	Eastern		Canada, Alaska		
	Mediterranean Isl.		Southeastern USA		*Indian Ocean Islands*
			Western USA		Mascarenes
	Africa		Remaining USA		Seychelles
	Northern		Mexico		Madagascar
X	Tropical				
●	Southern		*Southern America*		*Pacific Islands*
		N	Tropical		Micronesia
	Asia		Caribbean	●	Galapagos
	Temperate	N	Chile, Argentina	●	Hawaii
	Tropical				Melanesia, Polynesia

Invaded Habitats
Forests and forest edges.

Description
A large, glabrous vine growing to more than 8 m and with stout, stems. Leaves are large, 5-25 cm long and wide, and three-lobed. Petioles are 1.5-5 cm long and have two sessile glands at the apex. Stipules are linear-subulate and *c.* 10 mm long. Flowers are borne on peduncles of *c.* 6 cm length and are up to 8 cm in diameter. Sepals are green outside and white inside, petals white. Corona filaments are arranged in 4 or 5 ranks and purple at base. Fruits are ovoid to globose berries, 5-6 cm long and 4-5 cm wide, and purple when ripe. Seeds are ovoid, flattened and 4-6 mm long.

Ecology and Control
A widely cultivated species with many commercial varieties and hybrids. The plant tolerates slight frosts and grows in a wide range of soils. The vine grows quickly and smothers native vegetation where invasive, impeding growth and regeneration of native trees and shrubs. Invaded forests are species poor and host trees may be killed by the heavy weight of vines. Seedlings are unable to establish in stands of this plant and forest regeneration is prevented.

Specific control methods for this species are not available. The same methods as for *P. tripartita* may apply.

References
549, 725, 835, 1089, 1313.

Passiflora mixta L. f.
Passifloraceae

LF: Evergreen climber
CU: Food, ornamental
SN: *Passiflora tomentosa* Lam., *Tacsonia quitensis* Benth.

Geographic Distribution

Europe	*Australasia*	*Atlantic Islands*
Northern	Australia	Cape Verde
British Isles	● New Zealand	Canary, Madeira
Central, France		Azores
Southern	*Northern America*	South Atlantic Isl.
Eastern	Canada, Alaska	
Mediterranean Isl.	Southeastern USA	*Indian Ocean Islands*
	Western USA	Mascarenes
Africa	Remaining USA	Seychelles
Northern	Mexico	Madagascar
X Tropical		
Southern	*Southern America*	*Pacific Islands*
	N Tropical	Micronesia
Asia	Caribbean	Galapagos
Temperate	Chile, Argentina	Hawaii
Tropical		Melanesia, Polynesia

Invaded Habitats
Forests and forest edges.

Description
A glabrous or greyish pubescent vine with 3-4 m long slender, angular stems. Leaves are deeply three-lobed, serrated, 5-10 cm long and 6-15 cm wide, with the lobes being ovate-oblong. There are 4-10 petiole glands. Stipules are serrate or dentate, up to 75 mm long and 50 mm wide. Flowers are pink or pinky orange, up to 11 cm in diameter, and borne on stout peduncles of 4-6 cm length. The greenish white and pale pink calyx tube is 8-11 cm long. Corona filaments are short, deep mauve or purple, and arranged in one or two series. Fruits are ovoid berries, yellow when ripe, 4-7 cm long and 2-3.5 cm wide. Seeds are dark brown, 3-5 mm long and 2-4 mm wide.

Ecology and Control
Where native, this vine grows in mountain forests between 2,500-3,600 m elevation. It is a highly variable species with regard to degree of pubescence and leaf shape. There are several varieties in Ecuador. Where invasive, the fast growing vine smothers native vegetation and forms extensive mats on the floor. Dense infestations outcompete all underground species and lead to pure stands with reduced faunal and floral diversity. Since seedlings of shrubs and trees are unable to establish in invaded areas, natural forest regeneration is prevented.

Specific control methods for this species are not available. The same methods as for *P. tripartita* may apply.

References
1415.

Passiflora tripartita var. *mollissima* Niels   Passifloraceae

LF: Evergreen climber   CU: Ornamental
SN: *Passiflora mollissima* (Kunth) Bailey

Geographic Distribution

Europe		*Australasia*		*Atlantic Islands*	
	Northern	●	Australia		Cape Verde
	British Isles	●	New Zealand		Canary, Madeira
	Central, France				Azores
	Southern	*Northern America*			Southern
	Eastern		Canada, Alaska		South Atlantic Isl.
	Mediterranean Isl.		Southeastern USA		
			Western USA	*Indian Ocean Islands*	
Africa			Remaining USA		Mascarenes
	Northern		Mexico		Seychelles
X	Tropical				Madagascar
X	Southern	*Southern America*			
		N	Tropical	*Pacific Islands*	
Asia			Caribbean		Micronesia
	Temperate		Chile, Argentina		Galapagos
X	Tropical			●	Hawaii
					Melanesia, Polynesia

Invaded Habitats
Forests and forest edges, riparian habitats, woodland.

Description
A tall vine reaching 20 m, with striate and densely or softly downy stems. Leaves are sharply serrated, have three lobes and up to 12 cm long and 15 cm wide. Lobes are ovate-oblong and 3-4 cm wide. Petioles are 2-3 cm long. Flowers are pink or coral pink, 6-9 cm in diameter, and have a pinkish purple band of tubercles instead of corona filaments. Fruits are oblong-ovoid, softly pubescent berries, up to 10 cm long and 35 mm wide, becoming yellowish when ripe. Each fruit contains 100-200 dark brown seeds of *c.* 6 mm length.

Ecology and Control
In the native range, this vine grows from 2,000-3,600 m elevation. Where invasive, the plant forms dense curtains of trailing and climbing stems, completely smothering trees, shrubs and understorey plants. The altered structure and reduced species richness of invaded forests prevents forest regeneration and affects wildlife by reducing the abundance of food plants. Pieces of stems easily root. Fruits are abundantly produced and seeds dipsersed by birds and mammals. The shade tolerant seedlings reach high densities and grow rapidly if a canopy gap occurs, enabling the vine to enter closed forests.

Small plants and seedlings can be pulled or dug out. Larger vines are cut and the root system dug out, or sprayed with glyphosate. Vines can also be cut at breast height, the lower parts laid to the ground and sprayed with herbicide.

References
57, 121, 215, 284, 717, 718, 719, 924, 1415, 1417.

Pastinaca sativa L. — Apiaceae

LF: Biennial herb
SN: -
CU: Ornamental

Geographic Distribution

Europe	*Australasia*	*Atlantic Islands*
X Northern	X Australia	Cape Verde
N British Isles	X New Zealand	Canary, Madeira
N Central, France		Azores
N Southern	*Northern America*	South Atlantic Isl.
N Eastern	X Canada, Alaska	
N Mediterranean Isl.	Southeastern USA	*Indian Ocean Islands*
	● Western USA	Mascarenes
Africa	● Remaining USA	Seychelles
Northern	Mexico	Madagascar
Tropical		
X Southern	*Southern America*	*Pacific Islands*
	Tropical	Micronesia
Asia	Caribbean	Galapagos
N Temperate	X Chile, Argentina	Hawaii
Tropical		Melanesia, Polynesia

Invaded Habitats
Grassland, forests and forest edges, disturbed sites.

Description
A pubescent herb up to 100 cm tall, with pinnately compound leaves. The basal leaves are usually simply pinnate with 4-11 segments and have slender petioles. Leaflets are large, ovate. Inflorescences are umbels with 5-20 rays each. Flowers are yellow to yellowish green and 3-4 mm in diameter. Fruits are broadly elliptical, strongly compressed, 5-7 mm long, and have narrow wings along lateral ridges.

Ecology and Control
A variable species with at least four subspecies in the native range. The plant is invasive because it grows in dense stands that displace native vegetation. The tall size and vigorous growth makes it a strong competitor to native grasses and forbs and shades them out. Dense stands prevent the regeneration of shrubs and trees.

Specific control methods for this species are not available. Seedlings can be pulled or dug out, larger plants cut before flowering. Rosettes can be sprayed with herbicide.

References
77, 552, 571, 754.

Pennisetum clandestinum Hochst.   Poaceae

LF: Perennial herb
SN: -

CU: Erosion control, forage, ornamental

Geographic Distribution

Europe
    Northern
    British Isles
    Central, France
    Southern
    Eastern
    Mediterranean Isl.

Africa
● X Northern
  N Tropical
  ● Southern

Asia
    Temperate
  X Tropical

Australasia
● Australia
● New Zealand

Northern America
    Canada, Alaska
    Southeastern USA
  X Western USA
    Remaining USA
  X Mexico

Southern America
● Tropical
    Caribbean
  X Chile, Argentina

Atlantic Islands
    Cape Verde
    Canary, Madeira
    Azores
    South Atlantic Isl.

Indian Ocean Islands
  X Mascarenes
    Seychelles
    Madagascar

Pacific Islands
    Micronesia
● Galapagos
● Hawaii
  X Melanesia, Polynesia

Invaded Habitats
Forests, grass- and heathland, riparian habitats, coastal beaches.

Description
A prostrate grass, forming loose swards, with culms of 30-120 cm height, and with profusely branched rhizomes and stolons. Leaf blades are 3-10 cm long, 3-5 mm wide, sparsely and softly hairy above. Ligules are rings of hairs. The inflorescence is a spike with 2-4 almost sessile spikelets and partly enclosed within the sheath of the upper most leaf. Seeds are dark brown and *c.* 2 mm long, flat or ellipsoid.

Ecology and Control
A native of highland grassland and forest margins, occurring up to 2,700 m elevation in east Africa. Several cultivars have been developed and are widely used. The grass rarely produces seeds but spreads vigorously from rhizomes and stolons which root easily. The grass is mat-forming and smothers native plants, eliminating all other species. It tolerates occasional frosts and flooding. The thick rhizomes spread through the soil to a depth of 30-40 cm. The basal layer of dead leaves may support creeping fires in fire prone regions, thus increasing fire hazards.

Runners at the edge of infestations can be pulled out to stop the spread. Small patches can be dug out, all rhizomes and creeping stems must be removed. An effective control is solarization by covering infestations with plastic sheeting for 4-12 weeks. Herbicides used to control this grass include glyphosate, dalapon or 2,2-DPA.

References
215, 284, 580, 794, 835, 984, 1089, 1126, 1190, 1199, 1200, 1313, 1415, 1418.

Pennisetum macrourum Trin.   Poaceae

LF: Perennial herb   CU: Ornamental
SN: *Pennisetum angolense* Rendle, *Pennisetum giganteum* Rich.

Geographic Distribution

Europe	*Australasia*	*Atlantic Islands*
Northern	● Australia	Cape Verde
British Isles	● New Zealand	Canary, Madeira
Central, France		Azores
Southern	*Northern America*	South Atlantic Isl.
Eastern	Canada, Alaska	
Mediterranean Isl.	Southeastern USA	*Indian Ocean Islands*
	Western USA	X Mascarenes
Africa	Remaining USA	Seychelles
Northern	Mexico	Madagascar
N Tropical		
N Southern	*Southern America*	*Pacific Islands*
	Tropical	Micronesia
Asia	Caribbean	Galapagos
N Temperate	Chile, Argentina	Hawaii
Tropical		Melanesia, Polynesia

Invaded Habitats
Grassland, riparian habitats, freshwater wetlands.

Description
An erect grass up to 2 m tall, usually smaller, with glabrous and unbranched culms. Leaves are strongly ribbed, up to 120 cm long and *c.* 13 mm wide, light green above and grey-green below. Numerous spikelets are borne in slender spikes of 10-30 cm length. Spikelets have serrated awns of 10-13 mm length. The plant has rhizomes of 1 m length or more.

Ecology and Control
The strong rhizome growth increases rapidly the size of individual clumps and leads to a dense mat that eliminates all other vegetation. Root pieces and rhizome fragments can easily regenerate new plants. Once established, the grass is relatively drought resistant, persistant and individual clones expand by vegetative growth. Seeds are dispersed by water and by adhering to animals.

    Individual plants or small clumps can be removed manually, all rhizomes must be removed. Cutting before fruits ripen prevents seed dispersal but may enhance spread by dispersing rhizome fragments. Effective herbicides are glyphosate, tetrapion, 2,2-DPA, or bromacil, best applied in autumn or late spring.

References
121, 215, 528, 529, 985, 986, 1415.

Pennisetum polystachion (L.) Schult.   Poaceae

LF: Annual, perennial herb   CU: -
SN: *Panicum polystachion* L.

Geographic Distribution

Europe	*Australasia*	*Atlantic Islands*
Northern	● Australia	N Cape Verde
British Isles	New Zealand	Canary, Madeira
Central, France		Azores
Southern	*Northern America*	South Atlantic Isl.
Eastern	Canada, Alaska	
Mediterranean Isl.	Southeastern USA	*Indian Ocean Islands*
	Western USA	Mascarenes
Africa	Remaining USA	Seychelles
Northern	Mexico	Madagascar
N Tropical		
Southern	*Southern America*	*Pacific Islands*
	Tropical	X Micronesia
Asia	Caribbean	Galapagos
Temperate	Chile, Argentina	X Hawaii
X Tropical		X Melanesia, Polynesia

Invaded Habitats
Forest gaps and edges, woodland, disturbed sites.

Description
A tufted grass with few simple or branched culms of 0.5-3 m height, the branches often bearing inflorescences. Leaf blades are narrow, 5-45 cm long and 5-18 mm wide, smooth or hairy. Inflorescences are dense, yellow-brown spikes of 5-25 cm length and 10-25 mm width. Spikelets have two florets and are 3-5 mm long. They are surrounded by numerous bristles of 15-25 mm length.

Ecology and Control
This grass is well adapted to soils of low fertility and occurs mainly in grasslands on sandy soils where native. It grows quickly and is fairly shade tolerant. It becomes dominant in cleared forests and spreads quickly after fires, forming dense tussocks that may cover large areas. Native plants are displaced and natural regeneration of trees and shrubs is prevented. A soil seed bank may accumulate in dense stands of this grass.

    Control should aim at preventing seed formation and destroy the soil seed bank. Scattered plants can be hand pulled or dug out. Seedlings are best sprayed with paraquat. Dense stands can be treated with glyphosate.

References
580, 607, 623, 986, 1190.

Pennisetum purpureum Schumach.

Poaceae

LF: Annual, perennial herb
SN: -
CU: -

Geographic Distribution

Europe	*Australasia*	*Atlantic Islands*
Northern	X Australia	Cape Verde
British Isles	New Zealand	X Canary, Madeira
Central, France		Azores
Southern	*Northern America*	South Atlantic Isl.
Eastern	Canada, Alaska	
Mediterranean Isl.	● Southeastern USA	*Indian Ocean Islands*
	Western USA	Mascarenes
Africa	Remaining USA	Seychelles
Northern	X Mexico	Madagascar
N Tropical		
● Southern	*Southern America*	*Pacific Islands*
	X Tropical	X Micronesia
Asia	X Caribbean	● Galapagos
Temperate	X Chile, Argentina	X Hawaii
X Tropical		Melanesia, Polynesia

Invaded Habitats
Forests, grass- and heathland, riparian habitats, coastal beaches.

Description
A stout grass with a vigorous root system and a creeping rhizome of 15-25 cm length. Culms are smooth, 2-5 m tall and branched upwards. Leaf blades are finely pointed, 30-90 cm long and 1-3 cm wide, with the margins being thickened. Inflorescences are compact, bristly spikes of 8-30 cm length and 15-30 mm width, yellow-brown or rarely purplish. Each spikelet is surrounded by several bristles of 5-15 mm length and one bristle up to 4 cm length. Spikelets have two florets each.

Ecology and Control
This grass spreads mainly by vegetative growth and is found in a wide range of habitats. Where invasive, it forms dense reeds 3 m tall or more in moist and rich soils, displacing native vegetation and preventing any regeneration of native plants. The grass persists due to a deep root system. Seeds are rarely produced. Established plants are drought and fire tolerant, and vigorously resprout if cut.

    Individual plants can be dug out, all rhizomes must be removed. Larger infestations are controlled by cutting or burning off and treating any regrowth with 2,2-DPA.

References
284, 518, 549, 580, 715, 725, 835, 1190, 1313.

Pennisetum setaceum (Forssk.) Chiov.   Poaceae

LF: Perennial herb
CU: Ornamental
SN: *Cenchrus asperifolius* Desf., *Pennisetum ruppelii* Steud.

Geographic Distribution

Europe
- Northern
- British Isles
- Central, France
- Southern
- Eastern
- Mediterranean Isl.

Africa
- N Northern
- N Tropical
- X Southern

Asia
- N Temperate
- Tropical

Australasia
- ● Australia
- X New Zealand

Northern America
- Canada, Alaska
- Southeastern USA
- ● Western USA
- Remaining USA
- Mexico

Southern America
- Tropical
- Caribbean
- Chile, Argentina

Atlantic Islands
- Cape Verde
- Canary, Madeira
- Azores
- X South Atlantic Isl.

Indian Ocean Islands
- Mascarenes
- Seychelles
- Madagascar

Pacific Islands
- X Micronesia
- Galapagos
- ● Hawaii
- Melanesia, Polynesia

Invaded Habitats
Grass- and woodland, coastal dunes, desert scrub.

Description
A large bunch grass with stems of 20-100 cm height. The rigid leaves are up to 30 cm long and *c.* 3 mm wide. Inflorescences are 10-30 cm long, linear, with the spikelets being enclosed by an involucre of slender bristles of 1.5-4 mm length; these fall together with the spikelets. Spikelets are lanceolate to oblong, 4.5-6.5 mm long, and have two florets each. The lower glume is often minute.

Ecology and Control
A very fire and drought tolerant bunch grass that displaces native grassland communities where invasive. It forms thick stands that impede growth and regeneration of native plants, completely eliminating the native vegetation in time. It promotes fires by accumulation of large quantities of dead biomass, and the grass spreads as a result of fires because it rapidly colonizes burned areas. Seeds are wind dispersed.

Small patches can be uprooted or cut at ground level. Larger infestations can be chemically treated with hexazinone.

References
133, 284, 766, 1180, 1199, 1200, 1303, 1410.

Pereskia aculeata Mill.

Cactaceae

LF: Succulent perennial
SN: *Cactus pereskia* L.

CU: Ornamental

Geographic Distribution

Europe	*Australasia*	*Atlantic Islands*
Northern	Australia	Cape Verde
British Isles	New Zealand	Canary, Madeira
Central, France		Azores
Southern	*Northern America*	South Atlantic Isl.
Eastern	Canada, Alaska	
Mediterranean Isl.	X Southeastern USA	*Indian Ocean Islands*
	Western USA	Mascarenes
Africa	Remaining USA	Seychelles
Northern	X Mexico	Madagascar
Tropical		
● Southern	*Southern America*	*Pacific Islands*
	N Tropical	Micronesia
Asia	N Caribbean	Galapagos
N Temperate	N Chile, Argentina	Hawaii
Tropical		Melanesia, Polynesia

Invaded Habitats
Grass- and scrubland, dry forests.

Description
A large and slender, succulent shrub or climber, up to 20 m tall, bearing normal, fleshy and entire leaves, with the trunk and older branches armed with spines. Leaves are lanceolate to oblong-elliptic, 4-9 cm long, and have short petioles. Leaves are often reduced to thorns. Spines are solitary or in clusters of 2-3, and 2.5-5 cm long. The white, yellowish or pinkish flowers are borne in corymbose panicles and are 2.5-4.5 cm in diameter. Fruits are bright yellow to orange, globose berries of 1.5-2 cm diameter, crowned by the perianth. They contain few seeds each. Seeds are almost black, globose, 4-5 mm in diameter, and shiny.

Ecology and Control
In the native range, this plant grows in tropical hammocks, savanna forests, and coastal dunes. It is a variable species with many varieties and forms. The plant often grows upright first and changes into a climbing growth habit. It has a vigorous vegetative reproduction by means of rooting stem fragments. The plant forms dense and spiny thickets that displace native plants and affect wildlife habitats.

    Specific control methods for this species are not available. Small plants may be pulled or dug out. Larger thickets may be controlled in similar ways as *Opuntia* species.

References
205, 549, 617, 998, 1089.

Petrorhagia prolifera Ball & Heywood   Caryophyllaceae

LF: Annual, biennial herb   CU: -
SN: *Tunica prolifera* (L.) Scop.

Geographic Distribution

Europe
N   Northern
N   British Isles
N   Central, France
N   Southern
N   Eastern
N   Mediterranean Isl.

Africa
N   Northern
    Tropical
    Southern

Asia
N   Temperate
    Tropical

Australasia
●   Australia
X   New Zealand

Northern America
    Canada, Alaska
    Southeastern USA
X   Western USA
    Remaining USA
    Mexico

Southern America
    Tropical
    Caribbean
    Chile, Argentina

Atlantic Islands
    Cape Verde
N   Canary, Madeira
    Azores
    South Atlantic Isl.

Indian Ocean Islands
    Mascarenes
    Seychelles
    Madagascar

Pacific Islands
    Micronesia
    Galapagos
    Hawaii
    Melanesia, Polynesia

Invaded Habitats
Grassland, rock outcrops, riverbeds.

Description
A herb reaching 50 cm height, with pubescent or glabrous stems, and leaf sheaths about as long as wide. Basal leaves are linear-oblanceolate, 2-5 cm long and 1-4 mm wide. Stem leaves are smaller and linear. Pink flowers with petals of 10-14 mm length are borne in cymes of 1-6, enclosed in papery bract-like leaves. Fruits are dehiscent capsules with four teeth. Seeds are black, 1.3-1.9 mm long and 0.8-1.1 mm wide.

Ecology and Control
This herb establishes in vegetation openings and grows in dense patches displacing native plants. It competes for space and resources and prevents the establishment of native species. The plant spreads rapidly by seeds.
    Specific control methods for this species are not available. Single plants can be hand pulled. Cutting before fruits ripen prevents seed dispersal, larger populations can be treated with herbicide.

References
215, 1059.

Phalaris aquatica L.

Poaceae

LF: Perennial herb
SN: *Phalaris stenoptera* Hack., *Phalaris tuberosa* L.
CU: Fodder, forage

Geographic Distribution

	Europe		Australasia		Atlantic Islands
	Northern	•	Australia		Cape Verde
X	British Isles		New Zealand	X	Canary, Madeira
N	Central, France				Azores
N	Southern		Northern America		South Atlantic Isl.
N	Eastern		Canada, Alaska		
N	Mediterranean Isl.		Southeastern USA		Indian Ocean Islands
		•	Western USA		Mascarenes
	Africa		Remaining USA		Seychelles
N	Northern		Mexico		Madagascar
	Tropical				
X	Southern		Southern America		Pacific Islands
			Tropical		Micronesia
	Asia		Caribbean		Galapagos
N	Temperate		Chile, Argentina		Hawaii
	Tropical				Melanesia, Polynesia

Invaded Habitats
Grass- and heathland, riparian habitats, freshwater wetlands, coastal beaches.

Description
A perennial grass up to 150 cm tall, with short rhizomes and with smooth and erect to ascending culms. Leaves are 20-40 cm long and 10-20 mm wide, long-acute, and have ligules of *c.* 3 mm length. Inflorescences are contracted spike-like panicles of 5-15 cm length and 10-20 mm width, oblong-cylindrical, and occasionally purplish. Spikelets are 4.5-7 mm long and have 2-3 florets. The obovate and acute glumes have winged keels. Seeds are 2-3 mm long.

Ecology and Control
Mature plants of this grass develop an extensive and deep root system, thereby out-competing native species. It forms dense stands that exclude all other vegetation and impede recruitment of native shrubs and trees. Seeds are dispersed by water and animals. The grass accumulates large amounts of dead biomass and increases fire hazards.

Small plants can be dug out, rhizomes must be removed. Larger stands are regularly slashed or burned to reduce seed production, a follow-up programme is necessary to control regrowth and seedlings. Close mowing or clipping late in the growing season reduces the vigour of the grass. Effective herbicides are glyphosate or fluazifop applied as a foliar spray to actively growing plants. A pre-emergence herbicide to control this grass is sulfumeturon.

References
133, 215, 765, 924.

Phalaris arundinacea L.   Poaceae

LF: Perennial herb   CU: Erosion control, forage, ornamental
SN: -

Geographic Distribution

Europe		*Australasia*		*Atlantic Islands*	
N	Northern	●	Australia		Cape Verde
N	British Isles	X	New Zealand		Canary, Madeira
N	Central, France				Azores
N	Southern	*Northern America*			South Atlantic Isl.
N	Eastern	●	Canada, Alaska		
N	Mediterranean Isl.	●	Southeastern USA	*Indian Ocean Islands*	
		X	Western USA	X	Mascarenes
Africa		●	Remaining USA		Seychelles
N	Northern		Mexico		Madagascar
	Tropical				
X	Southern	*Southern America*		*Pacific Islands*	
		X	Tropical		Micronesia
Asia		X	Caribbean		Galapagos
N	Temperate	X	Chile, Argentina	X	Hawaii
X	Tropical				Melanesia, Polynesia

Invaded Habitats
Forests, freshwater wetlands, riparian habitats.

Description
A stout grass with long rhizomes and erect, smooth stems of 0.5-2 m height. The long-acute leaves are 10-35 cm long and 6-18 mm wide, with smooth sheaths and ligules of 6-10 mm length. Inflorescences are branched panicles of 5-25 cm length, lanceolate to oblong and often purplish. Spikelets have very short pedicels and are 4.5-6.5 mm long; they bear 2-3 florets. Glumes are not winged on the keel and are narrowly lanceolate. The upper lemma is 3-4 mm long and broadly lanceolate.

Ecology and Control
This grass is found along lakeshores, in riparian habitats, wet meadows, marshes, disturbed ground, and in shallow waters within the native range. It is a highly variable species with two subspecies in Europe. Many varieties are used in horticulture, most of them with light green and white-striped leaves. The species spreads vegetatively and forms dense monospecific mats often covering large areas and impeding water flow. Native wetland species quickly decline in areas colonized by this grass. The grass accumulates a persistant soil seed bank. Although native to the USA and Canada, invasive plants originate from Europe.
    Control measures include mowing, prescribed burning, or chemical control. Restoration by planting native grasses and sedges reduces recolonization by the grass. Seedlings are easily controlled by spraying glyphosate or amitrole.

References
44, 70, 215, 487, 571, 903, 1396.

Phalaris minor Retz.   Poaceae

LF: Annual herb
SN: -
CU: Fodder, forage

Geographic Distribution

Europe
- Northern
- X British Isles
- Central, France
- N Southern
- Eastern
- N Mediterranean Isl.

Africa
- N Northern
- Tropical
- X Southern

Asia
- X Temperate
- N Tropical

Australasia
- • Australia
- New Zealand

Northern America
- Canada, Alaska
- • Southeastern USA
- X Western USA
- Remaining USA
- X Mexico

Southern America
- X Tropical
- Caribbean
- X Chile, Argentina

Atlantic Islands
- Cape Verde
- X Canary, Madeira
- N Azores
- South Atlantic Isl.

Indian Ocean Islands
- Mascarenes
- Seychelles
- Madagascar

Pacific Islands
- Micronesia
- Galapagos
- X Hawaii
- Melanesia, Polynesia

Invaded Habitats
Forests, riparian habitats, freshwater wetlands, coastal beaches.

Description
A grass with rigid and smooth, usually branched stems of 20-120 cm height. Leaves are long-acute, up to 20 cm long and *c.* 10 mm wide, with ligules of 3-8 mm length. Inflorescences are ovoid to ovoid-oblong, contracted spike-like panicles of 2-5 cm length. Spikelets have 2-3 florets and are 4.5-5.5 mm long. Glumes have a toothed wing each. The lower lemma is minute.

Ecology and Control
A native of dry, open places that grows generally on sandy soils. It is invasive because it forms dense swards and becomes dominant, displacing native plants and preventing overstorey regeneration. It is also a significant agricultural weed.

Scattered plants can be dug out. Cutting before fruits ripen prevents seed formation. Larger patches can be treated with herbicides.

References
215, 993, 1159, 1184, 1185.

Phleum pratense L.   Poaceae

LF: Perennial herb
SN: *Phleum nodosum* L.
CU: Fodder, forage, erosion control

Geographic Distribution

Europe		*Australasia*		*Atlantic Islands*	
N	Northern	●	Australia		Cape Verde
N	British Isles	X	New Zealand		Canary, Madeira
N	Central, France			N	Azores
N	Southern	*Northern America*			South Atlantic Isl.
N	Eastern		Canada, Alaska		
N	Mediterranean Isl.		Southeastern USA	*Indian Ocean Islands*	
		●	Western USA	X	Mascarenes
Africa		X	Remaining USA		Seychelles
N	Northern		Mexico		Madagascar
	Tropical				
X	Southern	*Southern America*		*Pacific Islands*	
		X	Tropical		Micronesia
Asia			Caribbean		Galapagos
N	Temperate		Chile, Argentina	X	Hawaii
N	Tropical				Melanesia, Polynesia

Invaded Habitats
Grassland, coastal scrub, disturbed sites.

Description
A 6-150 cm tall tufted grass with 1-6 nodes per stem and the lower nodes sometimes being swollen and tuberous. Leaves are up to 45 cm long and *c.* 10 mm wide, and acuminate, with ligules of 1-6 mm length. Inflorescences are strongly contracted cylindrical panicles of 2-11 cm length and 4-12 mm width, green, often tinged with purple. Glumes are strongly keeled, 4-4.5 mm long, and have three veins and a short awn. The oblong spikelets are 2-5.5 mm long.

Ecology and Control
A common grass of meadows and pastures in the native range with two subspecies in Europe. Although found mostly in lower elevations, occasionally it invades damp alpine grasslands in Australia. The grass is invasive because it spreads quickly and forms dense colonies that crowd out native plants and prevent their regeneration. Little is known on the ecology of this grass.

Specific control methods for this species are not available. Mowing close to the ground or clipping may reduce the plant's vigour. Larger populations can be treated with herbicide.

References
215, 238, 1314.

Phragmites australis (Cav.) Trin.   Poaceae

LF: Perennial herb   CU: Erosion control, shelter
SN: *Arundo phragmites* L., *Phragmites communis* Trin.

Geographic Distribution

Europe
- N   Northern
- N   British Isles
- N   Central, France
- N   Southern
- N   Eastern
- N   Mediterranean Isl.

Africa
- X   Northern
-       Tropical
- X   Southern

Asia
- X   Temperate
- X   Tropical

Australasia
- ●   Australia
- ●   New Zealand

Northern America
- ●   Canada, Alaska
- ●   Southeastern USA
-       Western USA
- ●   Remaining USA
-       Mexico

Southern America
- X   Tropical
- X   Caribbean
- X   Chile, Argentina

Atlantic Islands
-       Cape Verde
- N   Canary, Madeira
-       Azores
-       South Atlantic Isl.

Indian Ocean Islands
-       Mascarenes
-       Seychelles
- X   Madagascar

Pacific Islands
-       Micronesia
-       Galapagos
-       Hawaii
- X   Melanesia, Polynesia

Invaded Habitats
Freshwater wetlands, riparian habitats.

Description
A large and stout aquatic or semi-aquatic grass with unbranched culms of 2.5-4.5 m height, and with stout creeping rhizomes and stolons. The greyish-green leaves are up to 60 cm long and 10-60 mm wide, gradually tapering to a long apex, and have glabrous sheaths. Inflorescences are large, spreading and usually purplish, oblong to ovoid panicles of 20-60 cm length. Spikelets have 2-10 florets each, long hairs, and are 10-16 mm long. The glumes and lemmas are lanceolate and without awns.

Ecology and Control
This grass spreads mainly by vegetative growth and establishes frequently after disturbances. The creeping rhizomes may reach 20 m length. Dense stands of the grass are species poor, crowd out native vegetation and block waterways. Due to the large size and high evapotranspiration rate, it can have a lowering effect on the water table. The grass tolerates moderate salinity and grows also in brackish waters. The stems are not overwintering.

     Cutting in spring and autumn or burning in summer is used to reduce growth. Control is often linked to restoring the pre-disturbance conditions by altering the hydrology. Effective herbicides include dalapon applied to actively growing plants.

References
485, 522, 542, 569, 580, 818, 964, 1190, 1287, 1327, 1442.

Physalis peruviana L. Solanaceae

LF: Perennial herb
SN: *Physalis edulis* Sims
CU: Ornamental, food

Geographic Distribution

Europe		*Australasia*		*Atlantic Islands*	
	Northern	●	Australia	X	Cape Verde
X	British Isles	X	New Zealand	X	Canary, Madeira
X	Central, France			X	Azores
	Southern	*Northern America*			South Atlantic Isl.
	Eastern		Canada, Alaska		
	Mediterranean Isl.		Southeastern USA	*Indian Ocean Islands*	
			Western USA		Mascarenes
Africa			Remaining USA		Seychelles
	Northern		Mexico		Madagascar
X	Tropical				
X	Southern	*Southern America*		*Pacific Islands*	
		N	Tropical		Micronesia
Asia		X	Caribbean	X	Galapagos
X	Temperate		Chile, Argentina	X	Hawaii
X	Tropical			X	Melanesia, Polynesia

Invaded Habitats
Forests and forest edges, riparian habitats, disturbed sites.

Description
A densely pubescent herb of 30-100 cm height, simple or branched above. The broadly ovate leaves are 7-15 cm long and 4-9 cm wide, entire or slightly toothed. Petioles are 1-4 cm long. Flowers appear solitary in the axils of leaves on short pedicels, are yellow with dark purple markings, and have corollas of 15-25 mm diameter. The anthers are purple and 3.5-4 mm long. The bell-shaped calyx is 8-9 mm long in flower, extending to 3-5 cm length to enclose the fruit. Fruits are yellow and globose berries of 12-20 mm length, surrounded and exceeded by the inflated calyx. The yellow seeds are *c.* 2 mm in diameter.

Ecology and Control
This large herb forms dense thickets that crowd out native vegetation. Little is known on the ecology of this species.

Specific control methods for this species are not available. Single plants can be hand pulled or dug out, larger plants cut or treated with herbicide.

References
215, 1265.

Pinus banksiana Lamb.   Pinaceae

LF: Evergreen tree   CU: Ornamental, wood
SN: -

Geographic Distribution

Europe	*Australasia*	*Atlantic Islands*
Northern	Australia	Cape Verde
British Isles	● New Zealand	Canary, Madeira
Central, France		Azores
Southern	*Northern America*	South Atlantic Isl.
Eastern	N Canada, Alaska	
Mediterranean Isl.	Southeastern USA	*Indian Ocean Islands*
	Western USA	Mascarenes
Africa	N Remaining USA	Seychelles
Northern	Mexico	Madagascar
Tropical		
Southern	*Southern America*	*Pacific Islands*
	Tropical	Micronesia
Asia	Caribbean	Galapagos
Temperate	Chile, Argentina	Hawaii
Tropical		Melanesia, Polynesia

Invaded Habitats
Grass- and heathland, forest margins, scrub.

Description
A needle tree up to 30 m tall, with strongly curved cones, slender and spreading branches, and a reddish-brown bark that forms thick scales. Leaves are borne in clusters of two, rigid and twisted, 2-5 cm long and bright to dark green. Male flowers are yellow. Cones are erect, ovoid, 3-5 cm long and 1.5-2.5 cm wide, yellowish-brown and shining. Seeds are 3-6 mm long and have a wing of 8-12 mm length.

Ecology and Control
A shade intolerant pioneer tree that colonizes exposed soils after major disturbances such as fire, growing well on dry and sandy soils. Once established, it forms extensive and persistent stands that exclude most native species. The juvenile period lasts for *c.* 3 years.

Specific control methods for this species are not available. The same methods as for other *Pinus* species may apply.

References
182, 260, 555, 660, 756, 1088, 1115, 1119, 1402.

Pinus contorta Douglas ex Loudon — Pinaceae

LF: Evergreen tree
SN: -
CU: Wood, shelter

Geographic Distribution

Europe		*Australasia*		*Atlantic Islands*	
X	Northern	●	Australia		Cape Verde
X	British Isles	●	New Zealand		Canary, Madeira
X	Central, France				Azores
	Southern	*Northern America*			South Atlantic Isl.
	Eastern	N	Canada, Alaska		
	Mediterranean Isl.		Southeastern USA	*Indian Ocean Islands*	
		N	Western USA		Mascarenes
Africa			Remaining USA		Seychelles
	Northern		Mexico		Madagascar
	Tropical				
	Southern	*Southern America*		*Pacific Islands*	
			Tropical		Micronesia
Asia			Caribbean		Galapagos
	Temperate		Chile, Argentina		Hawaii
	Tropical				Melanesia, Polynesia

Invaded Habitats
Grass- and scrubland, open forests.

Description
A tree of 6-10 m height, with short and contorted branches forming a round and dense crown and with a deeply fissured, dark brown bark. Twigs are glabrous and green in the first year, and become orange to brown with age. Leaves are in pairs, 2.5-7 cm long, rigid and dark green. The sessile cones are yellowish brown, 2-6 cm long and 2-3 cm wide, narrowly ovoid to conical and shining. Seeds are 4-5 mm long and have a wing of *c.* 8 mm length.

Ecology and Control
A tree growing in a wide range of habitats and with four different varieties in the native range. Plants in Australia belong to subsp. *latifolia*. It is a prolific seed producer, and the closed cones persist and accumulate on trees for decades. Fires lead to mass release of seeds, and the tree regenerates in extremely dense stands after fire, preventing the recruitment and growth of native plants. The tree's establishment depends on disturbances, and seedlings are sensitive to shading and competition. The tree invades and transforms native tussock grassland in New Zealand into species poor woodland and forest.

Specific control methods for this species are not available. Seedlings and saplings may be hand pulled or dug out. Cutting trees may lead to seed release and requires follow-up programmes to treat seedlings.

References
14, 19, 182, 773, 1088, 1415.

Pinus elliottii Engelm.   Pinaceae

LF: Evergreen tree   CU: Wood, shelter
SN: -

Geographic Distribution

Europe	*Australasia*	*Atlantic Islands*
Northern	● Australia	Cape Verde
British Isles	New Zealand	Canary, Madeira
Central, France		Azores
Southern	*Northern America*	South Atlantic Isl.
Eastern	Canada, Alaska	
Mediterranean Isl.	N Southeastern USA	*Indian Ocean Islands*
	Western USA	Mascarenes
Africa	Remaining USA	Seychelles
Northern	Mexico	Madagascar
Tropical		
● Southern	*Southern America*	*Pacific Islands*
	Tropical	Micronesia
Asia	Caribbean	Galapagos
Temperate	Chile, Argentina	Hawaii
Tropical		Melanesia, Polynesia

Invaded Habitats
Grass- and heathland, scrub.

Description
A large tree up to 30 m tall, with a scaly bark and a dense, rounded crown. Needles are stiff, dark blue-green, 15-30 cm long, and mostly borne in bundles of two. Male cones are 3-5 cm long, female cones are on short peduncles, and have prominent scale appendages. Seeds are 6-8 mm long and have a wing of 2-3 cm length.

Ecology and Control
In the native range, this tree prefers relatively moist soils. At least two varieties occur in the native range. Invasions in South Africa are common in areas which have been protected from fires for long times. The fast growing tree establishes dense stands that shade out native plants and prevent their regeneration. Invaded grasslands are transformed over time into species poor shrubs and forests.

    Specific control methods for this species are not available. Seedlings and saplings may be pulled or dug out, larger trees cut. Follow-up programmes are necessary to treat seedlings.

References
61, 182, 421, 549, 662, 797, 1088, 1089, 1145, 1175.

Pinus halepensis Mill.   Pinaceae

LF: Evergreen tree
SN: -
CU: Erosion control, ornamental

Geographic Distribution

Europe	*Australasia*	*Atlantic Islands*
Northern	● Australia	Cape Verde
British Isles	X New Zealand	Canary, Madeira
Central, France		Azores
N Southern	*Northern America*	South Atlantic Isl.
N Eastern	Canada, Alaska	
N Mediterranean Isl.	Southeastern USA	*Indian Ocean Islands*
	Western USA	Mascarenes
Africa	Remaining USA	Seychelles
N Northern	Mexico	Madagascar
Tropical		
● Southern	*Southern America*	*Pacific Islands*
	Tropical	Micronesia
Asia	Caribbean	Galapagos
N Temperate	Chile, Argentina	Hawaii
Tropical		Melanesia, Polynesia

Invaded Habitats
Grass- and heathland, woodland.

Description
A tree up to 20 m tall, with an open crown and the trunk and branches often being twisted. The bark is silvery grey at first, becoming reddish brown and fissured. Leaves are in clusters of two, flexuous, pale green and 6-15 cm long. Cones are shiny reddish brown, 5-12 cm long and *c.* 4 cm wide, often borne in pairs or in clusters of a few on woody stalks of 1-2 cm length. Seeds are *c.* 7 mm long and have a wing of 2-3 cm length.

Ecology and Control
This tree is common along the coast in the native range. It is a highly drought tolerant tree, growing in many different habitats. It produces large numbers of seeds and fires can lead to mass establishment of seedlings. Where invasive, stands of this tree displace native vegetation, alter nutrient and water relations, and reduce light and species richness. The thick litter layer prevents any seedling establishment of native plants. Invaded communities are slowly transformed to species poor forests.

    Seedlings and saplings are easy to pull out. Larger trees can be cut. Stumps usually do not require herbicide treatment to prevent regrowth if cuts are made below any branches. Trees can also be killed by ringbarking. Fire is used to kill larger stands but may lead to mass release of seeds.

References
549, 924, 1088, 1112, 1156.

Pinus nigra Arnold   Pinaceae

LF: Evergreen tree   CU: Erosion control, ornamental, wood
SN: -

Geographic Distribution

Europe	*Australasia*	*Atlantic Islands*
Northern	• Australia	Cape Verde
X British Isles	• New Zealand	Canary, Madeira
Central, France		Azores
N Southern	*Northern America*	South Atlantic Isl.
N Eastern	Canada, Alaska	
N Mediterranean Isl.	Southeastern USA	*Indian Ocean Islands*
	Western USA	Mascarenes
Africa	• Remaining USA	Seychelles
N Northern	Mexico	Madagascar
Tropical		
Southern	*Southern America*	*Pacific Islands*
	Tropical	Micronesia
Asia	Caribbean	Galapagos
N Temperate	Chile, Argentina	Hawaii
Tropical		Melanesia, Polynesia

Invaded Habitats
Grassland, coastal dunes, dry forests.

Description
A tree up to 50 m height, with a grey to dark brown, deeply fissured bark, and spreading branches forming a broad pyramidal crown. Twigs are glabrous and pale brown to orange-brown. The stiff needles are in pairs, light to dark green and 4-19 cm long. The ovoid yellowish or light brown and shining cones are 3-8 cm long, 2-4 cm wide and almost sessile. Seeds are reddish brown, 5-7 mm long and have a wing of *c.* 15 mm length. Male flowers appear in clusters of 3-10 and are bright yellow.

Ecology and Control
A variable species with many variants and five subspecies in Europe. Naturalized plants in Australia belong to var. *corsicana*. The tree grows well on nutrient poor and sandy soils. In the native range, it is found from the coast up to the timberline in mountains. Where invasive, the tree forms dense stands that reduce light levels and shade out native plants. The tree becomes dominant and accelerates succession to woody communities on invaded sand dunes in central USA. It establishes also in native tussock grasslands of New Zealand and transforms them to woodland.

     Specific control methods for this species are not available. Seedlings and saplings may be pulled or dug out. Larger trees can be cut and the cut stumps treated with herbicide, or ringbarked.

References
19, 182, 279, 732, 733, 1088.

Pinus patula Schiede ex Schltdl. & Cham.   Pinaceae

LF: Evergreen tree   CU: Ornamental, shelter, wood
SN: -

Geographic Distribution

Europe	*Australasia*	*Atlantic Islands*
Northern	Australia	Cape Verde
British Isles	X New Zealand	Canary, Madeira
Central, France		Azores
Southern	*Northern America*	South Atlantic Isl.
Eastern	Canada, Alaska	
Mediterranean Isl.	Southeastern USA	*Indian Ocean Islands*
	Western USA	Mascarenes
Africa	Remaining USA	Seychelles
Northern	N Mexico	● Madagascar
● Tropical		
● Southern	*Southern America*	*Pacific Islands*
	N Tropical	Micronesia
Asia	Caribbean	Galapagos
Temperate	Chile, Argentina	● Hawaii
Tropical		Melanesia, Polynesia

Invaded Habitats
Forests, grass- and woodland.

Description
A tree reaching 35 m height, with a rather smooth, grey to dark brown bark, a light green foliage, a long straight trunk, and the branches more or less horizontal and turning up at the ends. Leaves are slender needles of 15-23 cm length and appear in bundles of three. Cones are shiny brown, sessile, up to 10 cm long, and borne in clusters of 2-5. Male catkins are short and yellow-brown. Seeds are *c.* 4 mm long and have a wing of *c.* 13 mm length.

Ecology and Control
A fast growing tree that is native to highlands from 1,500-3,000 m elevation. The tree is invasive because it forms dense pure stands that shade out native herbaceous plants and shrubs. Several endemic plant species are threatened by expanding populations of this tree in Malawi. Seed production is abundant, and seedlings establish especially well in disturbed sites. The juvenile period of this tree lasts for *c.* 12 years.

A control measure includes cutting the trees followed by burning once the seeds have germinated.

References
284, 369, 370, 549, 1088, 1089, 1199.

Pinus pinaster Aiton   Pinaceae

LF: Evergreen tree
SN: -
CU: Erosion control, shelter, wood

Geographic Distribution

Europe	Australasia	Atlantic Islands
Northern	● Australia	Cape Verde
X British Isles	● New Zealand	Canary, Madeira
Central, France		Azores
N Southern	*Northern America*	South Atlantic Isl.
Eastern	Canada, Alaska	
N Mediterranean Isl.	Southeastern USA	*Indian Ocean Islands*
	Western USA	Mascarenes
Africa	Remaining USA	Seychelles
N Northern	Mexico	Madagascar
Tropical		
● Southern	*Southern America*	*Pacific Islands*
	Tropical	Micronesia
Asia	Caribbean	Galapagos
Temperate	● Chile, Argentina	● Hawaii
Tropical		Melanesia, Polynesia

Invaded Habitats
Grass- and heathland, forests, coastal scrub and dunes.

Description
A tree reaching 40 m height with spreading branches forming a broad pyramidal crown, and a deeply fissured, brown bark. Branches are bright reddish brown. Needles are borne in pairs, rigid and bright green, and 10-25 cm long. Female cones are ovoid-conical, light brown, 8-22 cm long and 5-8 cm wide, and have short stalks. Seeds are 6-8 mm long and have a wing of 22-24 mm length.

Ecology and Control
Where native, this tree is common on sandy and poor soils near the coast. There are two subspecies in Europe. The tree is drought tolerant and adapted to fires. Seeds are abundantly produced and dispersed by wind. The tree is invasive because the dense stands displace native vegetation and alter nutrient and water relations. The thick litter layer that accumulates under the canopies prevents seedling establishment of native plants. The juvenile period of this tree lasts for *c.* 10 years.

Seedlings and saplings are easy to pull out. Larger trees can be cut. Stumps usually do not require herbicide treatment to prevent regrowth if cuts are made below any branches. Trees can also be killed by ringbarking. Fire is used to kill larger stands but may lead to mass release of seeds. Seedlings need to be controlled in follow-up programmes.

References
137, 215, 272, 284, 348, 549, 610, 702, 795, 924, 1089, 1112, 1199, 1225, 1415.

Pinus radiata D. Don   Pinaceae

LF: Evergreen tree   CU: Erosion control, shelter, wood
SN: -

Geographic Distribution

Europe		*Australasia*		*Atlantic Islands*	
	Northern	●	Australia		Cape Verde
	British Isles	●	New Zealand		Canary, Madeira
	Central, France				Azores
	Southern		*Northern America*		South Atlantic Isl.
	Eastern		Canada, Alaska		
	Mediterranean Isl.		Southeastern USA		*Indian Ocean Islands*
		N	Western USA		Mascarenes
Africa			Remaining USA		Seychelles
	Northern	N	Mexico		Madagascar
	Tropical				
●	Southern		*Southern America*		*Pacific Islands*
			Tropical		Micronesia
Asia			Caribbean	X	Galapagos
	Temperate	X	Chile, Argentina		Hawaii
	Tropical				Melanesia, Polynesia

Invaded Habitats
Grass- and heathland, riparian habitats, coastal dunes, scrub.

Description
A tree reaching 40 m height with a dark brown bark becoming thick and fissured, with glabrous and reddish-brown twigs, and with an irregular, open crown. The bright green needles are 10-15 cm long and borne in densely crowded clusters of 2-3. The asymmetrical cones are sessile or have short stalks, usually borne in clusters of 2-3, reflexed, conical-ovoid, 7-14 cm long and 5-8 cm wide. Seeds are *c.* 7 mm long and have a wing of *c.* 20 mm length.

Ecology and Control
A fast growing tree with a juvenile period of *c.* 5 years, adapted to fire and with cones that may remain closed on the tree for years. The native range consists of small areas in California and Mexico. Plants in Mexico are referred to var. *binata*. The tree is a prolific seed producer, and fires lead to mass release of seeds. It establishes well in burned areas and forms dense stands that may cover large areas. The native vegetation is eliminated and transformed into species poor woodland. A thick litter layer accumulates beneath stands of this tree, preventing establishment of native plants. The tree does not resprout from stumps.

Control in fire adapted plant communities is done by cutting trees followed by burning after seeds have germinated, usually a few months after tree felling. Scattered seedlings and saplings can be pulled out.

References
109, 176, 177, 182, 215, 247, 284, 652, 924, 1086, 1087, 1089, 1112, 1334, 1418.

Piper aduncum L.
Piperaceae

LF: Evergreen shrub, tree CU: -
SN: *Piper celtidifolium* Kunth, *Piper elongatum* Vahl

Geographic Distribution

Europe	*Australasia*	*Atlantic Islands*
Northern	Australia	Cape Verde
British Isles	New Zealand	Canary, Madeira
Central, France		Azores
Southern	*Northern America*	South Atlantic Isl.
Eastern	Canada, Alaska	
Mediterranean Isl.	X Southeastern USA	*Indian Ocean Islands*
	Western USA	Mascarenes
Africa	Remaining USA	Seychelles
Northern	N Mexico	Madagascar
Tropical		
Southern	*Southern America*	*Pacific Islands*
	N Tropical	Micronesia
Asia	N Caribbean	Galapagos
Temperate	Chile, Argentina	Hawaii
● Tropical		● Melanesia, Polynesia

Invaded Habitats
Tropical hammocks, forests and forest edges, slopes, disturbed sites.

Description
A multi-stemmed shrub or small tree reaching 6-8 m height, with yellow-green, finely hairy stems and enlarged, ringed nodes. Leaves are alternate, broadly lanceolate to narrowly elliptic, tapering into long tips, 13-25 cm long and 3.5-8 cm wide, with the base being asymmetric. Tiny yellowish flowers are borne in conspicuous curving spikes opposite the leaves. Peduncles are 12-17 mm long. Fruits are small, ovoid pale green drupes, containing numerous minute seeds.

Ecology and Control
In the native range, this plant is found in evergreen forests, humid woodland and riparian habitats, occurring from sea level to 2,000 m elevation. It is a fast growing pioneer shrub that establishes well in disturbed sites and quickly builds up dense thickets that impede growth and regeneration of native species. The dense canopies reduce light and shade out herbaceous species. The invasion of secondary forests and cleared areas can suppress natural forest regeneration as native tree and shrub species are unable to establish seedlings in these dense thickets.

Specific control methods for this species are not available. Seedlings and saplings may be pulled or dug out. Larger shrubs can be cut and the cut stumps treated with herbicide.

References
538, 539, 667, 1098.

Pistia stratiotes L. Araceae

LF: Aquatic perennial
SN: -
CU: Ornamental

Geographic Distribution

Europe	*Australasia*	*Atlantic Islands*
Northern	X Australia	Cape Verde
British Isles	New Zealand	Canary, Madeira
Central, France		Azores
Southern	*Northern America*	South Atlantic Isl.
Eastern	Canada, Alaska	
Mediterranean Isl.	● Southeastern USA	*Indian Ocean Islands*
	X Western USA	Mascarenes
Africa	Remaining USA	X Seychelles
X Northern	Mexico	X Madagascar
● Tropical		
● Southern	*Southern America*	*Pacific Islands*
	X Tropical	Micronesia
Asia	● Caribbean	Galapagos
X Temperate	X Chile, Argentina	X Hawaii
● Tropical		Melanesia, Polynesia

Invaded Habitats
Freshwater wetlands, riparian habitats.

Description
A free-floating aquatic herb, nearly without a stem, sometimes producing stolons with new rosettes of leaves at their ends. Leaves have very short petioles, are pale green and form dense rosettes. The leaf blades are obovate, 5-17 cm long, 2-7 cm wide, thick and spongy, pubescent on both surfaces and with prominent veins beneath. Inflorescences are small and inconspicuous, almost sessile and borne among leaves. The white spathe is constricted in the middle, the spadix is shorter than the spathe and has a single female flower and 2-8 male flowers. There are two stamens in each male flower. Seeds are cylindrical and minute.

Ecology and Control
The place of origin of this plant is uncertain. It multiplies rapidly by vegetative growth and forms a compact continuous cover on the water surface, impeding water flow, displacing native water plants and lowering the oxygen content of the water. Dense mats may be colonized by emergent macrophytes. Seeds float on the water for a few days and are carried for long distances before they sink and germinate.

Plants may be harvested mechanically. Chemical control is done by applying a diquat herbicide approved for use in aquatic environments.

References
7, 229, 319, 331, 359, 549, 577, 580, 715, 986, 1162.

Pittosporum undulatum Vent.   Pittosporaceae

LF: Evergreen shrub, tree   CU: Ornamental
SN: -

Geographic Distribution

Europe	*Australasia*	*Atlantic Islands*
Northern	● N Australia	Cape Verde
British Isles	● New Zealand	X Canary, Madeira
Central, France		● Azores
X Southern	*Northern America*	X South Atlantic Isl.
Eastern	Canada, Alaska	
Mediterranean Isl.	Southeastern USA	*Indian Ocean Islands*
	X Western USA	Mascarenes
Africa	Remaining USA	Seychelles
Northern	● Mexico	Madagascar
Tropical		
● Southern	*Southern America*	*Pacific Islands*
	Tropical	Micronesia
Asia	● Caribbean	Galapagos
Temperate	Chile, Argentina	● Hawaii
Tropical		Melanesia, Polynesia

Invaded Habitats
Forests, grass- and heathland, riparian habitats, coastal dunes and scrub.

Description
A glabrous bushy tree of 4-14 m height, with a grey bark and a dense pyramidal crown. Leaves are radiating at the ends of branches, are elliptic-oblong, 6-14 cm long and 3-5 cm wide, dark green above and paler beneath. Fragrant and creamy-white flowers are borne in umbel-like cymes at the ends of branches. Fruits are globular or obovoid dehiscent capsules of 10-15 mm diameter, opening with two valves and becoming orange when ripe. They contain 12-22 red brown to black seeds.

Ecology and Control
This tree is a successful gap colonizer and eliminates native vegetation by the low and dense canopies, shading out almost all other species. The tree produces large quantities of sticky seeds that are dispersed by birds and mammals. Gaps are quickly colonized and seedlings form a dense ground cover. The nutrient rich litter leads to an increase of soil fertility levels. The tree displaces native laurel forests on the Azores with monospecific shrubland. The tree resprouts vigorously after damage.

    Seedlings are easy to hand pull. Small trees are cut, larger trees cut or girdled, often in combination with herbicide treatment. Effective herbicides are 2,4,5-T, picloram plus 2,4-D, or glyphosate. Follow-up programmes are necessary to treat regrowth and emerging seedlings.

References
215, 284, 460, 461, 462, 463, 471, 472, 473, 517, 549, 915, 916, 924, 1028, 1084, 1108, 1109, 1110.

Plantago coronopus L.   Plantaginaceae

LF: Annual, perennial herb   CU: -
SN: -

Geographic Distribution

Europe		*Australasia*		*Atlantic Islands*	
N	Northern	●	Australia		Cape Verde
N	British Isles	X	New Zealand	N	Canary, Madeira
N	Central, France			N	Azores
N	Southern	*Northern America*			South Atlantic Isl.
N	Eastern		Canada, Alaska		
N	Mediterranean Isl.		Southeastern USA	*Indian Ocean Islands*	
		X	Western USA		Mascarenes
Africa			Remaining USA		Seychelles
N	Northern		Mexico		Madagascar
	Tropical				
	Southern	*Southern America*		*Pacific Islands*	
			Tropical		Micronesia
Asia			Caribbean		Galapagos
N	Temperate		Chile, Argentina		Hawaii
N	Tropical				Melanesia, Polynesia

Invaded Habitats
Grass- and heathland, riparian habitats, freshwater wetlands, coastal dunes.

Description
A pubescent herb with one to many rosettes and flowering stems up to 20 cm tall. The numerous leaves are all basal, 3-20 cm long and 0.5-2 cm wide, linear and entire or toothed, or deeply divided into fine segments. Flowering stems are usually numerous. Flowers have four whitish and papery petals, and are crowded in a cylindrical spike of 2-5 cm length at the end of a long peduncle. Fruits are dehiscent capsules with 2-6 seeds of *c.* 1 mm length.

Ecology and Control
A variable complex in Europe with many closely related taxa differing in size and growth form. Native habitats include mainly coastal marshes and cliffs. It is invasive because it forms dense mats of small rosettes, displacing native vegetation and preventing regeneration of native plants. Seeds are dispersed by wind and water.

    Specific control methods for this species are not available. Scattered plants may be pulled or dug out, larger infestations treated with herbicide.

References
215, 346, 398, 1197, 1356, 1450.

Plantago lanceolata L.   Plantaginaceae

LF: Perennial herb
SN: -
CU: Fodder

Geographic Distribution

Europe	*Australasia*	*Atlantic Islands*
N Northern	● Australia	Cape Verde
N British Isles	X New Zealand	N Canary, Madeira
N Central, France		N Azores
N Southern	*Northern America*	South Atlantic Isl.
N Eastern	X Canada, Alaska	
N Mediterranean Isl.	X Southeastern USA	*Indian Ocean Islands*
	X Western USA	X Mascarenes
Africa	X Remaining USA	Seychelles
N Northern	Mexico	X Madagascar
X Tropical		
X Southern	*Southern America*	*Pacific Islands*
	X Tropical	Micronesia
Asia	X Caribbean	Galapagos
N Temperate	X Chile, Argentina	X Hawaii
N Tropical		Melanesia, Polynesia

Invaded Habitats
Grass- and heathland, riparian habitats, freshwater wetlands, coastal dunes.

Description
A glabrous to pubescent herb with one to many rosettes and flowering stems about twice as long as the leaves. Rhizomes are thick and short. Leaves are linear-lanceolate to lanceolate, 2-30 cm long and 5-35 mm wide, mostly entire, gradually tapering to the petiole, and are all basal. Flowering stems are up to 50 cm tall and bear one dense spike of 1-6 cm length each. Flowers are subtended by dark green to blackish bracts of 2.5-3.5 mm length. The brownish-white corolla is *c.* 4 mm in diameter. Fruits are capsules of 3-4 mm length, each having two seeds of *c.* 2 mm length.

Ecology and Control
This is an extremely variable species in the native range, with many varieties and subspecies ranging in growth habit from prostrate to erect. The plant is a widespread weed and colonizes rapidly open areas. It forms dense swards that crowd out native vegetation and prevent the establishment of native species. Seeds are sticky and adhere to animals. They remain viable in the soil for a long time.

Grazing or mowing may reduce the plant's growth. Effective herbicides are 2,4-D, MCPA, dicamba, or fenoprop.

References
215, 226, 580, 1120.

Poa pratensis L. — Poaceae

LF: Perennial herb
SN: -
CU: Erosion control, ornamental

Geographic Distribution

Europe		*Australasia*		*Atlantic Islands*	
N	Northern	●	Australia		Cape Verde
N	British Isles	X	New Zealand	N	Canary, Madeira
N	Central, France			X	Azores
N	Southern	*Northern America*			South Atlantic Isl.
N	Eastern	●	Canada, Alaska		
N	Mediterranean Isl.		Southeastern USA	*Indian Ocean Islands*	
		X	Western USA	X	Mascarenes
Africa		●	Remaining USA		Seychelles
N	Northern		Mexico		Madagascar
	Tropical				
X	Southern	*Southern America*		*Pacific Islands*	
			Tropical		Micronesia
Asia			Caribbean		Galapagos
N	Temperate		Chile, Argentina	X	Hawaii
N	Tropical				Melanesia, Polynesia

Invaded Habitats
Grassland, riparian habitats, freshwater wetlands.

Description
A perennial grass with a strong short-creeping rhizome, erect culms of 20-180 cm height that have 2-4 nodes, and leaves that are shorter than the stems. Leaves are green or greyish-green, up to 30 cm long and 2-5 mm wide. The papery ligules are 1-3 mm long. Inflorescences are purplish green panicles of 6-15 cm length and 1-12 cm width, with the branches mostly in clusters of 3-5. Spikelets are ovate to oblong, 4-6 mm long and have 2-5 florets. Glumes are 2-4 mm long, lemmas 3-4 mm long. Fruits are narrowly ovoid caryopses.

Ecology and Control
Where native, this grass is commonly found in meadows, pastures and disturbed sites. It is invasive because it spreads rapidly and forms dense swards that crowd out native vegetation. Populations reach high shoot densities and the large size makes it highly competitive to native grasses and forbs.

Single plants can be dug out, rhizomes must be removed to prevent regrowth. Larger patches can be treated with herbicides.

References
135, 215, 551, 571, 642, 828, 1069, 1454.

Polygala myrtifolia L. Polygalaceae

LF: Evergreen shrub
SN: -
CU: Ornamental

Geographic Distribution

Europe
- Northern
- British Isles
- Central, France
- Southern
- Eastern
- Mediterranean Isl.

Africa
- Northern
- Tropical
- N Southern

Asia
- Temperate
- Tropical

Australasia
- ● Australia
- X New Zealand

Northern America
- Canada, Alaska
- Southeastern USA
- Western USA
- Remaining USA
- Mexico

Southern America
- Tropical
- Caribbean
- Chile, Argentina

Atlantic Islands
- Cape Verde
- Canary, Madeira
- Azores
- South Atlantic Isl.

Indian Ocean Islands
- Mascarenes
- Seychelles
- Madagascar

Pacific Islands
- Micronesia
- Galapagos
- Hawaii
- Melanesia, Polynesia

Invaded Habitats
Heathland, riparian habitats, coastal dunes and bluffs.

Description
A much-branched shrub or small tree of 1-2.5 m height, with a dense foliage, a rounded crown and a pale green to grey brown bark. Leaves are spirally arranged and closely crowded along the branchlets. They are simple, narrowly oblong, 2-5 cm long and 8-15 mm wide, almost sessile and glabrous. Flowers grow in short and few-flowered racemes at the ends of branches. The pinkish to purple or lilac corolla is 13-18 mm long. The lower petal has a brush-like appendage, the two lateral sepals resemble petals. Fruits are small, oval and dehiscent capsules that are slightly winged, containing usually two seeds.

Ecology and Control
A highly variable species in the native range with regard to size and shape of leaves. Where invasive, the plant spreads rapidly and forms dense thickets, impeding growth and regeneration of native plants. Trees and shrubs are unable to establish seedlings under canopies of this shrub and forest regeneration is suppressed. The plant establishes well in disturbed sites and seedling density may exceed 2,000 m^{-2}. Seeds can remain viable in the soil for several years.

Plants can be killed by cutting close to the ground. Fires are used in fire adapted communities to kill topgrowth and stimulate seed germination. Seedlings are then sprayed with herbicide.

References
121, 215, 607, 924.

Polygala virgata Thunb. Polygalaceae

LF: Evergreen shrub
SN: *Polygala speciosa* Sims
CU: Ornamental

Geographic Distribution

Europe	Australasia	Atlantic Islands
Northern	● Australia	Cape Verde
British Isles	X New Zealand	Canary, Madeira
Central, France		Azores
Southern	*Northern America*	South Atlantic Isl.
Eastern	Canada, Alaska	
Mediterranean Isl.	Southeastern USA	*Indian Ocean Islands*
	Western USA	Mascarenes
Africa	Remaining USA	Seychelles
Northern	Mexico	Madagascar
N Tropical		
N Southern	*Southern America*	*Pacific Islands*
	Tropical	Micronesia
Asia	Caribbean	Galapagos
Temperate	Chile, Argentina	Hawaii
Tropical		Melanesia, Polynesia

Invaded Habitats
Grass- and scrubland, disturbed sites.

Description
A slender erect shrub of 1-3 m height, with sparsely hairy stems. Leaves are linear or oblanceolate to narrow elliptic, 10-50 mm long and 1-5 mm wide, and sparsely hairy. Inflorescences are racemes of 4-12 cm length borne at the ends of branches and in the upper leaf axils. The purple to pale lilac zygomorphic flowers are 12-15 mm long. Fruits are obovate capsules of *c.* 10 mm length surrounded by the persistent calyx. Seeds are blackish, oblong, *c.* 6 mm long and 2.5 mm wide.

Ecology and Control
A variable species in the native range with regard to degree of pubescence and leaf size. The plant grows to dense thickets that crowd out native plants and impede their regeneration. Little is known on the ecology of this species.

Specific control methods for this species are not available. The same methods as for the previous species may apply.

References
215.

Polygonum aviculare L.
Polygonaceae

LF: Annual herb
SN: *Polygonum heterophyllum* Lindm.
CU: Ornamental

Geographic Distribution

Europe		*Australasia*		*Atlantic Islands*	
N	Northern	X	Australia		Cape Verde
N	British Isles	X	New Zealand		Canary, Madeira
N	Central, France			N	Azores
N	Southern	*Northern America*			South Atlantic Isl.
N	Eastern	X	Canada, Alaska		
N	Mediterranean Isl.	●	Southeastern USA	*Indian Ocean Islands*	
		X	Western USA		Mascarenes
Africa		X	Remaining USA		Seychelles
X	Northern	X	Mexico		Madagascar
X	Tropical				
●	Southern	*Southern America*		*Pacific Islands*	
		X	Tropical		Micronesia
Asia			Caribbean		Galapagos
N	Temperate	X	Chile, Argentina	X	Hawaii
	Tropical				Melanesia, Polynesia

Invaded Habitats
Heath- and scrubland, disturbed sites.

Description
A low, mat-forming to scrambling herb with strongly branched stems of 10-100 cm length, and with a strong taproot. Leaves have short petioles, are lanceolate to ovate, 15-50 mm long and 5-18 mm wide. Leaves of the main stem are much larger than those of branches. Inflorescences are axillary, small spike-like clusters of 1-6 flowers. Flowers are greenish to red and 2-3 mm long. Fruits are nuts of 2.5-3.5 mm length, surrounded by the perianth remains.

Ecology and Control
A variable species in the native range, with creeping and erect growing subspecies. The plant's sprawling growth habit leads to dense mats that smother herbaceous species and small shrubs. Seeds are dispersed by birds, mammals and water. The plant grows best in rich soils but is found also on infertile soils. It is a troublesome agricultural weed and also invades natural plant communities.

Specific control methods for this species are not available. Plants may be cut before fruit formation to prevent seed dispersal. Seedlings can be hand pulled, larger plants treated with herbicide.

References
581, 851, 857, 858, 1089, 1241.

Polygonum perfoliatum L.   Polygonaceae

LF: Annual vine   CU: Ornamental
SN: *Persicaria perfoliata* (L.) Gross, *Tracaulon perfoliatum* (L.) Greene

Geographic Distribution

Europe		*Australasia*		*Atlantic Islands*	
	Northern		Australia		Cape Verde
	British Isles		New Zealand		Canary, Madeira
	Central, France				Azores
	Southern	*Northern America*			South Atlantic Isl.
	Eastern		Canada, Alaska		
	Mediterranean Isl.	●	Southeastern USA	*Indian Ocean Islands*	
			Western USA		Mascarenes
Africa		X	Remaining USA		Seychelles
	Northern		Mexico		Madagascar
	Tropical				
	Southern	*Southern America*		*Pacific Islands*	
			Tropical		Micronesia
Asia		X	Caribbean		Galapagos
N	Temperate		Chile, Argentina		Hawaii
N	Tropical				Melanesia, Polynesia

Invaded Habitats
Riparian habitats, hedges, disturbed sites.

Description
A prickly vine with elongated and branched stems, becoming woody at the base. Stems, petioles and leaf veins are covered with spikes of 1-2 mm length. Leaves are triangular, pale green and 2.5-7.5 cm long and wide. Flowers are 3-4 mm long. The persistent calyx thickens during fruit formation to produce a blue berry-like fruit of *c.* 5 mm diameter, containing a single black shiny achene of *c.* 3 mm diameter.

Ecology and Control
A rapidly growing plant that smothers native herbs and small shrubs. Plants of this vine form tangled mats that climb 6-8 m over shrubs and understorey trees, shading out the native vegetation. Seeds are dispersed by water, birds and small mammals.

    Physical control is done by mowing or cutting close to the ground. An effective herbicide to control dense stands is glyphosate. Follow-up programmes are necessary to treat seedlings. Pre-emergence herbicides include imazapyr or atrazine.

References
293, 843, 960, 1393.

Polypogon monspeliensis (L.) Desf.   Poaceae

LF: Annual herb   CU: -
SN: *Alopecurus monspeliensis* L.

Geographic Distribution

Europe
    Northern
N   British Isles
N   Central, France
N   Southern
N   Eastern
N   Mediterranean Isl.

Africa
N   Northern
    Tropical
X   Southern

Asia
N   Temperate
N   Tropical

Australasia
●   Australia
X   New Zealand

Northern America
X   Canada, Alaska
    Southeastern USA
●   Western USA
    Remaining USA
    Mexico

Southern America
X   Tropical
    Caribbean
X   Chile, Argentina

Atlantic Islands
N   Cape Verde
N   Canary, Madeira
N   Azores
    South Atlantic Isl.

Indian Ocean Islands
X   Mascarenes
    Seychelles
    Madagascar

Pacific Islands
    Micronesia
    Galapagos
X   Hawaii
    Melanesia, Polynesia

Invaded Habitats
Riparian habitats, freshwater wetlands, coastal marshes.

Description
A smooth and glabrous grass with stems of 10-90 cm height and with 3-6 nodes per stem, usually branched near the base. Leaves are 5-15 cm long and 2-8 mm wide. Ligules are up to 10 mm long. The inflorescence is a contracted and dense, cylindrical panicle of 1.5-16 cm length, covered with fine bristles. Spikelets are 2-3 mm long and have two glumes, each with an awn of 3.5-7 mm length. Spikelets have one floret each.

Ecology and Control
Where native, this grass is found mainly near the coast at the edges of salt marshes, other damp sites, and in disturbed sites. It is invasive because it forms dense swards that crowd out native plants and prevent their regeneration.

    Specific control methods for this species are not available. Scattered plants can be hand pulled or cut, larger populations treated with herbicide.

References
215, 618.

Populus alba L. Salicaceae

LF: Deciduous tree
SN: -
CU: Ornamental, wood

Geographic Distribution

Europe		*Australasia*		*Atlantic Islands*	
X	Northern	●	Australia		Cape Verde
X	British Isles	X	New Zealand	N	Canary, Madeira
N	Central, France			X	Azores
N	Southern	*Northern America*			South Atlantic Isl.
N	Eastern	X	Canada, Alaska		
N	Mediterranean Isl.		Southeastern USA	*Indian Ocean Islands*	
		X	Western USA		Mascarenes
Africa			Remaining USA		Seychelles
N	Northern		Mexico		Madagascar
	Tropical				
●	Southern	*Southern America*		*Pacific Islands*	
			Tropical		Micronesia
Asia			Caribbean		Galapagos
N	Temperate		Chile, Argentina		Hawaii
	Tropical				Melanesia, Polynesia

Invaded Habitats
Grass- and heathland, scrub, riparian habitats.

Description
A tree reaching 30 m height with a smooth and whitish grey bark, a broad crown, and white-tomentose young branchlets and buds. Leaves on long shoots are ovate, lobed, coarsely toothed, 6-12 cm long, dark green above and white-tomentose beneath. Leaves on short shoots are much smaller and ovate to elliptic-oblong. Petioles are shorter than the length of the leaf blades. Female catkins are *c.* 5 cm long, male catkins longer and yellow to orange or red. Fruiting catkins elongate to 8-10 cm length.

Ecology and Control
A native of riparian forests, this fast growing tree reproduces rapidly and forms many suckers from the roots. Where invasive, dense stands of this tree crowd out native vegetation and strongly reduce species diversity due to the shading effect. The tree hybridizes with the native *P. grandidentata* and *P. tremuloides* in the USA. It spreads mainly by root sprouting.

Trees and suckers can be controlled with repeated and frequent cutting. Cut surfaces can be treated with a glyphosate herbicide; the same herbicide can be used as a foliar application.

References
549, 1089, 1222, 1223.

Prosopis glandulosa Torr.   Fabaceae

LF: Deciduous shrub, tree   CU: Forage
SN: -

Geographic Distribution

Europe	*Australasia*	*Atlantic Islands*
Northern	● Australia	Cape Verde
British Isles	New Zealand	Canary, Madeira
Central, France		Azores
Southern	*Northern America*	South Atlantic Isl.
Eastern	Canada, Alaska	
Mediterranean Isl.	Southeastern USA	*Indian Ocean Islands*
	N Western USA	Mascarenes
Africa	Remaining USA	Seychelles
Northern	N Mexico	Madagascar
Tropical		
● Southern	*Southern America*	*Pacific Islands*
	X Tropical	Micronesia
Asia	Caribbean	Galapagos
Temperate	Chile, Argentina	Hawaii
Tropical		Melanesia, Polynesia

Invaded Habitats
Grass- and heathland, rangeland, riparian habitats.

Description
A multi-stemmed shrub or small tree of 3-15 m height, with reddish brown branches and pairs of straight spines at the nodes. The dark green leaves are compound and have one pair of pinnae, each of which has 6-25 pairs of widely spaced glabrous leaflets. These are 10-50 mm long and 20-40 mm wide. Inflorescences are axillary spikes of 6-10 cm length with numerous yellow flowers. Fruits are woody and slender pods of 10-25 cm length, cylindrical and slightly constricted between the seeds. Each pod contains 5-18 oblong seeds of 6-7 mm length.

Ecology and Control
This nitrogen-fixing tree is well adapted to dry conditions and has roots penetrating to 15 m depth. There are two naturalized varieties in Australia and a number of hybrids. The tree is a prolific seed producer; seeds are dispersed by water and animals and remain viable in the soil for many years. The tree resprouts vigorously after damage. It transforms invaded grasslands into woodlands and forms thorny thickets that tolerate only few species and reduce wildlife habitat. Loss in grass cover under its canopies cause soil erosion and sand dune formation.

    Seedlings and saplings can be pulled or dug out, roots must be removed. Cutting trees below the ground surface may prevent resprouting. Effective herbicides are clopyralid, picloram, triclopyr, or 2,4-D amine. Follow-up programmes are necessary to treat seedlings and regrowth.

References
137, 142, 143, 185, 189, 190, 284, 524, 549, 936, 986, 1068, 1232, 1332, 1459.

Prosopis juliflora (Sw.) DC.   Fabaceae

LF: Evergreen shrub, tree   CU: Shelter, wood
SN: -

Geographic Distribution

Europe		*Australasia*		*Atlantic Islands*	
	Northern	X	Australia		Cape Verde
	British Isles		New Zealand		Canary, Madeira
	Central, France				Azores
	Southern	*Northern America*			South Atlantic Isl.
	Eastern		Canada, Alaska		
	Mediterranean Isl.		Southeastern USA	*Indian Ocean Islands*	
		N	Western USA	●	Mascarenes
Africa			Remaining USA		Seychelles
	Northern	N	Mexico		Madagascar
X	Tropical				
	Southern	*Southern America*		*Pacific Islands*	
		N	Tropical		Micronesia
Asia		X	Caribbean		Galapagos
X	Temperate		Chile, Argentina	X	Hawaii
X	Tropical				Melanesia, Polynesia

Invaded Habitats
Grass- and shrubland, dry forests.

Description
A shrub or tree up to 8 m tall, sometimes more, with a gnarled trunk and branchlets that are armed with 1-4 cm long spines. Leaves are bipinnately compound with mostly two, sometimes more pairs of pinnae of 6-8 cm length. Leaflets are in 15-20 pairs per pinna and 8-10 mm long. Flowers are greenish white to yellow, and densely crowded in spikes of 5-10 cm length. Fruits are yellow to brown pods of 10-25 cm length and 8-15 mm width. Seeds are *c.* 7 mm long and 4-5 mm wide.

Ecology and Control
This salt and drought tolerant tree is nitrogen-fixing and has deep reaching roots and tolerates dry as well as waterlogged soils. Seed production is prolific, and seeds are dispersed by water and animals. Seeds accumulate in a soil seed bank. The tree rapidly forms dense thorny thickets that reduce native species richness and wildlife habitats. Invaded grasslands are transformed to woodland and forests. Loss of grass cover under canopies of this tree may promote soil erosion. The tree resprouts easily after damage.

     Seedlings and saplings can be pulled or dug out, roots must be removed. Cutting trees below the ground surface may prevent resprouting. Burning is used to kill plants aboveground. Chemical control includes basal bark spraying, treating cut stumps and foliar spraying. Effective herbicides are triclopyr or picloram. Follow-up programmes are necessary to treat seedlings and regrowth.

References
524, 986, 1165, 1166, 1278, 1332, 1459.

Prunus cerasifera Ehrh.   Rosaceae

LF: Deciduous tree
SN: -
CU: Ornamental, food

Geographic Distribution

Europe	*Australasia*	*Atlantic Islands*
Northern	● Australia	Cape Verde
X British Isles	X New Zealand	Canary, Madeira
X Central, France		Azores
N Southern	*Northern America*	South Atlantic Isl.
N Eastern	Canada, Alaska	
Mediterranean Isl.	Southeastern USA	*Indian Ocean Islands*
	X Western USA	Mascarenes
Africa	Remaining USA	Seychelles
Northern	Mexico	Madagascar
Tropical		
Southern	*Southern America*	*Pacific Islands*
	Tropical	Micronesia
Asia	Caribbean	Galapagos
N Temperate	Chile, Argentina	Hawaii
X Tropical		Melanesia, Polynesia

Invaded Habitats
Grassland, riparian habitats, forest edges, coastal scrub and dunes.

Description
A low much-branched tree or shrub up to 8 m height, with sometimes spiny branches, and the young twigs glabrous and glossy. The elliptic to ovate leaves are 4-7 cm long and 2-3.5 cm wide, pubescent on the veins beneath, finely serrated and almost rounded at the base. Flowers are borne solitary or in sessile lateral umbels, are white, 2-2.5 cm in diameter, with 25-30 stamens and pedicels of 5-15 mm length. Flowers appear with or slightly before the leaves. Fruits are globose, red to yellow drupes of 2-3cm diameter, containing one stone seed.

Ecology and Control
The exact native range of this tree is obscure due to its wide cultivation. The tree becomes abundant and shades out native vegetation, preventing the regeneration of native shrubs and trees and strongly reducing species richness under its canopies. Seedlings develop a strong root system before significant vertical growth commences and are thus highly competitive to native species. Seeds are dispersed by birds and mammals.

    Seedlings and saplings can be pulled or dug out. Larger trees are cut or ringbarked. Dense seedling growth can be treated with herbicide. Fruiting trees should be removed first to prevent seed dispersal.

References
121, 215, 607.

Prunus laurocerasus L.   Rosaceae

LF: Deciduous tree   CU: Ornamental, food
SN: *P. grandifolia* Salisb., *Cerasus laurocerasus* Loisel., *Laurocerasus officinalis* Roem.

Geographic Distribution

Europe		*Australasia*		*Atlantic Islands*	
X	Northern	●	Australia		Cape Verde
X	British Isles	X	New Zealand		Canary, Madeira
●	Central, France				Azores
N	Southern	*Northern America*			South Atlantic Isl.
N	Eastern		Canada, Alaska		
	Mediterranean Isl.		Southeastern USA	*Indian Ocean Islands*	
		X	Western USA		Mascarenes
Africa			Remaining USA		Seychelles
	Northern		Mexico		Madagascar
	Tropical				
	Southern	*Southern America*		*Pacific Islands*	
			Tropical		Micronesia
Asia			Caribbean		Galapagos
N	Temperate		Chile, Argentina		Hawaii
X	Tropical				Melanesia, Polynesia

Invaded Habitats
Forests and forest edges, riparian habitats.

Description
A wide spreading shrub or a tree up to 10 m tall with pale green young twigs. Leaves are dark green shining, elliptic-ovate to oblong-ovate, 5-15 cm long and 2.5-7 cm wide. Dull white flowers of *c.* 8 mm diameter grow in racemes of 7-13 cm length in leaf axils and at the ends of branches. Fruits are purplish-black, ovoid to subglobose drupes of 10-12 mm diameter, containing one stone seed.

Ecology and Control
This fast growing tree tolerates shade and full sun and is able to germinate in dense shade. A number of varieties and cultivars are used as ornamentals. The spreading growth habit and the dense foliage shade out native vegetation and strongly reduce species richness. Once established, it remains dominant and forms pure stands. Larger plants resprout after damage, and lower branches touching the soil may become rooted.

Seedlings and saplings may be pulled or dug out. Larger plants can be cut and the cut stumps treated with herbicide, or ringbarked.

References
121, 215, 607, 855, 1359.

Prunus serotina Ehrh.   Rosaceae

LF: Deciduous tree   CU: Ornamental, wood
SN: -

Geographic Distribution

Europe		*Australasia*		*Atlantic Islands*	
X	Northern		Australia		Cape Verde
X	British Isles		New Zealand		Canary, Madeira
●	Central, France				Azores
X	Southern	*Northern America*			South Atlantic Isl.
●	Eastern	N	Canada, Alaska		
	Mediterranean Isl.	N	Southeastern USA	*Indian Ocean Islands*	
			Western USA		Mascarenes
Africa		N	Remaining USA		Seychelles
	Northern	N	Mexico		Madagascar
	Tropical				
	Southern	*Southern America*		*Pacific Islands*	
		X	Tropical		Micronesia
Asia			Caribbean		Galapagos
	Temperate		Chile, Argentina		Hawaii
	Tropical				Melanesia, Polynesia

Invaded Habitats
Forests and forest gaps, woodland.

Description
A tree reaching 30 m height, with a dark brown bark and glabrous young branches. Leaves are oblong-ovate to lanceolate-oblong, 5-12 cm long, bright green above and pale green beneath, and slightly toothed at the margins. Petioles are 6-25 mm long. White flowers of 8-10 mm diameter grow in cylindrical racemes of 6-15 cm length, each having *c.* 30 flowers. Pedicels are 3-8 mm long. Fruits are globose purplish-black drupes of 8-10 mm diameter, containing one stone seed of 7-8 mm diameter.

Ecology and Control
A fast growing tree that colonizes forest gaps and edges. Saplings can survive for long periods of time under shade and then sprout rapidly if a canopy gap occurs. The short-lived seeds are mainly dispersed by birds. The tree vigorously resprouts if damaged. The fast growth and persistence leads to dense stands that eliminate native shrubs and trees.
    Seedlings and saplings can be pulled or cut. Larger trees are cut and the cut stumps treated with herbicide, or the basal bark is sprayed. Effective herbicides are 2,4-D amine, 2,4,5-T amine, or 2,4-D plus dicamba.

References
54, 284, 824, 919, 1203, 1228, 1229, 1324.

Psidium cattleianum Sabine Myrtaceae

LF: Evergreen shrub, tree
SN: *Psidium littorale* Raddi
CU: Food

Geographic Distribution

Europe		*Australasia*		*Atlantic Islands*	
	Northern	●	Australia		Cape Verde
	British Isles	X	New Zealand		Canary, Madeira
	Central, France			X	Azores
	Southern	*Northern America*			South Atlantic Isl.
	Eastern		Canada, Alaska		
	Mediterranean Isl.	●	Southeastern USA	*Indian Ocean Islands*	
			Western USA	●	Mascarenes
Africa			Remaining USA	●	Seychelles
	Northern		Mexico		Madagascar
X	Tropical				
●	Southern	*Southern America*		*Pacific Islands*	
		N	Tropical	X	Micronesia
Asia			Caribbean		Galapagos
	Temperate		Chile, Argentina	●	Hawaii
	Tropical			●	Melanesia, Polynesia

Invaded Habitats
Forests and forest openings, mountain slopes.

Description
A large shrub or small tree up to 6 m tall with a pale brown and smooth bark. Young shoots are pubescent. The simple and glabrous leaves are 4-8 cm long, 2.5-4.5 cm wide and pointed at the ends. Margins are often recurved. White flowers with petals of *c.* 5 mm length are borne solitary in leaf axils. Fruits are 20-25 mm in diameter, red, having several seeds of *c.* 3 mm length and 2.5 mm width.

Ecology and Control
A thicket forming, shade tolerant tree that is able to invade intact and undisturbed rainforest. It is fast growing and produces dense populations of root suckers and seedlings. The dense foliage shades out all other plants. The soil surface is often covered by mats of feeder roots. Seed production is prolific and seeds are dispersed by birds and mammals. The tree accumulates a large amount of litter that suppresses the establishment and growth of native tree seedlings.

Control should aim at removing fruit bearing trees first and containing small infestations to prevent spread. Seedlings and saplings may be pulled or dug out, roots must be removed. Larger trees can be cut and the cut stumps treated with herbicide.

References
284, 416, 417, 549, 599, 715, 769, 770, 1199, 1239.

Psidium guajava L.   Myrtaceae

LF: Evergreen shrub, tree   CU: Food, shelter, wood
SN: *Psidium pomiferum* L., *Psidium pumilum* Vahl

Geographic Distribution

Europe		*Australasia*		*Atlantic Islands*	
	Northern	●	Australia	X	Cape Verde
	British Isles	●	New Zealand	X	Canary, Madeira
	Central, France				Azores
	Southern	*Northern America*			South Atlantic Isl.
	Eastern		Canada, Alaska		
	Mediterranean Isl.	●	Southeastern USA	*Indian Ocean Islands*	
			Western USA		Mascarenes
Africa			Remaining USA		Seychelles
X	Northern	N	Mexico		Madagascar
X	Tropical				
●	Southern	*Southern America*		*Pacific Islands*	
		N	Tropical		Micronesia
Asia		N	Caribbean	●	Galapagos
X	Temperate		Chile, Argentina	●	Hawaii
X	Tropical			X	Melanesia, Polynesia

Invaded Habitats
Forests and forest margins, grassland, riparian habitats.

Description
A sparingly branched tree or shrub up to 9 m tall, with a smooth reddish brown bark that is mottled with reddish brown, grey and whitish patches. Branchlets are hairy and quadrangular in cross-section. The ovate to oblong-elliptic leaves are stiff, 4-15 cm long, with prominent veins and pubescent beneath. White flowers of 3-4 cm diameter grow solitary or in clusters in the axils of leaves, and become yellow at maturity. Fruits are yellow or pink berries of 3-6 cm diameter, containing numerous yellow seeds of *c.* 3 mm length.

Ecology and Control
A variable species with many cultivated varieties differing in fruit characters. The tree is drought tolerant, fast growing and extensively suckering from roots. It forms dense thickets that reduce native species richness and impede regeneration of native trees. The tree grows quickly in forest gaps.

Chemical control is difficult as the leaves have a waxy cuticle impeding penetration of herbicides. Seedlings and saplings may be pulled or dug out, roots must be removed. Larger trees can be cut and the cut stumps treated with herbicide.

References
284, 549, 550, 599, 607, 715, 725, 792, 794, 1089, 1142, 1313.

Psoralea pinnata L. — Fabaceae

LF: Evergreen shrub, tree
SN: -
CU: Ornamental

Geographic Distribution

Europe	*Australasia*	*Atlantic Islands*
Northern	● Australia	Cape Verde
British Isles	● New Zealand	Canary, Madeira
Central, France		Azores
Southern	*Northern America*	South Atlantic Isl.
Eastern	Canada, Alaska	
Mediterranean Isl.	Southeastern USA	*Indian Ocean Islands*
	Western USA	Mascarenes
Africa	Remaining USA	Seychelles
Northern	Mexico	Madagascar
Tropical		
N Southern	*Southern America*	*Pacific Islands*
	Tropical	Micronesia
Asia	Caribbean	Galapagos
Temperate	Chile, Argentina	Hawaii
Tropical		Melanesia, Polynesia

Invaded Habitats
Grass- and heathland, riparian habitats, coastal dunes and beaches.

Description
A small tree up to 4 m tall, with the leaves divided into 5-11 narrow leaflets. These are 2-3 cm long and 1-2 mm wide. Flowers are mauve-blue with white centres, 12-15 mm in diameter, fragrant and borne solitary or in axillary clusters. Fruits are black pods of 3-4 mm length, containing one seed.

Ecology and Control
A nitrogen-fixing shrub that grows naturally on riverbanks, in forest margins and scrubland. It is a prolific seed producer, and seeds can remain viable in the soil for more than 8 years. The shrub forms dense thickets that eliminate native vegetation and prevent any regeneration of native species.

Small plants and seedlings can be dug out. Larger plants can be ringbarked or cut at ground level. Burning is used to clear stands and to stimulate seed germination. Seedlings are then treated with herbicide.

References
121, 215, 607, 924, 1415.

Pueraria montana (Lour.) Merr.   Fabaceae

LF: Deciduous climber   CU: Erosion control, food, ornamental
SN: *Pueraria triloba* (Lour.) Makino, *Pueraria hirsuta* Schnied.

Geographic Distribution

Europe
    Northern
    British Isles
    Central, France
●   Southern
    Eastern
    Mediterranean Isl.

Africa
    Northern
    Tropical
●   Southern

Asia
N   Temperate
N   Tropical

Australasia
N   Australia
    New Zealand

Northern America
    Canada, Alaska
●   Southeastern USA
X   Western USA
●   Remaining USA
●   Mexico

Southern America
    Tropical
    Caribbean
    Chile, Argentina

Atlantic Islands
    Cape Verde
    Canary, Madeira
    Azores
    South Atlantic Isl.

Indian Ocean Islands
    Mascarenes
    Seychelles
    Madagascar

Pacific Islands
    Micronesia
    Galapagos
●   Hawaii
N   Melanesia, Polynesia

Invaded Habitats
Riparian habitats, forest edges, woodland.

Description
A hairy, trailing or high climbing vine with the older stems being woody and reaching 30 m length. The alternate leaves are compound with three leaflets. The base of the terminal leaflet has usually two lobes. Leaflets are up to 20 cm long and 12 cm wide, with petioles ranging from 3-30 cm in length. Purple and fragrant flowers of 2-2.5 cm length are borne in racemes of 10-20 cm length. Fruits are woolly, flattened and brownish pods of 4-8 cm length.

Ecology and Control
A native of forest margins, occurring up to 2,000 m elevation. This nitrogen-fixing and fast growing vine can quickly cover shrubs and trees with a dense tangle of stems, smothering them and shading out all other vegetation. Stems easily root at nodes and form mats that may be more than 2 m thick. The vine is able to smother trees up to 35 m tall, eventually killing them. Seeds are dispersed by birds and mammals.

Control includes grazing by goats, persistent weeding or mowing, and chemical control. An effective herbicide is triclopyr. Once established, the vine is difficult to control. If manually controlled, all roots should be removed to prevent regrowth.

References
184, 216, 284, 424, 567, 692, 715, 765, 878, 889, 1095, 1249.

Pyracantha angustifolia (Franch.) Schneid. Rosaceae

LF: Evergreen shrub, tree   CU: Ornamental
SN: *Cotoneaster angustifolius* Franch.

Geographic Distribution

Europe	*Australasia*	*Atlantic Islands*
Northern	● Australia	Cape Verde
British Isles	X New Zealand	Canary, Madeira
Central, France		Azores
Southern	*Northern America*	South Atlantic Isl.
Eastern	Canada, Alaska	
Mediterranean Isl.	Southeastern USA	*Indian Ocean Islands*
	Western USA	Mascarenes
Africa	Remaining USA	Seychelles
Northern	Mexico	Madagascar
Tropical		
● Southern	*Southern America*	*Pacific Islands*
	Tropical	Micronesia
Asia	Caribbean	Galapagos
N Temperate	Chile, Argentina	● Hawaii
Tropical		Melanesia, Polynesia

Invaded Habitats
Grass- and heathland, rocky ridges, riparian habitats.

Description
A dense and spreading shrub up to 4 m tall, sometimes prostrate, with stiff and spiny branches. Leaves are narrow-oblong to lanceolate-oblong, 15-50 mm long, with the margins entire or slightly toothed, dark green above and greyish-tomentose beneath. White flowers of *c.* 8 mm diameter grow in rather dense corymbs of 2-4 cm diameter. Fruits are short-stalked, bright orange to red and globose pomes of 6-8 mm diameter, containing five seeds.

Ecology and Control
This shrub forms dense thickets where invasive, shading out native plants and impeding the growth and regeneration of shrubs and trees. It invades high-altitude grasslands in Africa. The shrub produces fruits abundantly and seeds are dispersed by birds. Once established, the plant is fairly shade tolerant. The shrub suckers from roots, enabling populations to expand rapidly.
   Seedlings and small plants can be pulled or dug out. Larger plants are cut and the cut stumps treated with herbicide.

References
121, 215, 548, 549, 607, 924, 1089.

Pyracantha crenulata (Don) Roem. Rosaceae

LF: Evergreen shrub CU: Ornamental
SN: *Cotoneaster crenulatus* (Don) Koch

Geographic Distribution

Europe		*Australasia*		*Atlantic Islands*
	Northern	●	Australia	Cape Verde
	British Isles	X	New Zealand	Canary, Madeira
	Central, France			Azores
	Southern	*Northern America*		South Atlantic Isl.
	Eastern		Canada, Alaska	
	Mediterranean Isl.		Southeastern USA	*Indian Ocean Islands*
			Western USA	Mascarenes
Africa			Remaining USA	Seychelles
	Northern		Mexico	Madagascar
	Tropical			
●	Southern	*Southern America*		*Pacific Islands*
			Tropical	Micronesia
Asia			Caribbean	Galapagos
N	Temperate		Chile, Argentina	Hawaii
	Tropical			Melanesia, Polynesia

Invaded Habitats
Grassland, riparian habitats.

Description
A more or less thorny shrub or small tree of 2-4 m height, with young branches being rusty-pubescent and with spreading stems. Leaves are oblong to oblanceolate, usually acute, glabrous, 2-5 cm long and 5-15 mm wide, dark green above and pale green beneath. Petioles are 3-5 mm long. Creamy white flowers of *c.* 8 mm diameter are borne in rather loose corymbs of 2-3 cm diameter. Fruits are globose orange-red berries of 3-6 mm diameter.

Ecology and Control
A native of the temperate Himalaya, this shrub invades high-altitude grasslands in Africa and similar habitats in Australia. It forms dense and spiny thickets that crowd out native vegetation. Fruits are abundantly produced and seeds dispersed mainly by birds.

Specific control methods for this species are not available. Seedlings and small plants can be pulled or dug out. Larger plants may be cut and the cut stumps treated with herbicide.

References
215, 549, 607.

Ranunculus repens L.   Ranunculaceae

LF: Perennial herb
SN: -
CU: Ornamental

Geographic Distribution

Europe
- N   Northern
- N   British Isles
- N   Central, France
- N   Southern
- N   Eastern
- N   Mediterranean Isl.

Africa
- N   Northern
-      Tropical
- X   Southern

Asia
- N   Temperate
-      Tropical

Australasia
- ●   Australia
- X   New Zealand

Northern America
- X   Canada, Alaska
-      Southeastern USA
- X   Western USA
- X   Remaining USA
-      Mexico

Southern America
- X   Tropical
-      Caribbean
- X   Chile, Argentina

Atlantic Islands
-      Cape Verde
- N   Canary, Madeira
- N   Azores
-      South Atlantic Isl.

Indian Ocean Islands
- X   Mascarenes
-      Seychelles
-      Madagascar

Pacific Islands
-      Micronesia
-      Galapagos
-      Hawaii
-      Melanesia, Polynesia

Invaded Habitats
Forests, grassland, riparian habitats, freshwater wetlands.

Description
A nearly glabrous rhizomatous herb with long creeping stems rooting at nodes, and erect flowering stems up to 60 cm tall. Basal leaves have long petioles, are triangular-ovate, with three main segments, the middle one with a long stalk. Leaf blades are 1.5-6 cm long and 2-8 cm wide. Upper leaves are sessile and smaller. Flowers are yellow and 20-30 mm in diameter. Fruits are glabrous and smooth achenes of 2.5-4 mm length, with a short and curved beak.

Ecology and Control
Where native, this plant grows in wet grassland, woods, marshes, streamsides, and dune slacks. It is a highly variable species in the native range. The plant spreads rapidly where competition is low and forms dense swards that eliminate native vegetation. Seedlings establish readily on bare ground. The plant has a large seed bank and seeds remain viable for several years. It is an important agricultural weed and also invades natural plant communities.

     Small patches may be removed manually, roots must be removed. Repeated cutting may reduce the plant's vigour. Effective herbicides are 2,4-D or MCPA.

References
215, 357, 527, 581, 778, 790.

Ravenala madagascariensis Sonn. Strelitziaceae

LF: Evergreen tree
SN: -
CU: Ornamental

Geographic Distribution

Europe	*Australasia*	*Atlantic Islands*
Northern	Australia	Cape Verde
British Isles	New Zealand	Canary, Madeira
Central, France		Azores
Southern	*Northern America*	South Atlantic Isl.
Eastern	Canada, Alaska	
Mediterranean Isl.	Southeastern USA	*Indian Ocean Islands*
	Western USA	● Mascarenes
Africa	Remaining USA	Seychelles
Northern	Mexico	N Madagascar
Tropical		
Southern	*Southern America*	*Pacific Islands*
	Tropical	Micronesia
Asia	Caribbean	Galapagos
Temperate	Chile, Argentina	Hawaii
Tropical		Melanesia, Polynesia

Invaded Habitats
Forests, freshwater wetlands, marshes, mountain slopes.

Description
A large tree reaching 20 m height with stems formed by the sheathing leaf bases. The alternate leaves are arranged in two ranks and 2-4 m long. Inflorescences have 5-12 creamy white flowers of 20-25 cm length. Fruits are dehiscent capsules containing dark brown seeds. These are 10-12 mm long and 7-8 mm wide, and surrounded by a blue aril.

Ecology and Control
A native of secondary forests and riparian habitats, this tree is invasive because it spreads rapidly and forms dense stands that shade out all other plant species with the large leaves. It can become the dominant species in later stage forest successions. The tree spreads vegetatively by suckering from the roots. The species is variable in size and shape, depending on the habitat where it grows. It is an upright tree in forests and more a shrub in open sites. Little is known on the ecology of this species as an invader.

Specific control methods for this species are not available. Small plants may be pulled or dug out. Larger trees can be cut and the cut stumps treated with herbicide.

References
194, 284, 768, 769, 1239.

Rhamnus alaternus L.   Rhamnaceae

LF: Evergreen shrub   CU: Ornamental
SN: -

Geographic Distribution

Europe	*Australasia*	*Atlantic Islands*
Northern	● Australia	Cape Verde
British Isles	● New Zealand	N Canary, Madeira
Central, France		Azores
N Southern	*Northern America*	South Atlantic Isl.
N Eastern	Canada, Alaska	
Mediterranean Isl.	Southeastern USA	*Indian Ocean Islands*
	Western USA	Mascarenes
Africa	Remaining USA	Seychelles
N Northern	Mexico	Madagascar
Tropical		
Southern	*Southern America*	*Pacific Islands*
	Tropical	Micronesia
Asia	Caribbean	Galapagos
N Temperate	Chile, Argentina	Hawaii
Tropical		Melanesia, Polynesia

Invaded Habitats
Grass- and heathland, riparian habitats, forests, coastal dunes and beaches.

Description
A nearly glabrous shrub of 4-6 m height, usually with a rounded bushy habit. Leaves are dark green above and yellowish green beneath, lanceolate to ovate or elliptic, 2-6 cm long and 12-24 mm wide, entire or slightly toothed, and with petioles up to 8 mm length. Small yellowish green flowers grow in dense and mostly pubescent racemes. Flowers have a yellow calyx and no petals. Fruits are obovoid, reddish drupes of 4-6 mm length, becoming black at maturity. Each fruit has three seeds.

Ecology and Control
In the native range, this shrub is common in dry rocky places, mediterranean scrub, and dry forests. It is fast growing and resprouts vigorously from the base after damage. It spreads both by seeds and by root suckering. Fruits are abundantly produced and seeds dispersed by birds. The shrub forms dense thickets that shade out native plants and lead to species poor stands.
    Smaller plants and seedlings can be pulled or dug out. Larger plants are cut and the cut stumps treated with herbicide. Follow-up programmes may be necessary to treat seedlings and regrowth.

References
121, 215, 607, 924, 1415.

Rhamnus cathartica L.　　　　　　　　　　　　　　　　Rhamnaceae

LF: Deciduous shrub
SN: -
CU: Shelter

Geographic Distribution

Europe	*Australasia*	*Atlantic Islands*
N　Northern	Australia	Cape Verde
N　British Isles	New Zealand	Canary, Madeira
N　Central, France		Azores
N　Southern	*Northern America*	South Atlantic Isl.
N　Eastern	●　Canada, Alaska	
N　Mediterranean Isl.	Southeastern USA	*Indian Ocean Islands*
	Western USA	Mascarenes
Africa	●　Remaining USA	Seychelles
N　Northern	Mexico	Madagascar
Tropical		
Southern	*Southern America*	*Pacific Islands*
	Tropical	Micronesia
Asia	Caribbean	Galapagos
N　Temperate	Chile, Argentina	Hawaii
Tropical		Melanesia, Polynesia

Invaded Habitats
Forests and forest edges, riparian habitats.

Description
A shrub or small tree of 1-3 m height, with sharp thorns at the ends of most branches and with a grey to blackish brown bark becoming rough with age. Twigs and leaves are opposite. The oblong to elliptical leaves have toothed margins and are 3-6 cm long, with 2-3 pairs of lateral veins. The petioles are 2-4 times as long as the stipules. Flowers are yellowish-green, borne in axillary clusters of 2-8, and have four petals. Fruits are black berries of 6-8 mm diameter, containing 3-4 seeds.

Ecology and Control
This shrub grows in a wide range of different habitats within the native range. Growth habits range from multi-stemmed shrubs to single-stemmed trees. Where invasive, the shrub can form dense and impenetrable thickets that displace native vegetation. Seedlings are shade tolerant and may reach high densities in the understorey. Seeds are dispersed by birds. The shrub resprouts vigorously if damaged.

　　Seedlings are easy to pull by hand, larger plants should be girdled or cut at the base. To prevent resprouting, freshly cut stumps can be painted with a glyphosate or triclopyr herbicide. The best time for felling is autumn.

References
46, 466, 533, 534, 546, 571, 1396.

Rhododendron ponticum L.   Ericaceae

LF: Evergreen shrub, tree   CU: Ornamental, shelter
SN: *Rhododendron lancifolium* Moench

Geographic Distribution

Europe		*Australasia*		*Atlantic Islands*
	Northern		Australia	Cape Verde
●	British Isles	X	New Zealand	Canary, Madeira
X	Central, France			Azores
N	Southern		*Northern America*	South Atlantic Isl.
N	Eastern		Canada, Alaska	
	Mediterranean Isl.		Southeastern USA	*Indian Ocean Islands*
			Western USA	Mascarenes
Africa			Remaining USA	Seychelles
	Northern		Mexico	Madagascar
	Tropical			
	Southern		*Southern America*	*Pacific Islands*
			Tropical	Micronesia
Asia			Caribbean	Galapagos
N	Temperate		Chile, Argentina	Hawaii
	Tropical			Melanesia, Polynesia

Invaded Habitats
Forests, woodland, riparian habitats, coastal areas.

Description
A densely branched large shrub or tree of 2-8 m height, with spreading branches. The elliptic to oblong leaves are dark green and shining above, paler beneath, glabrous, and 6-25 cm long. Petioles are short. Flowers are violet-purple or mauvish-purple, 4-6 cm in diameter, and have ten stamens. They grow in racemes with 8-15 flowers, the pedicels are 2-6 cm long. Fruits are woody capsules containing numerous small seeds of *c.* 1.5 mm length.

Ecology and Control
This shrub grows mostly on sandy and peaty acid soils, and is highly variable in size and habit. The shrub forms large and extensive thickets that eliminate all other vegetation due to the shading effect of the dense foliage. Native shrubs and trees are unable to establish under canopies of this species. Stands accumulate a thick litter layer, further impeding seedling growth and establishment of native species. The shrub produces large quantities of wind dispersed seeds and suckers from roots. Damaged plants resprout vigorously.

Seedlings and saplings can be hand pulled but need to be removed to prevent adventitious root formation. Larger individuals can be cut and the cut stumps treated with herbicides such as amide glyphosate, or 2,4,5-T.

References
284, 285, 286, 287, 430, 503, 1167, 1269, 1270.

Rhodomyrtus tomentosa (Ait.) Hassk. Myrtaceae

LF: Evergreen shrub CU: Ornamental
SN: *Myrtus tomentosa* Ait., *Rhodomyrtus parviflora* Alston

Geographic Distribution

Europe	*Australasia*	*Atlantic Islands*
Northern	Australia	Cape Verde
British Isles	New Zealand	Canary, Madeira
Central, France		Azores
Southern	*Northern America*	South Atlantic Isl.
Eastern	Canada, Alaska	
Mediterranean Isl.	● Southeastern USA	*Indian Ocean Islands*
	Western USA	Mascarenes
Africa	Remaining USA	Seychelles
Northern	Mexico	Madagascar
Tropical		
Southern	*Southern America*	*Pacific Islands*
	Tropical	X Micronesia
Asia	Caribbean	Galapagos
N Temperate	Chile, Argentina	● Hawaii
N Tropical		Melanesia, Polynesia

Invaded Habitats
Open forests, woodland, pineland.

Description
A bushy shrub up to 3 m tall, with opposite leaves and the young branches whitish pubescent. Leaves are oblong, 2.5-10 cm long and 1.5-3 cm wide, white woolly on the lower side, and have short petioles. They have three conspicuous longitudinal veins. The flowers are 38-44 mm in diameter, have pedicels of 13-25 mm length, and magenta to pink petals fading white. Each flower is subtended by a pair of bracts. Fruits are oblong berries of 10-15 mm diameter, green at first and becoming purplish, containing numerous minute seeds.

Ecology and Control
This shrub is common in open sandy ground, sea shores, and riverbanks where native. It is well adapted to dry and hot places, and tolerates saline soils. It invades pinelands in Florida and displaces the native understorey vegetation with dense and impenetrable, pure thickets. Seeds are probably dispersed by birds and mammals. Little is known on the ecology of this plant.
   Once established, control is difficult. Seedlings and smaller plants may be hand pulled or dug out. Larger shrubs can be cut and the cut stumps treated with herbicide.

References
715, 1245.

Ricinus communis L.   Euphorbiaceae

LF: Evergreen shrub   CU: Ornamental
SN: -

Geographic Distribution

Europe		*Australasia*		*Atlantic Islands*	
	Northern	●	Australia	X	Cape Verde
	British Isles	X	New Zealand	X	Canary, Madeira
	Central, France			X	Azores
X	Southern	*Northern America*			South Atlantic Isl.
	Eastern		Canada, Alaska		
	Mediterranean Isl.		Southeastern USA	*Indian Ocean Islands*	
		●	Western USA		Mascarenes
Africa		X	Remaining USA		Seychelles
X	Northern	●	Mexico		Madagascar
N	Tropical				
●	Southern	*Southern America*		*Pacific Islands*	
		X	Tropical	X	Micronesia
Asia		X	Caribbean	●	Galapagos
X	Temperate	X	Chile, Argentina	●	Hawaii
X	Tropical			X	Melanesia, Polynesia

Invaded Habitats
Grass- and heathland, riparian habitats, disturbed sites.

Description
A much-branched woody herb or shrub with stout hollow stems up to 4 m tall. The peltate and palmately lobed leaves have long petioles of 7-35 cm length, with the blades being 15-60 cm long. Inflorescences are panicles of 10-30 cm length, terminal or opposite the leaves, with male flowers above and female flowers below within the same inflorescence. Petals are inconspicuous, male flowers have yellow stamens, female flowers conspicuously red styles. Fruits are capsules of 10-20 mm length, covered with soft spines, splitting into six parts and releasing reddish-brown to blackish, smooth seeds of 9-17 mm length.

Ecology and Control
This highly variable plant occurs with many races and probably originates in tropical Africa. It is a fast growing and short-lived thicket forming plant that shades out native species. Seeds are dispersed by birds and mammals and may remain dormant in the soil for several years. Establishment depends on disturbance.

Small plants can be hand pulled. Larger plants are cut and the stumps treated with herbicide. Chemical control is most effective before fruits develop, an effective herbicide is glyphosate or picloram plus 2,4-D. Burning is used to kill larger infestations.

References
133, 284, 792, 794, 924, 986, 1089, 1199, 1313.

Robinia pseudoacacia L.   Fabaceae

LF: Deciduous tree
SN: -
CU: Ornamental, wood

Geographic Distribution

Europe
- Northern
- British Isles
- ● Central, France
- ● Southern
- ● Eastern
- Mediterranean Isl.

Africa
- X Northern
- Tropical
- ● Southern

Asia
- X Temperate
- Tropical

Australasia
- ● Australia
- X New Zealand

Northern America
- X Canada, Alaska
- N Southeastern USA
- X Western USA
- N Remaining USA
- Mexico

Southern America
- Tropical
- Caribbean
- X Chile, Argentina

Atlantic Islands
- Cape Verde
- Canary, Madeira
- Azores
- South Atlantic Isl.

Indian Ocean Islands
- Mascarenes
- Seychelles
- Madagascar

Pacific Islands
- Micronesia
- Galapagos
- Hawaii
- Melanesia, Polynesia

Invaded Habitats
Forests, riparian habitats, grassland, rocky places.

Description
A tree up to 30 m tall with a brown and deeply furrowed bark, glabrous branches, and stipular spines on the twigs. The pinnately compound leaves are almost glabrous and have 3-10 pairs of leaflets. These are elliptic to ovate, 25-45 mm long and 12-25 mm wide, rounded at the apex. The very fragrant and white flowers grow in dense, drooping racemes of 10-20 cm length. Corollas are 15-20 mm long. Fruits are linear-oblong and smooth pods of 5-10 cm length and *c.* 10 mm width, containing 3-10 seeds of 4-5 mm length.

Ecology and Control
This shade intolerant pioneer tree grows in a wide range of soils. It is a variable species and many cultivars have been developed. Seedlings establish in forest gaps and grow rapidly to reach the canopy. The tree then spreads quickly by lateral growth and root sprouts, leading to dense pure stands that displace native vegetation. The tree is nitrogen-fixing and increases soil fertility levels which may affect the floristic composition of invaded sites. In Europe, it invades species rich dry grasslands and reduces the abundance of species adapted to nutrient poor soils. Damaged trees resprout vigorously from the base.

Seedlings and saplings can be pulled or dug out, roots must be removed. Repeated cutting can eventually kill the tree. Effective chemical control is achieved by treating freshly cut stumps with glyphosate.

References
105, 129, 215, 340, 549, 568, 612, 661, 672, 697, 1446.

Romulea rosea (L.) Eckl.   Iridaceae

LF: Perennial herb   CU: Ornamental
SN: *Trichonema purpurascens* Sweet, *Ixia rosea* L.

Geographic Distribution

Europe
       Northern
X   British Isles
       Central, France
       Southern
       Eastern
       Mediterranean Isl.

Africa
       Northern
       Tropical
N   Southern

Asia
       Temperate
       Tropical

Australasia
●   Australia
X   New Zealand

Northern America
       Canada, Alaska
       Southeastern USA
       Western USA
       Remaining USA
       Mexico

Southern America
       Tropical
       Caribbean
       Chile, Argentina

Atlantic Islands
       Cape Verde
       Canary, Madeira
       Azores
       South Atlantic Isl.

Indian Ocean Islands
       Mascarenes
       Seychelles
       Madagascar

Pacific Islands
       Micronesia
       Galapagos
       Hawaii
       Melanesia, Polynesia

Invaded Habitats
Grassland, riparian habitats, coastal areas.

Description
A glabrous cormous herb of 10-40 cm height, with a rounded corm and several basal leaves of 15-25 cm length and 1-2.5 mm width, appearing before the flowers. Flowers are magenta-pink, pale lilac or white with yellow inside at the base, and have perianths of 15-45 mm length. Filaments are hairy at the base. Pedicels are 3-8 cm long. Fruits are ovate to roundish and membranous capsules with numerous globose seeds.

Ecology and Control
A native of sandy and clay slopes and flats. Two varieties have become naturalized: var. *rosea* and var. *australis*, plants in Britain belong to var. *australis*. The plant spreads both by seeds and by vegetative growth and forms dense swards that crowd out native plants. It establishes well in disturbed sites and populations expand rapidly to form a continuous cover.

     Specific control methods for this species are not available. Plants may be dug out, corms must be removed to prevent regrowth. The same control methods as for *Moraea flaccida* or *Sparaxis bulbifera* may apply.

References
215, 337.

Rorippa palustris (L.) Besser — Brassicaceae

LF: Annual, perennial herb   CU: -
SN: *Rorippa islandica* (Oeder) Borb.

Geographic Distribution

Europe		*Australasia*		*Atlantic Islands*
N	Northern	●	Australia	Cape Verde
N	British Isles	X	New Zealand	Canary, Madeira
N	Central, France			Azores
N	Southern	*Northern America*		South Atlantic Isl.
N	Eastern	N	Canada, Alaska	
	Mediterranean Isl.		Southeastern USA	*Indian Ocean Islands*
		N	Western USA	Mascarenes
Africa		N	Remaining USA	Seychelles
	Northern		Mexico	Madagascar
	Tropical			
	Southern	*Southern America*		*Pacific Islands*
			Tropical	Micronesia
Asia			Caribbean	Galapagos
N	Temperate	X	Chile, Argentina	Hawaii
N	Tropical			Melanesia, Polynesia

Invaded Habitats
Forests, riparian habitats, freshwater wetlands.

Description
An annual to short-lived perennial, usually glabrous herb with ascending to erect stems of 10-110 cm height that are branched above. The petioled leaves are deeply lobed with 2-6 pairs of serrated lobes. Flowers are pale yellow and have four sepals. Petals and sepals are 1.5-2.8 mm long. Fruits are 4-12 mm long, cylindrical to oblong-elliptical, often curved, and 1.5-3 mm wide.

Ecology and Control
A frequent species of open damp places, ditches, streamsides, riparian habitats, and disturbed sites within the native range. The plant is variable in size and shape and grows in dense patches that displace native species by shading out. Its large size makes it a strong competitor to native herbaceous species.

Specific control methods for this species are not available. Scattered plants may be pulled or dug out, larger stands cut treated with herbicide.

References
215, 233.

Rosa canina L.   Rosaceae

LF: Deciduous shrub   CU: Ornamental, food
SN: *Rosa lutetiana* Léman

Geographic Distribution

Europe		*Australasia*		*Atlantic Islands*	
N	Northern	●	Australia		Cape Verde
N	British Isles	X	New Zealand	N	Canary, Madeira
N	Central, France				Azores
N	Southern	*Northern America*			South Atlantic Isl.
N	Eastern		Canada, Alaska		
N	Mediterranean Isl.		Southeastern USA	*Indian Ocean Islands*	
			Western USA		Mascarenes
Africa		X	Remaining USA		Seychelles
N	Northern		Mexico		Madagascar
	Tropical				
	Southern	*Southern America*		*Pacific Islands*	
			Tropical		Micronesia
Asia			Caribbean		Galapagos
N	Temperate		Chile, Argentina		Hawaii
N	Tropical				Melanesia, Polynesia

Invaded Habitats
Forests and forest edges, grassland, swamp and stream edges.

Description
A prickly shrub with erect or arching stems up to 5 m tall, with long internodes and stout, curved or hooked prickles. The compound leaves have 5-7 ovate to elliptic leaflets of 15-40 mm length and 12-20 mm width, dark to glaucous green, and sometimes slightly pubescent beneath. Flowers grow at the ends of short branchlets, are light pink or white, fragrant, 4-5 cm in diameter, and with 1-3 cm long pedicels. The sepals are usually lobed and become reflexed and deciduous. Fruits are ovoid to globose red berries of 10-20 mm length. The yellowish seeds are 4.5-6 mm long.

Ecology and Control
In the native range, this shrub is common in hedges, forest edges, disturbed and rocky places. It is a highly variable species, often considered as a complex with many microspecies. The strong growth, the climbing and trailing stems smother native vegetation and prevent any regeneration of native shrubs and trees. Thickets of this shrub hinder wildlife movement and access to water. Stems touching the ground may become rooted. Seeds are dispersed by birds and by attaching to animals.

    Shrubs can be cut at ground level but regrowth is fast. An effective chemical control is spraying the lower 50 cm of the stems with triclopyr during the flowering and early fruiting stage. Whole plants are sprayed with picloram plus 2,4-D, glyphosate, or triclopyr.

References
932, 986, 1325.

Rosa multiflora Thunb. — Rosaceae

LF: Deciduous shrub
SN: *Rosa polyantha* Siebold & Zucc.
CU: Ornamental, shelter

Geographic Distribution

Europe		*Australasia*		*Atlantic Islands*	
	Northern		Australia		Cape Verde
X	British Isles	X	New Zealand		Canary, Madeira
	Central, France				Azores
	Southern		*Northern America*		South Atlantic Isl.
	Eastern	●	Canada, Alaska		
	Mediterranean Isl.		Southeastern USA		*Indian Ocean Islands*
		●	Western USA		Mascarenes
Africa		●	Remaining USA		Seychelles
	Northern		Mexico		Madagascar
	Tropical				
X	Southern		*Southern America*		*Pacific Islands*
			Tropical		Micronesia
Asia			Caribbean		Galapagos
N	Temperate		Chile, Argentina		Hawaii
	Tropical				Melanesia, Polynesia

Invaded Habitats
Forests and forest margins, grassland.

Description
A large and wide spreading, prickly, climbing shrub of 1-2 m height with long and arching stems. Leaves are pinnately compound with 5-9 obovate to oblong leaflets of 15-30 mm length and 8-20 mm width. Margins are sharply serrated, with the petiole and rachis often glandular-hairy. Inflorescences are many-flowered corymbs with softly hairy pedicels. The white and fragrant flowers are 15-20 mm in diameter. Sepals are lanceolate, petals obovate. Fruits are red, small and globose berries *c.* 6 mm length.

Ecology and Control
Where native, this climbing shrub grows in open forests and ravines up to 2,500 m elevation. The shrub forms impenetrable and large thickets outcompeting native species, preventing forest regeneration and degrading grasslands. Seeds are dispersed by birds and may remain viable in the soil for several years.

In grasslands, regular mowing or burning is effective in preventing seedling establishment. Larger plants can be dug out or removed with a weed wrench. If cut, stumps should be treated with herbicide to prevent resprouting. Effective herbicides are glyphosate or chlor-flurenol.

References
329, 389, 571, 837, 1396.

Rosa rubiginosa L. — Rosaceae

LF: Deciduous shrub
SN: *Rosa eglanteria* L.
CU: Ornamental

Geographic Distribution

Europe	*Australasia*	*Atlantic Islands*
N Northern	● Australia	Cape Verde
N British Isles	● New Zealand	Canary, Madeira
N Central, France		Azores
N Southern	*Northern America*	South Atlantic Isl.
N Eastern	Canada, Alaska	
N Mediterranean Isl.	Southeastern USA	*Indian Ocean Islands*
	X Western USA	Mascarenes
Africa	● Remaining USA	Seychelles
Northern	Mexico	Madagascar
Tropical		
● Southern	*Southern America*	*Pacific Islands*
	Tropical	Micronesia
Asia	Caribbean	Galapagos
N Temperate	X Chile, Argentina	Hawaii
Tropical		Melanesia, Polynesia

Invaded Habitats
Forests and forest edges, grassland, riparian habitats, rock outcrops.

Description
A much-branched shrub up to 3 m tall, with arching branches, strong and hooked prickles, sometimes mixed with bristles, and pinnately compound leaves. These have 5-7 orbicular to ovate shortly stalked leaflets of 10-30 mm length and 8-15 mm width that are pubescent and densely glandular beneath. Flowers grow in clusters of 1-3, are white to bright pink, 3-5 cm in diameter, and have pedicels of 10-15 mm length. Fruits are globose to ovoid, orange to bright red berries of 10-15 mm length. Seeds are yellow and 4-7 mm long.

Ecology and Control
This shrub tolerates a wide range of soil conditions and establishes in sites with low competition by other species. The dense and often tall thickets affect wildlife movement, displace native plants and prevent the establishment of tree and shrub seedlings. Scattered clumps of this shrub may enlarge and merge, covering extensive areas by time. The shrub suckers freely from roots. The shrub spreads by seeds and by vegetative growth, even small pieces of roots may regenerate new plants. Seeds are dispersed by birds and small mammals.

Hand grubbing of smaller plants is effective but roots must be removed as far as possible to prevent recolonization. Grazing by goats or sheep is used to destroy seedlings. Effective herbicides are picloram, glyphosate or triclopyr applied as an overall spray or to the lower parts of stems.

References
121, 215, 299, 543, 549, 607, 924, 985, 986, 1266, 1415.

Rosa rugosa Thunb.   Rosaceae

LF: Deciduous shrub
SN: -
CU: Ornamental

Geographic Distribution

Europe	*Australasia*	*Atlantic Islands*
● Northern	Australia	Cape Verde
X British Isles	X New Zealand	Canary, Madeira
Central, France		Azores
Southern	*Northern America*	South Atlantic Isl.
Eastern	Canada, Alaska	
Mediterranean Isl.	Southeastern USA	*Indian Ocean Islands*
	Western USA	Mascarenes
Africa	Remaining USA	Seychelles
Northern	Mexico	Madagascar
Tropical		
Southern	*Southern America*	*Pacific Islands*
	Tropical	Micronesia
Asia	Caribbean	Galapagos
N Temperate	Chile, Argentina	Hawaii
Tropical		Melanesia, Polynesia

Invaded Habitats
Grassland, coastal scrub and dunes, disturbed sites.

Description
An upright shrub reaching 2.5 m height, with stems densely tomentose when young and densely covered with prickles and bristles. Leaves are pinnately compound and have 5-9 elliptic to elliptic-obovate leaflets of 2-5 cm length and 8-30 mm width, with tomentose and bristly petioles of 2-5 cm length. Flowers are borne solitary or in clusters of a few, purple to white, 6-8 cm in diameter, with short and bristly pedicels. Fruits are smooth, depressed-globose scarlet berries of 20-25 mm diameter.

Ecology and Control
A variable species with numerous varieties and hybrids. The shrub is freely suckering from roots and forms dense thickets that expand rapidly and displace native vegetation. Seeds are dispersed by birds and small mammals.

Specific control methods for this species are not available. The same methods as for other *Rosa* species may apply.

References
425.

Rubus argutus Link.   Rosaceae

LF: Deciduous shrub   CU: Food
SN: -

Geographic Distribution

Europe	Australasia	Atlantic Islands
Northern	Australia	Cape Verde
British Isles	X New Zealand	Canary, Madeira
Central, France		Azores
Southern	*Northern America*	South Atlantic Isl.
Eastern	Canada, Alaska	
Mediterranean Isl.	Southeastern USA	*Indian Ocean Islands*
	Western USA	Mascarenes
Africa	N Remaining USA	Seychelles
Northern	Mexico	Madagascar
Tropical		
Southern	*Southern America*	*Pacific Islands*
	Tropical	Micronesia
Asia	Caribbean	Galapagos
Temperate	Chile, Argentina	● Hawaii
Tropical		Melanesia, Polynesia

Invaded Habitats
Grassland, forests and forest edges.

Description
A thorny shrub reaching 3 m height, with erect or arching stems, with sharp, straight or hooked prickles. Leaves are compound with five leaflets and have petioles of 6-10 cm length. The stalk of the terminal leaflet is longest; the lower leaflets are sessile or almost so. Blades of leaflets are elliptic to lanceolate-ovate and have sharply toothed margins. The blade of the terminal leaflet is 6-10 cm long, the lower leaflets are 4-6 cm long. Leaves of flower bearing shoots are much smaller. Flowers grow in cymes, with the axis of the cyme and the pedicels bearing long and soft hairs. Petals are white to pink and 2-2.5 cm long.

Ecology and Control
Native habitats of this shrub include wet or swampy woodlands, often in standing water, streambanks, shoreline thickets, but also dry places. It forms impenetrable thickets by its arching stems that become rooted at the tips. Native plants are eliminated and wildlife movement is affected. Seedlings of shrubs and trees are unable to establish in areas invaded by this shrub. The shrub easily resprouts from basal and belowground buds after damage. Seeds are dispersed by birds.

Scattered shrubs may be dug out, roots must be removed. Larger thickets can be slashed but regrowth is fast. Effective herbicides are glyphosate, 2,4-D plus dicamba, or 2,4-D plus triclopyr.

References
840, 1063.

Rubus cuneifolius Pursh   Rosaceae

LF: Deciduous shrub   CU: -
SN: -

Geographic Distribution

Europe	Australasia	Atlantic Islands
Northern	• Australia	Cape Verde
British Isles	New Zealand	Canary, Madeira
Central, France		Azores
Southern	*Northern America*	South Atlantic Isl.
Eastern	Canada, Alaska	
Mediterranean Isl.	N Southeastern USA	*Indian Ocean Islands*
	Western USA	Mascarenes
Africa	N Remaining USA	Seychelles
Northern	Mexico	Madagascar
Tropical		
• Southern	*Southern America*	*Pacific Islands*
	Tropical	Micronesia
Asia	Caribbean	Galapagos
N Temperate	Chile, Argentina	Hawaii
Tropical		Melanesia, Polynesia

Invaded Habitats
Grass- and heathland.

Description
A densely pubescent shrub with erect or sprawling stems of 30-150 cm height, suckering from below ground parts, and with stiff and sharp prickles. The leaves are compound with five leaflets in the lower leaves and three leaflets in the upper leaves. Flower bearing stems have much smaller leaves. Leaflets are greyish-green, densely greyish-tomentose beneath, and have prickly petioles. Terminal leaflets have stalks of 5-20 mm length, the lateral leaflets are sessile or almost so. Flowers have white petals of 10-15 mm length.

Ecology and Control
In the native range, this plant occurs in a wide range of habitats, e.g. well-drained and wet places, old fields, open woodland, wet pine savanna, disturbed sites, and sandy places. The plant coppices vigorously and forms dense thickets, impeding wildlife and displacing native vegetation. Scattered clumps merge and may cover large areas over time.

Manual control includes cutting, but this stimulates strong regrowth. Stems can also be flattened to the ground to suppress growth. Herbicides used to control this plant include glyphosate, triclopyr, or 2,4,5-T plus dicamba.

References
323, 549, 882, 1089.

Rubus ellipticus Sm.   Rosaceae

LF: Deciduous shrub   CU: Ornamental
SN: *Rubus flavus* Hamilton

Geographic Distribution

Europe		*Australasia*		*Atlantic Islands*	
	Northern		Australia		Cape Verde
	British Isles		New Zealand		Canary, Madeira
	Central, France				Azores
	Southern	*Northern America*			South Atlantic Isl.
	Eastern		Canada, Alaska		
	Mediterranean Isl.		Southeastern USA	*Indian Ocean Islands*	
			Western USA		Mascarenes
Africa			Remaining USA		Seychelles
	Northern		Mexico		Madagascar
●	Tropical				
	Southern	*Southern America*		*Pacific Islands*	
		X	Tropical		Micronesia
Asia		X	Caribbean		Galapagos
N	Temperate		Chile, Argentina	●	Hawaii
N	Tropical				Melanesia, Polynesia

Invaded Habitats
Forests and forest edges, wet places.

Description
A scandent shrub with prickly and sparsely hairy stems, the prickles being 4-5 mm long. Petioles are prickly and 15-40 mm long. Leaves are compound with three broadly obovate leaflets of 2-7 cm length and 2-6 cm width, the margins being toothed. Inflorescences are more or less compact panicles of 5-10 cm length, containing about 20 white flowers of 8-10 mm diameter. Pedicels are 3-5 mm long. Fruits are ovoid to globose, yellow berries of 15-25 mm length and 15-20 mm width.

Ecology and Control
This shrub forms impenetrable thickets that displace native vegetation and affect wildlife by impeding movement and reducing habitats. It spreads rapidly by root suckers and regenerates from underground shoots after fire or cutting. Scattered clumps expand and form in time a continuous cover on the floor.

    Manual removal of individual plants should include removal of roots. Shrubs can be cut and the cut stumps treated with herbicide. Effective herbicides are glyphosate or 2,4,5-T plus 2,4-D.

References
284, 369, 370.

Rubus fruticosus L. agg.   Rosaceae

LF: Deciduous shrub   CU: Food, shelter
SN: -

Geographic Distribution

Europe	Australasia	Atlantic Islands
Northern	● Australia	Cape Verde
N British Isles	● New Zealand	Canary, Madeira
N Central, France		Azores
N Southern	*Northern America*	South Atlantic Isl.
N Eastern	Canada, Alaska	
Mediterranean Isl.	X Southeastern USA	*Indian Ocean Islands*
	X Western USA	Mascarenes
Africa	X Remaining USA	Seychelles
Northern	Mexico	Madagascar
Tropical		
● Southern	*Southern America*	*Pacific Islands*
	Tropical	Micronesia
Asia	Caribbean	Galapagos
X Temperate	X Chile, Argentina	X Hawaii
X Tropical		Melanesia, Polynesia

Invaded Habitats
Grass- and scrubland, forests, riparian habitats, disturbed sites.

Description
A 2-6 m tall shrub with arching, glabrous stems and hooked prickles of 3-12 mm length. Leaves are palmately compound and have 3-7 green and pubescent leaflets, the terminal leaflet being broad-ovate, cordate, 3-16 cm long and 1.5-12 cm wide. The lateral leaflets are narrower and smaller. Inflorescences are many-flowered pubescent and somewhat prickly panicles. Flowers are white or pink and 15-60 mm in diameter. Fruits are purple-black, globose berries of 10-30 mm diameter. The light to dark brown seeds are 2-3 mm long and deeply pitted.

Ecology and Control
This plant belongs to a complex group of many closely related taxa. The shrub suckers from roots and stems touching the soil become rooted at the tips, forming daughter plants. It forms extensive and dense impenetrable thickets, shading out all other vegetation, displacing it and affecting wildlife habitats. The plant can completely dominate invaded areas within a short time, as individual clumps expand laterally due to root suckering and stem rooting. Seeds are dispersed by birds and mammals.

    Cutting and burning is used but does not prevent regrowth. If individual plants are grubbed, the crowns and much of the roots must be removed. Chemical treatment is done by treating with glyphosate, picloram, amitrole-T, often applied to regrowth after burning or slashing. Grazing by goats proved to be effective as well.

References
25, 26, 28, 30, 121, 388, 476, 549, 607, 924, 939, 985, 986, 1143, 1415.

Rubus niveus Thunb.   Rosaceae

LF: Deciduous shrub   CU: Food
SN: *Rubus albescens* Roxb., *Rubus lasiocarpus* Sm., *Rubus micranthus* Don

Geographic Distribution

Europe	*Australasia*	*Atlantic Islands*
Northern	Australia	Cape Verde
British Isles	New Zealand	Canary, Madeira
Central, France		Azores
Southern	*Northern America*	South Atlantic Isl.
Eastern	Canada, Alaska	
Mediterranean Isl.	Southeastern USA	*Indian Ocean Islands*
	Western USA	Mascarenes
Africa	Remaining USA	Seychelles
Northern	Mexico	Madagascar
X Tropical		
Southern	*Southern America*	*Pacific Islands*
	X Tropical	Micronesia
Asia	Caribbean	● Galapagos
N Temperate	Chile, Argentina	● Hawaii
N Tropical		Melanesia, Polynesia

Invaded Habitats
Forest edges, riparian habitats, woodland, disturbed sites.

Description
A sparsely prickly scrambling shrub of 1.5-3 m height with glabrous stems covered with a whitish bloom. Prickles are 4-7 mm long. Leaves are palmately compound with 5-7 ovate and toothed leaflets. Petioles are 15-50 mm long. The terminal leaflet is 5-17 cm long and 4-8 cm wide, green above and white-tomentose beneath. Flowers are pink or mauve, 10-15 mm in diameter, and borne in dense clusters of 2-5 cm length. Fruits are ovoid-globose white, pink or deep purplish berries of 10-15 mm diameter.

Ecology and Control
As other species of the genus, this shrub forms dense and impenetrable thickets due to the arching and intertwining stems. It displaces native vegetation, impedes regeneration of native shrubs and trees and affects wildlife habitats.

Specific control methods for this species are not available. The same methods as for other *Rubus* species may apply.

References
835, 1313.

Rubus ulmifolius Schott Rosaceae

LF: Deciduous shrub CU: Ornamental
SN: *Rubus discolor* Weihe & Nees, *Rubus inermis* Willd.

Geographic Distribution

Europe		*Australasia*		*Atlantic Islands*	
	Northern	●	Australia		Cape Verde
N	British Isles	X	New Zealand	N	Canary, Madeira
N	Central, France			N	Azores
N	Southern	*Northern America*			South Atlantic Isl.
N	Eastern	X	Canada, Alaska		
N	Mediterranean Isl.		Southeastern USA	*Indian Ocean Islands*	
		●	Western USA		Mascarenes
Africa			Remaining USA		Seychelles
N	Northern		Mexico		Madagascar
	Tropical				
	Southern	*Southern America*		*Pacific Islands*	
			Tropical		Micronesia
Asia			Caribbean		Galapagos
	Temperate	X	Chile, Argentina		Hawaii
	Tropical			●	Melanesia, Polynesia

Invaded Habitats
Forests, riparian habitats, freshwater wetlands.

Description
A shrub with stout and arching, light brown to purplish stems, robust prickles and palmately compound leaves. These have 3-5 stalked leaflets that are dark green and glabrous above, white-tomentose beneath, sharply toothed, and with hooked prickles on the petioles. The terminal leaflet is broad-ovate to obovate and 3-8 cm long. Flowers are white or pink and grow in long, tomentose and prickly panicles. Fruits are small, ovoid or subglobose berries of 8-10 mm diameter.

Ecology and Control
A variable species with several cultivars. It spreads both by seeds and by vegetative growth. Stems root freely at the tips, and pieces of stems are carried by streams to form new infestations. The shrub grows in dense patches that displace native vegetation and prevent regeneration of shrubs and trees due to the shading effect. Seeds are dispersed by birds.

Specific control methods for this species are not available. The same methods as for other *Rubus* species may apply.

References
133, 215, 640, 641, 894.

Rumex acetosella L. Polygonaceae

LF: Perennial herb
SN: *Acetosella vulgaris* L.
CU: -

Geographic Distribution

Europe	*Australasia*	*Atlantic Islands*
N Northern	● Australia	Cape Verde
N British Isles	X New Zealand	N Canary, Madeira
N Central, France		N Azores
N Southern	*Northern America*	South Atlantic Isl.
N Eastern	X Canada, Alaska	
N Mediterranean Isl.	Southeastern USA	*Indian Ocean Islands*
	● Western USA	X Mascarenes
Africa	X Remaining USA	Seychelles
N Northern	X Mexico	Madagascar
X Tropical		
X Southern	*Southern America*	*Pacific Islands*
	X Tropical	Micronesia
Asia	Caribbean	Galapagos
N Temperate	X Chile, Argentina	X Hawaii
X Tropical		X Melanesia, Polynesia

Invaded Habitats
Grass- and heathland, seasonal freshwater wetlands, coastal beaches, rock outcrops.

Description
A 10-70 cm tall herb with slender glabrous erect or ascending and much-branched stems. Leaves are alternate, variable in shape and size, 3-15 times as long as wide, and sagittate with the basal lobes turning outwards. Flowers are borne in clusters. The plant is normally dioecious, male flowers are yellow-orange and female flowers red-orange. Fruits are achenes of 1-1.5 mm length and width. The plant has a long and deeply growing rhizome.

Ecology and Control
A highly variable species with four subspecies in Europe. It grows in forest gaps, bogs, grass- and heathland, mostly on dry sandy or loamy soils. It forms large stands by vegetative growth, and individual clones may last for decades or longer. The rosettes may cover large areas within a short time and shade out native species. Fragments of roots easily grow to new plants. Seeds are dispersed by wind, water, and by attaching to animals. The plant is shade intolerant and establishment of seedlings occurs in disturbed sites.

    Cutting or mowing results in quick replacement of new shoots. Seedlings and small patches can be dug out, the crowns and roots must be removed. Larger stands can be treated with herbicide, or cut and the regrowth treated chemically.

References
121, 215, 386, 566, 581, 676, 933.

Rumex conglomeratus Murray

Polygonaceae

LF: Perennial herb
SN: -
CU: -

Geographic Distribution

Europe		Australasia		Atlantic Islands	
	Northern	●	Australia		Cape Verde
N	British Isles	X	New Zealand	X	Canary, Madeira
N	Central, France			X	Azores
N	Southern	*Northern America*			South Atlantic Isl.
N	Eastern	X	Canada, Alaska		
N	Mediterranean Isl.		Southeastern USA	*Indian Ocean Islands*	
			Western USA	X	Mascarenes
Africa			Remaining USA		Seychelles
N	Northern		Mexico		Madagascar
	Tropical				
	Southern	*Southern America*		*Pacific Islands*	
			Tropical		Micronesia
Asia			Caribbean		Galapagos
N	Temperate	X	Chile, Argentina		Hawaii
	Tropical				Melanesia, Polynesia

Invaded Habitats
Floodplains, woodland, streambanks, marshy shores.

Description
A glabrous upright herb of 50-120 cm height, with slender sparsely branched stems. Leaves are dark green, ovate, entire, 3-15 cm long and 1-4 cm wide. The lower leaves have long petioles. Flowers are reddish-green, small and borne in many-flowered clusters arranged in an open panicle. The corolla is 2-3 mm long. The red-brown fruits are surrounded by three oblong valves of 2-3 mm length. Seeds are *c.* 1 mm long and 0.5 mm wide. The plant has a thick taproot.

Ecology and Control
Where native, this plant commonly grows in ditches, streamsides, meadows, damp places, and in forest gaps. It is invasive because it grows in dense patches that may merge to cover large areas, thereby displacing native vegetation and reducing species richness. The species reproduces by seeds and by vegetative growth from the crown. Once established, it is a persistant weed.

   Specific control methods for this species are not available. The same methods as for other *Rumex* species may apply.

References
215, 907, 986.

Rumex crispus L.   Polygonaceae

LF: Perennial herb   CU: -
SN: -

Geographic Distribution

Europe
N   Northern
N   British Isles
N   Central, France
N   Southern
N   Eastern
N   Mediterranean Isl.

Africa
N   Northern
X   Tropical
      Southern

Asia
N   Temperate
N   Tropical

Australasia
●   Australia
X   New Zealand

Northern America
X   Canada, Alaska
      Southeastern USA
X   Western USA
X   Remaining USA
X   Mexico

Southern America
X   Tropical
      Caribbean
X   Chile, Argentina

Atlantic Islands
N   Cape Verde
N   Canary, Madeira
X   Azores
      South Atlantic Isl.

Indian Ocean Islands
X   Mascarenes
      Seychelles
      Madagascar

Pacific Islands
X   Micronesia
      Galapagos
X   Hawaii
X   Melanesia, Polynesia

Invaded Habitats
Forests, grassland, riparian habitats, freshwater wetlands, coastal marshes.

Description
An upright herb of 30-150 cm height, branched in the upper part. The basal and lower stem leaves are lanceolate and up to 30 cm long. Upper leaves are narrowly oblong to elliptic-lanceolate and have undulate margins. Flowers are greenish red and have stalks of 5-10 mm length. Inflorescences are open to dense panicles with short branches. Fruits are achenes enclosed in the inner perianth segments, and 3.5-6 mm long. The dark brown seeds are *c.* 2 mm long. The plant has a fleshy deeply penetrating taproot.

Ecology and Control
A highly variable species with three subspecies in Europe and whose exact native range is obscure. It is a serious agricultural weed and invades natural plant communities where it persists and grows in dense patches that displace native vegetation. It produces large amounts of seeds that are dispersed by wind and water. Seeds may remain viable in the soil for decades. The species establishes mainly in disturbed sites with bare ground. Once established, the plant is persistent.

     Single plants or small groups can be dug out or deeply hoed. To prevent regrowth, the roots must be severed at least 20 cm below the soil surface. Effective herbicides include glyphosate, dicamba, picloram, or triclopyr.

References
215, 223, 224, 580, 986.

Rumex sagittatus Thunb.

Polygonaceae

LF: Perennial herb
CU: Ornamental
SN: *Acetosa sagittata* (Thunb.) Johnson & Briggs

Geographic Distribution

Europe	*Australasia*	*Atlantic Islands*
Northern	● Australia	Cape Verde
British Isles	X New Zealand	Canary, Madeira
Central, France		Azores
Southern	*Northern America*	South Atlantic Isl.
Eastern	Canada, Alaska	
Mediterranean Isl.	Southeastern USA	*Indian Ocean Islands*
	Western USA	Mascarenes
Africa	Remaining USA	Seychelles
Northern	Mexico	Madagascar
Tropical		
N Southern	*Southern America*	*Pacific Islands*
	Tropical	Micronesia
Asia	Caribbean	Galapagos
Temperate	Chile, Argentina	Hawaii
Tropical		Melanesia, Polynesia

Invaded Habitats
Forests, grassland, riparian habitats, freshwater wetlands, coastal bluffs and dunes.

Description
A climbing or scrambling herb with a tuberous woody rootstock and stems reaching 3 m height. Leaves have slender petioles as long as or longer than the blades. The leaf blade is 3-8 cm long and 2-6 cm wide, triangular or hastate with two long basal lobes. Flowers have long and thin pedicels. The green perianth is 1.5-2 mm long. Inflorescences are pyramidal panicles up to 30 cm length. Fruit valves are orbicular, 5-9 mm in diameter, and yellow or pink. The light brown seeds are *c.* 3 mm long. The plant has tuber bearing rhizomes reaching 5 m length.

Ecology and Control
This rapidly growing plant is invasive because it completely smothers herbs and shrubs, preventing any regeneration and reducing native species richness. Stems either trail over the ground or climb over supporting vegetation. The plant spreads vegetatively by tubers and by seeds. Seeds are dispersed by wind and water, tubers by water and soil movement. Although shade tolerant, growth is most vigorous in sunny conditions.

Control is difficult due to the many tubers that dislodge easily. Seedlings and small plants can be dug out, all tubers and rhizomes must be removed. Removing fruits before they ripen prevents seed dispersal. Chemical control is best done before fruits ripen. Repeated applications over several years are necessary to give effective control.

References
215, 924.

Salix babylonica L.   Salicaceae

LF: Deciduous tree   CU: Ornamental, wood
SN: *Salix elegantissima* Koch

Geographic Distribution

Europe	*Australasia*	*Atlantic Islands*
Northern	● Australia	Cape Verde
British Isles	X New Zealand	Canary, Madeira
Central, France		Azores
X Southern	*Northern America*	X South Atlantic Isl.
X Eastern	Canada, Alaska	
Mediterranean Isl.	Southeastern USA	*Indian Ocean Islands*
	X Western USA	Mascarenes
Africa	Remaining USA	Seychelles
Northern	Mexico	Madagascar
Tropical		
● Southern	*Southern America*	*Pacific Islands*
	Tropical	Micronesia
Asia	Caribbean	Galapagos
N Temperate	Chile, Argentina	Hawaii
Tropical		Melanesia, Polynesia

Invaded Habitats
Heath- and shrubland, riparian habitats, freshwater wetlands.

Description
A spreading tree reaching 20 m height, with brown, glabrous, long and pendent twigs, drooping almost to the ground. Older trees have a dark brown and furrowed bark. Leaves are bright green, narrowly lanceolate to linear-lanceolate, glabrous, 8-16 cm long and 8-15 mm wide, and have short petioles. Margins are toothed. Flowers are borne in drooping spike-like catkins. These are curved, sessile or on short stalks. The tree is dioecious, and male catkins are up to 4 cm long, female catkins up to 2 cm long. Fruits are sessile, ovoid-conic and glabrous capsules.

Ecology and Control
This is a common tree of riparian habitats. In South Africa, only female trees are present. The tree spreads mainly vegetatively by root sprouts and forms dense thickets along streams that shade out native riparian species and affect the invertebrate fauna of wetlands and rivers by changing and reducing the species composition and richness. Little is known on the ecology of this tree.

    Specific control methods for this species are not available. The same methods as for *Salix cinerea* may apply.

References
215, 548, 549, 1089.

Salix cinerea L.   Salicaceae

LF: Deciduous shrub, tree   CU: Erosion control, ornamental
SN: *Salix acuminata* Mill., *Salix aquatica* Sm.

Geographic Distribution

Europe	*Australasia*	*Atlantic Islands*
N Northern	● Australia	Cape Verde
N British Isles	● New Zealand	Canary, Madeira
N Central, France		Azores
N Southern	*Northern America*	South Atlantic Isl.
N Eastern	Canada, Alaska	
Mediterranean Isl.	Southeastern USA	*Indian Ocean Islands*
	Western USA	Mascarenes
Africa	Remaining USA	Seychelles
Northern	Mexico	Madagascar
Tropical		
Southern	*Southern America*	*Pacific Islands*
	Tropical	Micronesia
Asia	Caribbean	Galapagos
N Temperate	Chile, Argentina	Hawaii
Tropical		Melanesia, Polynesia

Invaded Habitats
Riparian habitats, freshwater wetlands, wet coastal places.

Description
A multi-stemmed shrub or small tree reaching 7 m height, sometimes more than 10 m, with young twigs being densely pubescent and dark reddish-brown, and a rather smooth bark. Leaves are elliptic to obovate-lanceolate, 2-10 cm long and 1-3 cm wide, bluish grey and densely pubescent beneath, dull green and almost glabrous above. Petioles are *c.* 10 mm long. Catkins are almost sessile, very silky, male catkins being up to 5 cm long and female catkins up to 8 cm long.

Ecology and Control
There are two subspecies of this tree in Europe. The spreading growth habit leads to dense thickets along rivers, competing for space, water and nutrients. The dense foliage of stands strongly reduce light levels and eliminate almost all native vegetation. They can alter the shape of riverbanks and streambeds through sediment accumulation and the shading affects aquatic invertebrates by reducing their richness and abundance. The short-lived seeds are dispersed by wind and water. Hybridization with other *Salix* species is common in Australia.

     Seedlings and small plants can be pulled or dug out, roots must be removed. Smaller trees are cut and the cut stumps treated with herbicide. Herbicide can also applied by the drill-fill method. Follow-up treatments are necessary to control regrowth.

References
22, 121, 215, 924, 1415.

Salix fragilis L. — Salicaceae

LF: Deciduous tree
SN: -
CU: Erosion control, ornamental, wood

Geographic Distribution

Europe	*Australasia*	*Atlantic Islands*
Northern	● Australia	Cape Verde
N British Isles	● New Zealand	Canary, Madeira
N Central, France		X Azores
N Southern	*Northern America*	South Atlantic Isl.
N Eastern	● Canada, Alaska	
N Mediterranean Isl.	Southeastern USA	*Indian Ocean Islands*
	X Western USA	Mascarenes
Africa	Remaining USA	Seychelles
Northern	Mexico	Madagascar
Tropical		
● Southern	*Southern America*	*Pacific Islands*
	Tropical	Micronesia
Asia	Caribbean	Galapagos
N Temperate	Chile, Argentina	Hawaii
Tropical		Melanesia, Polynesia

Invaded Habitats
Riparian habitats, lakesides, wet places.

Description
A sometimes shrubby tree reaching 30 m height, with a greyish and thick bark becoming fissured with age, and glabrous olive to brown twigs. Leaves are narrow-lanceolate to oblong-lanceolate, 6-16 cm long and 1.5-4 cm wide, long-acuminate, light green beneath, and with petioles of 6-25 mm length. Catkins grow on leafy stalks, are drooping, the male catkins being 2-4.5 cm long, the female catkins 3-6 cm long. The tree is dioecious.

Ecology and Control
This tree is invasive because it can become the dominant species in riparian vegetation and forms a dense canopy, reducing light levels and thereby shading out native plants and reducing macroinvertebrate abundance. The tree reproduces vigorously by vegetative growth. Twigs and broken branches become rooted, and pieces of twigs, stems and roots are carried by streams, leading to new infestations. In Australia, most plants of this species are male clones. In the USA, it hybridizes with the native *Salix alba*.

Specific control methods for this species are not available. The same methods as for *Salix cinerea* may apply.

References
121, 258, 549, 739, 924, 1415.

Salpichroa origanifolia (Lam.) Thell.   Solanaceae

LF: Perennial herb   CU: Ornamental, wood
SN: *Salpichroa rhomboidea* (Gill. & Hook.) Miers, *Atropa rhomboidea* Gill. & Hook.

Geographic Distribution

Europe
- Northern
- X British Isles
- Central, France
- Southern
- Eastern
- Mediterranean Isl.

Africa
- Northern
- Tropical
- Southern

Asia
- Temperate
- Tropical

Australasia
- ● Australia
- X New Zealand

Northern America
- Canada, Alaska
- Southeastern USA
- X Western USA
- Remaining USA
- Mexico

Southern America
- N Tropical
- Caribbean
- X Chile, Argentina

Atlantic Islands
- Cape Verde
- X Canary, Madeira
- X Azores
- South Atlantic Isl.

Indian Ocean Islands
- Mascarenes
- Seychelles
- Madagascar

Pacific Islands
- Micronesia
- Galapagos
- Hawaii
- Melanesia, Polynesia

Invaded Habitats
Grassland, riparian habitats, coastal dunes and beaches.

Description
A stout herb reaching 1.5 m height, with much-branched and sprawling stems, sometimes woody at the base. The simple leaves are suborbicular to ovate-rhombic and entire. Leaf blades are 5-50 mm long and 5-35 mm wide. The petiole is of about the same length as the blade. The whitish or cream flowers grow solitary in the axils of leaves, are 6-10 mm in diameter and have short corolla lobes. Fruits are whitish to yellow and ovoid berries of 10-15 mm diameter. Seeds are strongly compressed, hairy and 1.5-2 mm long.

Ecology and Control
This plant spreads by root segments and seeds. Once established, it has a sprawling growth habit and forms dense thickets, smothering all other vegetation, impeding growth and regeneration of native shrubs and trees. The extensive root system allows it to survive long dry periods. Seeds are dispersed by birds and small mammals.

Digging and removing most of the roots is possible but labour intensive. Herbicides to control this plant include 2,4-D, MCPA, or picloram.

References
215, 433, 985, 986, 1265.

Salsola kali L. — Chenopodiaceae

LF: Annual herb
SN: *Salsola australis* R.Br.
CU: -

Geographic Distribution

Europe		*Australasia*		*Atlantic Islands*	
N	Northern	X	Australia		Cape Verde
N	British Isles	X	New Zealand	N	Canary, Madeira
N	Central, France			N	Azores
X	Southern	*Northern America*			South Atlantic Isl.
N	Eastern	X	Canada, Alaska		
X	Mediterranean Isl.		Southeastern USA	*Indian Ocean Islands*	
		X	Western USA		Mascarenes
Africa		X	Remaining USA		Seychelles
X	Northern	X	Mexico	X	Madagascar
	Tropical				
●	Southern	*Southern America*		*Pacific Islands*	
			Tropical		Micronesia
Asia			Caribbean		Galapagos
X	Temperate	X	Chile, Argentina	X	Hawaii
X	Tropical				Melanesia, Polynesia

Invaded Habitats
Heath- and shrubland, coastal beaches.

Description
An upright much-branched herb with glabrous or pubescent stems of 5-120 cm height, becoming woody at maturity. The succulent alternate leaves are linear to linear-subulate, 1-4 cm long and 1-2 mm wide, usually with a spine at the end. The small pink to greenish-white flowers are borne solitary in leaf axils or in short axillary spikes, and subtended by two conspicuous leaf-like bracteoles. The fruiting calyx is enclosing the fruit, leathery and up to 10 mm wide. Seeds are *c.* 2 mm in diameter.

Ecology and Control
In the native range, this highly variable plant is common on coastal beaches but grows also on non-saline soils. There are many subspecies and varieties in Europe. In the Himalayas, it ascends to 4,500 m elevation. Where invasive, it displaces native plants and competes for space, water and nutrients. Some forms break off at soil level and disperse the seeds by rolling across the landscape. Seed production is prolific and seeds may remain viable for 6-12 months.

Specific control methods for this species are not available. Plants can be hand pulled or dug out. Cutting before fruits ripen prevents seed dispersal. Larger stands may be treated with herbicide.

References
391, 549, 581, 1089.

Salvia verbenaca L. Lamiaceae

LF: Perennial herb CU: -
SN: *Salvia cleistogama* de Bary & Paul, *Salvia clandestina* L.

Geographic Distribution

Europe		*Australasia*		*Atlantic Islands*	
	Northern	●	Australia		Cape Verde
N	British Isles	X	New Zealand	N	Canary, Madeira
N	Central, France				Azores
N	Southern	*Northern America*			South Atlantic Isl.
N	Eastern		Canada, Alaska		
N	Mediterranean Isl.		Southeastern USA	*Indian Ocean Islands*	
		X	Western USA		Mascarenes
Africa			Remaining USA		Seychelles
N	Northern		Mexico		Madagascar
	Tropical				
	Southern	*Southern America*		*Pacific Islands*	
			Tropical		Micronesia
Asia			Caribbean		Galapagos
N	Temperate		Chile, Argentina		Hawaii
	Tropical				Melanesia, Polynesia

Invaded Habitats
Grassland, riverbanks, disturbed sites.

Description
A moderately to densely hairy herb with erect, simple or branched stems of 10-80 cm height. Leaves are ovate to ovate-oblong, often pinnately lobed with wide lobes. The basal leaves have long petioles, are more or less lobed, 5-10 cm long and 2-4 cm wide. Stem leaves are sessile or have short petioles only. Flowers are blue, lilac or purple, 10-15 mm in diameter, growing in axillary clusters with 5-10 flowers each. Fruits are broad-ellipsoid nutlets of *c.* 2 mm length.

Ecology and Control
This plant is a polymorphic species in Europe with numerous local variants. Native habitats include dry grassy places, rough ground, dunes, and disturbed sites. It is invasive because it forms dense swards that crowd out native plants and reduce species richness. Little is known on the ecology of this species.

Specific control methods for this species are not available. Smaller plants may be pulled or dug out, larger stands treated with herbicide.

References
215, 929.

Salvinia molesta Mitch.   Salviniaceae

LF: Aquatic fern
SN: *Salvinia auriculata* auct.
CU: Ornamental

Geographic Distribution

Europe	*Australasia*	*Atlantic Islands*
Northern	● Australia	Cape Verde
British Isles	● New Zealand	Canary, Madeira
Central, France		Azores
Southern	*Northern America*	South Atlantic Isl.
Eastern	Canada, Alaska	
Mediterranean Isl.	X Southeastern USA	*Indian Ocean Islands*
	Western USA	Mascarenes
Africa	X Remaining USA	Seychelles
Northern	Mexico	Madagascar
● Tropical		
● Southern	*Southern America*	*Pacific Islands*
	N Tropical	Micronesia
Asia	X Caribbean	Galapagos
Temperate	X Chile, Argentina	Hawaii
● Tropical		● Melanesia, Polynesia

Invaded Habitats
Freshwater wetlands, ponds, streams.

Description
A free floating small perennial fern consisting of a branched rhizome and leaves. The plant has two types of leaves: strongly dissected ones hanging into the water and serving as roots, and floating leaves. The latter are elliptic, entire, variable in size, 1-5 cm wide, and densely covered with papillae on the upper surface.

Ecology and Control
This fern is a sterile hybrid incapable of sexual reproduction and spreading solely by vegetative growth and fragmentation. It is fast growing in nutrient rich waters and forms dense mats of tightly packed leaves. Such mats may completely cover the water surface and become up to 1 m thick. Native aquatic plants are eliminated, the water flow is affected and oxygen levels are reduced. Dead plants release large amounts of nutrients into the water, thereby increasing eutrophication.

Large infestations are mechanically harvested but this may promote fragmentation and further spread. Herbicides used are diquat formulated for use in running waters, hexazinone, chlorsulfuron, or fluridone.

References
251, 269, 284, 420, 508, 525, 607, 644, 793, 885, 886, 924, 1089, 1102, 1103, 1104, 1105, 1106, 1122, 1282, 1283, 1284, 1415.

Sambucus nigra L.   Caprifoliaceae

LF: Deciduous shrub
SN: -
CU: Ornamental

Geographic Distribution

Europe
- N Northern
- N British Isles
- N Central, France
- N Southern
- N Eastern
- N Mediterranean Isl.

Africa
- N Northern
-    Tropical
-    Southern

Asia
- N Temperate
-    Tropical

Australasia
-    Australia
- ● New Zealand

Northern America
-    Canada, Alaska
-    Southeastern USA
-    Western USA
-    Remaining USA
-    Mexico

Southern America
-    Tropical
-    Caribbean
-    Chile, Argentina

Atlantic Islands
-    Cape Verde
- X Canary, Madeira
- N Azores
-    South Atlantic Isl.

Indian Ocean Islands
-    Mascarenes
-    Seychelles
-    Madagascar

Pacific Islands
-    Micronesia
-    Galapagos
-    Hawaii
-    Melanesia, Polynesia

Invaded Habitats
Forests and forest edges, scrubland, disturbed sites.

Description
A shrub or small tree reaching 10 m height, with a deeply furrowed bark and grey branches that have numerous lenticels. Leaves are 10-30 cm long and pinnately compound with 3-7 elliptic to elliptic-ovate, sharply toothed leaflets of 4-12 cm length and 2-6 cm width. Flowers are yellowish white, very fragrant, 6-8 mm in diameter, and grow in large flat cymes of 12-24 cm diameter. Fruits are globose and black, berry-like drupes of 6-8 mm diameter, containing seeds of 3-4.5 mm length.

Ecology and Control
In the native range, this fast growing shrub grows in open areas and woodland edges, riparian forests, and is associated with nutrient rich soils. Where invasive, it forms dense stands that crowd out native plants by shading and competing for space, water and nutrients. The shrub is capable of establishing in a closed shrub canopy. New shoots grow readily from cut or burned stumps, and pieces of shoots may become rooted under suitable conditions.

Seedlings and saplings can be hand pulled or dug out. Larger shrubs are cut and the cut stumps treated with herbicide. Effective herbicides for chemical control are glyphosate, 2,4-D, fluroxypyr, or mecoprop.

References
51, 1415.

Scaevola taccada (Gaertn.) Roxb. Goodeniaceae

LF: Evergreen shrub
CU: Erosion control
SN: *Scaevola frutescens* Krause, *Scaevola sericea* Vahl

Geographic Distribution

Europe
- Northern
- British Isles
- Central, France
- Southern
- Eastern
- Mediterranean Isl.

Africa
- Northern
- N Tropical
- Southern

Asia
- Temperate
- N Tropical

Australasia
- N Australia
- New Zealand

Northern America
- Canada, Alaska
- ● Southeastern USA
- Western USA
- Remaining USA
- Mexico

Southern America
- Tropical
- X Caribbean
- Chile, Argentina

Atlantic Islands
- Cape Verde
- Canary, Madeira
- Azores
- South Atlantic Isl.

Indian Ocean Islands
- N Mascarenes
- N Seychelles
- N Madagascar

Pacific Islands
- N Micronesia
- Galapagos
- N Hawaii
- N Melanesia, Polynesia

Invaded Habitats
Coastal dunes and hammocks, beaches.

Description
A multi-stemmed shrub or small tree of 0.5-2 m height, sometimes taller, with a round growth form and succulent, light green leaves. Leaf axils are conspicuously white hairy. The simple and entire leaves are glabrous, glossy, elliptic to obovate, 5-23 cm long, 2-9 cm wide, and gradually tapering to the base. The white to pinkish flowers are zygomorphic, 18-20 mm long and borne in clusters of 2-4. Corollas are 20-25 mm in diameter. Fruits are whitish and indehiscent subglobose to ellipsoid berries of 7-13 mm length, containing 1-2 seeds of *c.* 8 mm length and 7 mm width.

Ecology and Control
Where native, this salt tolerant shrub occurs commonly along the high tidemark of sandy beaches, on rocky and sandy coasts, and in edges of coastal beach forests. Where invasive, the shrub competes with native coastal vegetation and can quickly form extensive and dense populations. In Florida, it competes directly with an endangered native congener, *Scaevola plumieri*. The species flowers throughout the year. Stems touching the soil become rooted. Seeds remain viable in sea water for a year or more, and since the fruits easily float on the water, they are carried by ocean currents to new sites.

The species roots easily from cuttings. Young plants can be hand pulled and larger plants dug out. Chemical control is done by cutting the shrub to the ground and treating the cut stumps with triclopyr.

References
564, 715, 930, 1281.

Schefflera actinophylla (Endl.) Harms   Araliaceae

LF: Evergreen tree
SN: -
CU: Ornamental

Geographic Distribution

Europe	*Australasia*	*Atlantic Islands*
Northern	N Australia	Cape Verde
British Isles	New Zealand	Canary, Madeira
Central, France		Azores
Southern	*Northern America*	South Atlantic Isl.
Eastern	Canada, Alaska	
Mediterranean Isl.	● Southeastern USA	*Indian Ocean Islands*
	Western USA	Mascarenes
Africa	Remaining USA	Seychelles
Northern	Mexico	Madagascar
Tropical		
Southern	*Southern America*	*Pacific Islands*
	Tropical	Micronesia
Asia	Caribbean	Galapagos
Temperate	Chile, Argentina	● Hawaii
N Tropical		Melanesia, Polynesia

Invaded Habitats
Hammocks, disturbed sites.

Description
A sparingly branched tree of 10-15 m height, usually with several trunks. Leaves are borne in large rosettes at the ends of branches, forming leafy umbrellas. Leaves are palmately compound with 5-18 large leaflets. These are 8-30 cm long, elliptic to obovate, and have petioles of 30-60 cm length. Bright red flowers are borne in flowerheads with 8-12 flowers each; these appear numerous on stiff and spreading inflorescences of 40-80 cm length at the ends of branches. Fruits are small, roundish and dark purplish to red drupes of 6-8 mm diameter. Fruits are aggregated in heads of 20-25 mm diameter. Each fruit head contains 10-12 seeds of 5-6 mm length and *c.* 3 mm width.

Ecology and Control
This plant grows mostly as a tree but occasionally as an epiphyte in lowland to highland rainforests, wet sclerophyll, and open forests. Flowers and fruits are attractive to birds which disperse seeds to new sites. Seeds germinate readily. Once established, the tree forms dense and shady thickets, outcompeting native plant species. The tree grows also as an epiphyte if seeds germinate in the crotches of large trees or in the old leaf bases of palms. The plant then sends aerial roots down to the ground.

Seedlings and young saplings can be pulled by hand. Older trees can be treated with triclopyr mixed with an oil diluent and applied to the base of the tree. If trees are cut, the stumps should be treated with triclopyr. Caution must be taken not to affect host trees.

References
715.

Schinus molle L.   Anacardiaceae

LF: Evergreen tree
SN: *Schinus areira* L.
CU: Ornamental, wood

Geographic Distribution

Europe		*Australasia*		*Atlantic Islands*	
	Northern	●	Australia		Cape Verde
	British Isles	X	New Zealand		Canary, Madeira
	Central, France				Azores
	Southern	*Northern America*			South Atlantic Isl.
	Eastern		Canada, Alaska		
	Mediterranean Isl.		Southeastern USA	*Indian Ocean Islands*	
		X	Western USA		Mascarenes
Africa			Remaining USA		Seychelles
	Northern		Mexico		Madagascar
X	Tropical				
X	Southern	*Southern America*		*Pacific Islands*	
		N	Tropical		Micronesia
Asia			Caribbean	X	Galapagos
X	Temperate		Chile, Argentina	X	Hawaii
	Tropical				Melanesia, Polynesia

Invaded Habitats
Forests, grass- and shrubland, riverbanks, coastal dunes and beaches.

Description
A 5-15 m tall tree with a short trunk, slender drooping branches, a large spreading crown, and a deeply fissured flaking bark. Leaves are alternate, pinnately compound, up to 25 cm long, and have 7-20 pairs of narrowly lanceolate to linear-lanceolate leaflets of 1.5-6 cm length and 2-7 mm width. The terminal leaflet is smaller than the lateral ones. Margins are entire or toothed. Flowers are yellowish white, small, and borne in axillary and terminal drooping panicles. Fruits are lavender to pinkish red, globose drupes of 6-8 mm diameter, each containing one seed. The tree is dioecious.

Ecology and Control
This fast growing and drought resistant tree grows in elevations up to 2,100 m in Kenya. It is freely coppicing and forms extensive species poor stands that shade out all native vegetation. Since the soil under the canopy remains bare and lacks a herbaceous ground flora, erosion can be accelerated in stands growing on slopes or near streams. Seeds are dispersed by birds.

Seedlings and saplings can be dug out. Larger trees are cut and the cut stumps treated with herbicide. The drill-fill method is also effective. Follow-up treatments are necessary to control regrowth. Fruit bearing trees should be removed first to prevent seed dispersal.

References
121, 215, 591, 924, 934, 935.

Schinus terebinthifolius Raddi — Anacardiaceae

LF: Evergreen tree
SN: -
CU: Ornamental

Geographic Distribution

Europe	*Australasia*	*Atlantic Islands*
Northern	● Australia	Cape Verde
British Isles	X New Zealand	Canary, Madeira
Central, France		Azores
X Southern	*Northern America*	X South Atlantic Isl.
Eastern	Canada, Alaska	
Mediterranean Isl.	● Southeastern USA	*Indian Ocean Islands*
	X Western USA	● Mascarenes
Africa	Remaining USA	Seychelles
X Northern	Mexico	Madagascar
Tropical		
● Southern	*Southern America*	*Pacific Islands*
	N Tropical	Micronesia
Asia	X Caribbean	Galapagos
Temperate	N Chile, Argentina	● Hawaii
X Tropical		Melanesia, Polynesia

Invaded Habitats
Forests, grassland, mangrove swamps, riparian habitats, coastal wetlands.

Description
A tree or shrub with several trunks, 1-7 m tall, and arching but not drooping branches. Leaves are pinnately compound, up to 40 cm long, and have 2-8 pairs of elliptic to lanceolate leaflets, and an additional leaflet at the end. Leaflets are glabrous, 1.5-7.5 cm long and 7-32 mm wide, the terminal leaflet being larger than the lateral ones. Margins are entire to serrated, and glabrous. The white flowers form large, terminal panicles. Petals are oblong to ovate and 1.2-2.5 mm long. Fruits are globose, bright red drupes of 4-5 mm diameter. The plant is dioecious, e.g. male and female flowers are borne on different individuals.

Ecology and Control
This shade and drought resistant tree can become dominant in the understorey and outcompetes native species for light and nutrients. The tree forms dense thickets that completely shade out and displace native vegetation with a species poor shrubland. Seeds are spread by birds and mammals and germinate readily; recruitment from seed is rapid. The tree easily sprouts from the trunk after damage. The spread of this tree is promoted by disturbance.

    Seedlings and saplings can be pulled by hand. Prescribed fire is used as a control measure in fire adapted communities. Chemical control includes basal application of bromacil, hexazinone, or triclopyr mixed with an oil diluent. Removing female trees eliminates future seed sources.

References
56, 133, 284, 353, 354, 355, 394, 549, 634, 715, 905, 926, 934, 935, 980, 1089, 1155, 1239, 1278.

Schismus arabicus Nees — Poaceae

LF: Annual herb
SN: -
CU: Forage

Geographic Distribution

Europe
- Northern
- British Isles
- Central, France
- Southern
- Eastern
- N Mediterranean Isl.

Africa
- N Northern
- Tropical
- Southern

Asia
- N Temperate
- N Tropical

Australasia
- X Australia
- New Zealand

Northern America
- Canada, Alaska
- Southeastern USA
- ● Western USA
- Remaining USA
- Mexico

Southern America
- Tropical
- Caribbean
- Chile, Argentina

Atlantic Islands
- Cape Verde
- Canary, Madeira
- Azores
- South Atlantic Isl.

Indian Ocean Islands
- Mascarenes
- Seychelles
- Madagascar

Pacific Islands
- Micronesia
- Galapagos
- Hawaii
- Melanesia, Polynesia

Invaded Habitats
Arid and semi-arid grassland, rangeland, desert scrub, coastal shrubland.

Description
A tufted grass with erect to spreading smooth culms of 10-20 cm height. Leaves are usually inrolled, smooth, and *c.* 2 mm wide. Ligules consist of rings of rigid hairs. Inflorescences are green to purplish dense and narrow panicles. Spikelets are laterally compressed and contain 5-10 florets. Glumes are lanceolate and persistent. Seeds are minute.

Ecology and Control
This winter annual invades open spaces between shrubs in desert communities and produces a dense carpet of grass. It becomes dominant and displaces native annuals due to competition for water and nutrients, and prevents establishment of native species. The roots form an extensive mat near the soil surface. The long-standing dead stems readily carry fires across the landscape, increasing fire frequency and intensity. Seeds are dispersed by wind and water and accumulate in a soil seed bank. The grass may become reproductive within a few weeks.

Once established, the grass is difficult to control because both grazing and burning promote its spread. Small patches can be removed manually or cut. An effective herbicide is glyphosate.

References
133, 403, 513, 514.

Schizachyrium condensatum (Kunth) Nees

Poaceae

LF: Perennial herb CU: -
SN: *Andropogon condensatus* Kunth, *Andropogon glomeratus* (Walter) Britton

Geographic Distribution

Europe
  Northern
    British Isles
    Central, France
  Southern
  Eastern
  Mediterranean Isl.

Africa
  Northern
  Tropical
  Southern

Asia
  Temperate
  Tropical

Australasia
  Australia
  New Zealand

Northern America
  Canada, Alaska
  Southeastern USA
  Western USA
  Remaining USA
  N Mexico

Southern America
  N Tropical
  N Caribbean
  Chile, Argentina

Atlantic Islands
  Cape Verde
  Canary, Madeira
  Azores
  South Atlantic Isl.

Indian Ocean Islands
  Mascarenes
  Seychelles
  Madagascar

Pacific Islands
  Micronesia
  Galapagos
  ● Hawaii
  Melanesia, Polynesia

Invaded Habitats
Wood- and shrubland, grassland, disturbed sites.

Description
A stout glabrous erect bunchgrass with culms of 1-1.8 m height and 2-3.5 mm diameter, unbranched in the lower part. The glabrous leaf sheaths are keeled, the ligules consist of firm membranes of 0.7-2 mm length. Leaf blades are up to 40 cm long and 3-8 mm wide, the lower surface being keeled. The loose inflorescences are 20-40 cm long and composed of numerous solitary racemes of 15-35 mm length. Spikelets are 4.5-5 mm long. Fruits are linear-cylindrical caryopses of 2.5-3 mm length.

Ecology and Control
This fast growing grass sends up new tillers each year from a small root crown. The species is closely related to *Schizachyrium microstachyum* and is regarded by some authors as belonging to this species complex. The grass is invasive because it promotes the spread of fires and displaces native vegetation with pure stands. Such stands accumulate large quantities of dead and flammable biomass, increasing fire frequency and intensity. The grass forms dense swards that crowd out native plant species and prevent their regeneration.

Specific control methods for this species are not available. The same methods as for *Pennisetum* species may apply.

References
304, 306, 600.

Schkuhria pinnata (Lam.) Kuntze   Asteraceae

LF: Annual herb   CU: -
SN: *Pectis pinnata* Lam.

Geographic Distribution

	Europe		Australasia		Atlantic Islands
	Northern	X	Australia		Cape Verde
X	British Isles		New Zealand		Canary, Madeira
	Central, France				Azores
	Southern		*Northern America*		South Atlantic Isl.
	Eastern		Canada, Alaska		
	Mediterranean Isl.		Southeastern USA		*Indian Ocean Islands*
			Western USA		Mascarenes
	Africa		Remaining USA		Seychelles
	Northern	N	Mexico		Madagascar
X	Tropical				
●	Southern		*Southern America*		*Pacific Islands*
		N	Tropical		Micronesia
	Asia		Caribbean		Galapagos
	Temperate	N	Chile, Argentina		Hawaii
	Tropical				Melanesia, Polynesia

Invaded Habitats
Grass- and woodland, disturbed sites.

Description
A branched herb with stems 25-75 cm tall and numerous branches forming a loose compound inflorescence. Leaves and phyllaries are densely covered with glands. Leaves are usually alternate, 2-5 cm long, pinnately or bipinnately lobed, with linear to filiform segments. Numerous flowerheads grow in corymbose panicles with strongly ascending flowering branches of 10-25 cm length. The involucres are 5-6 mm long and 3-4 mm wide and have two bracts at their bases. There is only one short yellow ray floret per flowerhead. Fruits are achenes with a pappus consisting of scales.

Ecology and Control
This plant is found in grasslands, oak woodland and in tropical deciduous forests within the native range. There are two varieties in Mexico: var. *guatemalensis* and var. *virgata*, the latter is often weedy. Where invasive, the plant grows in dense clumps that crowd out native plants and prevent their regeneration. Little is known on the ecology of this species.

   Specific control methods for this species are not available. Scattered plants may be hand pulled. Cutting prevents seed dispersal. Larger stands can be treated with herbicide.

References
1089, 1308.

Securigera varia (L.) Lassen
Fabaceae

LF: Perennial herb
SN: *Coronilla varia* L.
CU: Erosion control, ornamental

Geographic Distribution

Europe	*Australasia*	*Atlantic Islands*
Northern	Australia	Cape Verde
British Isles	X New Zealand	Canary, Madeira
N Central, France		Azores
N Southern	*Northern America*	South Atlantic Isl.
N Eastern	● Canada, Alaska	
Mediterranean Isl.	● Southeastern USA	*Indian Ocean Islands*
	Western USA	Mascarenes
Africa	● Remaining USA	Seychelles
Northern	Mexico	Madagascar
Tropical		
Southern	*Southern America*	*Pacific Islands*
	Tropical	Micronesia
Asia	Caribbean	Galapagos
N Temperate	Chile, Argentina	Hawaii
Tropical		Melanesia, Polynesia

Invaded Habitats
Grassland, riparian habitats.

Description
A variable low-growing perennial glabrous to pubescent herb with ascending or trailing stems of 30-120 cm length. Leaves are pinnately compound with 5-10 pairs of leaflets. Flowers are borne in umbels with 12-15 flowers each, are 10-15 mm long, and vary in colour from pale pink or lavender to white.

Ecology and Control
Native habitats of this plant include meadows, grassland and disturbed places. The species is nitrogen-fixing due to root nodulation and increases soil fertility levels which may change the composition of the associated flora. The plant climbs over shrubs and small trees and shades them out. It can rapidly establish solid, single-species stands that eliminate native vegetation and degrade wildlife habitats.

Single plants can be hand pulled or dug out. Small infestations can be contained by covering with mulch or shade cloth. Effective chemical control is done by spraying plants with 2,4-D plus dicamba.

References
571, 1065.

Sedum acre L.   Crassulaceae

LF: Perennial herb   CU: Ornamental
SN: -

Geographic Distribution

Europe	*Australasia*	*Atlantic Islands*
N Northern	Australia	Cape Verde
N British Isles	● New Zealand	Canary, Madeira
N Central, France		Azores
N Southern	*Northern America*	South Atlantic Isl.
N Eastern	Canada, Alaska	
N Mediterranean Isl.	Southeastern USA	*Indian Ocean Islands*
	Western USA	Mascarenes
Africa	Remaining USA	Seychelles
N Northern	Mexico	Madagascar
Tropical		
Southern	*Southern America*	*Pacific Islands*
	Tropical	Micronesia
Asia	Caribbean	Galapagos
N Temperate	Chile, Argentina	Hawaii
Tropical		Melanesia, Polynesia

Invaded Habitats
Coastal cliffs, riverbeds, rock outcrops, disturbed sites.

Description
A glabrous herb with trailing stems rooting at the nodes, with short non-flowering shoots and ascending flowering stems of 5-12 cm height. The alternate and succulent leaves are 3-6 mm long, 2.5-3 mm wide and elliptical in cross-section. Flowers grow in small cymes up to 15 mm length; they have five bright yellow petals of 6-8 mm length and ten stamens. Fruits are yellowish-white follicles. Seeds are narrowly-ellipsoid to obovoid and 0.5-0.8 mm long.

Ecology and Control
In the native range, this plant is found on walls and rocks, in open grassland and coastal sandy places. It is a highly variable species in Europe with regard to size, leaf shape and flower size. It spreads by seeds and vegetative growth as the stems easily break and root at the nodes. The creeping growth habit of this plant leads to dense mats that can completely carpet the ground, displacing native vegetation and impeding the establishment of native species.

    Scattered clumps can be removed manually, all stems must be removed to prevent re-growth. Larger patches can be sprayed with herbicides.

References
1415.

Senecio angulatus L.f.   Asteraceae

LF: Perennial herb
SN: -
CU: Ornamental

Geographic Distribution

Europe	*Australasia*	*Atlantic Islands*
Northern	● Australia	Cape Verde
British Isles	● New Zealand	Canary, Madeira
Central, France		Azores
Southern	*Northern America*	South Atlantic Isl.
Eastern	Canada, Alaska	
Mediterranean Isl.	Southeastern USA	*Indian Ocean Islands*
	Western USA	Mascarenes
Africa	Remaining USA	Seychelles
Northern	Mexico	Madagascar
Tropical		
N Southern	*Southern America*	*Pacific Islands*
	Tropical	Micronesia
Asia	Caribbean	Galapagos
Temperate	Chile, Argentina	Hawaii
Tropical		Melanesia, Polynesia

Invaded Habitats
Grass- and scrubland, dry forests, coastal wetlands.

Description
A scrambling and glabrous herb with thick leaves and sparingly branched stems reaching 2 m length. Leaves are thick and fleshy, coarsely lobed, ovate to lanceolate, the blade being 3-6 cm long and 2.5-4 cm wide, with petioles as long as the blade. Flowerheads of 12-25 mm diameter grow in compound corymbs or panicles, each flowerhead having 4-6 yellow florets of 5-10 mm length. Fruits are hairy achenes of 2-2.5 mm length and have a pappus of 5-7 mm length.

Ecology and Control
Where native, this herb or half-climbing shrub is commonly found in forest margins. It is a rapidly growing vine, smothering the ground cover and native shrubs up to 2 m height with a dense tangle of stems, especially in habitats near the coast. The plant spreads both by seeds and by vegetative growth. Branches touching the ground and stem fragments root at the tips. The plant also forms long aboveground runners. Seeds are dispersed by wind.

    Mechanical control is done by slashing, mowing or grubbing. An effective method proved to be cutting at soil level prior to flowering. Regrowth can then be sprayed with herbicide or manually removed.

References
121, 215, 1415.

Senecio elegans L. — Asteraceae

LF: Annual herb
SN: -
CU: Ornamental

Geographic Distribution

Europe	*Australasia*	*Atlantic Islands*
Northern	● Australia	Cape Verde
British Isles	X New Zealand	Canary, Madeira
Central, France		X Azores
Southern	*Northern America*	South Atlantic Isl.
Eastern	Canada, Alaska	
Mediterranean Isl.	Southeastern USA	*Indian Ocean Islands*
	X Western USA	Mascarenes
Africa	Remaining USA	Seychelles
Northern	Mexico	Madagascar
Tropical		
N Southern	*Southern America*	*Pacific Islands*
	Tropical	Micronesia
Asia	Caribbean	Galapagos
Temperate	Chile, Argentina	Hawaii
Tropical		Melanesia, Polynesia

Invaded Habitats
Coastal sand dunes and beaches.

Description
A 60-100 cm tall herb with ridged and glabrous to densely hairy stems usually branched only in the inflorescence. The fleshy leaves are deeply divided into 2-4 pairs of segments. The lower leaves have petioles, the upper ones are sessile. The open inflorescence is a dense corymb and has numerous flowerheads on long peduncles. Flowerheads are 20-25 mm in diameter and have purple ray florets with 6-8 mm long limbs. Fruits are usually hairy achenes of *c.* 2.5 mm length, with a pappus of 5-8 mm length.

Ecology and Control
Where native, this plant grows mostly in coastal sandy soils. It is invasive because it spreads rapidly and forms a continuous ground cover of dense growth, displacing native species and preventing their regeneration. Little is known on the ecology of this species.

    Specific control methods for this species are not available. Plants can be hand pulled or cut to prevent seed dispersal. Larger stands can be treated with herbicide.

References
215.

Senna alata (L.) Roxb.   Fabaceae

LF: Evergreen shrub   CU: Erosion control, ornamental
SN: *Cassia alata* L., *Herpetica alata* (L.) Raf.

Geographic Distribution

Europe	*Australasia*	*Atlantic Islands*
Northern	● Australia	Cape Verde
British Isles	New Zealand	Canary, Madeira
Central, France		Azores
Southern	*Northern America*	South Atlantic Isl.
Eastern	Canada, Alaska	
Mediterranean Isl.	Southeastern USA	*Indian Ocean Islands*
	Western USA	Mascarenes
Africa	Remaining USA	Seychelles
Northern	X Mexico	Madagascar
X Tropical		
Southern	*Southern America*	*Pacific Islands*
	N Tropical	Micronesia
Asia	X Caribbean	Galapagos
Temperate	Chile, Argentina	X Hawaii
X Tropical		X Melanesia, Polynesia

Invaded Habitats
Forests and forest edges, humid ravines, riverbanks, wood- and grassland.

Description
A glabrous shrub or small tree of 2-5 m height, with few rather coarse branches. Leaves are 45-75 cm long, pinnately compound and have 7-14 pairs of leaflets. These are broadly oblong to obovate, 7-20 cm long and 3-12 cm wide. Flowers are borne in racemes of 15-60 cm length, and have bright yellow petals of 16-24 mm length, and yellow to orange calyx lobes. The flowerbuds are covered with orange bracts that fall off when the flowers open. Fruits are ascending, linear-oblong and winged pods of 11-19 cm length and 2-3 cm width. Seeds are tan or dark brown, flattened and 5.5-7 mm long.

Ecology and Control
This short-lived shrub grows best in sunny locations on moist soils and forms dense thickets where invasive. The large leaves shade out most native plants so that species richness is strongly reduced under canopies of this shrub. The plant is not nitrogen-fixing. Seedlings establish an extensive root system in the first year and compete for space and nutrients. Established plants sucker from roots if damaged. The plant is also a significant agricultural weed.

    Seedlings can be pulled or dug out, roots must be removed. Larger stands are slashed close to the ground and cut surfaces treated with 2,4-D plus picloram or triclopyr.

References
607, 608, 986.

Senna bicapsularis (L.) Roxb.   Fabaceae

LF: Deciduous shrub, tree   CU: Ornamental
SN: *Cassia bicapsularis* L., *Cassia emarginata* L.

Geographic Distribution

Europe		*Australasia*		*Atlantic Islands*	
	Northern		Australia	X	Cape Verde
	British Isles		New Zealand	X	Canary, Madeira
	Central, France				Azores
	Southern	*Northern America*			South Atlantic Isl.
	Eastern		Canada, Alaska		
	Mediterranean Isl.	N	Southeastern USA	*Indian Ocean Islands*	
			Western USA		Mascarenes
Africa			Remaining USA		Seychelles
	Northern		Mexico		Madagascar
X	Tropical				
●	Southern	*Southern America*		*Pacific Islands*	
		N	Tropical		Micronesia
Asia		N	Caribbean	X	Galapagos
	Temperate		Chile, Argentina	X	Hawaii
	Tropical				Melanesia, Polynesia

Invaded Habitats
Woodland, riparian habitats, coastal bush.

Description
A spreading, scrambling or climbing shrub or small tree of 1.2-3 m height. Leaves are compound with usually three pairs of leaflets. A prominent gland is present on the leaf rachis between the lowest pair of leaflets. The obovate-elliptic to orbicular leaflets are 1-3.5 cm long and 7-24 mm wide. Inflorescences are racemes with 3-15 yellow or orange flowers each, and with a well developed peduncle. Petals are obovate and 10-14 mm long. The fruit is a cylindrical, oblong-linear pod of 8-15 cm length and 1-1.5 cm width.

Ecology and Control
This shrub forms extensive and dense thickets and climbs over native vegetation, impeding growth and regeneration of native species. Extensive thickets affect wildlife by reducing habitats and restricting access to water.

    Specific control methods for this species are not available. The same methods as for other *Senna* species may apply.

References
549, 792.

Senna didymobotrya (Fresen.) Irwin & Barneby   Fabaceae

LF: Evergreen shrub, tree
SN: *Cassia didymobotrya* Fresen.
CU: Ornamental, shelter

Geographic Distribution

Europe		*Australasia*		*Atlantic Islands*	
	Northern	●	Australia		Cape Verde
	British Isles		New Zealand		Canary, Madeira
	Central, France				Azores
	Southern	*Northern America*			South Atlantic Isl.
	Eastern		Canada, Alaska		
	Mediterranean Isl.	●	Southeastern USA	*Indian Ocean Islands*	
		X	Western USA		Mascarenes
Africa			Remaining USA		Seychelles
	Northern	X	Mexico		Madagascar
N	Tropical				
●	Southern	*Southern America*		*Pacific Islands*	
		N	Tropical		Micronesia
Asia			Caribbean		Galapagos
	Temperate		Chile, Argentina	X	Hawaii
	Tropical				Melanesia, Polynesia

Invaded Habitats
Grass- and woodland, forests, riparian habitats, coastal scrub.

Description
A multi-stemmed shrub or a tree of 1-7.5 m height. Leaves are 12-35 cm long and pinnately compound with 7-15 pairs of leaflets. These are oblong to elliptic, 15-60 mm long and 6-20 mm wide. The petioles are 2-8 cm long. Inflorescences are erect and spike-like racemes of 10-40 cm length, with many bright yellow flowers of 17-27 mm length and with dark, persisting bracts. Fruits are oblong and flat pods of 7-10 cm length and 15-20 mm width, constricted between the seeds. Seeds are compressed and 6-8 mm long.

Ecology and Control
A shrub that grows commonly in riverine habitats and damp forest edges in the native range. Where invasive, it forms dense impenetrable thickets that impede the growth and regeneration of native plants and affect wildlife movement.

Specific control methods for this species are not available. The same methods as for other *Senna* species may apply.

References
549, 792, 1089.

Senna obtusifolia (L.) Irwin & Barneby — Fabaceae

LF: Annual, perennial herb CU: -
SN: *Cassia obtusifolia* L., *Cassia tora* L.

Geographic Distribution

	Europe		*Australasia*		*Atlantic Islands*
	Northern	●	Australia		Cape Verde
	British Isles		New Zealand		Canary, Madeira
	Central, France				Azores
	Southern		*Northern America*		South Atlantic Isl.
	Eastern		Canada, Alaska		
	Mediterranean Isl.	X	Southeastern USA		*Indian Ocean Islands*
			Western USA	X	Mascarenes
Africa		X	Remaining USA	X	Seychelles
	Northern	X	Mexico	X	Madagascar
N	Tropical				
	Southern		*Southern America*		*Pacific Islands*
		X	Tropical	X	Micronesia
Asia		N	Caribbean	●	Galapagos
X	Temperate	X	Chile, Argentina	X	Hawaii
X	Tropical			X	Melanesia, Polynesia

Invaded Habitats
Grassland, forest edges, riparian habitats, floodplains, disturbed sites.

Description
Usually a shrubby herb of 0.5-2 m height with a stout taproot and smooth, branched stems. The alternate leaves are pinnately compound, 8-12 cm long, and have three pairs of linear to obovate leaflets of 25-50 mm length and 15-25 mm width. Leaflets of the terminal pair are larger than those of the basal pair. Stipules are prominent and linear-lanceolate. Yellow or orange flowers are borne in pairs in the leaf axils. The corolla is 8-15 mm in diameter and has five spreading petals. Fruits are dry, dehiscent, somewhat curved pods of 15-25 cm length and 3-6 mm width, containing 25-30 seeds each. Seeds are yellowish brown to tan, shiny, rhomboidal, and 4-5 mm long.

Ecology and Control
This plant is a significant agricultural weed that also invades natural plant communities. It grows in dense thickets, competing for light, water and nutrients and displacing native vegetation. The plant is a prolific seed producer, and seeds may remain viable in the soil for several years. The species is not nitrogen-fixing due to a lack of nodulation.

Single plants can be grubbed before flowering commences. Larger colonies are slashed to reduce the plant's vigour. Effective herbicides include 2,4-D plus picloram, glyphosate, or dichlorprop. Seedlings are best sprayed before they set flowers.

References
66, 581, 607, 608, 986, 1313.

Senna pendula (Willd.) Irwin & Barneby   Fabaceae

LF: Evergreen shrub
SN: -
CU: Ornamental

Geographic Distribution

Europe	*Australasia*	*Atlantic Islands*
Northern	● Australia	Cape Verde
British Isles	New Zealand	X Canary, Madeira
Central, France		Azores
Southern	*Northern America*	South Atlantic Isl.
Eastern	Canada, Alaska	
Mediterranean Isl.	● Southeastern USA	*Indian Ocean Islands*
	Western USA	Mascarenes
Africa	Remaining USA	Seychelles
Northern	X Mexico	Madagascar
Tropical		
X Southern	*Southern America*	*Pacific Islands*
	N Tropical	Micronesia
Asia	X Caribbean	Galapagos
Temperate	X Chile, Argentina	X Hawaii
Tropical		Melanesia, Polynesia

Invaded Habitats
Woodland, riparian habitats, tropical hammocks, coastal beaches.

Description
A spreading shrub up to 3 m tall, with pinnately compound leaves of 4-8 cm length. There is one gland on the rachis between the two lowest pairs of pinnae. Leaves have 3-6 pairs of pinnae, with broad-oblanceolate to obovate leaflets of 2-5 cm length and 10-15 mm width. Petioles of leaves are 2-4 cm long. The yellow to yellow-green flowers are 3-4 cm in diameter and borne in axillary racemes. The peduncles are 3-4 cm long, the pedicels 20-25 mm. Fruits are straw-coloured cylindrical pods of 10-14 cm length and 8-12 mm diameter. Seeds are 4-6 mm long.

Ecology and Control
This is a variable species, belonging to a complex with numerous varieties. The exact native range is obscure. Plants naturalized in Florida and Australia are referred to var. *glabrata*. It is a fast growing plant often becoming established in openings of hammocks and climbing over the adjacent canopies, suppressing the growth of native species and displacing them. It is quite salt tolerant and grows well in sandy soils. Seed production is prolific and the long-lived seeds are dispersed by water and soil movement. The plant resprouts from the base after damage.

    Seedlings and smaller plants can be removed manually. Larger plants are cut and the cut stumps treated with herbicide. Foliar sprays are most effective on seedlings and on fresh regrowth.

References
549, 715, 924.

Sesbania punicea (Cav.) Benth.   Fabaceae

LF: Deciduous shrub   CU: Ornamental
SN: *Daubentonia punicea* (Cav.) DC.

Geographic Distribution

Europe		*Australasia*		*Atlantic Islands*	
	Northern		Australia		Cape Verde
	British Isles		New Zealand		Canary, Madeira
	Central, France				Azores
	Southern	*Northern America*			South Atlantic Isl.
	Eastern		Canada, Alaska		
	Mediterranean Isl.	X	Southeastern USA	*Indian Ocean Islands*	
			Western USA		Mascarenes
Africa		X	Remaining USA		Seychelles
	Northern		Mexico		Madagascar
	Tropical				
•	Southern	*Southern America*		*Pacific Islands*	
		N	Tropical		Micronesia
Asia			Caribbean		Galapagos
	Temperate	X	Chile, Argentina		Hawaii
	Tropical				Melanesia, Polynesia

Invaded Habitats
Riparian habitats, freshwater wetlands, grassland, disturbed sites.

Description
A deciduous shrub or small tree, up to 4 m tall, with slender branches and drooping leaves. These are 10-20 cm long, have petioles of 1-2 cm length, and are pinnately compound with 10-14 pairs of leaflets. Leaflets are oblong, 1-3 cm long and 4-8 mm wide, entire and ending in a pointed tip. Red to orange flowers of 2-3 cm length are borne in axillary clusters up to 25 cm length. Fruits are oblong and cylindrical pods of 6-8 cm length and *c.* 10 mm width, with short wings and a sharp point at the end. They contain 4-10 seeds. These are somewhat flattened, reddish brown and 6-8 mm long.

Ecology and Control
This fast growing and short-lived, nitrogen-fixing shrub forms dense and impenetrable thickets along riverbanks and in wetlands, excluding native species and affecting wildlife. Both leaves and seeds are poisonous to humans and animals. Dense stands impede the flow of water and may increase soil erosion as the soil under canopies of this shrub lacks a herbaceous cover. Seeds are dispersed by water and remain viable for several years.

    Seedlings and saplings can be pulled or dug out. Larger shrubs are cut and the cut stumps treated with herbicide.

References
137, 284, 478, 549, 572, 573, 574, 1020, 1089.

Silybum marianum (L.) Gaertn.   Asteraceae

LF: Annual, biennial herb   CU: -
SN: *Carduus marianus* L.

Geographic Distribution

	Europe		Australasia		Atlantic Islands
X	Northern	●	Australia		Cape Verde
X	British Isles	X	New Zealand	N	Canary, Madeira
X	Central, France			X	Azores
N	Southern		*Northern America*		South Atlantic Isl.
N	Eastern	X	Canada, Alaska		
N	Mediterranean Isl.		Southeastern USA		*Indian Ocean Islands*
		●	Western USA		Mascarenes
	Africa	X	Remaining USA		Seychelles
N	Northern		Mexico		Madagascar
X	Tropical				
X	Southern		*Southern America*		*Pacific Islands*
		X	Tropical		Micronesia
	Asia		Caribbean		Galapagos
N	Temperate	●	Chile, Argentina		Hawaii
	Tropical				Melanesia, Polynesia

Invaded Habitats
Grassland, coastal scrub and dunes, riparian habitats.

Description
A stout herb with glabrous to slightly pubescent stems of 20-250 cm height, and spiny-dentate leaves. Leaves are alternate, glabrous, often coarsely lobed, and dark green with white blotches. The basal leaves are 25-50 cm long and 12-25 cm wide, and have a winged petiole. Stem leaves are sessile, smaller and less deeply divided, with yellowish white spines up to 8 mm length. Flowerheads are globose, 2-6 cm in diameter and borne on long peduncles. Involucral bracts are tapering into spines of 2-5 cm length. Florets are pink to purple. Fruits are black achenes of 6-8 mm length and 2.5-4 mm width. The pappus is 15-20 mm long.

Ecology and Control
This plant occurs frequently in disturbed sites and is invasive because the large rosettes shade out native plants and form extensive patches that crowd out native vegetation and impede wildlife. The plant spreads quickly once established and may form an impenetrable cover over large areas. Seedling establishment depends on local disturbance. Seeds are dispersed by attaching to animals, by water and in mud. Seeds may remain viable for 10 years. Dead plants leave spots of bare soil that may become colonized by this plant or other weeds.

Mowing before fruit formation prevents seed dispersal. Chemical control is most effective at the seedling and rosette stages. An effective herbicide is 2,4-D ester. Planting desirable species suppresses seedling establishment of this plant.

References
133, 215, 581, 985, 986, 1183.

Solanum laxum Spreng.   Solanaceae

LF: Evergreen climber   CU: Ornamental
SN: *Solanum jasminoides* Paxton

Geographic Distribution

Europe
    Northern
    British Isles
    Central, France
    Southern
    Eastern
    Mediterranean Isl.

Africa
    Northern
    Tropical
    Southern

Asia
    Temperate
    Tropical

Australasia
    Australia
    ● New Zealand

Northern America
    Canada, Alaska
    Southeastern USA
    Western USA
    Remaining USA
    Mexico

Southern America
    N   Tropical
    Caribbean
    Chile, Argentina

Atlantic Islands
    Cape Verde
    Canary, Madeira
    Azores
    South Atlantic Isl.

Indian Ocean Islands
    Mascarenes
    Seychelles
    Madagascar

Pacific Islands
    Micronesia
    Galapagos
    Hawaii
    Melanesia, Polynesia

Invaded Habitats
Forest edges, scrubland, riparian habitats.

Description
A spineless high climbing liana with stems up to 15 m long. Petioles are slender, twining and 2-3 cm long. The leaf blades are 2-6 cm long and 7-30 mm wide, lanceolate to ovate, often with tufts of white hairs on the lower side. Flowers are borne several to many in loose panicles of 4-6 cm length at the ends of branches. The white to pale blue corolla is 2-3.3 cm in diameter. Fruits are globose shining, dark blue to black berries of 4-6 mm diameter. Seeds are broad-obovoid and *c.* 1 mm wide.

Ecology and Control
This climber smothers native shrubs and trees and prevents their regeneration. Little is known on the ecology of this species as an invasive plant. Seeds are dispersed by birds.

    Specific control methods for this species are not available. Stems may be cut and the cut stumps treated with herbicide. Seedlings can be hand pulled or dug out.

References
1415.

Solanum linnaeanum Hepper & Jaeger   Solanaceae

LF: Evergreen shrub   CU: -
SN: *Solanum hermannii* Dunal

Geographic Distribution

Europe	Australasia	Atlantic Islands
Northern	● Australia	Cape Verde
British Isles	X New Zealand	X Canary, Madeira
Central, France		X Azores
Southern	*Northern America*	South Atlantic Isl.
Eastern	Canada, Alaska	
Mediterranean Isl.	● Southeastern USA	*Indian Ocean Islands*
	Western USA	Mascarenes
Africa	Remaining USA	Seychelles
Northern	Mexico	Madagascar
Tropical		
N Southern	*Southern America*	*Pacific Islands*
	Tropical	Micronesia
Asia	Caribbean	Galapagos
Temperate	Chile, Argentina	● Hawaii
Tropical		Melanesia, Polynesia

Invaded Habitats
Grassland, coastal beaches and scrub.

Description
A strongly armed spreading shrub of 1-2 m height, with straight and yellow spines up to 12 mm long. Leaves are deeply dissected and usually spiny. The prickly petioles are up to 5 cm long, the blades ovate to oblong, 8-18 cm long and 7-13 cm wide. Flowers are purplish to almost white, 20-40 mm in diameter, and grow solitary or in small clusters in the axils of leaves. Fruits are globose yellow berries of *c.* 25 mm diameter, usually mottled green and white. Seeds are broad-obovoid and *c.* 3 mm wide. The plant has a stout woody taproot.

Ecology and Control
Native habitats of this plant include rocky slopes, flats, and disturbed places. Where invasive, it forms dense thickets that crowd out native plants by competing for space, water and nutrients. Wildlife movement is restricted and dense patches provide harbour for rabbits in Australia.

Seedlings and small plants may be hand pulled or dug out. Larger shrubs are grubbed and heaped for burning. Seedlings and young plants can be controlled with 2,4-D, older plants with amitrole, picloram or tebuthiuron.

References
215, 607, 986.

Solanum mauritianum Scop.   Solanaceae

LF: Evergreen shrub, tree   CU: Ornamental
SN: *Solanum auriculatum* Ait.

Geographic Distribution

Europe	*Australasia*	*Atlantic Islands*
Northern	● Australia	Cape Verde
British Isles	X New Zealand	X Canary, Madeira
Central, France		● Azores
X Southern	*Northern America*	X South Atlantic Isl.
Eastern	Canada, Alaska	
Mediterranean Isl.	Southeastern USA	*Indian Ocean Islands*
	Western USA	X Mascarenes
Africa	Remaining USA	Seychelles
Northern	Mexico	Madagascar
● Tropical		
● Southern	*Southern America*	*Pacific Islands*
	N Tropical	Micronesia
Asia	Caribbean	Galapagos
Temperate	Chile, Argentina	X Hawaii
Tropical		X Melanesia, Polynesia

Invaded Habitats
Grass- and heathland, riparian habitats, wet forests, coastal beaches.

Description
A thornless densely-tomentose shrub or small tree of 1.5-10 m height, with a soft wood and a dense, greyish pubescence. Leaves are elliptic, 20-30 cm long and 6-11 cm wide, long-acuminate, with a petiole of 3-5 cm length, dull green above and whitish tomentose below. Purple flowers of *c.* 20 mm diameter grow in dense, terminal clusters. Berries are globose, greenish-yellow when ripe, 10-15 mm in diameter, and densely covered with stellate hairs. Each berry contains numerous seeds of 1.5-2 mm length.

Ecology and Control
This shade tolerant shrub grows to dense thickets shading out all other vegetation. It primarily colonizes disturbed sites and its spread is promoted by fire. Seed germination is enhanced after fire, and fires often lead to mass germination, resulting in a dense cover of seedlings. Seeds are dispersed by birds. The shrub resprouts prolifically from the base and trunk if damaged.

    Burning is used to kill the plants but may enhance seed germination. Scattered seedlings can be pulled or dug out. Larger shrubs and trees can be cut and the cut stumps treated with herbicide. Follow-up programmes are necessary to treat regrowth and seedlings.

References
121, 207, 208, 209, 324, 549, 792, 956, 957, 958, 1089, 1129.

Solanum nigrum L. Solanaceae

LF: Annual, biennial herb CU: -
SN: -

Geographic Distribution

Europe		*Australasia*		*Atlantic Islands*	
N	Northern	●	Australia	N	Cape Verde
N	British Isles	X	New Zealand	N	Canary, Madeira
N	Central, France			N	Azores
N	Southern	*Northern America*		X	South Atlantic Isl.
N	Eastern	X	Canada, Alaska		
N	Mediterranean Isl.		Southeastern USA	*Indian Ocean Islands*	
		X	Western USA	X	Mascarenes
Africa		X	Remaining USA		Seychelles
N	Northern	X	Mexico	X	Madagascar
X	Tropical				
X	Southern	*Southern America*		*Pacific Islands*	
		X	Tropical		Micronesia
Asia		X	Caribbean		Galapagos
N	Temperate	X	Chile, Argentina	X	Hawaii
N	Tropical			X	Melanesia, Polynesia

Invaded Habitats
Forests, grassland, riparian habitats, coastal beaches.

Description
A spreading or erect herb of 30-100 cm height, with sparsely to densely pubescent and sparsely branched stems. Leaves are ovate-rhombic to lanceolate and have petioles of 2-5 cm length. Leaf blades are 4-10 cm long and 3-7 cm wide, entire or toothed. Inflorescences are solitary cymes on peduncles of 14-30 mm length and have 5-10 white flowers. Corollas are 5-15 mm in diameter, deeply lobed, with five lobes, anthers are yellow. Fruits are black, rarely green or yellow, globose berries of 6-10 mm diameter. Seeds are *c.* 2 mm long.

Ecology and Control
This plant grows commonly in hedges, forest margins, and disturbed places in the native range, and is found up to 3,000 m elevation in China. It is a highly variable complex with many taxa sometimes granted species status. North American plants are referred to subsp. *nigrum*. The plant grows over herbs and small shrubs, forms dense thickets and crowds out native plants. Seeds are dispersed by birds and small mammals.

Seedlings and small plants can be hand pulled. Mowing or cutting is used to prevent seed formation. Effective herbicides are atrazine, dicamba or 2,4-D.

References
82, 215, 455, 580, 654, 953, 954, 1094, 1384.

Solanum pseudocapsicum L.   Solanaceae

LF: Perennial herb, shrub   CU: -
SN: *Solanum capsicastrum* Link ex Schauer, *Solanum diflorum* Vell.

Geographic Distribution

Europe		*Australasia*		*Atlantic Islands*	
	Northern	●	Australia		Cape Verde
	British Isles	X	New Zealand	X	Canary, Madeira
	Central, France			X	Azores
	Southern	*Northern America*			South Atlantic Isl.
	Eastern		Canada, Alaska		
	Mediterranean Isl.		Southeastern USA	*Indian Ocean Islands*	
			Western USA		Mascarenes
Africa			Remaining USA		Seychelles
	Northern	N	Mexico		Madagascar
	Tropical				
X	Southern	*Southern America*		*Pacific Islands*	
		N	Tropical		Micronesia
Asia			Caribbean		Galapagos
	Temperate	X	Chile, Argentina	X	Hawaii
	Tropical				Melanesia, Polynesia

Invaded Habitats
Forests and forest edges, riparian habitats.

Description
A herbaceous to somewhat woody perennial of 50-150 cm height, unarmed and with an unpleasant odour. Leaves are oblong- to linear-lanceolate, 1.5-10 cm long and 5-20 mm wide, acuminate, and with blades gradually tapering into a winged petiole. Inflorescences are solitary and sessile cymes with 1-3 flowers. Corollas are white or mauve and 10-15 mm in diameter, the anthers are orange. Fruits are globose, red and shining berries of 10-15 mm width, containing seeds of 2-3 mm diameter.

Ecology and Control
Where native, this plant is common in grassland, thickets, and disturbed places. Where invasive, it becomes abundant and forms dense patches that crowd out native plants and prevent the regeneration of shrubs and trees. Little is known on the ecology of this plant.
     Specific control methods for this species are not available. The same methods as for other herbaceous or shrubby *Solanum* species may apply.

References
215, 607.

Solanum tampicense Dunal

Solanaceae

LF: Evergreen shrub
SN: -
CU: -

Geographic Distribution

Europe
- Northern
- British Isles
- Central, France
- Southern
- Eastern
- Mediterranean Isl.

Africa
- Northern
- Tropical
- Southern

Asia
- Temperate
- Tropical

Australasia
- Australia
- New Zealand

Northern America
- Canada, Alaska
- ● Southeastern USA
- Western USA
- Remaining USA
- N Mexico

Southern America
- N Tropical
- Caribbean
- Chile, Argentina

Atlantic Islands
- Cape Verde
- Canary, Madeira
- Azores
- South Atlantic Isl.

Indian Ocean Islands
- Mascarenes
- Seychelles
- Madagascar

Pacific Islands
- Micronesia
- Galapagos
- Hawaii
- Melanesia, Polynesia

Invaded Habitats
Forests, riparian habitats, freshwater wetlands, swamps.

Description
A sprawling and prickly shrub, woody below, with stems reaching 5 m length and 1.5 cm diameter. Prickles on the stems are white to tan, recurved and up to 5 mm long. Leaves are simple, alternate, with petioles up to 3 cm long and blades of 15-25 cm length and 4-7 cm width. Leaves have deeply sinuate margins and straight prickles on the veins. Small white flowers are borne in axillary clusters of 3-11. Fruits are globose, orange to bright red berries of *c.* 1 cm diameter, each containing 10-60 yellowish and flat seeds.

Ecology and Control
The often interlocking stems of this shrub lead to dense thickets that may cover large areas and smother all native vegetation. The plant can become the dominant understorey species of cypress swamps. The plant grows both in full shade and full sun. Stems can form adventitious roots at the nodes, and the plant regrows from the crown if damaged. In sunny conditions, the plant produces fruits abundantly. Seeds as well as stem fragments are dispersed by streams.

Specific control methods for this species are not available. The same methods as for other *Solanum* species may apply.

References
422, 715, 1438.

Solanum viarum Dunal — Solanaceae

LF: Perennial herb
SN: -
CU: -

Geographic Distribution

	Europe		Australasia		Atlantic Islands
	Northern		Australia		Cape Verde
	British Isles		New Zealand		Canary, Madeira
	Central, France				Azores
	Southern		*Northern America*		South Atlantic Isl.
	Eastern		Canada, Alaska		
	Mediterranean Isl.	●	Southeastern USA		*Indian Ocean Islands*
			Western USA		Mascarenes
	Africa	X	Remaining USA		Seychelles
	Northern	X	Mexico		Madagascar
X	Tropical				
	Southern		*Southern America*		*Pacific Islands*
		N	Tropical		Micronesia
	Asia	X	Caribbean		Galapagos
X	Temperate		Chile, Argentina		Hawaii
X	Tropical				Melanesia, Polynesia

Invaded Habitats
Grassland, forests and hammocks, riparian habitats.

Description
An erect herb of 50-150 cm height, with densely pubescent stems and branches and straight prickles of 2-5 mm length. Leaf blades are broadly ovate, 6-20 cm long and 6-15 cm wide, with prickles and hairs on both sides. Petioles are stout, 3-7 cm long, and have prickles. Inflorescences are axillary racemes with 1-5 flowers, peduncles are short or absent. Corollas are white or green and pubescent. Fruits are globose, pale yellow berries of 2-3 cm diameter, containing brown, lens-shaped seeds of 2-3 mm diameter.

Ecology and Control
Where native, this fast growing plant grows commonly in grassland, thickets, and disturbed places. Where invasive, it forms dense stands that outcompete native species. The plant can regenerate new shoots from its extensive root system and can quickly build up large populations. Seed production is prolific, and a single individual may produce up to 50,000 seeds. These are dispersed by birds and mammals.

Eradication is difficult due to the extensive root system. Seedlings and small plants may be pulled or dug out, roots must be removed to prevent regrowth. Larger patches may be controlled by herbicides.

References
9, 715, 910, 911, 912, 913, 914, 992, 1344, 1438.

Solidago canadensis L.   Asteraceae

LF: Perennial herb   CU: Ornamental, bee plant
SN: -

Geographic Distribution

Europe
- X Northern
- ● British Isles
- ● Central, France
- ● Southern
- ● Eastern
-    Mediterranean Isl.

Africa
-    Northern
-    Tropical
-    Southern

Asia
- ● Temperate
-    Tropical

Australasia
- X Australia
- X New Zealand

Northern America
- N Canada, Alaska
- N Southeastern USA
- X Western USA
- N Remaining USA
-    Mexico

Southern America
-    Tropical
-    Caribbean
-    Chile, Argentina

Atlantic Islands
-    Cape Verde
-    Canary, Madeira
-    Azores
-    South Atlantic Isl.

Indian Ocean Islands
-    Mascarenes
-    Seychelles
-    Madagascar

Pacific Islands
-    Micronesia
-    Galapagos
- X Hawaii
-    Melanesia, Polynesia

Invaded Habitats
Grassland, forest edges, riparian habitats, disturbed sites.

Description
A tall herb of 50-200 cm height, with erect, sparsely to densely pubescent stems that are branched only in the inflorescence. Leaves are lanceolate, slightly to sharply toothed, 5-20 cm long, 15-40 mm wide, and pubescent beneath. Inflorescences are large panicles at the ends of stems with recurved-secund branches. Flowerheads are 3-5 mm in diameter and have short ray florets and numerous disk florets. Florets are yellow. Fruits are achenes of 0.5-1.2 mm length with a white pappus of *c.* 3 mm length. Rhizomes arise from the base and are 5-12 cm long.

Ecology and Control
This plant belongs to a highly variable species complex. Plants in Europe are referred to var. *scabra* (*Solidago altissima* L.) It forms extensive colonies with a high shoot density and covering large areas, eliminating almost all other species. Individual clones expand rapidly by vegetative lateral growth. The annual stems die in autumn and new shoots arise from the rhizomes in spring. Once established, the plant remains dominant for a long period of time. Seeds are abundantly produced and dispersed by wind.

    Control measures include repeated mowing, or mulching and reseeding of native species. Single clumps may be grubbed, rhizomes must be removed. Effective herbicides are 2,4-D or picloram.

References
2, 372, 469, 540, 631, 860, 868, 869, 870, 1140, 1236, 1321, 1351, 1370, 1372, 1373, 1388.

Solidago gigantea Ait. Asteraceae

LF: Perennial herb
SN: *Solidago serotina* Ait.
CU: Ornamental, bee plant

Geographic Distribution

Europe		*Australasia*		*Atlantic Islands*	
X	Northern		Australia		Cape Verde
X	British Isles		New Zealand		Canary, Madeira
●	Central, France			X	Azores
X	Southern	*Northern America*			South Atlantic Isl.
●	Eastern	N	Canada, Alaska		
	Mediterranean Isl.	N	Southeastern USA	*Indian Ocean Islands*	
		N	Western USA		Mascarenes
Africa		N	Remaining USA		Seychelles
	Northern		Mexico		Madagascar
	Tropical				
	Southern	*Southern America*		*Pacific Islands*	
			Tropical		Micronesia
Asia			Caribbean		Galapagos
X	Temperate		Chile, Argentina		Hawaii
	Tropical				Melanesia, Polynesia

Invaded Habitats
Grassland, forest edges, riparian habitats, wet meadows, disturbed sites.

Description
An erect herb of 50-150 cm height, with glabrous and often glaucous stems, branched only in the inflorescence. Leaves are glabrous, variably toothed, 10-20 cm long and 1.5-4 cm wide. Inflorescences are large panicles with pubescent branches at the ends of stems. Flowerheads are 4-8 mm in diameter, with the ray florets being slightly longer than the disk florets. Florets are yellow. Fruits are achenes of 1-1.3 mm length, with a yellowish-brown pappus of 3-4 mm length. Rhizomes are branched and may exceed 50 cm in length.

Ecology and Control
This perennial forms extensive and dense, pure populations where invasive, displacing native vegetation and preventing establishment of native species. Shoot densities may exceed 300 shoots m^{-2}. Once established, the species spreads mainly by rhizomes which easily break off and are carried by streams to new sites. Individual clones rapidly expand due to vegetative growth and merge into a complete cover.

Once established the plant is difficult to control. Scattered stems may be pulled or dug out, all rhizomes must be removed. Repeated mowing reduces the plant's vigour. Stands can be treated with 2,4-D or glyphosate applied during active growth.

References
136, 516, 540, 1371, 1372.

Sonchus oleraceus L.   Asteraceae

LF: Annual herb
SN: -
CU: -

Geographic Distribution

Europe
- N Northern
- N British Isles
- N Central, France
- N Southern
- N Eastern
- N Mediterranean Isl.

Africa
- N Northern
- X Tropical
- X Southern

Asia
- N Temperate
- N Tropical

Australasia
- ● Australia
- X New Zealand

Northern America
- X Canada, Alaska
- X Southeastern USA
- X Western USA
-   Remaining USA
- X Mexico

Southern America
- X Tropical
- X Caribbean
- X Chile, Argentina

Atlantic Islands
- X Cape Verde
- N Canary, Madeira
- N Azores
- X South Atlantic Isl.

Indian Ocean Islands
- X Mascarenes
-   Seychelles
-   Madagascar

Pacific Islands
-   Micronesia
- X Galapagos
- X Hawaii
-   Melanesia, Polynesia

Invaded Habitats
Forests, grassland, riparian habitats, freshwater wetlands, coastal estuaries and dunes.

Description
An erect herb of 10-140 cm height, with simple or branched stems, sometimes white-tomentose in the upper part. Leaves are glabrous, the lower ones have narrowly winged petioles, the upper ones are larger and lobed. Flowerheads are numerous, pale yellow and 2-3 cm wide. Ligules are about as long as the corolla-tube. Fruits are oblanceolate and weakly compressed achenes of 2.5-4 mm length and *c.* 1 mm width. The pappus is 5-8 mm long and persistent.

Ecology and Control
This widespread weed is invasive in natural habitats because it grows in dense patches that crowd out native plants. It is a shade intolerant pioneer species establishing in disturbed sites. The plant's large size and its high nutrient uptake may result in impoverishment of heavily infested sites. Seeds are dispersed by wind and water.

Since plants do not regenerate from root fragments, physical control by cutting or mowing may control the plant. Effective pre-emergence herbicides are simazine or atrazine. Seedlings and rosettes are sprayed with MCPA or 2,4-D.

References
215, 580, 614, 741.

Sparaxis bulbifera (L.) Ker Gawl.   Iridaceae

LF: Perennial herb
CU: Ornamental
SN: *Ixia bulbifera* L., *Ixia alba* (L.) Ker.

Geographic Distribution

Europe	*Australasia*	*Atlantic Islands*
Northern	● Australia	Cape Verde
British Isles	New Zealand	Canary, Madeira
Central, France		X Azores
Southern	*Northern America*	South Atlantic Isl.
Eastern	Canada, Alaska	
Mediterranean Isl.	Southeastern USA	*Indian Ocean Islands*
	X Western USA	Mascarenes
Africa	Remaining USA	Seychelles
Northern	Mexico	Madagascar
Tropical		
N Southern	*Southern America*	*Pacific Islands*
	Tropical	Micronesia
Asia	Caribbean	Galapagos
Temperate	Chile, Argentina	Hawaii
Tropical		Melanesia, Polynesia

Invaded Habitats
Grass- and heathland, woodland, riparian habitats.

Description
A cormous herb of 15-45 cm height, with papery to fibrous corms and branched stems that produce axillary bulbils of *c.* 5 mm diameter after flowering. Leaves are linear and grass-like, 10-30 cm long and 3-10 mm wide. Flowers are almost symmetrical, 2-3 cm long and have a white to cream perianth that is often purplish on the lower side. Flowering stems are usually branched.

Ecology and Control
Native habitats of this plant include wet, sandy or clay flats near the coast. Where invasive, the plant forms extensive populations, especially after disturbance, and impedes the growth and regeneration of native plants. The species reproduces by seeds, corms and bulbils. Seeds and bulbils are dispersed by water, wind and mud.

Plants can be dug out easily, but all corms must be removed to prevent recolonization. Cutting is done before leaf bulbils or seeds are formed, otherwise spread is enhanced. Stands may also be treated with herbicide.

References
121, 215, 924.

Spartina alterniflora Loisel.   Poaceae

LF: Perennial herb   CU: Erosion control
SN: -

Geographic Distribution

Europe		*Australasia*		*Atlantic Islands*	
	Northern		Australia		Cape Verde
X	British Isles	●	New Zealand		Canary, Madeira
X	Central, France				Azores
X	Southern	*Northern America*			South Atlantic Isl.
	Eastern	N	Canada, Alaska		
	Mediterranean Isl.	N	Southeastern USA	*Indian Ocean Islands*	
		●	Western USA		Mascarenes
Africa		N	Remaining USA		Seychelles
	Northern		Mexico		Madagascar
	Tropical				
	Southern	*Southern America*		*Pacific Islands*	
			Tropical		Micronesia
Asia			Caribbean		Galapagos
	Temperate		Chile, Argentina		Hawaii
	Tropical				Melanesia, Polynesia

Invaded Habitats
Coastal estuaries, salt marshes, mudflats.

Description
A tall grass with fleshy rhizomes and culms of 40-120 cm height and 10 mm or more in diameter. Leaf blades are flat and usually glabrous, 10-40 cm long and 5-10 mm wide. Ligules are 1-2 mm long and hairy. The inflorescence consists of a number of spikes of 5-15 cm length, arranged in a panicle of 10-25 cm length and 3-5 cm width. The sessile spikelets have 1-2 florets each and are 10-15 mm long, they fall entirely at maturity. The lower glume is linear, the upper glume ovate-lanceolate.

Ecology and Control
Where native, this grass is found in brackish waters of coastal lagoons and marshes. It is invasive because it forms dense pure stands that displace native vegetation, thus reducing native plant and invertebrate species richness. Once established, individual clumps increase annually by peripheral rhizome growth. The grass transforms open intertidal habitats into monospecific stands of tall growth, reducing shorebird feeding areas. The dense growth traps sediments and alters the hydrology of invaded areas.

     Small plants can be hand pulled, all rhizomes and roots must be removed to prevent regrowth. Solarization by covering infestations with black plastic sheeting can be used for smaller infestations. Repeated mowing close to the ground may be effective. Chemical control is done by spraying glyphosate.

References
106, 133, 193, 294, 295, 296, 297, 758, 814, 817, 1038, 1221, 1415.

Spartina anglica Hubb.   Poaceae

LF: Perennial herb
SN: -
CU: -

Geographic Distribution

Europe	*Australasia*	*Atlantic Islands*
Northern	• Australia	Cape Verde
N British Isles	• New Zealand	Canary, Madeira
X Central, France		Azores
Southern	*Northern America*	South Atlantic Isl.
Eastern	Canada, Alaska	
Mediterranean Isl.	Southeastern USA	*Indian Ocean Islands*
	• Western USA	Mascarenes
Africa	Remaining USA	Seychelles
Northern	Mexico	Madagascar
Tropical		
Southern	*Southern America*	*Pacific Islands*
	Tropical	Micronesia
Asia	Caribbean	Galapagos
X Temperate	Chile, Argentina	Hawaii
Tropical		Melanesia, Polynesia

Invaded Habitats
Coastal estuaries, salt marshes, mudflats, mangrove swamps.

Description
A deep rooted grass of 30-130 cm height with stout rhizomes. The yellowish-green leaves are 10-45 cm long and 6-15 mm wide. Ligules consist of silky hairs of 2-3 mm length. The erect and stiff inflorescence is a panicle, 12-35 cm long and consists of 2-8 or more spikes of 7-23 cm length. Spikelets are closely overlapping, 17-25 mm long and have soft-pubescent, awnless glumes.

Ecology and Control
This grass is a native of brackish waters of coastal lagoons, salt marshes and tidal mudflats. As the previous species, it forms extensive and dense pure meadows that crowd out native vegetation. It completely transforms intertidal mudflats and beaches into pure stands, eliminating habitats for fish, birds and invertebrates. By accumulating sediments, estuaries are widened and the water flow is changed. Seeds are dispersed by water and in mud.

Seedlings and scattered plants can be dug out but all rhizomes must be removed. Small infestations can be killed by solarization: plants are slashed and covered with black plastic sheeting for 4-6 months. Stands can also be treated with glyphosate.

References
121, 133, 215, 295, 352, 475, 598, 729, 924, 1289, 1415.

Spartina maritima (Curtis) Fernald   Poaceae

LF: Perennial herb   CU: -
SN: *Dactylis maritima* Curtis

Geographic Distribution

Europe	*Australasia*	*Atlantic Islands*
Northern	● Australia	Cape Verde
N British Isles	New Zealand	Canary, Madeira
N Central, France		N Azores
N Southern	*Northern America*	South Atlantic Isl.
Eastern	Canada, Alaska	
Mediterranean Isl.	Southeastern USA	*Indian Ocean Islands*
	Western USA	Mascarenes
Africa	Remaining USA	Seychelles
N Northern	Mexico	Madagascar
Tropical		
X Southern	*Southern America*	*Pacific Islands*
	Tropical	Micronesia
Asia	Caribbean	Galapagos
Temperate	Chile, Argentina	Hawaii
Tropical		Melanesia, Polynesia

Invaded Habitats
Coastal estuaries, salt marshes, intertidal mudflats.

Description
A rather robust stiff grass with stems of 15-80 cm height and green, often purplish-tinged leaves and rhizomes of 5-20 cm length. Leaves are 2-18 cm long and 3.5-5 mm wide. Ligules are short and consist of hairs of 0.2-0.6 mm length. The inflorescence is a 4-10 cm long panicle and consists of 2-5 spikes of 3-8 cm length. Spikelets are 10-15 mm long. Glumes are linear-lanceolate and densely hairy, awnless. Lemmas are narrowly oblong-lanceolate.

Ecology and Control
As the previous species of the genus, this grass grows in dense tufts of very rigid shoots, forming a complete cover over large areas, eliminating native plants and wildlife habitats. The plant spreads by vegetative fragments that are carried by streams and in mud. It builds extensive meadows that reduce wildlife habitats and impede feeding birds. Once established, the grass expands laterally by rhizome growth.

    Specific control methods for this species are not available. The same methods as for other *Spartina* species may apply.

References
3, 214, 220, 815, 1021, 1043, 1130, 1131.

Spathodea campanulata Beauv.

Bignoniaceae

LF: Evergreen tree
SN: *Spathodea nilotica* Seem.
CU: Ornamental

Geographic Distribution

Europe	*Australasia*	*Atlantic Islands*
Northern	Australia	Cape Verde
British Isles	New Zealand	Canary, Madeira
Central, France		Azores
Southern	*Northern America*	South Atlantic Isl.
Eastern	Canada, Alaska	
Mediterranean Isl.	Southeastern USA	*Indian Ocean Islands*
	Western USA	Mascarenes
Africa	Remaining USA	X Seychelles
Northern	Mexico	Madagascar
N Tropical		
Southern	*Southern America*	*Pacific Islands*
	Tropical	Micronesia
Asia	Caribbean	Galapagos
Temperate	Chile, Argentina	● Hawaii
Tropical		Melanesia, Polynesia

Invaded Habitats
Forests and forest edges.

Description
A tree of 7-25 m height with a pale grey bark and opposite leaves. These are up to 50 cm long, pinnately compound with 3-19 elliptic leaflets of 7-13 cm length and 4-7 cm width. There are pseudostipules of *c.* 3 cm length at the leaf bases. Flowers are borne in terminal racemes, are orange red, cup-shaped, 8-10 cm long and 4.5-5 cm in diameter. Fruits are hard blackish brown capsules of 17-25 cm length and 3.5-7 cm diameter, containing winged seeds.

Ecology and Control
A tree that spreads rapidly in mesic to wet areas and forms dense thickets. Native plants are eliminated by the shading effect of the large leaves, resulting in reduced species richness under the canopies of this tree. Little is known on the ecology of this species.
    Seedlings and saplings can be hand pulled or dug out. Larger trees are cut and the cut stumps treated with herbicide.

References
752, 1092.

Sphagneticola trilobata (L.) Pruski   Asteraceae

LF: Perennial herb   CU: Ornamental
SN: *Wedelia trilobata* (L.) Hitchc., *Thelechitonia trilobata* (L.) Rob. & Cuatrec.

Geographic Distribution

Europe	*Australasia*	*Atlantic Islands*
Northern	Australia	Cape Verde
British Isles	New Zealand	Canary, Madeira
Central, France		Azores
Southern	*Northern America*	South Atlantic Isl.
Eastern	Canada, Alaska	
Mediterranean Isl.	Southeastern USA	*Indian Ocean Islands*
	Western USA	Mascarenes
Africa	Remaining USA	Seychelles
Northern	N Mexico	Madagascar
Tropical		
● Southern	*Southern America*	*Pacific Islands*
	N Tropical	Micronesia
Asia	N Caribbean	Galapagos
Temperate	Chile, Argentina	● Hawaii
X Tropical		● Melanesia, Polynesia

Invaded Habitats
Pinelands, riparian habitats, wetland edges, coastal dunes, disturbed sites.

Description
A glabrous to hairy creeping herb with decumbent and often branching stems of 10-40 cm length, rooting at the lower nodes. Leaves are somewhat fleshy, opposite, sessile or with short petioles, 4-10 cm long and 2-5 cm wide, three to five-lobed or irregularly dentate. The central lobe is broad and the lateral lobes are narrower. Flowerheads of 2-4 cm diameter are borne solitary on axillary peduncles of 3-11 cm length. Flowerheads have *c.* 10 yellow ray florets and numerous yellow disk florets. The involucral scales are elliptic-oblong and 8-12 mm long. Ligules of ray florets are *c.* 10 mm long. Fruits are achenes of 3.5-5 mm length, ovoid, with a pappus consisting of irregular and short scales.

Ecology and Control
This creeping and mat-forming herb grows in dense patches suppressing native plants and preventing their regeneration. It establishes especially well in disturbed sites and prevents the establishment of native plants. Seeds are dispersed by wind and water. Little is known on the ecology of this plant.

    Specific control methods for this species are not available. Seedlings and small plants may be hand pulled or dug out, larger stands treated with herbicides.

References
549, 1132.

Sporobolus indicus (L.) R.Br.   Poaceae

LF: Perennial herb CU: -
SN: *Agrostis indica* L., *Sporobolus africanus* (Poir.) Rob. & Tournay

Geographic Distribution

Europe
    Northern
    British Isles
    Central, France
    Southern
    Eastern
    Mediterranean Isl.

Africa
    Northern
    Tropical
N   Southern

Asia
X   Temperate
X   Tropical

Australasia
●   Australia
●   New Zealand

Northern America
    Canada, Alaska
X   Southeastern USA
X   Western USA
    Remaining USA
X   Mexico

Southern America
X   Tropical
X   Caribbean
X   Chile, Argentina

Atlantic Islands
    Cape Verde
    Canary, Madeira
●   Azores
    South Atlantic Isl.

Indian Ocean Islands
    Mascarenes
    Seychelles
    Madagascar

Pacific Islands
    Micronesia
    Galapagos
●   Hawaii
X   Melanesia, Polynesia

Invaded Habitats
Grass- and heathland, pine forests, riparian habitats, coastal scrub and dunes.

Description
A tufted grass with one to several dark green stems of 30-100 cm height and with numerous leafy shoots at the base. Stem leaves are 20-30 cm long and 3-5 mm wide; ligules are short and hairy. Inflorescences are stiff and linear-cylindrical, dark greyish green panicles of 10-80 cm length. Spikelets are green to brownish, 1.5-2.5 mm long, and have one floret. Glumes are unequal and up to 1 mm long, lemmas are ovate, acute and 2-2.5 mm long. Fruits are caryopses of *c.* 1 mm length.

Ecology and Control
This grass is a highly variable species with several varieties within the native range. Plants naturalized in Hawaii are referred to as *S. indicus* var. *capensis*. It reproduces from seeds. Where invasive, the grass forms a dense cover eliminating native vegetation and preventing regeneration of shrubs and trees. The grass produces large amounts of dead biomass promoting the occurrence of wildfires. Seed production is prolific and seeds are mainly dispersed by attaching to animals. A large seed bank accumulates in the soil.

    Small individuals can be removed manually. Chemical control is done by spraying with hexazinone, 2,2-DPA, or amitrole plus atrazine. Spraying with flupropanate before seed heads appear gives good selective control.

References
215, 883, 938, 986.

Stachytarpheta cayennensis (Rich.) Vahl   Verbenaceae

LF: Perennial herb   CU: -
SN: *Stachytarpheta urticifolia* Sims, *Verbena cayennensis* Rich.

Geographic Distribution

Europe	*Australasia*	*Atlantic Islands*
Northern	Australia	Cape Verde
British Isles	New Zealand	Canary, Madeira
Central, France		Azores
Southern	*Northern America*	South Atlantic Isl.
Eastern	Canada, Alaska	
Mediterranean Isl.	X Southeastern USA	*Indian Ocean Islands*
	Western USA	Mascarenes
Africa	Remaining USA	X Seychelles
Northern	N Mexico	Madagascar
● Tropical		
Southern	*Southern America*	*Pacific Islands*
	N Tropical	Micronesia
Asia	N Caribbean	Galapagos
Temperate	Chile, Argentina	X Hawaii
N Tropical		Melanesia, Polynesia

Invaded Habitats
Forests and forest edges, moist gullies, disturbed sites.

Description
A large herb of 80-200 cm height, sometimes woody at the base, and with branched and sparsely pubescent stems. Leaves are opposite, membranous, elliptic to broadly elliptic or ovate, 4-8 cm long and 2-4.5 cm wide, with the margins sharply and coarsely toothed. Petioles are 5-20 mm long. Flowers are borne in slender spikes of 14-40 cm length and *c.* 3 mm width. They are dark purplish blue with a pale centre and have tubes of 7-8 mm length.

Ecology and Control
The exact native range of this species is obscure. It is invasive because it grows rapidly and forms dense thickets, shading out native plants and preventing establishment of shrubs and trees. In Hawaii, the plant hybridizes with other species of the same genus. Little is known on the ecology of this species.

    Single plants and seedlings can be pulled or dug out. Larger plants are cut and treated with herbicide. Larger patches are best sprayed as slashing does not give effective control. Herbicides recommended include 2,4-D amine, 2,4-D ester, or glyphosate.

References
986, 1168, 1273.

Stenotaphrum secundatum (Walter) Kuntze Poaceae

LF: Perennial herb
SN: *Ischaemum secundatum* Walter
CU: Erosion control, ornamental

Geographic Distribution

Europe		*Australasia*		*Atlantic Islands*	
	Northern	•	Australia		Cape Verde
	British Isles		New Zealand	X	Canary, Madeira
	Central, France			•	Azores
•	Southern		*Northern America*		South Atlantic Isl.
	Eastern		Canada, Alaska		
	Mediterranean Isl.	X	Southeastern USA		*Indian Ocean Islands*
		X	Western USA		Mascarenes
Africa			Remaining USA		Seychelles
	Northern	N	Mexico		Madagascar
N	Tropical				
N	Southern		*Southern America*		*Pacific Islands*
		N	Tropical		Micronesia
Asia		N	Caribbean		Galapagos
	Temperate		Chile, Argentina	X	Hawaii
	Tropical			X	Melanesia, Polynesia

Invaded Habitats
Wetland forests, coastal grass- and heathland, coastal dunes, disturbed sites.

Description
A creeping much-branched grass with prostrate vegetative stems and decumbent flowering stems of 10-40 cm height, and a branched rhizome system. Leaves are broadly linear, 5-20 cm long and 5-15 mm wide. Sheaths are glabrous and 1.2-10 cm long. Ligules consist of fringes of short hairs. The inflorescence is 3-15 cm long and consists of very short racemes embedded in a thickened central axis. Racemes are 5-10 mm long, bearing 1-3 spikelets with two florets each. Spikelets are 4-5 mm long, pale green and narrowly ovate. The lower glume is 1-2 mm long, the upper one 4-5 mm. Fruits are caryopses of *c.* 2 mm length.

Ecology and Control
Where native, this grass is found in moist swampy soils, mostly near the coast. The plant forms long stolons and stems root easily at the lower nodes. It is frost tolerant and withstands salt spray. It spreads quickly by means of vegetative growth, forming dense mats that eliminate all other species. Stems can climb 1-2 m tall and smother small shrubs. Growth and regeneration of native shrubs and trees is impeded in areas invaded by this grass. It usually does not set seeds where introduced.

Trailing stems at the edge of infestations can be pulled out. Small patches are dug out, all rhizomes and creeping stems must be removed. An effective control is solarization by covering infestations with plastic sheeting for 4-12 weeks. Herbicides used to control this grass include glyphosate, dalapon or 2,2-DPA.

References
187, 188, 215, 1126, 1190.

Swietenia macrophylla King Meliaceae

LF: Deciduous tree
SN: *Swietenia candollei* Pittier
CU: Shelter, wood

Geographic Distribution

Europe	*Australasia*	*Atlantic Islands*
Northern	Australia	Cape Verde
British Isles	New Zealand	Canary, Madeira
Central, France		Azores
Southern	*Northern America*	South Atlantic Isl.
Eastern	Canada, Alaska	
Mediterranean Isl.	Southeastern USA	*Indian Ocean Islands*
	Western USA	X Mascarenes
Africa	Remaining USA	Seychelles
Northern	N Mexico	Madagascar
Tropical		
Southern	*Southern America*	*Pacific Islands*
	N Tropical	Micronesia
Asia	Caribbean	X Galapagos
Temperate	Chile, Argentina	Hawaii
● Tropical		Melanesia, Polynesia

Invaded Habitats
Forests and forest edges.

Description
A tree reaching 30 m height with a dark grey, ridged and fissured bark, a dense crown and a dark green foliage. The greyish brown twigs are covered with numerous pale brown and thick scales. The glabrous leaves are 20-50 cm long, pinnately compound, and have 3-6 pairs of leaflets. These are 10-18 cm long and 4-7.5 cm wide, elliptical, asymmetrical at the base, entire and dark glossy green. The pale greenish yellow flowers are *c.* 8 mm in diameter and borne in panicles. Fruits are large, cylindrical and woody capsules of *c.* 15 cm length and 6.5 cm width, greyish brown and opening with five valves. Seeds are numerous, winged and 5-9 cm long.

Ecology and Control
A fast growing and shade tolerant tree that withstands pronounced periods of dry weather. It establishes well in disturbed sites and in secondary forests, becoming the dominant species and will suppress native plants. Seed production is prolific and seeds are dispersed by wind. The tree has some ability to sprout after cutting.

Specific control methods for this species are not available. Seedlings and saplings may be hand pulled or dug out. Larger trees are cut and the cut stumps treated with herbicide.

References
24, 69, 284, 333, 445, 509, 1123.

Syzygium cumini (L.) Skeels   Myrtaceae

LF: Evergreen tree CU: Ornamental, food
SN: *Eugenia cumini* (L.) Druce, *Eugenia jambolana* Lam., *Syzygium jambolanum* DC.

Geographic Distribution

Europe		*Australasia*		*Atlantic Islands*	
	Northern	X	Australia		Cape Verde
	British Isles		New Zealand		Canary, Madeira
	Central, France				Azores
	Southern	*Northern America*			South Atlantic Isl.
	Eastern		Canada, Alaska		
	Mediterranean Isl.	●	Southeastern USA	*Indian Ocean Islands*	
			Western USA		Mascarenes
Africa			Remaining USA		Seychelles
	Northern		Mexico		Madagascar
X	Tropical				
●	Southern	*Southern America*		*Pacific Islands*	
		X	Tropical		Micronesia
Asia			Caribbean		Galapagos
N	Temperate		Chile, Argentina	●	Hawaii
N	Tropical			X	Melanesia, Polynesia

Invaded Habitats
Tropical hammocks and forests, wet pinelands, coastal bush.

Description
A small to medium-sized tree reaching 25 m height, with a greyish white and rather scaly bark. Leaves are broadly oblong to ovate and have long petioles of 1-3 cm length. Leaf blades are 5-15 cm long and 2.5-9 cm wide. White to pink flowers of *c.* 13 mm diameter are borne in branched cymes in the axils of leaves. The petals are fused into a cup. Fruits are deep purplish-red to black and ovoid berries of 15-25 mm length and 12-15 mm width, each containing usually one ellipsoid seed of 10-15 mm length.

Ecology and Control
This rapidly spreading tree forms dense canopies that shade out young native trees and prevent their regeneration. The tree grows in both wet and well drained soils, and withstands prolonged flooding but not highly saline soils. In the native range, it is common in forests, especially along margins of streams and ponds. Seeds are dispersed by birds and mammals. The tree resprouts vigorously after damage.

Seedlings and saplings can be removed manually. Larger trees are cut and the cut stumps treated with herbicide to prevent regrowth.

References
549, 715, 988.

Syzygium jambos (L.) Alston — Myrtaceae

LF: Evergreen tree
SN: *Eugenia jambos* L.

CU: Food

Geographic Distribution

Europe	*Australasia*	*Atlantic Islands*
Northern	Australia	Cape Verde
British Isles	New Zealand	Canary, Madeira
Central, France		Azores
Southern	*Northern America*	South Atlantic Isl.
Eastern	Canada, Alaska	
Mediterranean Isl.	X Southeastern USA	*Indian Ocean Islands*
	Western USA	● Mascarenes
Africa	Remaining USA	● Seychelles
Northern	X Mexico	Madagascar
Tropical		
X Southern	*Southern America*	*Pacific Islands*
	X Tropical	Micronesia
Asia	● Caribbean	● Galapagos
Temperate	Chile, Argentina	● Hawaii
N Tropical		Melanesia, Polynesia

Invaded Habitats
Forests and forest edges, riparian habitats, coastal bush.

Description
A tree of 9-12 m height, with several stems arising from a single base and with a dense foliage. Leaves are oblong-lanceolate, shiny green above and paler beneath, 10-20 cm long and 2.5-5 cm wide, entire, and acuminate. Petioles are *c.* 10 mm long. The white-greenish flowers are up to 8 cm in diameter, have four petals and numerous conspicuous stamens up to 5 cm long. Flowers are borne in terminal cymes. Fruits are globose and yellowish berries of 2.5-4 cm diameter, containing 1-2 seeds of 2-2.5 cm diameter.

Ecology and Control
This tree grows on mountain slopes and in riparian forests where native. The exact native range is unknown but it probably originates in Malesia. It is invasive because it forms dense impenetrable thickets that expand rapidly. The dense canopies shade out almost all native species and lead to monospecific stands. The tree resprouts vigorously after damage.

    Seedlings and saplings can be removed manually. Larger trees are cut and the cut stumps treated with herbicide to prevent regrowth.

References
380, 725, 835, 1239, 1313.

Tabebuia pallida (Lindl.) Miers   Bignoniaceae

LF: Evergreen tree   CU: Wood
SN: *Bignonia pallida* Lindl., *Tabebuia pentaphylla* Hemsl.

Geographic Distribution

Europe
    Northern
    British Isles
    Central, France
    Southern
    Eastern
    Mediterranean Isl.

Africa
    Northern
    Tropical
    Southern

Asia
    Temperate
    Tropical

Australasia
    Australia
    New Zealand

Northern America
    Canada, Alaska
    Southeastern USA
    Western USA
    Remaining USA
  N Mexico

Southern America
  N Tropical
  N Caribbean
    Chile, Argentina

Atlantic Islands
    Cape Verde
    Canary, Madeira
    Azores
    South Atlantic Isl.

Indian Ocean Islands
    Mascarenes
● Seychelles
    Madagascar

Pacific Islands
    Micronesia
    Galapagos
    Hawaii
    Melanesia, Polynesia

Invaded Habitats
Forests and forest edges, tropical hammocks.

Description
A tree of 5-35 m height with a greyish deeply fissured bark. Leaves are 8-15 cm long and palmately compound with 3-5 spreading leaflets. Petioles are 5-25 cm long. The leaflets are broadly elliptic, 5-22 cm long and 2-11 cm wide. Inflorescences are terminal panicles and consist of many pink rose or white flowers with corollas of 5-7 cm length. Fruits are cylindrical and dehiscent capsules of 10-20 cm length and *c.* 15 mm width, containing numerous winged seeds.

Ecology and Control
This tree is a characteristic species of tropical dry and moist forests in the native range. Where invasive, it establishes well in disturbed sites and forms dense thickets that shade out native plants and strongly reduce species richness. Regeneration of native shrubs and trees is prevented, hindering natural succession and forest regeneration. Little is known on the ecology of this tree.
    Specific control methods for this species are not available. Seedlings and saplings can be hand pulled or dug out. Larger stems are cut and the cut stumps treated with herbicide.

References
412, 416, 417, 1239.

Tagetes minuta L.   Asteraceae

LF: Annual herb   CU: Ornamental
SN: *Tagetes glandulifera* Schrank, *Tagetes glandulosa* Link.

Geographic Distribution

Europe		*Australasia*		*Atlantic Islands*	
	Northern	X	Australia	X	Cape Verde
X	British Isles	X	New Zealand	X	Canary, Madeira
X	Central, France				Azores
X	Southern	*Northern America*			South Atlantic Isl.
	Eastern		Canada, Alaska		
X	Mediterranean Isl.	X	Southeastern USA	*Indian Ocean Islands*	
		X	Western USA		Mascarenes
Africa		X	Remaining USA		Seychelles
X	Northern		Mexico	X	Madagascar
●	Tropical				
●	Southern	*Southern America*		*Pacific Islands*	
		N	Tropical		Micronesia
Asia			Caribbean		Galapagos
X	Temperate	N	Chile, Argentina	●	Hawaii
X	Tropical				Melanesia, Polynesia

Invaded Habitats
Coastal areas, disturbed places.

Description
A strongly scented glabrous and erect herb of 50-200 cm height, with short branches. Leaves are 3-15 cm long and 3-10 cm wide, and deeply dissected into 9-17 lanceolate to narrowly elliptic lobes. Leaf margins are sharply toothed. Numerous flowerheads grow in dense terminal corymbs. Involucres are cylindrical, 8-12 mm long and *c.* 2 mm wide, and have 3-4 yellowish-green bracts. The yellowish-green ligules are 1-3 mm long. Each flowerhead has a few yellow disc florets and a few creamy white ray florets. Fruits are linear and black achenes of 4-6 mm length, with appressed hairs and a scaly pappus.

Ecology and Control
The species establishes readily in disturbed sites and can form dense populations after a fire or when forests are cut and burned. It is a prolific seed producer and seeds are dispersed by attaching to animals. The large size and dense growth makes it highly competitive to native plant species.

    Plants can be removed manually. An effective herbicide to control large infestations is paraquat. Cutting before flowers open prevents seed formation.

References
581, 808, 1089, 1445.

Tamarix aphylla (L.) H. Karst.   Tamaricaceae

LF: Evergreen shrub, tree
SN: *Tamarix orientalis* Forssk.
CU: Erosion control, ornamental, shelter

Geographic Distribution

Europe
   Northern
   British Isles
   Central, France
   Southern
   Eastern
   Mediterranean Isl.

Africa
N   Northern
N   Tropical
     Southern

Asia
N   Temperate
N   Tropical

Australasia
●   Australia
    New Zealand

Northern America
    Canada, Alaska
    Southeastern USA
X   Western USA
    Remaining USA
    Mexico

Southern America
    Tropical
    Caribbean
    Chile, Argentina

Atlantic Islands
    Cape Verde
    Canary, Madeira
    Azores
    South Atlantic Isl.

Indian Ocean Islands
    Mascarenes
    Seychelles
    Madagascar

Pacific Islands
    Micronesia
    Galapagos
X   Hawaii
    Melanesia, Polynesia

Invaded Habitats
Desert scrub and washes, riparian habitats, disturbed sites.

Description
A shrub or small tree up to 12 m tall, branching from the base, with a dark reddish-brown bark, grey-green branchlets and scale-like leaves. These are 0.5-2 mm long, strongly clasping the twig but not overlapping. Inflorescences are spike-like racemes of 2-6 cm length with numerous white to reddish flowers, appearing at the ends of the previous year's branches. Petals are oblong to elliptic and *c.* 2 mm long. Fruits are bell-shaped capsules containing numerous minute seeds. The plant has a strong rootstock and deeply penetrating roots.

Ecology and Control
Where native, this fast growing and salt tolerant tree grows commonly in oases and along ephemeral streams. The plant forms extensive and dense thickets and the deeply penetrating roots often lower the water table while surface soil salinity increases due to salt excretion. Native plants are eliminated under stands of this tree and wildlife habitats are reduced. In Australia, the tree displaces native *Eucalyptus camaldulensis* stands. Stands of this tree increase sedimentation rates by trapping and stabilizing sediment during floods. Seeds are dispersed mainly by flood waters but also by animals. The tree suckers from roots.

    Small plants can be dug out but as much of the root system as possible must be removed to prevent resprouting. Larger trees are cut at ground level and the cut stumps treated with picloram plus 2,4-D, or triclopyr plus picloram plus 2,4-D. The same herbicides can be used to control regrowth.

References
115, 282, 341, 499, 607, 986.

Tamarix ramosissima Ledeb.   Tamaricaceae

LF: Evergreen shrub, tree
SN: *Tamarix pentandra* Pall.
CU: Erosion control, ornamental

Geographic Distribution

Europe		*Australasia*		*Atlantic Islands*	
	Northern	●	Australia		Cape Verde
	British Isles		New Zealand		Canary, Madeira
	Central, France				Azores
	Southern		*Northern America*		South Atlantic Isl.
N	Eastern		Canada, Alaska		
	Mediterranean Isl.		Southeastern USA		*Indian Ocean Islands*
		●	Western USA		Mascarenes
Africa		X	Remaining USA		Seychelles
	Northern		Mexico		Madagascar
	Tropical				
●	Southern		*Southern America*		*Pacific Islands*
			Tropical		Micronesia
Asia			Caribbean		Galapagos
N	Temperate		Chile, Argentina		Hawaii
N	Tropical				Melanesia, Polynesia

Invaded Habitats
Desert washes and streambanks, desert scrub, ditches.

Description
A glabrous shrub or small tree reaching 8 m height, with a reddish-brown bark. The scaly leaves are sessile and 1.5-3.5 mm long. Inflorescences are dense racemes of 1.5-7 cm length and 3-4 mm width, arranged in panicles. Bracts are exceeding the pedicels. Flowers are pink and have petals of 1-2 mm length. Fruits are capsules with numerous minute seeds. The plant has deeply penetrating roots.

Ecology and Control
This tree grows commonly in damp places, especially on saline and alkaline soils. It forms dense and impenetrable thickets over large areas that lower the water table due to a high water consumption and increase salinity of the upper soil levels. They displace native riparian vegetation, reduce species richness and reduce food sources available for wildlife. Thickets also increase sediment deposition because the extensive root system is more stable than that of native trees and shrubs, thus trapping sediment and affecting the severity of floods. Since the dead material burns easily and the tree quickly resprouts after fire, wildfires are more frequent in areas dominated by this species.

Once established, tamarisks are difficult to control because they rapidly resprout from roots. Small seedlings can be pulled by hand, older individuals grubbed but roots must be removed. Larger trees are cut and the stumps treated with herbicide such as triclopyr.

References
115, 133, 158, 162, 163, 186, 212, 255, 282, 341, 365, 379, 393, 483, 578, 740, 819, 1161, 1174, 1360.

Tecoma stans (L.) Juss. — Bignoniaceae

LF: Evergreen shrub, tree CU: Ornamental
SN: *Bignonia stans* L., *Stenolobium stans* (L.) seem.

Geographic Distribution

Europe
   Northern
   British Isles
   Central, France
   Southern
   Eastern
   Mediterranean Isl.

Africa
   Northern
X Tropical
● Southern

Asia
   Temperate
   Tropical

Australasia
   Australia
   New Zealand

Northern America
   Canada, Alaska
N Southeastern USA
   Western USA
   Remaining USA
N Mexico

Southern America
N Tropical
N Caribbean
●N Chile, Argentina

Atlantic Islands
   Cape Verde
   Canary, Madeira
   Azores
   South Atlantic Isl.

Indian Ocean Islands
● Mascarenes
   Seychelles
   Madagascar

Pacific Islands
   Micronesia
   Galapagos
X Hawaii
   Melanesia, Polynesia

Invaded Habitats
Tropical hammocks, rock outcrops, riparian habitats, forest edges.

Description
A shrub or small tree of 2-6 m height with a dense foliage. Leaves are 10-20 cm long, bright green above and paler below, pinnately compound with 5-13 sharply toothed leaflets. Flowers are bright yellow, *c.* 5 cm long and 12-24 mm in diameter, trumpet-shaped, and borne in terminal inflorescences. Fruits are shiny brown capsules of 12-20 cm length and 5-10 mm width, releasing numerous papery and winged seeds of 3-5 mm length.

Ecology and Control
This tree spreads rapidly in dry to mesic areas. It forms dense stands that displace native plants and prevents their regeneration. It establishes well in disturbed sites. Seeds are abundantly produced and wind dispersed. Little is known on the ecology of this species.

Specific control methods for this species are not available. Seedlings and saplings may be removed manually, larger trees cut and the cut stumps treated with herbicide.

References
274, 549, 891, 1218, 1261.

Tectaria incisa Cav.

Dryopteridaceae

LF: Fern
CU: Ornamental
SN: *Aspidium macrophyllum* Sw., *Tectaria martinicensis* Spreng.

Geographic Distribution

Europe	*Australasia*	*Atlantic Islands*
Northern	Australia	Cape Verde
British Isles	New Zealand	Canary, Madeira
Central, France		Azores
Southern	*Northern America*	South Atlantic Isl.
N Eastern	Canada, Alaska	
Mediterranean Isl.	● Southeastern USA	*Indian Ocean Islands*
	Western USA	Mascarenes
Africa	Remaining USA	Seychelles
Northern	N Mexico	Madagascar
Tropical		
Southern	*Southern America*	*Pacific Islands*
	N Tropical	Micronesia
Asia	N Caribbean	Galapagos
Temperate	Chile, Argentina	Hawaii
Tropical		Melanesia, Polynesia

Invaded Habitats
Tropical hammocks, forest edges, rocky shady places.

Description
A stout and erect, rhizomatous fern with pale green fronds. Rhizomes are stout, 15-20 mm in diameter, short-creeping and have brownish black scales. Fertile and sterile fronds are similar in shape and size, once pinnate, and have petioles as long or longer than blades. Blades reach 90 cm length and 60 cm width and have a large, deeply lobed terminal leaflet. Each frond has 3-10 pairs of pinnae. Sori appear in one to several rows on the lower surface of the leaflets along lateral veins.

Ecology and Control
This large fern grows on the floor or on rocks and becomes dominant in the understorey of tropical hammocks. It is a variable species with at least eight varieties in the native range. It forms dense patches that threat rare native ferns in Florida by the shading effect of the large fronds. The fern is fertile all year and spreads by natural dispersal of spores. It has the potential to hybridize with native ferns.

Specific control methods for this species are not available. Plants may be hand pulled or dug out but rhizomes must be removed to prevent resprouting.

References
715, 904.

Thespesia populnea (L.) Soland. Malvaceae

LF: Evergeen shrub, tree
SN: *Thespesia macrophylla* Blume
CU: Erosion control, ornamental, wood

Geographic Distribution

Europe	*Australasia*	*Atlantic Islands*
Northern	Australia	Cape Verde
British Isles	New Zealand	Canary, Madeira
Central, France		Azores
Southern	*Northern America*	South Atlantic Isl.
Eastern	Canada, Alaska	
Mediterranean Isl.	• Southeastern USA	*Indian Ocean Islands*
	Western USA	Mascarenes
Africa	Remaining USA	N Seychelles
Northern	Mexico	Madagascar
N Tropical		
Southern	*Southern America*	*Pacific Islands*
	X Tropical	Micronesia
Asia	X Caribbean	Galapagos
Temperate	Chile, Argentina	N Hawaii
N Tropical		Melanesia, Polynesia

Invaded Habitats
Coastal hammocks and sand dunes, mangrove swamps.

Description
A shrub or tree of 3-18 m height, with a light grey to brown bark becoming fissured, and with a rather dense crown. Leaves are alternate, simple and entire, and have petioles of 5-12 cm length. The leaf blades are broad-ovate to triangular, 12-15 cm long and 5-12 cm wide, both surfaces being dotted with small whitish scales. The pale yellow flowers grow solitary in the axils of leaves, are *c.* 8 cm in diameter and have a dark purple centre. Pedicels are 2-3 cm long and petals *c.* 6 cm long. Fruits are leathery globose yellowish to brownish-green capsules of 20-35 mm diameter, containing a few brown hairy seeds of *c.* 10 mm length.

Ecology and Control
Where native, this shrub grows at the edges of mangrove swamps, along tidal waters, usually on sandy and rocky coasts. The shrub is resistant to salt spray and strong winds. The shrub's spreading lower branches leads to dense and impenetrable thickets that affect wildlife and crowd out native vegetation. Sometimes it forms forests of seedlings at the high-tide line of beaches. It produces large fruit crops and seeds are dispersed by tides and ocean currents. The small seeds can withstand extended periods of floating and easily germinate in sand.

    Specific control methods for this species are not available. Seedlings and saplings may be hand pulled or dug out, larger trees cut and the cut stumps treated with herbicide.

References
715.

Thunbergia grandiflora Roxb.   Acanthaceae

LF: Evergreen climber   CU: Ornamental
SN: -

Geographic Distribution

Europe
     Northern
     British Isles
     Central, France
     Southern
     Eastern
     Mediterranean Isl.

Africa
     Northern
     Tropical
     Southern

Asia
     Temperate
N   Tropical

Australasia
●   Australia
     New Zealand

Northern America
     Canada, Alaska
     Southeastern USA
     Western USA
     Remaining USA
     Mexico

Southern America
X   Tropical
     Caribbean
     Chile, Argentina

Atlantic Islands
     Cape Verde
     Canary, Madeira
     Azores
     South Atlantic Isl.

Indian Ocean Islands
     Mascarenes
X   Seychelles
     Madagascar

Pacific Islands
     Micronesia
     Galapagos
●   Hawaii
     Melanesia, Polynesia

Invaded Habitats
Forests and forest edges, riparian habitats, coastal areas.

Description
A woody climber with stems reaching 2-3 m length, with swollen roots and branches that are covered with yellow-brown bristles. The simple leaves are opposite, broad-ovate, acuminate, broadly toothed, 10-20 cm long and 3-9 cm wide. Petioles are 2-6 cm long. Large flowers with trumpet-shaped, blue or whitish corollas are borne in axillary clusters of 1-2 flowers, or in pendulous racemes; the pedicels are 4-10 cm long. The corolla is 5-9 cm long and *c.* 7 cm in diameter. Fruits are globose capsules of 3-5 cm length, containing four flattened seeds. Each fruit has a conspicuous, long and woody beak at the end. Seeds are 5-10 mm wide.

Ecology and Control
A native of subtropical scrub of moist gullies and steep slopes, this vigorous climber spreads mainly vegetatively by stolons. The plant develops an extensive tuberous root system with some tubers reaching 50 cm diameter and 3 m length. The fast growing climber smothers all vegetation up to 12 m aboveground, especially along rivers, and becomes the dominant ground cover. Dense infestations can kill host trees by the weight of the stems. The plant sprouts repeatedly from the roots. It does not usually produce seeds but spreads by stem and tuber pieces carried by streams.

     Control is difficult once established. Young plants may be grubbed, all roots and tubers must be removed. Larger vines are cut and the cut stumps treated with herbicide. An effective herbicide is imazapyr applied to the foliage during active growth.

References
607, 986, 1331.

Toona ciliata Roem.   Meliaceae

LF: Semi-deciduous tree CU: Ornamental, wood
SN: *Cedrela toona* Rottler, *Cedrela australis* Muell.

Geographic Distribution

Europe		*Australasia*		*Atlantic Islands*	
	Northern	N	Australia		Cape Verde
	British Isles		New Zealand		Canary, Madeira
	Central, France				Azores
	Southern	*Northern America*			South Atlantic Isl.
	Eastern		Canada, Alaska		
	Mediterranean Isl.		Southeastern USA	*Indian Ocean Islands*	
			Western USA		Mascarenes
Africa			Remaining USA	X	Seychelles
	Northern		Mexico		Madagascar
X	Tropical				
●	Southern	*Southern America*		*Pacific Islands*	
		X	Tropical		Micronesia
Asia			Caribbean		Galapagos
N	Temperate		Chile, Argentina	●	Hawaii
N	Tropical				Melanesia, Polynesia

Invaded Habitats
Forests and forest gaps, riparian habitats.

Description
A tree of 10-35 m height, with a wide and rounded crown and a drooping foliage. The bright green leaves are glabrous, pinnately compound with 8-12 pairs of entire leaflets. Flowers are white or yellowish, fragrant, *c.* 5 mm long, and borne in hanging inflorescences. Fruits are woody capsules of *c.* 2 cm length, splitting open. Seeds are winged at both ends, thin, the embryo portion being 1-2 mm long.

Ecology and Control
A fast growing, light-demanding, early successional pioneer tree that spreads rapidly in disturbed forests and cleared areas. Plants in Hawaii are referred to var. *australis*, distinguished from other varieties by the glabrous leaflets. The tree is shade intolerant and becomes persistent once established. Little is known on the ecology of this tree as an invasive species.

Specific control methods for this species are not available. Seedlings and small plants can be hand pulled, larger trees cut and the cut stumps treated with herbicide.

References
47, 104, 549, 556, 1186.

Tradescantia albiflora Kunth — Commelinaceae

LF: Perennial herb
SN: -
CU: -

Geographic Distribution

Europe	*Australasia*	*Atlantic Islands*
Northern	● Australia	Cape Verde
British Isles	New Zealand	Canary, Madeira
Central, France		Azores
Southern	*Northern America*	South Atlantic Isl.
Eastern	Canada, Alaska	
Mediterranean Isl.	Southeastern USA	*Indian Ocean Islands*
	Western USA	Mascarenes
Africa	Remaining USA	Seychelles
Northern	Mexico	Madagascar
Tropical		
Southern	*Southern America*	*Pacific Islands*
	N Tropical	Micronesia
Asia	Caribbean	Galapagos
Temperate	Chile, Argentina	Hawaii
Tropical		Melanesia, Polynesia

Invaded Habitats
Forests, riparian habitats, scrubland.

Description
A more or less succulent herb with fibrous roots and branching, creeping or slightly ascending stems rooting at the nodes. The glabrous leaves are almost sessile, ovate to lanceolate, 2.5-5.5 cm long and 1-2.5 cm wide, and acuminate at the end. The ciliate leaf sheaths are 5-8 mm long. Inflorescences have 15-20 flowers, each with a pedicel of 1-2 cm length. The outer perianth parts are green and 5-8 mm long, the inner perianth parts are white and 7-10 mm long. Filaments are covered with long white hairs. The fruits are papery capsules.

Ecology and Control
Once established, this plant spreads mainly by vegetative growth. The creeping and branching stems lead to a dense ground cover, outcompeting native plants and preventing the establishment of trees and shrubs. The species is closely related to *Tradescantia fluminensis*, which has leaves that are green above and purple beneath.

Specific control methods for this species are not available. The same methods as for *Tradescantia fluminensis* may apply.

References
4, 215.

Tradescantia fluminensis Vell. Commelinaceae

LF: Perennial herb
SN: -
CU: Ornamental

Geographic Distribution

Europe	*Australasia*	*Atlantic Islands*
Northern	● Australia	Cape Verde
British Isles	● New Zealand	X Canary, Madeira
Central, France		X Azores
● Southern	*Northern America*	South Atlantic Isl.
Eastern	Canada, Alaska	
Mediterranean Isl.	● Southeastern USA	*Indian Ocean Islands*
	X Western USA	Mascarenes
Africa	Remaining USA	Seychelles
Northern	Mexico	Madagascar
Tropical		
Southern	*Southern America*	*Pacific Islands*
	N Tropical	Micronesia
Asia	Caribbean	Galapagos
X Temperate	Chile, Argentina	Hawaii
Tropical		Melanesia, Polynesia

Invaded Habitats
Forests and forest edges, woodland, riparian habitats.

Description
A herb with trailing stems of 1 m length or more, rooting at the nodes. The alternate leaves are ovate-oblong, glabrous, purplish beneath, 2-6 cm long and 1-3 cm wide. Inflorescences are paired cymes at the ends of stems. Flowers are white to pale lilac, 10-15 mm in diameter, and grow on slender pedicels of 1-2 cm length. Petals are narrowly ovate and up to 12 mm long. The white filaments are densely covered with hairs. Inflorescences are subtended by paired leaf-like bracts. Fruits are papery capsules.

Ecology and Control
This plant spreads rapidly by vegetative growth and forms dense mats up to 60 cm deep, impeding the growth and regeneration of native shrubs and trees. It is a smothering creeper becoming dominant on the floor and tolerating heavy shade. It rapidly colonizes vegetation openings and the dense ground cover prevents seeds of native plants from reaching the ground. Plants in Australia fail to set seeds, they spread solely by vegetative growth.

Small infestations can be removed by hand. Larger infestations are removed by raking or rolling up the plants. Killing by solarization is used in warmer locations, the plastic sheeting should last for 2-6 weeks. An effective herbicide is paraquat. Follow-up treatments are necessary to control regrowth.

References
121, 196, 658, 659, 715, 836, 924, 1296, 1415.

Tradescantia spathacea Sw.   Commelinaceae

LF: Perennial herb   CU: Ornamental
SN: *Tradescantia discolor* L'Hér, *Rhoeo spathacea* (Sw.) Stearn

Geographic Distribution

Europe	*Australasia*	*Atlantic Islands*
Northern	Australia	Cape Verde
British Isles	New Zealand	Canary, Madeira
Central, France		Azores
Southern	*Northern America*	South Atlantic Isl.
Eastern	Canada, Alaska	
Mediterranean Isl.	● Southeastern USA	*Indian Ocean Islands*
	Western USA	Mascarenes
Africa	Remaining USA	X Seychelles
Northern	N Mexico	Madagascar
Tropical		
Southern	*Southern America*	*Pacific Islands*
	N Tropical	X Micronesia
Asia	N Caribbean	Galapagos
X Temperate	Chile, Argentina	Hawaii
Tropical		Melanesia, Polynesia

Invaded Habitats
Coastal tropical hammocks and scrub, pinelands, disturbed sites.

Description
A purplish herb with a short stout stem hidden by overlapping leaf bases. Leaves are spreading to erect, broadly-linear, stiff and somewhat fleshy, 15-30 cm long and 2.5-8 cm wide. The upper leaf surfaces are dark green or green with pale stripes, lower surfaces are usually purple. Small white flowers are clustered within a folded bract of 3-4 cm length. Flowers have three petals and six stamens. Fruits are capsules with two seeds each. The plant has a fleshy rootstock.

Ecology and Control
This plant spreads readily by seeds and by producing numerous offshoots. Its dense clumps form a continuous cover on the floor, preventing growth and establishment of native plants. Tree seedlings are unable to grow in these stands. Damaged plants easily resprout from the root.

Seedlings and scattered plants can be pulled or dug out, roots must be removed. Larger infestations can be cut and regrowth treated with herbicide.

References
715.

Triadica sebifera (L.) Small   Euphorbiaceae

LF: Deciduous tree   CU: Food, ornamental, shelter
SN: *Sapium sebiferum* (L.) Roxb.

Geographic Distribution

Europe	*Australasia*	*Atlantic Islands*
Northern	Australia	Cape Verde
British Isles	New Zealand	Canary, Madeira
Central, France		Azores
Southern	*Northern America*	South Atlantic Isl.
Eastern	Canada, Alaska	
Mediterranean Isl.	● Southeastern USA	*Indian Ocean Islands*
	Western USA	Mascarenes
Africa	● Remaining USA	Seychelles
X Northern	Mexico	Madagascar
Tropical		
Southern	*Southern America*	*Pacific Islands*
	Tropical	Micronesia
Asia	Caribbean	Galapagos
N Temperate	Chile, Argentina	Hawaii
X Tropical		Melanesia, Polynesia

Invaded Habitats
Wet forests and grassland, freshwater wetlands, riparian habitats.

Description
A small to medium-sized tree up to 10 m tall, sometimes more, with a milky sap and alternate leaves. The simple leaves have petioles of 2-5 cm length and broadly ovate blades of 3-6 cm width, tapering to a slender point. Small yellowish green flowers are borne on narrowly cylindrical spikes of 10-20 cm length. The rounded and three-lobed fruits are capsules of 7-13 mm diameter, exposing 2-3 waxy and dull white seeds. Seeds are 7-9 mm long and *c.* 6 mm wide.

Ecology and Control
This fast growing and salt tolerant tree spreads rapidly and transforms native wetland prairies into woodland dominated by this species. Once established, it forms pure stands that exclude almost all other plants and affect wildlife by reducing food sources. Extensive stands alter nutrient cycling and the species composition of decomposers due to a rapid leaf decay. Such stands reduce fuel loads in invaded areas and prevent the spread of fires. The tree is a prolific seed producer and seeds are mainly dispersed by birds and water. Seedlings are able to establish in a wide range of environmental conditions including closed-canopy forests. The tree suckers from roots and resprouts vigorously if damaged.

     Whereas seedlings can easily be hand pulled, mature trees should be treated with a triclopyr herbicide applied to the base of the tree. Cutting trees older than one year leads to root and stump suckering. Follow-up programmes are necessary to treat regrowth and seedlings.

References
73, 167, 168, 172, 196, 261, 262, 263, 480, 632, 636, 637, 643, 715, 746, 1076.

Trifolium arvense L.   Fabaceae

LF: Annual herb   CU: Forage
SN: -

Geographic Distribution

Europe		*Australasia*		*Atlantic Islands*	
N	Northern	●	Australia		Cape Verde
N	British Isles	X	New Zealand	N	Canary, Madeira
N	Central, France			N	Azores
N	Southern	*Northern America*			South Atlantic Isl.
N	Eastern		Canada, Alaska		
N	Mediterranean Isl.		Southeastern USA	*Indian Ocean Islands*	
		X	Western USA		Mascarenes
Africa			Remaining USA		Seychelles
N	Northern		Mexico		Madagascar
	Tropical				
	Southern	*Southern America*		*Pacific Islands*	
			Tropical		Micronesia
Asia			Caribbean		Galapagos
N	Temperate		Chile, Argentina	X	Hawaii
	Tropical				Melanesia, Polynesia

Invaded Habitats
Grassland, riverbeds, coastal dunes.

Description
An erect or ascending, glabrous herb with branched stems of 4-40 cm height. The compound leaves have three leaflets. The upper leaves are sessile and have linear-oblong leaflets of 5-20 mm length. The lower leaves have petioles of 5-20 mm length. Flowerheads are numerous, axillary or terminal, ovoid or oblong, up to 20 mm in diameter, and contain numerous flowers. These are white to pink and 3-6 mm long. Flowerheads are becoming elongated during fruit. Fruits are glabrous straight pods of *c.* 2 mm length, containing one seed of *c.* 1 mm diameter.

Ecology and Control
A highly variable species in the native range growing generally in sandy soils. The plant is nitrogen-fixing and increases soil fertility levels which may alter the species composition of invaded areas. The plant grows in dense patches that crowd out native plants.

Specific control methods for this species are not available. Scattered plants can be hand pulled. Cutting prevents seed formation. The same methods as for other *Trifolium* species may apply.

References
215.

Trifolium repens L.   Fabaceae

LF: Perennial herb
SN: -
CU: Forage

Geographic Distribution

Europe		*Australasia*		*Atlantic Islands*	
N	Northern	●	Australia		Cape Verde
N	British Isles	X	New Zealand	N	Canary, Madeira
N	Central, France			N	Azores
N	Southern	*Northern America*			South Atlantic Isl.
N	Eastern	X	Canada, Alaska		
N	Mediterranean Isl.	X	Southeastern USA	*Indian Ocean Islands*	
		X	Western USA		Mascarenes
Africa		X	Remaining USA		Seychelles
N	Northern	X	Mexico		Madagascar
	Tropical				
X	Southern	*Southern America*		*Pacific Islands*	
		X	Tropical		Micronesia
Asia		X	Caribbean		Galapagos
N	Temperate	X	Chile, Argentina	X	Hawaii
X	Tropical				Melanesia, Polynesia

Invaded Habitats
Grass- and heathland, riparian habitats, coastal beaches.

Description
A glabrous, creeping and mat-forming perennial with trailing stems up to 50 cm long, rooting at nodes. The bright green leaves are compound with three leaflets of 10-25 mm length that have light or dark marks along the veins. Stipules are large and membranous. Globose flower-heads are borne on erect, axillary peduncles of up to 20 cm length. Flowers are white, the corolla is 8-13 mm long, becoming light brown. Fruits are yellowish brown linear, compressed pods of 4-5 mm length that are constricted between seeds, containing 3-4 seeds of *c.* 1 mm diameter.

Ecology and Control
This is a highly variable species with many varieties and cultivars. It is mat-forming due to the extensively creeping and rooting stems. The plant is nitrogen-fixing and increases soil fertility levels. The mats may cover large areas and reduce native species richness. It establishes well in disturbed and open sites but is not shade tolerant. Seeds are dispersed by adhering to animals. In Australia, the plant is host to the red-legged earth mite, which causes considerable damage to native plants.

Light grazing can reduce the cover of this plant, frequent and intense grazing encourages the growth of it. Small infestations can be removed manually. Effective herbicides are diuron, simazine, atrazine, or 2,4-D plus dicamba.

References
121, 175, 215, 731, 1305, 1361.

Trifolium subterraneum L.   Fabaceae

LF: Annual herb
SN: -
CU: Erosion control, forage

Geographic Distribution

Europe		*Australasia*		*Atlantic Islands*	
	Northern	●	Australia		Cape Verde
N	British Isles	X	New Zealand	N	Canary, Madeira
N	Central, France			N	Azores
N	Southern	*Northern America*			South Atlantic Isl.
N	Eastern		Canada, Alaska		
N	Mediterranean Isl.	X	Southeastern USA	*Indian Ocean Islands*	
		X	Western USA		Mascarenes
Africa			Remaining USA		Seychelles
N	Northern		Mexico		Madagascar
	Tropical				
	Southern	*Southern America*		*Pacific Islands*	
			Tropical		Micronesia
Asia			Caribbean		Galapagos
N	Temperate		Chile, Argentina	X	Hawaii
	Tropical				Melanesia, Polynesia

Invaded Habitats
Grass- and woodland, seasonal freshwater wetlands, coastal beaches, rock outcrops.

Description
A hairy herb with numerous procumbent stems up to 30 cm tall. Leaves have petioles of 1-5 cm length and are compound with three broadly obovate leaflets of 5-15 mm length. Flowerheads are globose, and appressed to or buried in the soil during fruit by the long peduncles. Corollas are white, sometimes pink, and 10-15 mm long. Fruits are ovoid glabrous pods of 2-3 mm length, with one seed of *c.* 2 mm diameter each.

Ecology and Control
A highly variable species, growing usually in dry grassy places and sandy soils. There are many different strains that became naturalized in Australia, most of these belong to subsp. *subterraneum*. The plant is invasive because it grows in dense patches that eliminate native vegetation. It establishes readily in disturbed sites.

    Specific control methods for this species are not available. The same methods as for other *Trifolium* species may apply.

References
215, 457, 458, 651, 1029, 1030.

Turbina corymbosa (L.) Raf. Convolvulaceae

LF: Perennial vine
SN: *Ipomoea burmannii* Choisy
CU: Ornamental

Geographic Distribution

Europe	*Australasia*	*Atlantic Islands*
Northern	● Australia	Cape Verde
British Isles	New Zealand	Canary, Madeira
Central, France		Azores
Southern	*Northern America*	South Atlantic Isl.
Eastern	Canada, Alaska	
Mediterranean Isl.	N Southeastern USA	*Indian Ocean Islands*
	Western USA	Mascarenes
Africa	Remaining USA	Seychelles
Northern	N Mexico	Madagascar
Tropical		
Southern	*Southern America*	*Pacific Islands*
	N Tropical	Micronesia
Asia	N Caribbean	Galapagos
Temperate	Chile, Argentina	Hawaii
Tropical		Melanesia, Polynesia

Invaded Habitats
Hammocks, forest edges.

Description
A high-climbing or trailing vine with stems reaching several metres in length that are woody at the base. The alternate leaves are simple, entire, glabrous or rarely pubescent, and have petioles of 4-7 cm length. Leaf blades are cordate to ovate and 4-10 cm long. Flowers are borne in axillary clusters on peduncles as long as or longer than the leaves. They are bell-shaped, tubular, 3-4 cm long and the white corolla has greenish bands. Fruits are indehiscent capsules of 8-10 mm length, containing mostly one ovoid seed.

Ecology and Control
Native habitats of this vine include tropical hammocks and shrubby areas. It is invasive because it is fast growing and smothers native plants, preventing regeneration of shrubs and trees. Little is known on the ecology of this species.

Specific control methods for this species are not available. The same methods as for *Ipomoea* species may apply.

References
607, 823.

Typha latifolia L. Typhaceae

LF: Perennial herb
SN: -
CU: Ornamental

Geographic Distribution

Europe		*Australasia*		*Atlantic Islands*	
N	Northern	●	Australia		Cape Verde
N	British Isles	X	New Zealand		Canary, Madeira
N	Central, France				Azores
N	Southern	*Northern America*			South Atlantic Isl.
N	Eastern	●N	Canada, Alaska		
	Mediterranean Isl.	N	Southeastern USA	*Indian Ocean Islands*	
		N	Western USA		Mascarenes
Africa		N	Remaining USA		Seychelles
N	Northern	N	Mexico		Madagascar
N	Tropical				
X	Southern	*Southern America*		*Pacific Islands*	
		N	Tropical		Micronesia
Asia		X	Caribbean		Galapagos
N	Temperate	X	Chile, Argentina	X	Hawaii
N	Tropical				Melanesia, Polynesia

Invaded Habitats
Freshwater wetlands, marshes, riparian habitats, coastal estuaries.

Description
A large glabrous herb with stout stems of 1-3 m height. The alternate leaves are single and entire, sessile and have a sheathing base. They are greyish green, 8-25 mm wide, and overtop the spikes. The plant has strong spreading rhizomes reaching 70 cm length and 30 mm diameter. Numerous unisexual flowers are borne in a dense, cylindrical spike. The male flowers are borne above the female flowers, the male and female parts of the inflorescence are contiguous or separated by not more than 2-3 cm. The female part is 8-15 cm long and 2-3 cm wide and dark brown, the male part 6-14 cm long. Fruits are small capsules, each containing one seed. Fruits are shed in hair tufts of *c.* 8 mm length. Seeds are 1-1.5 mm long.

Ecology and Control
This plant grows vigorously in shallow waters and forms dense monospecific stands that reduce species richness and eliminate native vegetation. Such stands impede the water flow and increase bank erosion and siltation. Through anaerobic decay of excess plant material the water may become polluted. Newly established plants spread rapidly by rhizome growth and expand laterally. Seeds are dispersed by water and remain viable for long periods of time.

Hand pulling is practicable for scattered plants and small infestations. Mechanical cutting below the water line proved to be effective in killing the plant, the best time is at the end of flowering. A number of herbicides are used for chemical control, e.g. glyphosate, amitrole, dalapon, 2,4-D, or TCA. Success depends on time of application and density of infestations.

References
62, 215, 409, 479, 482, 547, 581, 952, 986, 1378.

Ulex europaeus L. — Fabaceae

LF: Evergreen shrub
SN: -
CU: Erosion control, ornamental

Geographic Distribution

Europe
- X Northern
- N British Isles
- N Central, France
- N Southern
- X Eastern Mediterranean Isl.

Africa
- X Northern Tropical
- ● Southern

Asia
- ● Temperate
- ● Tropical

Australasia
- ● Australia
- ● New Zealand

Northern America
- X Canada, Alaska Southeastern USA
- ● Western USA
- X Remaining USA
- X Mexico

Southern America
- X Tropical
- X Caribbean
- X Chile, Argentina

Atlantic Islands
- Cape Verde
- X Canary, Madeira
- ● Azores
- X South Atlantic Isl.

Indian Ocean Islands
- ● Mascarenes
- Seychelles
- Madagascar

Pacific Islands
- Micronesia
- Galapagos
- ● Hawaii
- Melanesia, Polynesia

Invaded Habitats
Grass- and heathland, forests, riparian habitats, coastal scrub and cliffs.

Description
A very spiny and densely branched shrub of 0.5-2.5 m height, with numerous branched and strong spines of 12-25 mm length. Leaves on young plants are ternate and up to 25 mm long, on mature plants simple and often reduced to scales or small spines. Twigs are tomentose with grey to reddish-brown hairs. Flowers are bright yellow, 2-2.5 cm long, and borne in leaf axils and in terminal clusters. Fruits are dehiscent densely hairy pods of 10-20 mm length. Seeds are brownish, 2.5-3 mm long, and have a white aril each.

Ecology and Control
This nitrogen-fixing shrub establishes in disturbed sites and forms dense and spiny impenetrable thickets eliminating native vegetation and affecting wildlife. It regenerates rapidly in burned areas. It is a prolific seed producer and seeds are dispersed by water and animals. The shrub accumulates a large soil seed bank and the long-lived seeds may persist for 30 years or more. The shrub resprouts vigorously from the base after damage.

Burning removes the shrub but stimulates seed germination. Grazing is often used to control seedlings and regrowth. Chemical control includes spraying seedlings with picloram or triclopyr. Mature plants are cut and the cut stumps treated with glyphosate. Regrowth and whole plants can be sprayed with dicamba, triclopyr, or 2,4-D plus triclopyr.

References
64, 133, 215, 284, 373, 374, 410, 440, 532, 581, 607, 730, 924, 986, 1060, 1090, 1100, 1199, 1278, 1415, 1451.

Ulmus pumila L.   Ulmaceae

LF: Deciduous tree   CU: -
SN: -

Geographic Distribution

Europe	*Australasia*	*Atlantic Islands*
Northern	Australia	Cape Verde
British Isles	New Zealand	Canary, Madeira
Central, France		Azores
Southern	*Northern America*	South Atlantic Isl.
Eastern	Canada, Alaska	
Mediterranean Isl.	Southeastern USA	*Indian Ocean Islands*
	X Western USA	Mascarenes
Africa	● Remaining USA	Seychelles
Northern	Mexico	Madagascar
Tropical		
Southern	*Southern America*	*Pacific Islands*
	Tropical	Micronesia
Asia	Caribbean	Galapagos
N Temperate	Chile, Argentina	Hawaii
Tropical		Melanesia, Polynesia

Invaded Habitats
Grass- and woodland, disturbed sites.

Description
A small to medium-sized tree up to 20 m tall, with glabrous to sparsely pubescent twigs, with an open and round crown and a grey to brown bark with shallow furrows. The dark green leaves are ovate to ovate-lanceolate, coarsely toothed, 2-7 cm long and 8-25 mm wide. The small flowers appear before the leaves, are greenish and are borne in small, drooping clusters of 2-5. Fruits are winged, smooth samaras of 10-15 mm width, with translucent and whitish wings. Each fruit contains one seed.

Ecology and Control
This tree is fast growing and occurs in a wide range of habitats. It produces seeds abundantly which are dispersed by wind. Seedlings may form dense thickets with hundreds of plants. The tree can become dominant in prairies subjected to disturbances and displaces native forbs and grasses. It sprouts vigorously from the roots.

Seedlings can be hand pulled. Larger trees may be girdled but not too deeply in order to prevent resprouting from the roots. Stumps of cut trees can be treated with glyphosate to prevent resprouting. In fire-adapted communities, regular burning may control this tree.

References
442, 571, 747, 748, 1065, 1198.

Urochloa mutica (Forssk.) Nguyen — Poaceae

LF: Perennial herb
CU: Erosion control, fodder
SN: *Brachiaria mutica* Stapf, *B. purpurascens* (Raddi) Henr., *Panicum purpurascens* Raddi

Geographic Distribution

	Europe		*Australasia*		*Atlantic Islands*
	Northern		● Australia		Cape Verde
	British Isles		X New Zealand	X	Canary, Madeira
	Central, France				Azores
	Southern		*Northern America*		South Atlantic Isl.
	Eastern		Canada, Alaska		
	Mediterranean Isl.		● Southeastern USA		*Indian Ocean Islands*
			Western USA	X	Mascarenes
	Africa		Remaining USA		Seychelles
	Northern	X	Mexico		Madagascar
N	Tropical				
	Southern		*Southern America*		*Pacific Islands*
			● Tropical		Micronesia
	Asia	X	Caribbean	●	Galapagos
X	Temperate	X	Chile, Argentina	●	Hawaii
X	Tropical			X	Melanesia, Polynesia

Invaded Habitats
Riparian habitats, freshwater wetlands, swamps, disturbed sites.

Description
A perennial grass with widely creeping stolons and with creeping and ascending stems of 1-3 m length, rooting at the lower nodes. Leaf sheaths have dense stiff hairs below, the ligule is a densely ciliate membrane. Leaf blades are flat, 25-30 cm long and 10-15 mm wide, usually glabrous. Inflorescences are terminal panicles up to 20 cm long, with 8-20 ascending and alternate branches of 2-5 cm length. Spikelets are borne in pairs and *c.* 3 mm long, glabrous and often tinged purple.

Ecology and Control
This fast growing semi-aquatic grass competes with native plants and forms dense monocultural stands that displace all other vegetation. It overgrows shrubs and young trees, impeding their growth. The grass also flourishes in wet conditions and its mats may reach 1 m or more in depth. In wetlands, it sends floating stems of more than 6 m length across slowly moving waters. It tolerates drought and brackish water. The plant reproduces and spreads mainly by stem fragments, seed production is low.

Small infestations can be removed manually or by mechanical harvesters. Larger stands may be treated with herbicides approved for use in aquatic environments.

References
278, 580, 607, 715, 984, 1190, 1313.

Vellereophyton dealbatum (Thunb.) Hill.   Asteraceae

LF: Annual herb   CU: Ornamental
SN: *Gnaphalium candidissimum* Lam.

Geographic Distribution

Europe
     Northern
     British Isles
     Central, France
     Southern
     Eastern
     Mediterranean Isl.

Africa
     Northern
     Tropical
N   Southern

Asia
     Temperate
     Tropical

Australasia
●   Australia
X   New Zealand

Northern America
     Canada, Alaska
     Southeastern USA
     Western USA
     Remaining USA
     Mexico

Southern America
     Tropical
     Caribbean
     Chile, Argentina

Atlantic Islands
     Cape Verde
     Canary, Madeira
     Azores
     South Atlantic Isl.

Indian Ocean Islands
     Mascarenes
     Seychelles
     Madagascar

Pacific Islands
     Micronesia
     Galapagos
     Hawaii
     Melanesia, Polynesia

Invaded Habitats
Heaths, freshwater wetlands, coastal estuaries and beaches, salt marshes.

Description
A herb with prostrate to ascending and often branched, densely white-tomentose stems of 5-50 cm height. Leaves are 10-45 mm long and 2-6 mm wide, decreasing in size upwards, both surfaces being white-woolly. Numerous small flowerheads grow in dense clusters at the ends of branches. They are 2.5-3 mm long, 1-2 mm wide and have yellow florets tipped red-purple. Fruits are achenes of *c.* 0.5 mm length. The pappus consists of many bristles.

Ecology and Control
Where native, this plant grows mainly in damp and sandy places near the coast, e.g. stream-beds, sand dune slacks, marshes, and margins of salt pans. The plant is very variable in size and shape. It is invasive because it forms dense patches that crowd out native plants and prevent their establishment. Little is known on the ecology of this plant.

     Specific control methods for this species are not available. Plants may be hand pulled, larger stands treated with herbicides. Cutting prevents seed dispersal.

References
215.

Verbascum thapsus L.   Scrophulariaceae

LF: Biennial herb   CU: -
SN: -

Geographic Distribution

Europe
N   Northern
N   British Isles
N   Central, France
N   Southern
N   Eastern
N   Mediterranean Isl.

Africa
     Northern
     Tropical
     Southern

Asia
N   Temperate
X   Tropical

Australasia
●   Australia
X   New Zealand

Northern America
X   Canada, Alaska
X   Southeastern USA
●   Western USA
X   Remaining USA
     Mexico

Southern America
     Tropical
     Caribbean
X   Chile, Argentina

Atlantic Islands
     Cape Verde
N   Canary, Madeira
N   Azores
     South Atlantic Isl.

Indian Ocean Islands
     Mascarenes
     Seychelles
     Madagascar

Pacific Islands
     Micronesia
     Galapagos
●   Hawaii
     Melanesia, Polynesia

Invaded Habitats
Grassland, riparian habitats, disturbed sites.

Description
A large herb with densely greyish- or whitish-tomentose stems of 30-200 cm height and alternate leaves. The basal leaves are elliptic- to obovate-oblong, 8-50 cm long and 3-14 cm wide. The upper stem leaves are smaller and decurrent. Flowers are borne in terminal, usually simple spike-like racemes of 20-50 cm length. The pale yellow flowers are 15-30 mm in diameter and have five unequal stamens. The upper three ones have villous filaments, the lower ones glabrous or villous filaments. Fruits are elliptic to ovoid capsules of 7-10 mm length, containing numerous small seeds. The plant has a strong taproot.

Ecology and Control
This plant is found mainly on dry sandy soils and primarily colonizes sites of low fertility. It forms dense patches with the large rosettes that shade out native plants. The plant spreads rapidly after disturbances and forms a continuous cover, eliminating the native vegetation. The long-lived seeds may remain viable in the soil for several decades. Once exposed to light they germinate rapidly.

     Individual plants can be killed by cutting *c.* 8 cm below the soil surface to remove the top of the taproot. Cutting stands before seed set prevents dispersal and burial of seeds. Chemical control is done by spraying rosettes with 2,4-D, glyphosate or tebuthiuron. Repeated applications may be necessary to control regrowth.

References
133, 215, 504, 505, 985, 986, 1154.

Verbascum virgatum Stokes   Scrophulariaceae

LF: Biennial herb
SN: -
CU: Ornamental

Geographic Distribution

Europe		*Australasia*		*Atlantic Islands*	
	Northern	●	Australia		Cape Verde
N	British Isles	X	New Zealand	N	Canary, Madeira
	Central, France			N	Azores
N	Southern	*Northern America*			South Atlantic Isl.
	Eastern		Canada, Alaska		
N	Mediterranean Isl.		Southeastern USA	*Indian Ocean Islands*	
		X	Western USA		Mascarenes
Africa			Remaining USA		Seychelles
	Northern		Mexico		Madagascar
	Tropical				
X	Southern	*Southern America*		*Pacific Islands*	
			Tropical		Micronesia
Asia			Caribbean		Galapagos
	Temperate	X	Chile, Argentina	X	Hawaii
	Tropical			X	Melanesia, Polynesia

Invaded Habitats
Riverbanks, disturbed sites.

Description
A glandular-pubescent herb with stems of 30-150 cm height. The basal leaves are obovate to elliptic, 10-35 cm long and 3-10 cm wide, the margins being toothed. Upper leaves are smaller and stem-clasping. Flowers are borne in a spike-like raceme of 30-80 cm length, flowers in the lower part often being in clusters of 2-5. The yellow corollas are 3-4 cm in diameter and have a purple centre. Fruits are globose glandular-hairy capsules of 5-8 mm length, containing numerous small seeds.

Ecology and Control
A plant that rapidly colonizes disturbed sites and forms dense patches due to the large rosette leaves. Native plants are crowded out and their establishment is prevented. Little is known on the ecology of this species.

Specific control methods for this species are not available. The same methods as for the previous species may apply.

References
215.

Verbena bonariensis L. Verbenaceae

LF: Perennial herb
SN: -
CU: Ornamental

Geographic Distribution

Europe		*Australasia*		*Atlantic Islands*	
	Northern	●	Australia		Cape Verde
	British Isles	X	New Zealand	X	Canary, Madeira
	Central, France			X	Azores
	Southern	*Northern America*			South Atlantic Isl.
	Eastern		Canada, Alaska		
	Mediterranean Isl.	X	Southeastern USA	*Indian Ocean Islands*	
		X	Western USA	X	Mascarenes
Africa			Remaining USA		Seychelles
	Northern		Mexico		Madagascar
X	Tropical				
●	Southern	*Southern America*		*Pacific Islands*	
		N	Tropical		Micronesia
Asia		X	Caribbean		Galapagos
X	Temperate	N	Chile, Argentina	X	Hawaii
	Tropical				Melanesia, Polynesia

Invaded Habitats
Grass- and woodland, riparian habitats, forest edges, disturbed sites.

Description
A glabrous to sparsely hairy, sparsely branched herb of 50-150 cm height, with stiff stems squared in cross-section. The elliptic to elliptic-lanceolate leaves are opposite, sessile or clasping the stem, sharply serrated at least in the upper half, 7-15 cm long and 1-2.5 cm wide. The inflorescence consists of several cymes on long peduncles, each cyme having dense and sessile spikes. Spikes are 5-6 mm wide, and rarely exceeding 2 cm length but elongating in fruit. The white, purplish or blue flowers are 5-6 mm long and 3-4 mm across. Fruits are narrowly ellipsoid nutlets of 1-2 mm length.

Ecology and Control
Where invasive, this plant often forms dense swards that crowd out native plants. It establishes well in disturbed sites and may persist. The long-lived seeds are accumulated in a soil seed bank. Little is known on the ecology of this species.

Specific control methods for this species are not available. Scattered plants can be removed manually by pulling out, larger stands can be treated with herbicides.

References
215, 234, 436, 921, 1089.

Vinca major L.   Apocynaceae

LF: Perennial herb   CU: Ornamental
SN: -

Geographic Distribution

	Europe		*Australasia*		*Atlantic Islands*
	Northern	•	Australia		Cape Verde
X	British Isles	•	New Zealand	X	Canary, Madeira
N	Central, France				Azores
N	Southern		*Northern America*		South Atlantic Isl.
N	Eastern		Canada, Alaska		
N	Mediterranean Isl.		Southeastern USA		*Indian Ocean Islands*
		•	Western USA		Mascarenes
	Africa		Remaining USA		Seychelles
	Northern		Mexico		Madagascar
	Tropical				
X	Southern		*Southern America*		*Pacific Islands*
			Tropical		Micronesia
	Asia		Caribbean		Galapagos
	Temperate	X	Chile, Argentina		Hawaii
	Tropical				Melanesia, Polynesia

Invaded Habitats
Forests, grassland, riparian habitats, coastal dunes.

Description
A low, slightly woody perennial with arching to ascending, often trailing vegetative stems up to 1.5 m length and flowering stems up to 30 cm length. The opposite and entire leaves are ovate, 2.5-9 cm long and 2-6 cm wide, have short petioles and ciliate margins. Bluish-purple flowers of 3-5 cm diameter grow solitary in the axils of leaves. Pedicels are 3-5 cm long. Fruits are spreading follicles of *c.* 5 cm diameter and 3.5-5 cm length. Seeds are oblong and 7-8 mm long.

Ecology and Control
A fast growing plant that spreads mainly by vegetative growth. It is a vigorous creeper, occurring in large infestations in semi-shady conditions. The numerous intertwined stems form dense and thick mats that cover the ground, smother small plants and crowd out native species. Establishment of shrub and tree seedlings is prevented. Stems root at the tips and nodes when they touch the ground. Seed set is low in naturalized plants of New Zealand.

Seedlings are easy to hand pull, small infestations can be dug out. Solarization by plastic sheeting for 4-6 months is used to kill smaller infestations. Larger infestations can be mown or slashed and the regrowth treated with herbicide. Effective herbicides are glyphosate or triclopyr. Follow-up treatments are necessary to control seedlings and regrowth.

References
121, 133, 215, 1063, 1311.

Vulpia bromoides (L.) Gray Poaceae

LF: Annual herb
CU: -
SN: *Festuca bromoides* L., *Vulpia dertonensis* (All.) Gola

Geographic Distribution

Europe
-     Northern
- N British Isles
- N Central, France
- N Southern
- N Eastern
- N Mediterranean Isl.

Africa
- N Northern
- N Tropical
- X Southern

Asia
-     Temperate
-     Tropical

Australasia
- ● Australia
- ● New Zealand

Northern America
-     Canada, Alaska
-     Southeastern USA
- ● Western USA
-     Remaining USA
-     Mexico

Southern America
-     Tropical
- X Caribbean
-     Chile, Argentina

Atlantic Islands
-     Cape Verde
- N Canary, Madeira
- N Azores
- X South Atlantic Isl.

Indian Ocean Islands
- X Mascarenes
-     Seychelles
-     Madagascar

Pacific Islands
-     Micronesia
-     Galapagos
- X Hawaii
-     Melanesia, Polynesia

Invaded Habitats
Grass- and heathland, riparian habitats, freshwater wetlands, coastal beaches.

Description
An erect tufted grass with 6-50 cm tall stems. Leaf blades are 1-14 cm long, 0.5-3 mm wide and have white papery ligules of *c.* 1 mm length. Inflorescences are narrow and sparingly branched, green-purple panicles of 1-11 cm length. Spikelets are 7-16 mm long and have 5-10 florets. Lower glumes are 2.5-5 mm long, upper glumes 4.5-9 mm long and the lemmas 4.5-7.5 mm long.

Ecology and Control
This shallow-rooted grass grows well in soils of low fertility. Where invasive, it successfully competes with native grasses and forbs for water, space and nutrients. It forms dense swards crowding out native plants and reducing species richness.

    Scattered plants can be hand pulled or dug out. A combination of grazing in spring (to reduce seed set) and grazing in autumn (to reduce seedling establishment) has shown to reduce the density of the grass. Chemical control includes spraying flowering plants with glyphosate to prevent seed set, or treating with simazine, simazine plus paraquat, or dalapon.

References
215, 276, 339, 742, 1358.

Vulpia myuros (L.) Gmel.   Poaceae

LF: Annual herb   CU: Erosion control, revegetation
SN: -

Geographic Distribution

Europe		*Australasia*		*Atlantic Islands*	
	Northern	●	Australia	N	Cape Verde
N	British Isles	X	New Zealand	N	Canary, Madeira
N	Central, France			N	Azores
N	Southern	*Northern America*			South Atlantic Isl.
N	Eastern		Canada, Alaska		
N	Mediterranean Isl.		Southeastern USA	*Indian Ocean Islands*	
		X	Western USA		Mascarenes
Africa			Remaining USA		Seychelles
N	Northern	X	Mexico		Madagascar
	Tropical				
X	Southern	*Southern America*		*Pacific Islands*	
			Tropical		Micronesia
Asia			Caribbean		Galapagos
N	Temperate	X	Chile, Argentina	X	Hawaii
N	Tropical			X	Melanesia, Polynesia

Invaded Habitats
Grass- and woodland, coastal estuaries and dunes, salt marshes.

Description
A slender tufted grass with erect culms of 8-65 cm height. Leaf blades are 1-14 cm long and 0.5-3 mm wide. The white ligules are membranous and *c.* 1 mm long. The inflorescence is a narrow sparingly branched panicle or raceme of 5-35 cm length, erect or slightly nodding. Spikelets are 6-15 mm long and have 5-10 florets. The lower glume is 0.4-2.5 mm long, the upper glume 2.5-6.5 mm long, including a short awn. Lemmas are 4.5-7.5 mm long and have an awn each 1-2 times as long as the lemma.

Ecology and Control
A highly competitive grass that suppresses native perennial grasses and forbs and prevents their establishment. Dense swards of this grass displace native vegetation and strongly reduce species richness. The short-lived seeds are dispersed mainly by attaching to animals. The grass is shallow-rooted and well adapted to soils of low fertility.

     A combination of spring and autumn grazing can reduce the density of the grass. Spraying glyphosate during flowering prevents seed set. Other effective herbicides are simazine applied in autumn, propyzamide or dalapon.

References
165, 215, 276, 339, 742, 1358.

Watsonia meriana (L.) Mill.   Iridaceae

LF: Perennial herb   CU: Ornamental
SN: *Watsonia bulbillifera* Mathews & Bolus

Geographic Distribution

Europe	*Australasia*	*Atlantic Islands*
Northern	● Australia	Cape Verde
British Isles	X New Zealand	X Canary, Madeira
Central, France		Azores
Southern	*Northern America*	South Atlantic Isl.
Eastern	Canada, Alaska	
Mediterranean Isl.	Southeastern USA	*Indian Ocean Islands*
	Western USA	Mascarenes
Africa	Remaining USA	Seychelles
Northern	Mexico	Madagascar
Tropical		
N Southern	*Southern America*	*Pacific Islands*
	Tropical	Micronesia
Asia	Caribbean	Galapagos
Temperate	Chile, Argentina	Hawaii
Tropical		Melanesia, Polynesia

Invaded Habitats
Heath- and woodland, riparian habitats, freshwater wetlands, coastal dunes.

Description
A cormous herb of 60-200 cm height, sometimes forming cormlets at the nodes. The rounded corms are coated by fibrous roots and reach 7 cm diameter. Leaves are glabrous, rigid, up to 60 cm long and 6 cm wide, and 5-6 in number. Inflorescences are spikes with 10-15 flowers facing in two opposite directions. The flowers are asymmetric and sessile, reddish-purple to pink, and have sharply bent perianth tubes of 4-5 cm length. Fruits are oblong capsules.

Ecology and Control
This plant favours moist situations and will withstand flooding for several weeks. It grows to dense stands that prevent the regeneration of overstorey species and reduce native plant species richness. It does not set viable seeds but spreads rapidly by the numerous cormlets. These are dispersed by water and in soil.

    Isolated plants can be hand pulled or grubbed at any time before cormlets are forming. Mowing or slashing kills the topgrowth but does not affect the corms in the soil. Cutting at 10-15 cm height when the stems first emerge but before they elongate prevents formation of cormlets. An effective herbicide is 2,2-DPA applied before the first flowers are formed.

References
121, 215, 271, 709, 896, 985, 986, 1000, 1425.

Watsonia versfeldiae Mathews & L. Bolus Iridaceae

LF: Perennial herb
SN: -
CU: Ornamental

Geographic Distribution

Europe	*Australasia*	*Atlantic Islands*
Northern	● Australia	Cape Verde
British Isles	New Zealand	Canary, Madeira
Central, France		Azores
Southern	*Northern America*	South Atlantic Isl.
Eastern	Canada, Alaska	
Mediterranean Isl.	Southeastern USA	*Indian Ocean Islands*
	Western USA	Mascarenes
Africa	Remaining USA	Seychelles
Northern	Mexico	Madagascar
Tropical		
N Southern	*Southern America*	*Pacific Islands*
	Tropical	Micronesia
Asia	Caribbean	Galapagos
Temperate	Chile, Argentina	Hawaii
Tropical		Melanesia, Polynesia

Invaded Habitats
Grass- and woodland, disturbed sites.

Description
A robust cormous herb with stems of 1-2 m height. The rounded corms are coated by fibrous roots. The inflorescence is a dense spike with usually more than 30 flowers, up to 2 m tall, and with the flowers facing in two opposite directions. The white, pink or purple flowers are asymmetric, sessile, and have funnel-shaped perianths of 3.5-4.5 cm length. The corolla tube is gradually bent. Fruits are capsules containing *c.* 20 seeds. These have papery wings.

Ecology and Control
As the previous species, this plant competes vigorously with native species, eventually displacing them. The plant spreads rapidly after disturbances and fires. Vegetative propagation by corms leads to clumps that rapidly expand in size, forming a continuous cover. The above ground parts of the plant dry off during summer and may create fire hazards. Seeds are dispersed by wind. The species flowers abundantly after fires.

Small infestations can be contained by hand pulling and removing the corms, or painting individual plants with herbicide. Removing the flowering stems before seed set slows the spread. An effective herbicide is glyphosate applied to actively growing plants. Follow-up programmes are necessary to control seedlings.

References
215, 613.

Xanthium spinosum L.   Asteraceae

LF: Annual herb
SN: -
CU: -

Geographic Distribution

Europe	*Australasia*	*Atlantic Islands*
Northern	● Australia	Cape Verde
X British Isles	X New Zealand	Canary, Madeira
X Central, France		X Azores
X Southern	*Northern America*	South Atlantic Isl.
X Eastern	Canada, Alaska	
Mediterranean Isl.	X Southeastern USA	*Indian Ocean Islands*
	X Western USA	Mascarenes
Africa	Remaining USA	Seychelles
X Northern	Mexico	Madagascar
X Tropical		
● Southern	*Southern America*	*Pacific Islands*
	N Tropical	Micronesia
Asia	Caribbean	Galapagos
X Temperate	X Chile, Argentina	Hawaii
Tropical		Melanesia, Polynesia

Invaded Habitats
Grassland, riparian habitats, seasonal freshwater wetlands, rock outcrops, disturbed sites.

Description
A large, much-branched and spiny herb of 20-120 cm height, with 1-2 stout and yellow spines in the axils of leaves. Stems are smooth and often pubescent. Leaves are alternate, lanceolate to narrowly ovate, 6-10 cm long and 1.5-2 cm wide, sessile or with short petioles, dark green above and white-tomentose beneath. Each leaf base is armed with yellow spines of 2-5 cm length. Male flowerheads are borne in terminal inflorescences, female flowerheads in axillary ones. Involucres are 10-12 mm long, 6-8 mm wide, and covered with slender spines. Fruits are ovoid burrs, 10-13 mm long and *c.* 4 mm wide, pale yellow and covered with numerous hooked spines. Each burr contains two seeds.

Ecology and Control
This plant is invasive because it forms thorny and impenetrable thickets, displacing native vegetation and affecting wildlife by restricting movement. It establishes mainly in disturbed sites. Seeds are dispersed by water and by attaching to animals.
    Single plants can be hoed. Chemical control is done by spraying plants with 2,4-D, MCPA or imazaquin. Control measures should aim at preventing seed formation.

References
33, 215, 506, 580, 745, 776, 985, 986, 1089.

Xanthium strumarium L.   Asteraceae

LF: Annual herb   CU: -
SN: *Xanthium echinatum* Murr., *X. orientale* L., *X. pungens* Wallr., *X. vulgare* Hill

Geographic Distribution

Europe		*Australasia*		*Atlantic Islands*	
	Northern	●	Australia		Cape Verde
X	British Isles	X	New Zealand	X	Canary, Madeira
X	Central, France			X	Azores
X	Southern	*Northern America*			South Atlantic Isl.
X	Eastern	X	Canada, Alaska		
	Mediterranean Isl.	X	Southeastern USA	*Indian Ocean Islands*	
		N	Western USA		Mascarenes
Africa		X	Remaining USA		Seychelles
X	Northern	X	Mexico		Madagascar
X	Tropical				
●	Southern	*Southern America*		*Pacific Islands*	
		X	Tropical		Micronesia
Asia		N	Caribbean		Galapagos
X	Temperate	X	Chile, Argentina	X	Hawaii
X	Tropical			X	Melanesia, Polynesia

Invaded Habitats
Grass- and heathland, riparian habitats, floodplains, disturbed sites.

Description
A large herb, with erect to decumbent stems of 20-150 cm height, without spines and usually branched. Leaves are alternate and have petioles of 2-8 cm length, the leaf blades are broadly ovate to triangular, 2-12 cm long, entire or lobed, and with stiff hairs on both sides. Flowerheads grow in axillary clusters, occasionally in a terminal inflorescence. Male flowerheads are 5-8 mm wide and located above the female flowerheads. The involucre has spines and distinct beaks. Fruits are ovoid burrs covered with hooked spines and have two straight or curved beaks at the end.

Ecology and Control
A highly variable species with many local variants often granted species status. The exact native range is obscure. It is a widespread weed that is mainly associated with open and disturbed grounds. It forms dense and tall thickets that are species poor and crowd out native vegetation. Fruits are carried by streams and dispersed by adhering to animals.
    Control measures should aim at preventing seed formation. An effective chemical control method is spraying 2,4-D or MCPA before flowering commences. Small plants can be hand pulled or hoed. Prescribed burning is also effective if the fire is hot enough.

References
66, 71, 501, 580, 745, 776, 962, 985, 1089, 1458.

Zantedeschia aethiopica (L.) Spreng.   Araceae

LF: Perennial herb   CU: Ornamental
SN: *Calla aethiopica* L., *Richardia africana* Kunth

Geographic Distribution

Europe		*Australasia*		*Atlantic Islands*	
	Northern	●	Australia		Cape Verde
X	British Isles	X	New Zealand	X	Canary, Madeira
	Central, France			X	Azores
	Southern		*Northern America*		South Atlantic Isl.
	Eastern		Canada, Alaska		
	Mediterranean Isl.		Southeastern USA		*Indian Ocean Islands*
		X	Western USA	X	Mascarenes
Africa			Remaining USA		Seychelles
	Northern		Mexico		Madagascar
	Tropical				
N	Southern		*Southern America*		*Pacific Islands*
			Tropical		Micronesia
Asia			Caribbean		Galapagos
	Temperate		Chile, Argentina	X	Hawaii
	Tropical				Melanesia, Polynesia

Invaded Habitats
Forests, riparian habitats, freshwater wetlands, coastal scrub and grasslands.

Description
A stemless glabrous herb of 60-100 cm height with leaves being all basal and flower scapes somewhat exceeding the leaves. The rootstock is a short rhizome. Leaves have spongy petioles up to 80 cm long and arrow-shaped blades of 10-45 cm length and 10-25 cm width. The spathes are pure white, 10-25 cm long and rhombic-ovate. The flower-bearing spadix is about half as long as the spathe, bright yellow, the lower part being wrapped by the spathe. The small flowers are densely crowded along the spadix with the upper flowers being male and the lower ones female. Fruits are yellow to orange berries with several roundish seeds of *c.* 5 mm length.

Ecology and Control
Where native, this plant grows in sandy or rocky places that are usually sesonally damp. It is invasive because it forms dense patches that crowd out native plants and prevent their regeneration. Dense infestations completely displace the native vegetation. The short-lived seeds are dispersed by water, birds and mammals. Rhizome fragments are carried by streams. Plants in Britain rarely produce any seeds.

    Scattered plants can be hand pulled or dug out, the rootstock must be removed. Seed production can be prevented by spraying plants with 2,4-D, glyphosate or chlorsulfuron before fruits are becoming ripe. Follow-up programmes may be necessary to treat regrowth and seedlings.

References
121, 215, 978, 986, 1149, 1187.

Zizania latifolia (Griseb.) Turcz. ex Stapf   Poaceae

LF: Perennial herb
SN: -
CU: Food

Geographic Distribution

Europe
- Northern
- British Isles
- Central, France
- Southern
- Eastern
- Mediterranean Isl.

Africa
- Northern
- Tropical
- Southern

Asia
- N Temperate
- N Tropical

Australasia
- Australia
- ● New Zealand

Northern America
- Canada, Alaska
- Southeastern USA
- Western USA
- Remaining USA
- Mexico

Southern America
- Tropical
- Caribbean
- Chile, Argentina

Atlantic Islands
- Cape Verde
- Canary, Madeira
- Azores
- South Atlantic Isl.

Indian Ocean Islands
- Mascarenes
- Seychelles
- Madagascar

Pacific Islands
- Micronesia
- Galapagos
- Hawaii
- Melanesia, Polynesia

Invaded Habitats
Freshwater wetlands, riparian habitats.

Description
A 1.5-3 m tall grass with erect to decumbent and often branched stems up to 2 cm in diameter. Leaves are 50-125 cm long, 5-25 mm wide, linear-lanceolate, and have ligules up to 10 mm long. The inflorescence is a large panicle of 30-55 cm length. Spikelets have one floret each and are *c.* 10 mm long, the female spikelets have awns. Glumes are small or lacking. Grains are linear, 12-20 mm long and 1.5-2 mm wide, purplish-black and cylindrical.

Ecology and Control
This grass is an emergent aquatic perennial that spreads by seeds and rhizomes. It forms dense stands that displace native aquatic vegetation and reduce species richness. The grass is well adapted to grow in deep water. Little is known on the ecology of this species.

Specific control methods for this species are not available. Individual plants and small infestations may be removed manually. Larger stands can be treated with herbicides approved for use in aquatic environments.

References
1302, 1415, 1441, 1442.

Ziziphus mauritiana Lam.   Rhamnaceae

LF: Evergreen shrub, tree   CU: Food, ornamental
SN: *Rhamnus jujuba* L.

Geographic Distribution

Europe	*Australasia*	*Atlantic Islands*
Northern	● Australia	X Cape Verde
British Isles	New Zealand	Canary, Madeira
Central, France		Azores
Southern	*Northern America*	South Atlantic Isl.
Eastern	Canada, Alaska	
Mediterranean Isl.	Southeastern USA	*Indian Ocean Islands*
	Western USA	Mascarenes
Africa	Remaining USA	N Seychelles
Northern	Mexico	Madagascar
N Tropical		
X Southern	*Southern America*	*Pacific Islands*
	X Tropical	Micronesia
Asia	X Caribbean	Galapagos
Temperate	Chile, Argentina	Hawaii
N Tropical		Melanesia, Polynesia

Invaded Habitats
Bush- and woodland, coastal sites.

Description
A spiny shrub or small tree of 3-8 m height, occasionally taller, with a dull black and rough bark, and with a spreading, drooping crown. Leaves have stipules and are variable in shape, generally oblong-elliptic, glabrous above and white tomentose beneath, with three prominent veins, 3-8 cm long and 1.5-5 cm wide, and finely toothed. Stipular spines are often in pairs. Greenish yellow flowers of 5-6 mm diameter are borne in axillary clusters along the branches. The pedicels are 5-10 mm long. Fruits are smooth and globose to ellipsoid drupes of 20-50 mm diameter, yellow or red when ripe, with a white pulp and 1-2 compressed seeds.

Ecology and Control
A highly variable species with numerous varieties and cultivars and whose exact native range is obscure. The shrub forms dense and impenetrable thickets affecting wildife and displacing native plants. In Australia, it forms a continuous shrub layer within formerly open eucalypt woodlands. The plant suckers from roots and resprouts even after complete removal of the shoots, e.g. after fires. Seeds are dispersed by birds and mammals.

    Single plants can be slashed and the roots grubbed. Larger stands are slashed and the cut stumps treated with triclopyr. Basal bark sprays with triclopyr or picloram are also effective. Glyphosate can be used as an overall spray. Follow-up programmes are necessary to treat seedlings and regrowth.

References
438, 493, 494, 495, 496, 497, 498, 607, 986.

References

1) Aarssen, L.W. (1981) The biology of Canadian weeds. 50. *Hypochoeris radicata* L. *Canadian Journal of Plant Science* 61, 365-381.

2) Abrahamson, W.G., Anderson, S.S., and McCrea, K.D. (1991) Clonal integration: nutrient sharing between sister ramets of *Solidago altissima* (Compositae). *American Journal of Botany* 78, 1508-1514.

3) Adams, J.B. and Bate, G.C. (1995) Ecological implications of tolerance of salinity and inundation by *Spartina maritima*. *Aquatic Botany* 52, 183-191.

4) Adamson, H.Y., Chow, W.S., Anderson, J.M., Vesk, M., and Sutherland, M.W. (1991) Photosynthetic acclimation of *Tradescantia albiflora* to growth irradiance: morphological, ultrastructural and growth responses. *Physiologia Plantarum* 82, 353-359.

5) Adler, P.B., D'Antonio, C.M., and Tunison, J.T. (1998) Understory succession following a dieback of *Myrica faya* in Hawaii' Volcanoes National Park. *Pacific Science* 52, 69-78.

6) Adoki, A. (1993) Nitrogen fixation and biomass production in *Hyparrhenia rufa*. *Malaysian Agricultural Journal* 55, 31-40.

7) Agami, M. and Reddy, K.R. (1990) Competition for space between *Eichhornia crassipes* (Mart.) Solms and *Pistia stratiotes* L. cultured in nutrient-enriched water. *Aquatic Botany* 38, 195-208.

8) Aiken, S.G., Newroth, P.R., and Wile, I. (1979) The biology of Canadian weeds. 34. *Myriophyllum spicatum* L. *Canadian Journal of Plant Science* 59, 201-215.

9) Akanda, R.U., Mullahey, J.J., and Shilling, D.G. (1996) Environmental factors affecting germination of tropical soda apple (*Solanum viarum*). *Weed Science* 44, 570-574.

10) Akridge, R.E. and Fonteyn, P.J. (1981) Naturalization of *Colocasia esculenta* (Araceae) in the San Marcos river, Texas. *Southwestern Naturalist* 26, 210-211.

11) Albariño, R.J. and Balseiro, E.G. (2002) Leaf litter breakdown in Patagonian streams: native versus exotic trees and the effect of invertebrate size. *Aquatic Conservation* 12, 181-192.

12) Albert, M.E. (1995) Portrait of an invader II: the ecology and management of *Carpobrotus edulis*. *California Exotic Pest Plant Council, CalEPPC News* 1995, 4-6.

13) Albert, M.E., D'Antonio, C.M., and Schierenbeck, K.A. (1997) Hybridization and introgression in *Carpobrotus* spp. (Aizoaceae) in California. I. Morphological evidence. *American Journal of Botany* 84, 896-904.

14) Alexander, R.R. and Carleton, B.E. (1980) Lodgepole pine management in the central Rocky Mountains. *Journal of Forestry* 78, 196-201.

15) Al-Henaid, J.S., Ferrell, M.A., and Miller, S.D. (1993) Effect of 2,4-D on leafy spurge (*Euphorbia esula*) viable seed production. *Weed Technology* 7, 76-81.

16) Ali, S.I. and Faruqi, S.A. (1969) A taxonomic study of *Acacia nilotica* complex in W. Pakistan. *Pakistan Journal of Botany* 1, 1-8.

17) Allen, L.J.S., Allen, E.J., Kunst, C.R.G., and Sosebee, R.E. (1991) A diffusion model for dispersal of *Opuntia imbricata* (cholla) on rangeland. *Journal of Ecology* 79, 1123-1135.

18) Allen, R.B. (1991) A preliminary assessment of the establishment and persistence of *Berberis darwinii* Hook., a naturalised shrub in secondary vegetation near Dunedin, New Zealand. *New Zealand Journal of Botany* 29, 353-360.

19) Allen, R.B. and Lee, W.G. (1989) Seedling establishment microsites of exotic conifers in *Chionochloa rigida* tussock grassland, Otago, New Zealand. *New Zealand Journal of Botany* 27, 491-498.

20) Allen, R.B. and Lee, W.G. (1992) Fruit selection by birds in relation to fruit abundance and appearance in the naturalised shrub *Berberis darwinii*. *New Zealand Journal of Botany* 30, 121-124.

21) Allen, R.B. and Wilson, J.B. (1992) Fruit and seed production in *Berberis darwinii* Hook., a shrub recently naturalised in New Zealand. *New Zealand Journal of Botany* 30, 45-55.

22) Alliende, M.C. and Harper, J.L. (1989) Demographic studies of a dioecious tree. I. Colonization, sex and age structure of a population of *Salix cinerea*. *Journal of Ecology* 77, 1029-1047.

23) Alonso, S.I. and Okada, K.A. (1996) Capacidad de propagación vegetativa de *Alternanthera philoxeroides* en suelos agrícolas. *Ecología Austral* 6, 9-16. *(Spain)*

24) Alvarenga, S. and Flores, E.M. (1988) Morphology and germination of the seed of *Swietenia macrophylla* King (Meliaceae). *Revista de Biologia Tropical* 36, 261-268. *(Spanish)*

25) Amor, R.L. (1974) Ecology and control of blackberry (*Rubus fruticosus* L. agg.). *Weed Research* 14, 231-238.

26) Amor, R.L. (1975) Ecology and control of blackberry (*Rubus fruticosus* L. agg.). IV. Effect of single and repeated applications of 2,4,5-T, picloram and aminotriazole. *Weed Research* 15, 39-45.

27) Amor, R.L. and Harris, R.V. (1974) Distribution and seed production of *Cirsium arvense* (L.) Scop. in Victoria, Australia. *Weed Research* 14, 317-323.

28) Amor, R.L. and Miles, B.A. (1974) Taxonomy and distribution of *Rubus fruticosus* L. agg. (Rosaceae) naturalized in Victoria. *Muelleria* 3, 37-62.

29) Amor, R.L. and Harris, R.V. (1977) Control of *Cirsium arvense* (L.) Scop. by pesticides and mowing. *Weed Research* 17, 303-309.

30) Amor, R.L. and Harris, R.V. (1979) Survey of the distribution of blackberry (*Rubus fruticosus* L. agg.) in Victoria and the use and effectiveness of chemical control measures. *Journal of the Australian Institute of Agricultural Science* 45, 260-263.

31) Anable, M.E., McClaran, M.P., and Ruyle, G.B. (1992) Spread of introduced Lehmann lovegrass *Eragrostis lehmanniana* Nees. in southern Arizona, USA. *Biological Conservation* 61, 181-188.

32) Anderson, J.E. and Reznicek, A.A. (1994) *Glyceria maxima* (Poaceae) in New England. *Rhodora* 96, 97-101.

33) Andrews, J.A. (1993) Control of Bathurst burr (*Xanthium spinosum*) in irrigated soybeans in southern New South Wales. *Plant Protection Quarterly* 8, 15-18.

34) Aniszewski, T., Kupari, M.H., and Leinonen, A.J. (2001) Seed number, seed size and seed diversity in Washington lupin (*Lupinus polyphyllus* Lindl.). *Annals of Botany* 87, 77-82.

35) Andersen, U.V. (1994) Sheep grazing as a method of controlling *Heracleum mantegazzianum*. In: de Waal, L.C., Child, L.E., Wade, P.M., and Brock, J.H. (eds) *Ecology and management of invasive riverside plants*. John Wiley & Sons, Chichester, pp. 77-91.

36) Anderson, M.G. (1995) Interactions between *Lythrum salicaria* and native organisms: a critical review. *Environmental Management* 19, 225-231.

37) Anderson, R.C. and Gardner, D.E. (1999) An evaluation of the wilt-causing bacterium *Ralstonia solanacearum* as a potential biological control agent for the alien kahili ginger (*Hedychium gardnerianum*) in Hawaiian forests. *Biological Control* 15, 89-96.

38) Anderson, R.C., Dhillion, S.S., and Kelley, T.M. (1996) Aspects of the ecology of an invasive plant, garlic mustard (*Alliaria petiolata*), in central Illinois. *Restoration Ecology* 4, 181-191.

39) Anderson, S.J., Stone, C.P., and Higashino, P.K. (1992) Distribution and spread of alien plants in Kipahulu Valley, Haleakala National Park, Above 2,300 ft elevation. In: Stone, C.P., Smith, C.W., and Tunison, J.T. (eds) *Alien plant invasions in native ecosystems of Hawaii: management and research.* Cooperative National Park Resources Studies Unit, University of Hawaii. University of Hawaii Press, Honolulu, pp. 300-338.

40) Andrascik, R.J. (1994) Process for developing a leafy spurge strategic management plan within Theodore Roosevelt National Park. *Leafy Spurge News* 16, 5.

41) Angerstein, M.B. and Lemke, D.E. (1994) First record of the aquatic weed *Hygrophila polysperma* (Acanthaceae) from Texas. *SIDA Contributions to Botany* 16, 365-371.

42) Annecke, D.P. and Moran, V.C. (1978) Critical review of biological pest control in South Africa. 2. The prickly pear *Opuntia ficus-indica* (L.) Miller. *Journal of the Entomological Society of South Africa* 41, 161-188.

43) Anonymous (1999) *The national weeds strategy: a strategic approach to weed problems of national significance.* Agriculture and Resource Management Council of Australia and New Zealand, Australian and New Zealand Environment and Conservation Council and Forestry Ministers, 52 pp.

44) Apfelbaum, S.I. and Sams, C.E. (1987) Ecology and control of reed canarygrass. *Natural Areas Journal* 7, 69-74.

45) Apfelbaum, S.I., Ludwig, J.P., and Ludwig, C.E. (1983) Ecological problems associated with disruption of dune vegetation dynamics by *Casuarina equisetifolia* L. at Sand Island, Midway Atoll. *Atoll Research Bulletin* 261, 1-19.

46) Archibold, O.W., Brooks, D., and Delanoy, L. (1997) An investigation of the invasive shrub European buckthorn, *Rhamnus cathartica* L., near Saskatoon, Saskatchewan. *Canadian Field-Naturalist* 111, 617-621.

47) Ares, A. and Fownes, J.H. (2000) Productivity, nutrient and water-use efficiency of *Eucalyptus saligna* and *Toona ciliata* in Hawaii. *Forest Ecology and Management* 139, 227-236.

48) Ares, A. and Fownes, J.H. (2001) Productivity, resource use, and competitive interactions of *Fraxinus uhdei* in Hawaii uplands. *Canadian Journal of Forest Research* 31, 132-142.

49) Ashenden, T.W., Stewart, W.S., and Williams, W. (1975) Growth responses of sand dune populations of *Dactylis glomerata* L. to different levels of water stress. *Journal of Ecology* 63, 97-107.

50) Asner, G.P. and Beatty, S.W. (1996) Effects of an African grass invasion on Hawaiian shrubland nitrogen biogeochemistry. *Plant and Soil* 186, 205-211.

51) Atkinson, M.D. and Atkinson, E. (2002) Biological flora of the British Isles. *Sambucus nigra* L. *Journal of Ecology* 90, 895-923.

52) Atwal, B.S. and Gopal, R. (1972) *Oxalis latifolia* and its control by chemical and mechanical methods in the hills. *Indian Journal of Weed Science* 4, 74-80.

53) Auge, H. and Brandl, R. (1997) Seedling recruitment in the invasive clonal shrub, *Mahonia aquifolium* Pursh (Nutt.). *Oecologia* 110, 205-211.

54) Auclair, A.N. and Cottam, G. (1971) Dynamics of black cherry (*Prunus serotina* Erhr.) in southern Wisconsin oak forests. *Ecological Monographs* 41, 153-177.

55) Auld, B.A. and Coote, B.G. (1981) Prediction of pasture invasion by *Nassella trichotoma* (Gramineae) in south east Australia. *Protection Ecology* 3, 271-277.

56) Austin, D.F. (1978) Exotic plants and their effects in southeast Florida. *Environmental Conservation* 5, 25-34.

57) Baars, R. and Kelly, D. (1996) Survival and growth responses of native and introduced vines in New Zealand to light availability. *New Zealand Journal of Botany* 34, 389-400.

58) Bailey, J.P. (1994) Reproductive biology and fertility of *Fallopia japonica* (Japanese knotweed) and its hybrids in the British Isles. In: de Waal, L.C., Child, L.E., Wade, P.M., and Brock, J.H. (eds) *Ecology and management of invasive riverside plants.* John Wiley & Sons, Chichester, pp. 141-158.

59) Bailey, J.P. (1997) The Japanese knotweed invasion of Europe; the potential for further evolution in non-native regions. In: Yano, E., Matsuo, K., Shiyomi, M., and Andow, D.A. (eds) *Biological invasions of ecosystem by pests and beneficial organisms.* National Institute of Agro-Environmental Sciences (NIAES) Series 3, Tsukuba, Japan, pp. 27-37.

60) Bailey, J.P., Child, L.E., and Wade, M. (1995). Assessment of the genetic variation and spread of British populations of *Fallopia japonica* and its hybrid *Fallopia x bohemica*. In: Pysek, P., Prach, K., Rejmánek, M., and Wade, M. (eds) *Plant invasions - general aspects and special problems.* SPB Academic Publishing, Amsterdam, pp. 141-150.

61) Baker, J.B. (1973) Intensive cultural practices increases growth of juvenile slash pine in Florida sandhills. *Forest Science* 19, 197-202.

62) Ball, J.P. (1990) Influence of subsequent flooding depth on cattail control by burning and mowing. *Journal of Aquatic Plant Management* 28, 32-36.

63) Ballard, R. (1986) *Bidens pilosa* complex (Asteraceae) in North and Central America. *American Journal of Botany* 73, 1452-1465.

64) Balneaves, J.M. and Perry, C. (1982) Long-term control of gorse/bracken mixtures for forest establishment in Nelson. *New Zealand Journal of Forestry* 27, 219-225.

65) Banting, J.D. (1966) Studies on the persistence of *Avena fatua*. *Canadian Journal of Plant Science* 46, 129-140.

66) Bararpour, M.T. and Oliver, L.R. (1998) Effect of tillage and interference on common cocklebur (*Xanthium strumarium*) and sicklepod (*Senna obtusifolia*) population, seed production, and seedbank. *Weed Science* 46, 424-431.

67) Barden, L.S. (1987) Invasion of *Microstegium vimineum* (Poaceae), an exotic, annual, shade-tolerant, C_4 grass, into a North Carolina floodplain. *The American Midland Naturalist* 118, 40-45.

68) Barden, L.S. and Matthews, J.F. (1980) Change in abundance of honeysuckle (*Lonicera japonica*) and other ground flora after prescribed burning of a Piedmont pine forest. *Castanea* 45, 257-260.

69) Barker, M.G. and Perez, S.D. (2000) Comparative water relations of mature mahogany (*Swietenia macrophylla*) trees with and without lianas in a subhumid, seasonally dry forest in Bolivia. *Tree Physiology* 20, 1167-1174.

70) Barnes, W.J. (1999) The rapid growth of a population of reed canarygrass (*Phalaris arundinacea* L.) and its impact on some riverbottom herbs. *Journal of the Torrey Botanical Society* 126, 133-138.

71) Barrentine, W.L. (1989) Minimum effective rate of chlorimuron and imazaquin applied to common cocklebur (*Xanthium strumarium*). *Weed Technology* 3, 126-130.

72) Barreto, R.W. and Evans, H.C. (1995) The mycobiota of the weed *Mikania micrantha* in southern Brazil with particular reference to fungal pathogens for biological control. *Mycological Research* 99, 343-352.

73) Barrilleaux, T.C. and Grace, J.B. (2000) Growth and invasive potential of *Sapium sebiferum* (Euphorbiaceae) within the coastal prairie region: effects of soil and moisture regime. *American Journal of Botany* 87, 1099-1106.

74) Baruch, Z. and Fernández, D.S. (1993) Water relations of native and introduced C_4 grasses in a neotropical savanna. *Oecologia* 96, 179-185.

75) Baruch, Z., Pattison, R.R., and Goldstein, G. (2000) Responses to light and water availability of four invasive Melastomataceae in the Hawaiian islands. *International Journal of Plant Sciences* 161, 107-118.

76) Basinger, M. (1998) *Aster subulatus* var. *ligulatus* (Asteraceae) and *Chloris virgata* (Poaceae) in Illinois. *Transactions of the Illinois State Academy of Science* 91, 119-121.

77) Baskin, J.M. and Baskin, C.C. (1979) Studies on the autecology and population biology of the weedy monocarpic perennial *Pastinaca sativa*. *Journal of Ecology* 67, 601-610.

78) Baskin, J.M. and Baskin, C.C. (1990) Seed germination ecology of poison hemlock, *Conium maculatum*. *Canadian Journal of Botany* 68, 2018-2024.

79) Baskin, J.M. and Baskin, C.C. (1992) Seed germination biology of the weedy biennial *Alliaria petiolata*. *Natural Areas Journal* 12, 191-197.

80) Bass, D.A. (1990) Dispersal of an introduced shrub (*Crataegus monogyna*) by the brush-tailed possum (*Trichosurus vulpecula*). *Australian Journal of Ecology* 15, 227-229.

81) Bassett, I.J. and Crompton, C.W. (1978) The biology of Canadian weeds. 32. *Chenopodium album* L. *Canadian Journal of Plant Science* 58, 1061-1072.

82) Bassett, I.J. and Munro, D.B. (1985) The biology of Canadian weeds. 67. *Solanum ptycanthum* Dun., *S. nigrum* L. and *S. sarrachoides* Sendt. *Canadian Journal of Plant Science* 65, 401-414.

83) Bastlová-Hanzélyová, D. (2001) Comparative study of native and invasive populations of *Lythrum salicaria*: population characteristics, site and community relationships. In: Brundu, G. *et al.* (eds) *Plant invasions: species ecology and ecosystem management.* Backhuys Publishers, Leiden, pp. 33-40.

84) Basu, D. and Chakraverty, R.K. (1986) Dormancy, viability and germination of *Adenanthera pavonina* seeds. *Acta Botanica Indica* 14, 68-72.

85) Baum, B.R. and Bailey, L.G. (1984) Taxonomic studies in wall berley (*Hordeum murinum* sensu lato) and sea barley (*Hordeum marinum* sensu lato). 2. Multivariate morphometrics. *Canadian Journal of Botany* 62, 2754-2764.

86) Bayer, E. von (1987) Die Gattung *Alstroemeria* in Chile. *Mitteilungen der Botanischen Staatssammlung München* 24, 1-362.

87) Beatley, J.C. (1966) Ecological status of introduced brome grasses (*Bromus* spp.) in desert vegetation of southern Nevada. *Ecology* 47, 548-554.

88) Beatty, S.W. and Licari, D.L. (1992) Invasion of fennel (*Foeniculum vulgare*) into shrub communities on Santa Cruz island, California. *Madroño* 39, 54-66.

89) Bebawi, F.F. and Campbell, S.D. (2000) Effects of fire on germination and viability of rubber vine (*Cryptostegia grandiflora*) seeds. *Australian Journal of Experimental Agriculture* 40, 949-957.

90) Bebawi, F.F. and McKenzie, J.R. (1999) Pod classification and its role in rubber vine (*Cryptostegia grandiflora*) germination and emergence. *Plant Protection Quarterly* 14, 30-34.

91) Bebawi, F.F. and Row, P.J. (2001) Effect of dry heat on germination and viability of *Cryptostegia grandiflora* seeds. *Plant Protection Quarterly* 16, 108-110.

92) Bebawi, F.F., Campbell, S.D., Lindsay, A.M., and Grice, A.G. (2000) Impact of fire on rubber vine (*Cryptostegia grandiflora* R.Br.) and associated pasture and germinable seed bank in a sub-riparian habitat of north Queensland. *Plant Protection Quarterly* 15, 62-66.

93) Beddows, A.R. (1959) Biological flora of the British Isles. *Dactylis glomerata*. *Journal of Ecology* 47, 223-239.

94) Beddows, A.R. (1961) Biological flora of the British Isles. *Holcus lanatus* L. *Journal of Ecology* 49, 421-430.

95) Beerling, D.J. (1991) The effect of riparian land use on the occurrence and abundance of Japanese knotweed (*Reynoutria japonica*) on selected rivers in South Wales. *Biological Conservation* 55, 329-337.

96) Beerling, D.J. (1993) The impact of temperature on the northern distribution limits of the introduced species *Fallopia japonica* and *Impatiens glandulifera* in north-west Europe. *Journal of Biogeography* 20, 45-53.

97) Beerling, D.J. and Perrins, J.M. (1993) Biological flora of the British Isles. *Impatiens glandulifera* Royle (*Impatiens roylei* Walp.). *Journal of Ecology* 81, 367-382.

98) Beerling, D.J. and Palmer, J.P. (1994) Status of *Fallopia japonica* (Japanese knotweed) in Wales. In: de Waal, L.C., Child, L.E., Wade, P.M., and Brock, J.H. (eds) *Ecology and management of invasive riverside plants*. John Wiley & Sons, Chichester, pp. 199-211.

99) Beerling, D.J., Bailey, J.P., and Conolly, A.P. (1994) Biological flora of the British Isles. *Fallopia japonica* (Houtt.) Ronse Decraene. *Journal of Ecology* 82, 959-979.

100) Belcher, J.W. and Wilson, S.D. (1989) Leafy spurge and the species composition of a mixed-grass prairie. *Journal of Range Management* 42, 172-175.

101) Bell, G.P. (1997) Ecology and management of *Arundo donax*, and approaches to riparian habitat restoration in southern California. In: Brock, J.H., Wade, M., Pysek, P., and Green, D. (eds) *Plant invasions: studies from North America and Europe*. Backhuys Publishers, Leiden, pp. 103-113.

102) Bellingham, P.J. (1998) Shrub succession and invasibility in an New Zealand montane grassland. *Australian Journal of Ecology* 23, 562-573.

103) Benefield, C.B., Tomaso, J.M., Kyser, G.B., Orloff, S.B., Churches, K.R., Marcum, D.B., and Nader, G.A. (1999) Success of mowing to control yellow starthistle depends on timing and plants branching form. *California Agriculture* 53, 17-21.

104) Beniwal, B.S., Dhawan, V.K., and Joshi, S.R. (1990) Effect of shade and mulch on germination of *Toona ciliata* Roem. *Indian Forester* 116, 942-945.

105) Bertacchi, A., Lombardi, T., and Onnis, A. (2001) *Robinia pseudoacacia* in the forested agricultural landscape of the Pisan hills (Italy). In: Brundu, G., Brock, J., Camarda, I., Child, L., and Wade, M. (eds) *Plant invasions: species ecology and ecosystem management*. Backhuys Publishers, Leiden, pp. 41-46.

106) Bertness, M.D. (1991) Zonation of *Spartina patens* and *Spartina alterniflora* in a New England salt marsh. *Ecology* 72, 138-148.

107) Best, K.F., Bowes, G.G., Thomas, A.G., and Maw, M.G. (1980) The biology of Canadian weeds. 39. *Euphorbia esula* L. *Canadian Journal of Plant Science* 60, 651-663.

108) Bhatt, Y.D., Rawat, Y.S., and Singh, S.P. (1994) Changes in ecosystem functioning after replacement of forest by *Lantana* shrubland in Kumaun Himalaya. *Journal of Vegetation Science* 5, 67-70.

109) Bi, H. and Turvey, N.D. (1994) Interspecific competition between seedlings of *Pinus radiata*, *Eucalyptus regnans* and *Acacia melanoxylon*. *Australian Journal of Botany* 42, 61-70.

110) Bianchi, D.S., Senften, J.K., and Felber, F. (2002) Isozyme variation of *Hordeum murinum* in Switzerland and test of hybridization with cultivated barley. *Weed Research* 42, 325-333.

111) Bímová, K., Mandák, B., and Pysek, P. (2001) Experimental control of *Reynoutria* congeners: a comparative study of a hybrid and its parents. In: Brundu, G., Brock, J., Camarda, I., Child, L., and Wade, M. (eds) *Plant invasions: species ecology and ecosystem management*. Backhuys Publishers, Leiden, pp. 283-290.

112) Biedenbender, S.H. and Roundy, B.A. (1996) Establishment of native semidesert grasses into existing stands of *Eragrostis lehmanniana* in southeastern Arizona. *Restoration Ecology* 4, 155-162.

113) Bishop, G.F. and Davy, A.J. (1984) Significance of rabbits for the population regulation of *Hieracium pilosella* in Breckland (East Anglia, England, UK). *Journal of Ecology* 72, 273-284.

114) Bishop, G.F. and Davy, A.J. (1994) Biological flora of the British Isles. *Hieracium pilosella* L. (*Pilosella officinarum* F. Schultz & Schultz-Bip.). *Journal of Ecology* 82, 195-210.

115) Blackburn, W.H., Knight, R.W., and Schuster, J.L. (1982) Saltcedar influence on sedimentation in the Brazos River. *Journal of Soil and Water Conservation* 37, 298-301.

116) Blair, R.M. (1990) *Gleditsia triacanthos* L. Honeylocust. In: Burns, R.M. and Honkala, B.H. (eds) *Silvics of North America, vol. 2. Hardwoods*. Agriculture Handbook 654, US Department of Agriculture, Washington DC, pp. 358-364.

117) Blanco, M.A., Schrauf, G., and Deregibus, V.A. (1995) Germination response of *Paspalum dilatatum* from the flooding Pampa of Argentina to different incubation and pre-incubation conditions. *Ecologia Austral* 5, 149-155.

118) Blank, R.R. and Young, J.A. (1997) *Lepidium latifolium*: influences on soil properties, rate of spread, and competitive stature. In: Brock, J.H., Wade, M., Pysek, P., and Green, D. (eds) *Plant invasions: studies from North America and Europe*. Backhuys Publishers, Leiden, pp. 69-80.

119) Blankespoor, G.W. and Larson, E.A. (1994) Response of smooth brome (*Bromus inermis* Leyss.) to burning under varying soil moisture conditions. *The American Midland Naturalist* 131, 266-272.

120) Bleakley, S. and Matheson, A.C. (1992) Patterns of morphological variation in seedlings of *Acacia mearnsii* De Wild. *Commonwealth Forestry Review* 71, 101-109.

121) Blood, K. (2001) *Environmental weeds: a field guide for SE Australia*. CRC Weed Management Systems, Adelaide.

122) Blossey, B. (1993) Herbivory below ground and biological weed control: life history of a root-boring weevil on purple loosestrife. *Oecologia* 94, 380-387.

123) Blumenthal, M.J. and Harris, C.A. (1998) Effects of photoperiod and temperature on shoot, root and rhizome growth in three *Lotus uliginosus* Schkuhr populations. *Annals of Botany* London 81, 55-59.

124) Blumenthal, M.J., Bowman, A.M., Cole, A., Jones, R.M., Kelman, W.M., Launders, T.E., and Nicol, H.I. (1999) Establishment, growth and persistence of greater lotus (*Lotus uliginosus*) at six sites in eastern Australia. *Australian Journal of Experimental Agriculture* 39, 819-827.

125) Bocchieri, E., de Martis, B., and Marchioni, A. (1980) Physioecology of germination of *Parapholis strigosa* at different conditions of salinity and temperature. *Bollettino della Societa' Sarda di Scienze Naturali* 20, 139-156. *(Italian)*

126) Bocchieri, E., de Martis, B., and Marchioni, A. (1980) *Parapholis incurva* (Gramineae). 2. The ecology of germination. *Bollettino della Societa' Sarda di Scienze Naturali* 20, 131-138. *(Italian)*

127) Böcker, R. and Kowarik, I. (1982) Der Götterbaum (*Ailanthus altissima*) in Berlin (West). *Berliner Naturschutzblätter* 26, 4-9. *(German)*

128) Boggs, K.W. and Story, J.M. (1987) The population age structure of spotted knapweed (*Centaurea maculosa*) in Montana. *Weed Science* 35, 194-198.

129) Boring, L.R. and Swank, W.T. (1984) The role of black locust (*Robinia pseudoacacia*) in forest succession. *Journal of Ecology* 72, 749-766.

130) Bossard, C.C. (1991) The role of habitat disturbance, seed predation and ant dispersal on establishment of the exotic shrub *Cytisus scoparius* in California. *The American Midland Naturalist* 126, 1-13.

131) Bossard, C.C. (1993) Seed germination in the exotic shrub *Cytisus scoparius* (Scotch broom) in California. *Madroño* 40, 47-61.

132) Bossard, C.C. and Rejmánek, M. (1994) Herbivory, growth, seed production, and resprouting of an exotic invasive shrub *Cytisus scoparius*. *Biological Conservation* 67, 193-200.

133) Bossard, C.C., Randall, J.M., and Hoshovsky, M.C. (2000) *Invasive plants of California's wildlands*. University of California Press, Berkeley, California.

134) Boswell, C.C. and Espie, P.R. (1998) Uptake of moisture and nutrients by *Hieracium pilosella* and effects on soil in a dry sub-humid grassland. *New Zealand Journal of Agricultural Research* 41, 251-261.

135) Bosy, J.L. and Reader, R.J. (1995) Mechanisms underlaying the suppression of forb seedling emergence by grass (*Poa pratensis*) litter. *Functional Ecology* 9, 635-639.

136) Botta-Dukát, Z. and Dancza, I. (2001) Effect of weather conditions on the growth of *Solidago gigantea*. In: Brundu, G., Brock, J., Camarda, I., Child, L., and Wade, M. (eds) *Plant invasions: species ecology and ecosystem management*. Backhuys Publishers, Leiden, pp. 47-54.

137) Boucher, C. and Stirton, C.H. (1980) *Plant invaders - beautiful but dangerous.* Dept. of Nature and Environmental Conservation, Cape Town.

138) Bourdôt, G.W. and Hurrell, G.A. (1989) Cover of *Stipa neesiana* Trin. & Rupr. (Chilean needle grass) on agricultural and pastoral land near Lake Grassmere, Marlborough. *New Zealand Journal of Botany* 27, 415-420.

139) Bourdôt, G.W. and Hurrell, G.A. (1989) Ingress of *Stipa neesiana* Trin. & Rupr. into swards of *Lolium perenne* L., *Dactylis glomerata* L. and *Phalaris aquatica* L., as affected by fertiliser and 2,2-DPA. *New Zealand Journal of Agricultural Research* 32, 317-326.

140) Bourdôt, G.W. and Hurrell, G.A. (1992) Aspects of the ecology of *Stipa neesiana* Trin. & Rupr. seeds. *New Zealand Journal of Agricultural Research* 35, 101-108.

141) Bovey, R.W. (1965) Control of russian olive by aerial application of herbicides. *Journal of Range Management* 18, 194-195.

142) Bovey, R.W. and Meyer, R.E. (1985) Herbicide mixtures for control of honey mesquite (*Prosopis glandulosa*). *Weed Science* 33, 349-352.

143) Bovey, R.W. and Meyer, R.E. (1989) Control of huisache and honey mesquite with a carpeted roller herbicide applicator. *Journal of Range Management* 42, 407-411.

144) Bowden, B.N. (1963) Studies on *Andropogon gayanus* Kunth. I. The use of *Andropogon gayanus* in agriculture. *Empire Journal of Experimental Agriculture* 31, 267-273.

145) Bowden, B.N. (1963) Studies on *Andropogon gayanus* Kunth. II. An outline of the morphology and anatomy of *Andropogon gayanus* var. *bisquamulatus* (Hochst.) Hack. *Journal of the Linnean Society* (Botany) 58, 509-519.

146) Bowden, B.N. (1964) Studies on *Andropogon gayanus* Kunth. III. An outline of its biology. *Journal of Ecology* 47, 255-271.

147) Bowden, D. and Rogers, R.W. (1996) *Protasparagus densiflorus:* an environmental weed of coastal vegetation reserves. *Pacific Conservation Biology* 2, 293-298.

148) Bowmer, K.H., Mitchell, D.S., and Short, D.L. (1984) Biology of *Elodea canadensis* Mich. and its management in Australian irrigation systems. *Aquatic Botany* 18, 231-238.

149) Boyd, R.S. (1992) Influence of *Ammophila arenaria* on foredune plant microdistributions at Point Reyes National Seashore, California. *Madroño* 39, 67-76.

150) Brabec, J. and Pysek, P. (2000) Establishment and survival of three invasive taxa of the genus *Reynoutria* (Polygonaceae) in mesic mown meadows: a field experimental study. *Folia Geobotanica* 35, 27-42.

151) Bradbury, M. (1990) The effect of water stress on growth and dry matter distribution in juvenile *Sesbania sesban* and *Acacia nilotica*. *Journal of Arid Environments* 18, 325-333.

152) Braithwaite, R.W., Lonsdale, W.M., and Esthbergs, J.A. (1989) Alien vegetation and native biota in tropical Australia: the impact of *Mimosa pigra*. *Biological Conser-vation* 48, 189-210.

153) Brandes, D. (1989) On the sociology of some neophytes of the Insubrian region. *Tuexenia* 9, 267-274. *(German)*

154) Brandes, D. (1991) Soziologie und Ökologie von *Oxalis pes-caprae* L. im Mittelmeer-gebiet unter besonderer Berücksichtigung von Malta. *Phytocoenologia* 19, 285-306. *(German)*

155) Brandt, L.A. and Black, D.W. (2001) Impacts of the introduced fern, *Lygodium micro-phyllum*, on the native vegetation of tree islands in the Arthur R. Marshall Loxahatchee National Wildlife Refuge. *Florida Scientist* 64, 191-196.

156) Breidahl, R. and Hewett, P.J. (1995) A review of silvicultural research in the karri (*Eucalyptus diversicolor*) forest. *CALMScience* 2, 51-100.

157) Brenton, R.K. and Klinger, R.C. (2002) Factors influencing the control of fennel (*Foe-niculum vulgare* Miller) using triclopyr on Santa Cruz Island, California, USA. *Natural Areas Journal* 22, 135-147.

158) Brock, J.H. (1994) *Tamarix* (saltcedar), an invasive exotic woody plant in arid and semi-arid riparian habitats of western USA. In: De Waal, L.C. *et al.* (eds) *Ecology and management of invasive riverside plants*. Wiley, New York, pp. 27-44.

159) Brock, J.H. (1998) Invasion, ecology and management of *Elaeagnus angustifolia* (Russian Olive) in the southwestern United States of America. In: Starfinger, U., Edwards, K., Kowarik, I., and Williamson, M. (eds) *Plant invasions - ecological mechanisms and human responses*. Backhuys Publishers, Leiden, pp. 123-136.

160) Brock, J.H., Child, L.E., de Waal, L.C., and Wade, M. (1995) The invasive nature of *Fallopia japonica* is enhanced by vegetative regeneration from stem tissues. In: Pysek, P. *et al.* (eds) *Plant invasions - general aspects and special problems*. SPB Academic Publishing, Amsterdam, pp. 131-140.

161) Brook, R.M. (1989) Review of literature on *Imperata cylindrica* (L.) Raeuschel with particular reference to South East Asia. *Tropical Pest Management* 35, 12-25.

162) Brotherson, J.D. and Winkel, V. (1986) Habitat relationships of saltcedar (*Tamarix ramosissima*) in central Utah. *Great Basin Naturalist* 46, 535-541.

163) Brotherson, J.D. and Field, D. (1987) *Tamarix*: impacts of a successful weed. *Range-lands* 9, 110-112.

164) Brown, C.L. and Whelan, R.J. (1999) Seasonal occurrence of fire and availability of germinable seeds in *Hakea sericea* and *Petrophile sessilis*. *Journal of Ecology* 87, 932-941.

165) Brown, C.S. and Rice, K.J. (2000) The mark of Zorro: effects of the exotic annual grass *Vulpia myuros* on California native perennial grasses. *Restoration Ecology* 8, 10-17.

166) Brown, J.R. and Carter, J. (1998) Spatial and temporal patterns of exotic shrub invasion in an Australian tropical grassland. *Landscape Ecology* 13, 93-102.

167) Bruce, K.A., Cameron, G.N., and Harcombe, P.A. (1995) Initiation of a new woodland type on the Texas coastal prairie by the Chinese tallow tree (*Sapium sebiferum* (L.) Roxb.). *Bulletin of the Torrey Botanical Club* 116, 371-377.

168) Bruce, K.A., Cameron, G.N., Harcombe, P.A., and Jubinsky, G. (1997) Introduction, impact on native habitats, and management of a woody invader, the Chinese tallow tree, *Sapium sebiferum* (L.) Roxb. *Natural Areas Journal* 17, 255-260.

169) Bryson, C.T. and Carter, R. (1993) Cogongrass, *Imperata cylindrica*, in the United States. *Weed Technology* 7, 1005-1009.

170) Buckingham, G.R. (1996) Biological control of alligatorweed, *Alternanthera philoxeroides*, the world's first aquatic weed success story. *Castanea* 61, 232-243.

171) Buell, A.C., Pickart, A.J., and Stuart, J.D. (1995) Introduction history and invasion patterns of *Ammophila arenaria* on the north coast of California. *Conservation Biology* 9, 1587-1593.

172) Bungard, R.A., Daly, G.T., McNeil, D.L., Jones, A.V., and Morton, J.D. (1997) *Clematis vitalba* in a New Zealand native forest remnant: does seed germination explain distribution? *New Zealand Journal of Botany* 35, 525-534.

173) Bungard, R.A., McNeil, D., and Morton, J.D. (1997) Effects of chilling, light and nitrogen-containing compounds on germination, rate of germination and seed imbibition of *Clematis vitalba* L. *Annals of Botany* 79, 643-650.

174) Bungard, R.A., Morton, J.D., McNeil, D.L., and Daly, G.T. (1998) Effects of irradiance and nitrogen on *Clematis vitalba* establishment in a New Zealand lowland podocarp forest remnant. *New Zealand Journal of Botany* 36, 661-670.

175) Burdon, J.J. (1983) Biological flora of the British Isles. *Trifolium repens* L. *Journal of Ecology* 71, 307-330.

176) Burdon, J.J. and Chilvers, G.A. (1977) Preliminary studies on a native Australian eucalypt forest invaded by exotic pines. *Oecologia* 31, 1-12.

177) Burdon, J.J. and Chilvers, G.A. (1994) Demographic changes and the development of competition in a native Australian eucalypt forest invaded by exotic pines. *Oecologia* 97, 419-423.

178) Burdon, J.J., Thrall, P.H., Groves, R.H., and Chaboudez, P. (2000) Biological control of *Carduus pycnocephalus* and *C. tenuiflorus* using the rust fungus *Puccinia cardui-pycnocephali*. *Plant Protection Quarterly* 15, 14-17.

179) Burgason, B.N. (1976) Prescribed burning for management of hawthorn and elder. *New York Fish and Game Journal* 23, 160-169.

180) Burgess, T.L., Bowers, J.E., and Turner, R.M. (1991) Exotic plants at the desert laboratory, Tucson, Arizona. *Madroño* 38, 96-114.

181) Bürki, C. and Nentwig, W. (1997) Comparison of herbivore insect communities of *Heracleum sphondylium* and *H. mantegazzianum* in Switzerland (Spermatophyta: Apiaceae). *Entomologia Generalis* 22, 147-155.

182) Burns, R.M. and Honkala, B.H. (1990) *Silvics of North America. Vol. 1. Conifers.* US Department of Agriculture, Agriculture Handbook 654. Washington, DC.

183) Burrell, J.P. (1981) Invasion of coastal heaths of Victoria by *Leptospermum laevigatum* (J. Gaertn.) F. Muell. *Australian Journal of Botany* 29, 747-764.

184) Burrows, J.E. (1989) Kudzu vine - a new plant invader of South Africa. *Veld & Flora* 75, 116-117.

185) Busby, F.E. and Schuster, J.L. (1971) Woody phreatophyte infestation of the middle Brazos River flood plain. *Journal of Range Management* 24, 285-287.

186) Busch, D.E. and Smith, S.D. (1995) Mechanisms associated with decline of woody species in riparian ecosystems of the southwestern US. *Ecological Monographs* 65, 347-370.

187) Busey, P. (1986) Morphological identification of St. Augustinegrass (*Stenotaphrum secundatum*) cultivars. *Crop Science* 26, 28-32.

188) Busey, P., Broschat, T.K., and Center, B.J. (1982) Classification of St. Augustinegrass (*Stenotaphrum secundatum*). *Crop Science* 22, 469-473.

189) Bush, J.K. and van Auken, O.W. (1990) Growth and survival of *Prosopis glandulosa* seedlings associated with shade and herbaceous competition. *Botanical Gazette* 151, 234-239.

190) Bush, J.K. and van Auken, O.W. (1991) Importance of time of germination and soil depth on growth of *Prosopis glandulosa* (Leguminosae) seedlings in the presence of a C_4 grass. *American Journal of Botany* 78, 1732-1739.

191) Caffrey, J.M. (1994) Spread and management of *Heracleum mantegazzianum* (giant hogweed) along Irish river corridors. In: de Waal, L.C. *et al.* (eds) *Ecology and management of invasive riverside plants.* John Wiley & Sons, Chichester, pp. 67-76.

192) Caffrey, J.M. (1999) Phenology and long-term control of *Heracleum mantegazzianum*. *Hydrobiologia* 415, 223-228.

193) Callaway, J.C. (1992) The introduction and spread of smooth cordgrass (*Spartina alterniflora*) in South San Francisco Bay. *Estuaries* 15, 218-226.

194) Calley, M., Braithwaite, B.W., and Ladd, P.G. (1993) Reproductive biology of *Ravenala madagascariensis* Gmel. as an alien species. *Biotropica* 25, 61-72.

195) Callihan, R.H., Prather, T.S., and Northam, F.E. (1993) Longevity of yellow starthistle (*Centaurea solstitialis*) achenes in soil. *Weed Technology* 7, 33-35.

196) Cameron, G. and Spencer, S.R. (1989) Rapid leaf decay and nutrient release in a Chinese tallow forest. *Oecologia* 80, 222-228.

197) Cameron, L. (1988) The biology of Australian weeds. 18. *Hypochoeris radicata* L. *Plant Protection Quarterly* 3, 156-162.

198) Campbell, M.H. (1974) Efficiency of aerial techniques for long term control of serrated tussock (*Nassella trichotoma*). *Australian Journal of Experimental Agriculture and Animal Husbandry* 14, 405-411.

199) Campbell, M.H. (1982) The biology of Australian weeds. 9. *Nassella trichotoma* (Nees) Arech. *Journal of the Australian Institute of Agricultural Science* 48, 76-84.

200) Campbell, M.H. (1985) Germination, emergence and seedling growth of *Hypericum perforatum* L. *Weed Research* 25, 259-266.

201) Campbell, M.H. and Delfosse, E.S. (1984) The biology of Australian weeds. 13. *Hypericum perforatum* L. *Journal of the Australian Institute of Agricultural Science* 50, 63-73.

202) Campbell, M.H., Miller, L.G., and Nicol, H.I. (1998) Effect of herbicides on seedhead production and control of serrated tussock (*Nassella trichotoma* (Nees) Arech.). *Plant Protection Quarterly* 13, 106-110.

203) Campbell, M.H. and Nicol, H.I. (1999) Seed dormancy in serrated tussock (*Nassella trichotoma* (Nees) Arech.) in New South Wales. *Plant Protection Quarterly* 14, 82-85.

204) Campbell, M.H. and Nicol, H.I. (2001) Herbicides to replace flupropanate for the control of serrated tussock (*Nassella trichotoma* (Nees) Arech.). *Plant Protection Quarterly* 16, 69-74.

205) Campbell, P.L. (1988) Seed germination of *Harrisia martinii* and *Pereskia aculeata* with reference to their potential spread in Natal. *Applied Plant Science* 2, 60-62.

206) Campbell, P.L., Bell, R.S., and Kluge, R.L. (1990) Identifying the research requirements for the control of silver wattle (*Acacia dealbata*) in Natal. *South African Forestry Journal* 155, 37-41.

207) Campbell, P.L. and van Staden, J. (1990) Utilization of solasodine from fruits for long-term control of *Solanum mauritianum*. *South African Forestry Journal* 155, 57-60.

208) Campbell, P.L. and van Staden, J. (1994) The viability and germination characteristics of exhumed *Solanum mauritianum* seeds buried for different periods of time. *Plant Growth Regulation* 14, 97-108.

209) Campbell, P.L., van Staden, J., Stevens, C., and Whitwell, M.I. (1992) The effects of locality, season and year of seed collection on the germination of bugweed (*Solanum mauritianum* Scop.) seeds. *South African Journal of Botany* 58, 310-316.

210) Campbell, P. (2000) *Wattle control.* Plant Protection Research Institute Handbook no. 3, Plant Protection Research Institute, Pretoria, South Africa.

211) Caple, D.R. (1971) Lehmann lovegrass on the Santa Rita experimental range, 1937-1968. *Journal of Range Management* 24, 17-21.

212) Carman, J.G. and Brotherson, J.D. (1982) Comparisons of sites infested and not infested with saltcedar (*Tamarix ramosissima*) and russian olive (*Elaeagnus angustifolia*). *Weed Science* 30, 360-364.

213) Carr, B. (1996) Bridal creeper at Woodman Point - its current status and recommended control strategies. *Plant Protection Quarterly* 11, 67-69.

214) Carr, G.W. (1993) Exotic flora of Victoria and its impact on indigenous biota. In: Foreman, D.B. and Walsh, N.G. (eds) *Flora of Victoria*. Inkata Press, Melbourne, pp. 256-297.

215) Carr, G.W., Yugovic, J.V., and Robinson, K.E. (1992) *Environmental weed invasions in Victoria. Conservation and management implications.* Department of Conservation and Environment, Melbourne, Victoria, Australia.

216) Carter, G.A. and Teramura, A.H. (1988) Vine photosynthesis and relationships to climbing mechanics in a forest understory. *American Journal of Botany* 75, 1011-1018.

217) Carter, J.O. (1994) *Acacia nilotica*: a tree legume out of control. In: Gutteridge, R.C. and Shelton, H.M. (eds) *Forage tree legumes in tropical agriculture*. CAB International, Wallingford, UK, pp. 338-351.

218) Carter, K. and Signor, A. (2000) Controlling broom (*Cytisus scoparius* (L.) Link) in native forest ecosystems. *Plant Protection Quarterly* 15, 165-166.

219) Cartwright, B. and Kok, L.T. (1985) Growth responses of musk and plumeless thistles (*Carduus nutans* and *Carduus acanthoides*) to damage by *Trichsirocalus horridus* (Coleoptera: Curculionidae). *Weed Science* 33, 57-62.

220) Castellanos, E.M., Heredia, C., Figueroa, M.E., and Davy, A.J. (1998) Tiller dynamics of *Spartina maritima* in successional and non-successional mediterranean salt marsh. *Plant Ecology* 137, 213-225.

221) Catling, P.M. and Porebski, Z.S. (1994) The history of invasion and current status of glossy buckthorn, *Rhamnus frangula*, in southern Ontario. *Canadian Field-Naturalist* 108, 305-310.

222) Catling, P.M., Oldham, M.J., Sutherland, D.A., Brownell, V.R., and Larson, B.M.H. (1997) The recent spread of autumn-olive, *Elaeagnus umbellata*, into southern Ontario and its current status. *Canadian Field-Naturalist* 111, 376-380.

223) Cavers, P.B. and Harper, J.L. (1964) Biological flora of the British Isles. *Rumex crispus* L. *Journal of Ecology* 52, 754-766.

224) Cavers, P.B. and Harper, J.L. (1966) Germination polymorphism in *Rumex crispus* and *R. obtusifolius*. *Journal of Ecology* 54, 367-382.

225) Cavers, P.B., Heagy, M.I., and Kokron, R.F. (1979) The biology of Canadian weeds. 35. *Alliaria petiolata* (M. Bieb.) Cavara and Grande. *Canadian Journal of Plant Science* 59, 217-229.

226) Cavers, P.B., Bassett, I.J., and Crompton, C.W. (1980) The biology of Canadian weeds. 47. *Plantago lanceolata* L. *Canadian Journal of Plant Science* 60, 1269-1282.

227) Center, T.D. and Van, T.K. (1989) Alteration of water hyacinth (*Eichhornia crassipes* (Mart.) Solms) leaf dynamics and phytochemistry by insect damage and plant density. *Aquatic Botany* 35, 181-195.

228) Chacalo, A., Watson, G., Bye, R., Ordaz, V., Aldama, A., and Vazquez, H.J. (2000) Root growth of *Quercus crassifolia*, *Q. crassipes*, and *Fraxinus uhdei* in two different soil types. *Journal of Arboriculture* 26, 30-37.

229) Chadwick, M.J. and Obeid, M. (1966) A comparative study of the growth of *Eichhornia crassipes* Solms and *Pistia stratiotes* L. in water-culture. *Journal of Ecology* 54, 563-575.

230) Chandler, K., Murphy, S.D., and Swanton, C.J. (1994) Effect of tillage and glyphosate on control of quackgrass (*Elytrigia repens*). *Weed Technology* 8, 450-456.

231) Chapman, H.M. and Bannister, P. (1990) The spread of heather, *Calluna vulgaris* (L.) Hull, into indigenous plant communities of Tongariro National Park. *New Zealand Journal of Ecology* 14, 7-16.

232) Charudattan, R. (1986) Integrated control of waterhyacinth (*Eichhornia crassipes*) with a pathogen, insects, and herbicides. *Weed Science* 34 (Suppl. 1), 26-30.

233) Chater, A.O. and Rich, T.C.G. (1995) *Rorippa islandica* (oeder ex Murray) Borbas (Brassicaceae) in Wales. *Watsonia* 20, 229-238.

234) Chaw, S.M., Peng, C.I., and Kao, M.T. (1986) *Verbena bonariensis*, new record L. (Verbenaceae): a newly naturalized plant in Taiwan. *Journal of the Taiwan Museum* 39, 123-126.

235) Chawdry, M.A. and Sagar, G.R. (1974) Control of *Oxalis latifolia* H.B.K. and *O. pes-caprae* L. by defoliation. *Weed Research* 14, 293-299.

236) Chawdry, M.A. and Sagar, G.R. (1974) Dormancy and sprouting of bulbs in *Oxalis latifolia* H.B.K. and *O. pes-caprae* L. *Weed Research* 14, 349-354.

237) Chen, H., Qualls, R.G., and Miller, M.C. (2002) Adaptive responses of *Lepidium latifolium* to soil flooding: biomass allocation, adventitious rooting, aerenchyma formation and ethylene production. *Environmental and Experimental Botany* 48, 119-128.

238) Cheplick, G.P. and Chui, T. (2001) Effects of competitive stress on vegetative growth, storage, and regrowth after defoliation in *Phleum pratense*. *Oikos* 95, 291-299.

239) Cherrill, A. (1995) Infestation of improved grasslands by *Juncus effusus* L. in the catchment of the river Tyne, Northern England: a field survey. *Grass and Forage Science* 50, 85-91.

240) Chikoye, D. and Ekeleme, F. (2001) Weed flora and soil seedbanks in fields dominated by *Imperata cylindrica* in the moist savannah of West Africa. *Weed Research* 41, 475-490.

241) Chikoye, D., Manyong, V.M., and Ekeleme, F. (2000) Characteristics of speargrass (*Imperata cylindrica*) dominated fields in West Africa: crops, soil properties, farmer perceptions and management strategies. *Crop Protection* 19, 481-487.

242) Child, L.E. and Spencer-Jones, D. (1995) Treatment of *Crassula helmsii* - a case study. In: Pysek, P. *et al.* (eds) *Plant invasions - general aspects and special problems*. SPB Academic Publishing, Amsterdam, pp. 195-202.

243) Child, L. and Wade, M. (1997) *Fallopia japonica* in the British Isles: the traits of an invasive species and implications for management. In: Yano, E., Matsuo, K., Shiyomi, M., and Andow, D.A. (eds) *Biological invasions of ecosystem by pests and beneficial organisms*. National Institute of Agro-Environmental Sciences (NIAES) Series 3, Tsukuba, Japan, pp. 200-210.

244) Child, L. and Wade, M. (2000) *The Japanese knotweed manual*. Packard Publishing, Chichester.

245) Child, L., Wade, M., and Wagner, M. (1998) Cost effective control of *Fallopia japonica* using combination treatments. In: Starfinger, U. *et al.* (eds) *Plant invasions - ecological mechanisms and human responses.* Backhuys Publishers, Leiden, pp. 143-154.

246) Child, L., Wade, M., and Hathaway, S. (2001) Strategic invasive plant management, linking policy and practice: a case study of *Fallopia japonica* in Swansea, South Wales (United Kingdom). In: Brundu, G. *et al.* (eds) *Plant invasions: species ecology and ecosystem management.* Backhuys Publishers, Leiden, pp. 291-302.

247) Chilvers, G.A. and Burdon, J.J. (1983) Further studies on a native Australian eucalypt forest invaded by exotic pines. *Oecologia* 59, 239-245.

248) Chivinge, O.A. (1996) Studies on the germination and seedling emergence of *Bidens pilosa* and its response to fertilizer application. *Transactions of the Zimbabwe Scientific Association* 70, 1-5.

249) Chow, P.N.P. (1982) Wild oat (*Avena fatua*) herbicide studies: I. Physiological response of wild oat to five postemergence herbicides. *Weed Science* 30, 1-6.

250) Christiansen, E.M. (1963) Naturalization of Russian olive (*Elaeagnus angustifolia* L.) in Utah. *American Midland Naturalist* 70, 133-137.

251) Cilliers, C.J. (1991) Biological control of water fern, *Salvinia molesta* (Salviniaceae), in South Africa. *Agriculture, Ecosystems and Environment* 37, 219-224.

252) Cilliers, C.J. and Neser, S. (1991) Biological control of *Lantana camara* (Verbenaceae), in South Africa. *Agriculture, Ecosystems and Environment* 37, 57-75.

253) Cintron, B.B. (1990) *Cedrela odorata* L. Cedro hembra, Spanish-cedar. In: Burns, R.M. and Honkala, B.H. (eds) *Silvics of North America, vol. 2. Hardwoods.* Agriculture Handbook 654, US Department of Agriculture, Washington DC, pp. 250-257.

254) Clarke, P.J. and Allaway, W.G. (1996) Litterfall in *Casuarina glauca* coastal wetland forests. *Australian Journal of Botany* 44, 373-380.

255) Cleverly, J.R., Smith, S.D., Sala, A., and Devitt, D.A. (1997) Invasive capacity of *Tamarix ramosissima* in a Mojave Desert floodplain: the role of drought. *Oecologia* 111, 12-18.

256) Coffey, B.T. and McNabb, C.D. (1974) Eurasian water-milfoil in Michigan. *The Michigan Botanist* 13, 159-165.

257) Cole, M.A.R. (1991) Vegetation management guideline: white and yellow sweet clover (*Melilotus alba* Desr. And *Melilotus officinalis* (L.) Lam.). *Natural Areas Journal* 11, 214-215.

258) Collier, K.J. and Winterbourn, M.J. (1986) Processing of willow leaves in two suburban streams in Christchurch, New Zealand. *New Zealand Journal of Marine and Freshwater Research* 20, 575-582.

259) Conant, P., Medeiros, A.C., and Loope, L.L. (1997) A multiagency containment program for Miconia (*Miconia calvescens*), an invasive tree in Hawaiian rain forests. In: Luken, J.O. and Thieret, J.W. (eds) *Assessment and management of plant invasions.* Springer, New York, pp. 249-254.

260) Conkey, L.E., Keifer, M., and Lloyd, A.H. (1995) Disjunct Jack pine (*Pinus banksiana* Lamb.) structure and dynamics, Acadia National Park, Maine. *Ecoscience* 2, 168-176.

261) Conner, W.H. (1994) The effect of salinity and waterlogging on growth and survival of baldcypress and Chinese tallow seedlings. *Journal of Coastal Research* 10, 1045-1049.

262) Conner, W.H. and Askew, G.R. (1993) Impact of saltwater flooding on red maple, redbay, and chinese tallow seedlings. *Castanea* 58, 214-219.

263) Conner, W.H., Inabinette, L.W., and Lucas, C.A. (2001) Effects of flooding on early growth and competitive ability of two native wetland tree species and an exotic. *Castanea* 66, 237-244.

264) Connor, H.E. (1965) Breeding systems in New Zealand grasses V. Naturalized species of *Cortaderia*. *New Zealand Journal of Botany* 3, 17-23.

265) Connor, H.E. (1992) Hawkweeds, *Hieracium* spp., in tussock grasslands of Canterbury, New Zealand, in 1960s. *New Zealand Journal of Botany* 30, 247-261.

266) Connor, H.E., Edgar, E., and Bourdôt, G.W. (1993) Ecology and distribution of naturalised species of *Stipa* in New Zealand. *New Zealand Journal of Agricultural Research* 36, 301-307.

267) Conolly, A.P. (1977) The distribution and history in the British Isles of some alien species of *Polygonum* and *Reynoutria*. *Watsonia* 11, 291-311.

268) Cook, B.A. (1929) Some notes on the plant associates and habitat of *Clidemia hirta* (L.) D.Don in Trinidad. *Fiji Agricultural Journal* 2, 92-93.

269) Cook, C.D.K. and Gut, B.J. (1971) *Salvinia* in the state of Kerala, India. *Pest Articles and News Summary (PANS)* 17, 438-447.

270) Cook, C.D.K. and Lüönd, R. (1982) A revision of the genus *Hydrilla* (Hydrocharitaceae). *Aquatic Botany* 13, 485-504.

271) Cooke, D.A. (1998) Bulbil Watsonia is a variety of *Watsonia meriana* (L.) Miller (Iridaceae). *Journal of the Adelaide Botanic Gardens* 18, 5-7.

272) Corbett, D.P. (1990) Control of cluster pine on French Island, Victoria. *Plant Protection Quarterly* 6, 128.

273) Corbineau, F., Defresne, S., and Come, D. (1985) Some characteristics of seed germination and seedling growth of *Cedrela odorata* (Meliaceae). *Bois et Forets des Tropiques* 0, 17-22. *(French)*

274) Cordero, S.R.A. and Di Stefano, G.J.F. (1991) Effect of osmotic stress on the germination of *Tecoma stans* (Bignoniaceae). *Revista de Biologia Tropical* 39, 107-110. *(Spanish)*

275) Corlett, R.T. (1984) The phenology of *Ficus benjamina* and *F. microcarpa* in Singapore. *Biotropica* 19, 122-124.

276) Cotton, R. and Stace, C.A. (1976) Taxonomy of the genus *Vulpia* (Gramineae). I. Chromosome numbers and geographical distribution of the Old World species. *Genetica* 46, 235-255.

277) Courtney, S.P. and Manzur, M.I. (1985) Fruiting and fitness in *Crataegus monogyna*: the effects of frugivores and seed predators. *Oikos* 44, 398-406.

278) Cowie, I.D. and Werner, P.A. (1993) Alien plant species invasive in Kakadu National Park, tropical northern Australia. *Biological Conservation* 63, 127-135.

279) Cown, D.J. (1974) Physical properties of Corsican pine grown in New Zealand. *New Zealand Journal of Forestry* 4, 76-93.

280) Cox, J.R., Martin, M.H., Ibarra, F.A., Fourie, J.H., Rethman, N.F.G., and Wilcox, D.G. (1988) The influence of climate and soils on the distribution of our African grasses. *Journal of Range Management* 41, 127-139.

281) Creager, R.A. (1992) Seed germination, physical and chemical control of catclaw mimosa (*Mimosa pigra* var. *pigra*). *Weed Technology* 6, 884-891.

282) Crins, W.J. (1989) The Tamaricaceae in the southwestern United States. *Journal of the Arnold Arboretum* 70, 403-425.

283) Crompton, C.W., Hall, I.V., Jensen, K.I.N., and Hildebrand, P.D. (1988) The biology of Canadian weeds. 83. *Hypericum perforatum* L. *Canadian Journal of Plant Science* 68, 149-162.

284) Cronk, Q.C.B. and Fuller, J.L. (1995) *Plant invaders*. Chapman & Hall, London.

285) Cross, J.R. (1975) Biological flora of the British Isles. *Rhododendron ponticum*. *Journal of Ecology* 63, 345-364.

286) Cross, J.R. (1981) The establishment of *Rhododendron ponticum* in the Killarney oakwoods, SW Ireland. *Journal of Ecology* 69, 807-824.

287) Cross, J.R. (1982) The invasion and impact of *Rhododendron ponticum* in native Irish vegetation. In: White, J. (ed.) *Studies on Irish Vegetation*. Royal Dublin Society, Dublin, pp. 209-220.

288) Cruden, R.W., McClain, A.M., and Shrivastava, G.P. (1996) Pollination biology and breeding system of *Alliaria petiolata* (Brassicaceae). *Bulletin of the Torrey Botanical Club* 123, 273-280.

289) Cruz, F.J. and Laweson, J. (1986) *Lantana camara* L., a threat to native plants and animals. *Noticias de Galapagos* 43, 10-11.

290) Csurhes, S.M. and Kriticos, D. (1994) *Gleditsia triacanthos* L. (Caesalpiniaceae), another thorny, exotic fodder tree gone wild. *Plant Protection Quarterly* 9, 101-105.

291) Csurhes, S.M. and Edwards, R. (1998) *Potential environmental weeds in Australia*. Environment Australia, Canberra, 208 pp.

292) Cullen, J.M., Briese, D.T., and Groves, R.H. (1997) Towards the integration of control methods for St. John's wort: workshop summary and recommendations. *Plant Protection Quarterly* 12, 103-106.

293) Cusik, A.W. and Ortt, M. (1987) *Polygonum perfoliatum* L. (Polygonaceae): a significant new weed in the Mississippi drainage. *SIDA Contributions to Botany* 12, 246-249.

294) Daehler, C.C. and Strong, D.R. (1994) Variable reproductive output among clones of *Spartina alterniflora* (Poaceae) invading San Francisco Bay, California: the influence of herbivory, pollination, and establishment site. *American Journal of Botany* 81, 307-313.

295) Daehler, C.C. and Strong, D.R. (1996) Status, prediction and prevention of introduced cordgrass *Spartina* spp. invasions in Pacific estuaries, USA. *Biological Conservation* 78, 51-58.

296) Daehler, C.C. and Strong, D.R. (1997) Hybridization between introduced smooth cordgrass (*Spartina alterniflora*; Poaceae) and native California cordgrass (*S. foliosa*) in San Francisco Bay, California, USA. *American Journal of Botany* 84, 607-611.

297) Daehler, C.C. and Strong, D.R. (1997) Reduced herbivore resistance in introduced smooth cordgrass (*Spartina alterniflora*) after a century of herbivore-free growth. *Oecologia* 110, 99-108.

298) Dajdok, Z., Aniol-Kwiatkowska, A., and Kacki, Z. (1998) *Impatiens glandulifera* Royle in the floodplain vegetation of the Odra river (west Poland). In: Starfinger, U. *et al.* (eds) *Plant invasions - ecological mechanisms and human responses*. Backhuys Publishers, Leiden, pp. 161-168.

299) Damascos, M.A. and Gallopin, G.G. (1992) Ecology of introduced shrub (*Rosa rubiginosa* L. equals *Rosa eglanteria* L.): invasion risks and effects on the plant communities of the Andean-Patagonian region of Argentina. *Revista Chilena de Historia Natural* 65, 395-407.

300) D'Antonio, C.M. (1990) Seed production and dispersal in the non-native, invasive succulent *Carpobrotus edulis* (Aizoaceae) in coastal strand communities of central California. *Journal of Applied Ecology* 27, 693-702.

301) D'Antonio, C.M. (1993) Mechanisms controlling invasion of coastal plant communities by the alien succulent *Carpobrotus edulis*. *Ecology* 74, 83-95.

302) D'Antonio, C.M. and Mahall, B.E. (1991) Root profiles and competition between the invasive, exotic perennial, *Carpobrotus edulis*, and two native shrub species in California coastal scrub. *American Journal of Botany* 78, 885-894.

303) D'Antonio, C.M., Dennis, D.C., and Tyler, C.M. (1993) Invasion of maritime chaparral by the introduced succulent *Carpobrotus edulis*. The roles of fire and herbivory. *Oecologia* 95, 14-21.

304) D'Antonio, C.M., Hughes, R.F., Mack, M., Hitchcock, D., and Vitousek, P.M. (1998) The response of native species to removal of invasive exotic grasses in a seasonally dry Hawaiian woodland. *Journal of Vegetation Science* 9, 699-712.

305) D'Antonio, C.M., Tunison, J.T., and Loh, R.K. (2000) Variation in the impact of exotic grasses on native plant composition in relation to fire across an elevation gradient. *Austral Ecology* 25, 507-522.

306) D'Antonio, C.M., Hughes, R.F., and Vitousek, P.M. (2001) Factors influencing dynamics of two introduced invasive C_4 grasses in seasonally dry Hawaiian woodlands. *Ecology* 82, 89-104.

307) Das, R.R. (1969) A study of reproduction in *Eichhornia crassipes* (Mart.) Solms. *Tropical Ecology* 10, 195-198.

308) Davidson, E.D. and Barbour, M.G. (1977) Germination, establishment, and demography of coastal bush lupine (*Lupinus arboreus*) at Bodega Head, California. *Ecology* 58, 592-600.

309) Davies, P.A. (1944) The root system of *Ailanthus altissima*. *Transactions of the Kentucky Academy of Sciences* 11, 33-35.

310) Davis, E.S., Fay, P.K., Chicoine, T.K., and Lacey, C.A. (1993) Persistence of spotted knapweed (*Centaurea maculosa*) seed in soil. *Weed Science* 41, 57-61.

311) Davis, O.H. (1927) Germination and early growth of *Cornus florida*, *Sambucus canadensis*, and *Berberis thunbergii*. *Botanical Gazette* 84, 225-263.

312) Dawson, F.H. (1994) Spread of *Crassula helmsii* in Britain. In: de Waal, L.C. *et al.* (eds) *Ecology and management of invasive riverside plants*. John Wiley & Sons, Chichester, pp. 1-14.

313) Dawson, F.H. and Hallows, H.B. (1983) Practical applications of a shading material for macrophyte control in watercourses. *Aquatic Botany* 17, 301-308.

314) Dawson, F.H. and Warman, E.A. (1987) *Crassula helmsii* (T. Kirk) Cockayne: is it an aggressive alien aquatic plant in Britain? *Biological Conservation* 42, 247-272.

315) Deering, R.H. and Vankat, J.L. (1999) Forest colonization and developmental growth of the invasive shrub *Lonicera maackii*. *The American Midland Naturalist* 141, 43-50.

316) Delaigue, J. (1987) Ecological and chorological contribution to the knowledge of *Erigeron karvinskianus*, new record in France. *Bulletin Mensuel de la Société Linnéenne de Lyon* 56, 42-56.

317) Delucchi, G. (1989) *Morus alba* (Moraceae) amplification of its geographical area in Argentina and observations about its naturalization. *Darwiniana* (San Isidro) 29, 405-406.

318) Del Moral, R. and Muller, C.H. (1969) Fog drip: a mechanism of toxin transport from *Eucalyptus globulus*. *Bulletin of the Torrey Botanical Club* 96, 467-475.

319) Den Hollander, N.G., Schenk, I.W., Diouf, S., Kropff, M.J., and Pieterse, A.H. (1999) Survival strategy of *Pistia stratiotes* L. in the Djoudj National Park in Senegal. *Hydrobiologia* 415, 21-27.

320) DeMars, B.G. (1994) Star-of-Bethlehem, *Ornithogalum umbellatum* L. (Liliaceae): an invasive, naturalized plant in woodlands of Ohio. *Natural Areas Journal* 14, 306-307.

321) Demos, E.K., Peterson, P., and Williams, G.J. (1973) Frost tolerance among populations of *Acer negundo* L. *American Midland Naturalist* 89, 223-228.

322) Dennill, G.B. and Donnelly, D. (1991) Biological control of *Acacia longifolia* and related weed species (Fabaceae) in South Africa. *Agriculture, Ecosystems and Environment* 37, 115-135.

323) Denny, R.P. and Goodall, J.M. (1991) Variable effects of glyphosate and triclopyr used for the control of American bramble, *Rubus cuneifolius* agg., in pine plantations. *South African Forestry Journal* 159, 11-15.

324) Denny, R.P. and Goodall, J.M. (1992) Herbicide treatments applied to stems and stumps of bugweed, *Solanum mauritianum*. *South African Forestry Journal* 161, 41-43.

325) Denny, R.P. and Goodall, J.M. (1992) Control of inkberry, *Cestrum laevigatum*, with triclopyr and imazapyr. *Applied Plant Science* 6, 52-54.

326) Denny, R.P. and Naudé, D.C. (1994) Imazapyr applied to cut-stumps kills *Chromolaena odorata*. *Applied Plant Science* 8, 43-45.

327) Dernoeden, P.H. (1986) Selective tall fescue (*Festuca arundinacea*) control in kentucky bluegrass (*Poa pratensis*) turf with diclofop. *Agronomy Journal* 78, 660-663.

328) De Rouw, A. (1991) The invasion of *Chromolaena odorata* (L.) King & Robinson (ex *Eupatorium odoratum*), and competition with the native flora, in a rain forest zone, south-west Côte d'Ivoire. *Journal of Biogeography* 18, 13-23.

329) Derr, J.F. (1989) Multiflora rose (*Rosa multiflora*) control with metsulfuron. *Weed Technology* 3, 381-384.

330) Desrochers, A.M., Bain, J.F., and Warwick, S.I. (1988) The biology of Canadian weeds. 89. *Carduus nutans* L. and *Carduus acanthoides* L. *Canadian Journal of Plant Science* 68, 1053-1068.

331) Dewald, L.B. and Lounibos, L.P. (1990) Seasonal growth of *Pistia stratiotes* L. in south Florida. *Aquatic Botany* 36, 263-275.

332) De Winton, M.D. and Clayton, J.S. (1996) The impact of invasive submerged weed species on seed banks in lake sediments. *Aquatic Botany* 53, 31-45.

333) De Zoysa, N.D., Gunatilleke, C.V.S., and Gunatilleke, I.A.U.N. (1986) Vegetation studies of a skid-trail planted with mahogany in Sinharaja. *The Sri Lanka Forester* 17, 142-147.

334) Diamond, P. (1999) *Paederia foetida* (Rubiaceae), new to the flora of North Carolina. *SIDA Contributions to Botany* 18, 1273-1276.

335) Dibbern, J.C. (1947) Vegetative responses of *Bromus inermis* to certain variations in environment. *Botanical Gazette* 109, 44-58.

336) Dick, M.A. (1994) Blight of *Lupinus arboreus* in New Zealand. *New Zealand Journal of Forestry Science* 24, 51-68.

337) Diez, S. and Clarke, M.F. (1996) The consumption of onion grass *Romulea rosea* corms by purple swamphens *Porphyrio porphyrio*: is there potential for native animals to help control weeds? *Victorian Naturalist Blackburn* 113, 247-255.

338) Dillenburg, L.R., Whigham, D.F., Teramura, A.H., and Forseth, I.N. (1993) Effects of below- and aboveground competition from the vines *Lonicera japonica* and *Parthenocissus quinquefolia* on the growth of the tree host *Liquidambar styraciflora*. *Oecologia* 93, 48-54.

339) Dillon, S.P. and Forcella, F. (1984) Germination, emergence, growth and flowering of two silvergrasses, *Vulpia bromoides* and *V. myuros*. *Australian Journal of Botany* 32, 165-175.

340) Dini-Papanastasi, O. and Panetsos, C.P. (2000) Relation between growth and morphological traits and genetic parameters of *Robinia pseudoacacia* var. *monophylla* D.C. in northern Greece. *Silvae Genetica* 49, 37-44.

341) DiTomaso, J.M. (1998) Impact, biology, and ecology of saltcedar (*Tamarix* spp.) in the southwestern United States. *Weed Technology* 12, 326-336.

342) DiTomaso, J.M., Kyser, G.B., and Hastings, M.S. (1999) Prescribed burning for control of yellow starthistle (*Centaurea solstitialis*) and enhanced native plant diversity. *Weed Science* 47, 233-242.

343) Dixon, I.R. (1996) Control of bridal creeper (*Asparagus asparagoides*) and the distribution of *Asparagus declinatus* in Kings Park bushland, 1991-1995. *Plant Protection Quarterly* 11, 61-63.

344) Dodd, A.P. (1961) Biological control of *Eupatorium adenophorum* in Queensland. *Australian Journal of Science* 23, 356-365.

345) Dodd, F.S., de Waal, L.C., Wade, P.M., and Tiley, G.E.D. (1994) Control and management of *Heracleum mantegazzianum* (giant hogweed). In: de Waal, L.C. *et al.* (eds) *Ecology and management of invasive riverside plants.* John Wiley & Sons, Chichester, pp. 111-126.

346) Dodds, J.G. (1953) Biological flora of the British Isles. *Plantago coronopus* L. *Journal of Ecology* 41, 467-478.

347) Doing, H., Biddiscombe, E.F., and Knedlhans, S. (1969) Ecology and distribution of the *Carduus nutans* group, nodding thistles, in Australia. *Vegetatio* 17, 313-351.

348) Donald, D.G.M. (1982) The control of *Pinus pinaster* in the fynbos biome. *South African Forestry Journal* 123, 3-7.

349) Donald, D.G.M. and Nel, F.P. (1989) Chemical control of *Hakea sericea* and *Hakea gibbosa* seedlings. *Applied Plant Science* 3, 107-109.

350) Donaldson, S. and Swensen, E. (2001) Community-based efforts to control early invasion of *Lepidium latifolium* in the Lake Tahoe basin (USA). In: Brundu, G. *et al.* (eds) Plant invasions: species *ecology and ecosystem management.* Backhuys Publishers, Leiden, pp. 303-310.

351) Dong, M. and de Kroon, H. (1994) Plasticity in morphology and biomass allocation in *Cynodon dactylon*, a grass species forming stolons and rhizomes. *Oikos* 70, 99-106.

352) Doody, J.P. (2001) *Coastal conservation and management: an ecological perspective.* Kluwer Academic Publishers, Boston, pp. 99-109.

353) Doren, R.F. and Whiteaker, L.D. (1990) Comparison of economic feasibility of chemical control strategies on differing age and density classes of *Schinus terebinthifolius*. *Natural Areas Journal* 10, 28-34.

354) Doren, R.F. and Whiteaker, L.D. (1990) Effects of fire on different size individuals of *Schinus terebinthifolius*. *Natural Areas Journal* 10, 107-113.

355) Doren, R.F., Whiteaker, L.D., and LaRosa, A.M. (1991) Evaluation of fire as a management tool for controlling *Schinus terebinthifolius* as secondary successional growth on abandoned agricultural land. *Environmental Management* 15, 121-129.

356) Douglas, B.J., Thomas, A.G., and Derksen, D.A. (1990) Downy brome (*Bromus tectorum*) invasion into southwestern Saskatchewan. *Canadian Journal of Plant Science* 70, 1143-1151.

357) Doust, L.L. (1981) Population dynamics and local specialization in a clonal perennial (*Ranunculus repens*): 1. The dynamics of ramets in contrasting habitats. *Journal of Ecology* 69, 743-756.

358) Downey, P.O. (2000) Broom (*Cytisus scoparius* (L.) Link) and fire: management implications. *Plant Protection Quarterly* 15, 178-183.

359) Dray, F.A. and Center, T.D. (1988) Seed production by *Pistia stratiotes* L. (water lettuce) in the United States. *Aquatic Botany* 33, 155-160.

360) Drayton, B. and Primack, R.B. (1999) Experimental extinction of garlic mustard (*Alliaria petiolata*) populations: implications for weed science and conservation biology. *Biological Invasions* 1, 159-167.

361) Drescher, A. and Prots, B. (2000) Warum breitet sich das Drüsen-Springkraut (*Impatiens glandulifera* Royle) in den Alpen aus? *Wulfenia* 7, 5-26. *(German)*

362) Dreyer, G.D. (1988) Efficacy of triclopyr in rootkilling Oriental bittersweet (*Celastrus orbiculatus* Thunb.) and certain other woody weeds. *Proceedings of the Northeastern Weed Science Society* 42, 120-121.

363) Dreyer, G.D., Baird, L.M., and Fickler, C. (1987) *Celastrus scandens* and *Celastrus orbiculatus*: Comparisons of reproductive potential between a native and introduced woody vine. *Bulletin of the Torrey Botanical Club* 114, 260-264.

364) Dunbabin, M.T. and Cocks, P.S. (1999) Ecotypic variation for seed dormancy contributes to the success of capeweed (*Arctotheca calendula*) in western Australia. *Australian Journal of Agricultural Research* 50, 1451-1458.

365) Duncan, K.W. (1997) A case study in *Tamarix ramosissima* control: Spring Lake, New Mexico. In: Brock, J.H., Wade, M., Pysek, P., and Green, D. (eds) *Plant invasions: studies from North America and Europe.* Backhuys Publishers, Leiden, pp. 115-121.

366) Dyer, C. and Richardson, D.M. (1992) Population genetics of the invasive Australian shrub *Hakea sericea* (Proteaceae) in South Africa. *South African Journal of Botany* 58, 117-124.

367) Ebinger, J. and Lehnen, L. (1981) Naturalized autumn olive (*Elaeagnus umbellata*) in Illinois, USA. *Transactions of the Illinois State Academy of Science* 74, 83-86.

368) Edgin, B. and Ebinger, J.E. (2001) Control of autumn olive (*Elaeagnus umbellata* Thunb.) at Beall Woods Nature Preserve, Illinois, USA. *Natural Areas Journal* 21, 386-388.

369) Edwards, I.D. (1982) Plant invaders on Mulanje Mountain. *Nyala* 8, 89-94.

370) Edwards, I.D. (1985) Conservation of plants on Mulanje Mountain, Malawi. *Oryx* 19, 86-90.

371) Edwards, K.R., Adams, M.S., and Kvet, J. (1995) Invasion history and ecology of *Lythrum salicaria* in North America. In: Pysek, P. *et al.* (eds) *Plant invasions: general aspects and special problems.* SPB Academic Publishing, Amsterdam, pp. 161-180.

372) Egli, P. and Schmid, B. (2000) Seasonal dynamics of biomass and nitrogen in canopies of *Solidago altissima* and effects of a yearly mowing treatment. *Acta Oecologica* 21, 63-77.

373) Egunjobi, J.K. (1969) Dry matter and nitrogen accumulation in secondary successions involving gorse (*Ulex europaeus* L.) and associated shrubs and trees. *New Zealand Journal of Science* 12, 175-193.

374) Egunjobi, J.K. (1971) Ecosystem processes in a stand of *Ulex europaeus* L. I. Dry matter production, litter fall and efficiency of solar energy utilization. *Journal of Ecology* 59, 31-38.

375) Ehrenfeld, J.G. (1997) Invasion of deciduous forest preserves in the New York metropolitan region by Japanese barberry (*Berberis thunbergii* DC.). *Journal of the Torrey Botanical Society* 124, 210-215.

376) Ehrenfeld, J.G. (1999) Structure and dynamics of populations of Japanese barberry (*Berberis thunbergii* DC.) in deciduous forests of New Jersey. *Biological Invasions* 1, 203-213.

377) Elhaak, M.A., Migahid, M.M., and Wegmann, K. (1997) Ecophysiological studies on *Euphorbia paralias* under soil salinity and sea water spray treatments. *Journal of Arid Environments* 35, 459-471.

378) Ellenberg, H. (1989) *Opuntia dillenii* als problematischer Neophyt im Nordjemen. *Flora* 182, 3-12. *(German)*

379) Ellis, L.M. (1995) Bird use of saltcedar and cottonwood vegetation in the Middle Rio Grande Valley of New Mexico, USA. *Journal of Arid Environments* 30, 339-349.

380) El-Siddig, K., Luedders, P., Ebert, G., and Adiku, S.G.K. (1998) Response of rose apple (*Eugenia jambos* L.) to water and nitrogen supply. *Journal of Applied Botany* 72, 203-206.

381) Epp, G.A. (1987) The seed bank of *Eupatorium odoratum* along a successional gradient in a tropical rain forest in Ghana. *Journal of Tropical Ecology* 3, 139-149.

382) Ernst, W. and Lugtenborg, T.F. (1980) Comparative ecophysiology of *Juncus articulatus* and *Holcus lanatus*. *Flora* 169, 121-134.

383) Ervin, G.N. and Wetzel, R.G. (1997) Shoot:root dynamics during growth stages of the rush *Juncus effusus* L. *Aquatic Botany* 59, 63-73.

384) Ervin, G.N. and Wetzel, R.G. (2001) Seed fall and field germination of needlerush, *Juncus effusus* L. *Aquatic Botany* 71, 233-237.

385) Ervin, G.N. and Wetzel, R.G. (2002) Influence of a dominant macrophyte, *Juncus effusus*, on wetland plant species richness, diversity, and community composition. *Oecologia* 130, 626-636.

386) Escarre, J., Houssard, C., and Thompson, J.D. (1994) An experimental study of the role of seedling density and neighbor relatedness in the persistence of *Rumex acetosella* in an old-field succession. *Canadian Journal of Botany* 72, 1273-1281.

387) Esler, D. (1990) Avian community responses to *Hydrilla* invasion. *Wilson Bulletin* 102, 427-440.

388) Evans, K.J., Symon, D.E., and Roush, R.T. (1998) Taxonomy and genotypes of the *Rubus fruticosus* L. aggregate in Australia. *Plant Protection Quarterly* 13, 152-156.

389) Evans, J.E. (1983) A literature review of management practices for multiflora rose (*Rosa multiflora*). *Natural Areas Journal* 3, 6-15.

390) Evans, J.E. (1984) Canada Thistle (*Cirsium arvense*): a literature review of management practices. *Natural Areas Journal* 4, 11-21.

391) Evans, R.A. and Young, J.A. (1972) Germination and establishment of *Salsola* in relation to seedbed environment. Part II. Seed distribution and seedling growth of *Salsola* and microenvironmental monitoring of the seedbed. *Agronomy Journal* 64, 219-224.

392) Evans, R.A., Young, Y.A., and Hawkes, R. (1979) Germination characteristics of Italian thistle (*Carduus pycnocephalus*) and Slenderflower thistle (*Carduus tenuiflorus*). *Weed Science* 27, 327-332.

393) Everitt, B.L. (1980) Ecology of saltcedar – a plea for research. *Environmental Geology* 3, 77-84.

394) Ewel, J.J. (1986) Invasibility: lessons from South Florida. In: Mooney, H.A. and Drake, J.A. (eds) *Ecology of biological invasions of North American and Hawaii.* Springer Verlag, New York, pp. 214-230.

395) Faensen, T.A. (1992) The ecology of *Melilotus alba* L. 3. The life cycle. *Flora* 186, 377-391. *(German)*

396) Fagg, P.C. (1989) Control of *Delairea odorata* (Cape ivy) in native forest with the herbicide clopyralid. *Plant Protection Quarterly* 4, 107-110.

397) Fairbrothers, D.E. and Gray, J.R. (1972) *Microstegium vimineum* (Trin.) A. Camus (Gramineae) in the United States. *Bulletin of the Torrey Botanical Club* 99, 97-100.

398) Fajer, E.D., Bowers, M.D., and Bazzaz, F.A. (1991) Performance and allocation patterns of the perennial herb, *Plantago lanceolata*, in response to simulated herbivory and elevated carbon dioxide environments. *Oecologia* 87, 37-42.

399) Falinski, J.B. (1998) Invasive alien plants and vegetation dynamics. In: Starfinger, U. *et al.* (eds) *Plant invasions - ecological mechanisms and human responses.* Backhuys Publishers, Leiden, pp. 3-21.

400) Fan, J. and Harris, W. (1996) Effects of soil fertility level and cutting frequency on interference among *Hieracium pilosella*, *H. praealtum*, *Rumex acetosella*, and *Festuca novae-zelandiae*. *New Zealand Journal of Agricultural Research* 39, 1-32.

401) Farnsworth, E.J. (1993) Ecology of semi-evergreen plant assemblages in the Guánica dry forest, Puerto Rico. *Caribbean Journal of Science* 29, 106-123.

402) Farrell, T.P. and Ashton, D.H. (1978) Population studies on *Acacia melanoxylon* R.Br. I: variation in seed and vegetative characteristics. *Australian Journal of Botany* 26, 365-379.

403) Faruqi, S.A. and Quraish, H.B. (1979) Studies on Libyan grasses: 5. Population variability and distribution of *Schismus arabicus* and *Schismus barbatus* in Libya. *Pakistan Journal of Botany* 11, 167-172.

404) Feijoo, C.S., Momo, F.R., Bonetto, C.A., and Tur, N.M. (1996) Factors influencing biomass and nutrient content of the submersed macrophyte *Egeria densa* Planch. in a pampasic stream. *Hydrobiologia* 341, 21-26.

405) Feldman, S.R. (1997) Biological control of plumeless thistle (*Carduus acanthoides* L.) in Argentina. *Weed Science* 45, 534-537.

406) Feldman, S.R., Vesprini, J.L., and Lewis, J.P. (1994) Survival and establishment of *Carduus acanthoides* L. *Weed Research* 34, 265-273.

407) Fenner, M. (1980) The inhibition of germination of *Bidens pilosa* seeds by leaf canopy shade in some natural vegetation types. *New Phytologist* 84, 95-102.

408) Fensham, R.J., Fairfax, R.J., and Cannell, R.J. (1994) The invasion of *Lantana camara* L. in Forty Mile Scrub National Park, north Queensland. *Australian Journal of Ecology* 19, 297-305.

409) Fiala, K. (1978) Underground organs of *Typha angustifolia* and *Typha latifolia*, their growth, propagation and production. *Acta Sciencia Naturalia Brno* 12, 1-43.

410) Field, R.J. and Hong, P. (1980) The preferential accumulation of picloram at sites of active growth in gorse (*Ulex europaeus* L.). *Weed Research* 20, 177-182.

411) Figueiredo, R.A., Motta, J.C., and Vasconcellos, L.A.S. (1995) Pollination, seed dispersal, seed germination and establishment of seedlings of *Ficus microcarpa*, Moraceae, in southeastern Brazil. *Revista Brasileira de Biologia* 55, 233-239.

412) Figueroa, P.G. and Fournier, O.L.A. (1995) Phenology and physiology in two populations of *Tabebuia rosea* in Costa Rica (Scrophulariales: Bignoniaceae). *Revista de Biologia Tropical* 44, 61-70.

413) Fike, J. and Niering, W.A. (1999) Four decades of old field vegetation development and the role of *Celastrus orbiculatus* in the northeastern United States. *Journal of Vegetation Science* 10, 483-492.

414) Firth, D.J. (1981) Camphor laurel (*Cinnamomum camphora*) – a new weed in northeastern New South Wales. *Australian Weeds* 1, 26-28.

415) Flanagan, G.J., Wilson, C.G., and Gillett, J.D. (1990) The abundance of native insects on the introduced weed *Mimosa pigra* in Northern Australia. *Journal of Tropical Ecology* 6, 219-230.

416) Fleischmann, K. (1997) Problems with invasive alien plants on the islands of Mahé and Silhouette, Seychelles. PhD thesis, Swiss Federal Institute of Technology, Zurich, Switzerland.

417) Fleischmann, K. (1997) Invasion of alien woody plants on the islands of Mahé and Silhouette, Seychelles. *Journal of Vegetation Science* 8, 5-12.

418) Fleming, T.H. and Williams, C.F. (1990) Phenology, seed dispersal, and recruitment in *Cecropia peltata* (Moraceae) in Costa Rican tropical dry forest. *Journal of Tropical Ecology* 6, 163-178.

419) Forcella, F., Wood, J.T., and Dillon, S.P. (1986) Characteristics distinguishing invasive weeds within *Echium* (Bugloss). *Weed Research* 26, 351-364.

420) Forno, I.W. (1983) Native distribution of the *Salvinia auriculata* complex and keys to species identification. *Aquatic Botany* 17, 71-83.

421) Foster, T.E. and Brooks, J.R. (2001) Long-term trends in growth of *Pinus palustris* and *Pinus elliottii* along a hydrological gradient in central Florida. *Canadian Journal of Forest Research* 31, 1661-1670.

422) Fox, A.M. and Bryson, C.T. (1998) Wetland nightshade (*Solanum tampicense*): a threat to wetlands in the United States. *Weed Technology* 12, 410-413.

423) Frandsen, P.R. (1997) Team Arundo: interagency cooperation to control giant cane (*Arundo donax*). In: Luken, J.O. and Thieret, J.W. (eds) *Assessment and management of plant invasions.* Springer, New York, pp. 244-248.

424) Frankel, E. (1989) Distribution of *Pueraria lobata* in and around New York City. *Bulletin of the Torrey Botanical Club* 116, 390-394.

425) Fremstad, E. (1997) Alien plants in Norway. Japanese Rose - *Rosa rugosa*. Fremmede planter i Norge. Rynkerose - *Rosa rugosa*. *Blyttia* 55, 115-121. *(Norwegian)*

426) Fremstad, E. and Elven R. (1996) Fremmede planter i Norge. Platanlønn (*Acer pseudoplatanus* L.). *Blyttia* 54, 61-78. *(Norwegian)*

427) Frey, H. (1984) Begegnung mit *Arctotheca calendula* (L.) Levyns. Eine Komposite vom Kapland erobert die südwesteuropäische Küste. *Dissertationes Botanicae* 72, 453-458. *(German)*

428) Frey, L. (1997) The eastern limit of European distribution of *Aira caryophyllea* (Poaceae). *Fragmenta Floristica et Geobotanica* 42, 255-263.

429) Fugler, S.R. (1983) The control of silky hakea in South Africa. *Bothalia* 14, 977-980.

430) Fuller, R.M. and Boorman, L.A. (1977) The spread and development of *Rhododendron ponticum* L. on dunes at Winterton, Norfolk, in comparison with invasion by *Hippophae rhamnoides* L. at Saltfleeby, Lincolnshire. *Biological Conservation* 12, 83-94.

431) Furch, K. and Junk, W.J. (1992) Nutrient dynamics of submersed decomposing amazonian herbaceous plant species *Paspalum fasciculatum* and *Echinochloa polystachya*. *Revue d'Hydrobiologie Tropicale* 25, 75-85.

432) Gabor, T.S. and Murkin, H.R. (1990) Effects of clipping purple loosestrife seedlings during a simulated wetland drawdown. *Journal of Aquatic Plant Management* 28, 98-100.

433) Galetto, L., Morales, C.L., and Torres, C. (1999) Reproductive biology of *Salpichroa origanifolia* (Solanaceae). *Kurtziana* 27, 211-224.

434) Gallagher, K.G., Schierenbeck, K.A., and D'Antonio, C.M. (1997) Hybridization and introgression in *Carpobrotus* spp. (Aizoaceae) in California II. Allozyme evidence. *American Journal of Botany* 84, 905-911.

435) Gann, G. and Gordon, D.R. (1998) *Paederia foetida* (skunk vine) *and P. cruddasiana* (sewer vine): threats and management strategies. *Natural Areas Journal* 18, 169-174.

436) Ganzaugh, N. (1980) The viability and germination capacity of *Verbena bonariensis* and *Verbena rigida* seeds. *Seed Science and Technology* 8, 615-624.

437) Gardener, M.R. and Sindel, B.M. (1998) The biology of *Nassella* and *Achnatherum* species naturalized in Australia and the implications for manageemnt on conservation lands. *Plant Protection Quarterly* 13, 76-79.

438) Gardiner, C.P. and Gardiner, S.P. (1996) The dissemination of chinese apple (*Ziziphus mauritiana*): a woody weed of the tropical subhumid savanna and urban fringe of north Queensland. *Tropical Grasslands* 30, 174.

439) Gautier, L. (1992) Taxonomy and distribution of a tropical weed: *Chromolaena odorata* (L.) R. King & H. Robinson. *Candollea* 47, 645-662.

440) Gaynor, D.L. and MacCarter, L.E. (1981) Biology, ecology, and control of gorse (*Ulex europaeus* L.): a bibliography. *New Zealand Journal of Agricultural Research* 24, 123-137.

441) Geary, T.F. and Woodall, S.L. (1990) *Melaleuca quinquenervia* (Cav.) S.T. Blake. Melaleuca. In: Burns, R.M. and Honkala, B.H. (eds) *Silvics of North America, vol. 2. Hardwoods.* Agriculture Handbook 654, US Department of Agriculture, Washington DC, pp. 461-465.

442) Geng, M.C. (1989) A provenance test of white elm (*Ulmus pumila* L.) in China. *Silvae Genetica* 38, 37-44.

443) Gentle, C.B. and Duggin, J.A. (1997) *Lantana camara* L. invasions in dry rainforest-open forest ecotones: the role of disturbances associated with fire and cattle grazing. *Australian Journal of Ecology* 22, 298-306.

444) Gentry, H.S. (1982) *Agaves of continental North America.* The University of Arizona Press, Tucson, Arizona.

445) Gerhardt, K. (1996) Germination and development of sown mahogany (*Swietenia macrophylla* King) in secondary tropical dry forest habitats in Costa Rica. *Journal of Tropical Ecology* 12, 275-289.

446) Gerlach, J. (1993) Invasive Melastomataceae in Seychelles. *Oryx* 27, 22-26.

447) Gerlach, J. (1996) The effects of habitat domination by invasive Melastomataceae. *Phelsuma* 4, 19-26.

448) Gerrath, J.M. and Posluszny, U. (1989) Morphological and anatomical development in the Vitaceae. V. Vegetative and floral development in *Ampelopsis brevipedunculata*. *Canadian Journal of Botany* 67, 2371-2386.

449) Ghuman, B.S., Lal, R., and Vanelslande, A. (1985) Effect of drought stress on water yam (*Dioscorea alata*). *International Journal of Tropical Agriculture* 3, 35-42.

450) Gibson, D.J. and Newman, J.A. (2001) Biological flora of the British Isles. *Festuca arundinacea* Schreber (*F. elatior* L. ssp. *arundinacea* (Schreber) Hackel). *Journal of Ecology* 89, 304-324.

451) Gibson, D.J., Spyreas, G., and Benedict, J. (2002) Life history of *Microstegium vimineum* (Poaceae), an invasive grass in southern Illinois. *Journal of the Torrey Botanical Society* 129, 207-219.

452) Gill, M.A. (1985) *Acacia cyclops* G. Don (Leguminosae – Mimosaceae) in Australia: Distribution and dispersal. *Journal of the Royal Society of Western Australia* 67, 59-65.

453) Gimingham, C.H. (1960) Biological flora of the British Isles. *Calluna vulgaris* (L.) Hull. *Journal of Ecology* 48, 455-483.

454) Gimingham, C.H. (1978) *Calluna* and its associated species: some aspects of co-existence in communities. *Vegetatio* 36, 179-186.

455) Givelberg, A. and Horowitz, M. (1984) Germination behavior of *Solanum nigrum* seeds. *Journal of Experimental Botany* 35, 588-598.

456) Glad, J.B. and Halse, R.R. (1993) Invasion of *Amorpha fruticosa* L. (Leguminosae) along the Columbia and Snake rivers in Oregon and Washington. *Madroño* 40, 62-65.

457) Gladstones, J.S. (1966) Naturalized subterranean clover (*Trifolium subterraneum* L.) in Western Australia: the strains, their distributions, characteristics and possible origins. *Australian Journal of Botany* 14, 329-354.

458) Gladstones, J.S. and Collins, W.J. (1983) Subterranean clover as a naturalized plant in Australia. *The Journal of the Australian Institute of Agricultural Science* 49, 191-202.

459) Glass, B. (1991) Vegetation management guideline: cut-leaved teasel (*Dipsacus laciniatus* L.) and common teasel (*Dipsacus sylvestris* Huds.). *Natural Areas Journal* 11, 213-214.

460) Gleadow, R.M. and Ashton, D.H. (1981) Invasion by *Pittosporum undulatum* of the forests of Central Victoria. I. Invasion patterns and morphology. *Australian Journal of Botany* 29, 705-720.

461) Gleadow, R.M. (1982) Invasion by *Pittosporum undulatum* of the forests of Central Victoria. II. Dispersal, germination and establishment. *Australian Journal of Botany* 30, 185-198.

462) Gleadow, R.M. and Rowan, K.S. (1982) Invasion by *Pittosporum undulatum* of the forests of Central Victoria. III. Effects of temperature and light on growth and drought resistance. *Australian Journal of Botany* 30, 347-357.

463) Gleadow, R.M., Rowan, K.S., and Ashton, D.H. (1983) Invasion by *Pittosporum undulatum* of the forests of Central Victoria. IV. Shade tolerance. *Australian Journal of Botany* 31, 151-160.

464) Gleason, H.A. (1939) The genus *Clidemia* in Mexico and Central America. *Brittonia* 3, 97-140.

465) Glyphis, J.P., Milton, S.J., and Siegfried, W.R. (1981) Dispersal of *Acacia cyclops* by birds. *Oecologia* 48, 138-141.

466) Godwin, H. (1943) Biological Flora of The British Isles. *Rhamnus cathartica* L. *Journal of Ecology* 31, 69-76.

467) Godwin, H. (1943) Biological Flora of The British Isles. *Frangula alnus* Miller (*Rhamnus frangula* L.). *Journal of Ecology* 31, 77-92.

468) Goeden, R.D. and Ricker, D.W. (1982) Poison hemlock, *Conium maculatum*, in southern California, USA: an alien weed attacked by few insects. *Annals of the Entomological Society of America* 75, 173-176.

469) Goldberg, D.E. (1988) Response of *Solidago canadensis* clones to competition. *Oecologia* 77, 357-364.

470) Goodall, J.M. and Erasmus, D.J. (1996) Review of the status and integrated control of the invasive alien weed, *Chromolaena odorata*, in South Africa. *Agriculture, Ecosystem and Environment* 56, 151-164.

471) Goodland, T. (1990) A report on the spread of an invasive tree species *Pittosporum undulatum* into the forests of the Blue Mountains, Jamaica. BSc report, University College of North Wales.

472) Goodland, T. and Healey, J.R. (1996) *The invasion of Jamaican montane rainforests by the Australian tree* Pittosporum undulatum. School of Agricultural and Forest Sciences, University of Wales, Bangor, UK.

473) Goodland, T. and Healey, J.R. (1997) *The effect of* Pittosporum undulatum *on the native vegetation of the Blue Mountains of Jamaica*. Research report, UK Overseas Development Administration and UK Department of the Environment.

474) Gopal, B. (1987) *Water Hyacinth. Aquatic Plant Studies 1*. Elsevier, Amsterdam.

475) Goss-Custard, J.D. and Moser, M.E. (1988) Rates of change in the numbers of dunlin *Calidris alpina* wintering in British estuaries in relation to the spread of *Spartina anglica*. *Journal of Applied Ecology* 25, 95-109.

476) Gouin, F.R. (1979) Controlling brambles in established christmas tree plantations. *HortScience* 14, 189-190.

477) Gould, A.M.A. and Gorchov, D.L. (2000) Effects of the exotic invasive shrub *Lonicera maackii* on the survival and fecundity of three species of native annuals. *American Midland Naturalist* 144, 36-50.

478) Graaff, J.L. and van Staden, J. (1984) The germination characteristics of two *Sesbania* species. *South African Journal of Botany* 3, 59-62.

479) Grace, J.B. (1989) Effects of water depth on *Typha latifolia* and *Typha domingensis*. *American Journal of Botany* 76, 762-768.

480) Grace, J.B. (1998) Can prescribed fire save the endangered coastal prairie ecosystem from Chinese tallow invasion? *Endangered Species Update* 15, 70-76.

481) Grace, J.B. and Wetzel, R.G. (1978) The production biology of eurasian watermilfoil (*Myriophyllum spicatum* L.): a review. *Journal of Aquatic Plant Management* 16, 1-11.

482) Grace, J.B. and Harrison, J.S. (1986) The biology of Canadian weeds. 73. *Typha latifolia* L., *Typha angustifolia* L. and *Typha x glauca* Godr. *Canadian Journal of Plant Science* 66, 361-379.

483) Graf, W.L. (1982) Tamarisk and river-channel management. *Environmental Management* 6, 283-296.

484) Graham, M.S. and Mitchell, M.D. (1996) Practical experiences in management for control of bridal creeper (*Asparagus asparagoides*) on nature reserves in the southern wheatbelt of Western Australia. *Plant Protection Quarterly* 11, 64-66.

485) Graneli, W. (1989) Influence of standing litter on shoot production in reed, *Phragmites australis* (Cav.) Trin. ex Steudel. *Aquatic Botany* 35, 99-109.

486) Gray, S.G. (1968) A review of research on *Leucaena leucocephala*. *Tropical Grasslands* 2, 19-30.

487) Green, E.K. and Galatowitsch, S.M. (2002) Effects of *Phalaris arundinacea* and nitrate-N addition on the establishment of wetland plant communities. *Journal of Applied Ecology* 39, 134-144.

488) Green, P.S. (1985) *Fraxinus rotundifolia*, *Fraxinus parvifolia* or *Fraxinus angustifolia*? *Kew Bulletin* 40, 131-134.

489) Greenberg, C.H., Smith, L.M., and Levey, D.J. (2001) Fruit fate, seed germination and growth of an invasive vine – an experimental test of 'sit and wait' strategy. *Biological Invasions* 3, 363-372.

490) Greenway, M. (1994) Litter accession and accumulation in a *Melaleuca quinquenervia* (Cav.) S.T. Blake wetland in south-eastern Queensland. *Australian Journal of Marine and Freshwater Research* 45, 1509-1519.

491) Gremmen, N.J.M. and Smith, V.R. (1981) *Agrostis stolonifera* on Marion Island (South Africa) (sub-Antarctic). *South African Journal of Antarctic Research* 10/11, 33-34.

492) Gremmen, N.J.M., Chown, S.L., and Marshall, D.J. (1998) Impact of the introduced grass *Agrostis stolonifera* on vegetation and soil fauna communities at Marion Island, sub-Antarctic. *Biological Conservation* 85, 223-231.

493) Grice, A.C. (1996) Aspects of the seed ecology of two invasive shrubs in a tropical woodland: *Cryptostegia grandiflora* and *Ziziphus mauritiana*. *Australian Journal of Ecology* 21, 324-331.

494) Grice, A.C. (1996) Seed production, dispersal and germination in *Cryptostegia grandiflora* and *Ziziphus mauritiana*, two invasive shrubs in tropical woodlands of northern Australia. *Australian Journal of Ecology* 21, 324-331.

495) Grice, A.C. (1997) Post-fire regrowth and survival of the invasive tropical shrubs *Cryptostegia grandiflora* and *Ziziphus mauritiana*. *Australian Journal of Ecology* 22, 49-55.

496) Grice, A.C. (1998) Ecology in the management of Indian jujube (*Ziziphus mauritiana*). *Weed Science* 46, 467-474.

497) Grice, A.C. (2002) The biology of Australian weeds. 39. *Ziziphus mauritiana* Lam. *Plant Protection Quarterly* 17, 2-11.

498) Grice, A.C., Radford, I.J., and Abbott, B.N. (2000) Regional and landscape-scale patterns of shrub invasion in tropical savannas. *Biological Invasions* 2, 187-205.

499) Griffin, G.F., Stafford Smith, D.M., Morton, S.R., Allan, G.E., and Masters, K.A. (1989) Status and implications of the invasion of Tamarisk (*Tamarix aphylla*) on the

Finke River, Northern Territory, Australia. *Journal of Environmental Management* 29, 297-315.

500) Griffin, J.L., Watson, V.H., and Strachan, W.F. (1988) Selective broomsedge (*Andropogon virginicus* L.) control in permanent pastures. *Crop Protection* 7, 80-83.

501) Griffin, J.L., Reynolds, D.B., Vidrine, P.R., and Saxton, A.M. (1992) Common cocklebur (*Xanthium strumarium*) control with reduced rates of soil and foliar-applied imazaquin. *Weed Technology* 6, 847-851.

502) Grilz, P.L. and Romo, J.T. (1995) Management considerations for controlling smooth brome in fescue prairie. *Natural Areas Journal* 15, 148-156.

503) Gritten, R.H. (1995) *Rhododendron ponticum* and some other invasive plants in the Snowdonia National Park. In: Pysek, P., Prach, K., Rejmánek, M., and Wade, M. (eds) *Plant invasions - general aspects and special problems*. SPB Academic Publishing, Amsterdam, pp. 213-222.

504) Gross, K.L. (1980) Colonization by *Verbascum thapsus* (Mullein) of an old-field in Michigan: experiments on the effects of vegetation. *Journal of Ecology* 68, 919-927.

505) Gross, K.L. and Werner, P.A. (1978) The biology of Canadian weeds. 28. *Verbascum thapsus* L. and *V. blattaria* L. *Canadian Journal of Plant Science* 58, 401-413.

506) Groves, R.H., Shepherd, R.C.H., and Richardson, R.G. (1995) *The Biology of Australian Weeds*. R.G. and F.J. Richardson, Melbourne.

507) Grubb, P.J. (1982) Control of relative abundance in roadside *Arrhenatheretum*: results of a long-term garden experiment. *Journal of Ecology* 70, 845-861.

508) Guillarmod, A.J. (1979) Water weeds in southern Africa. *Aquatic Botany* 6, 377-391.

509) Gullison, R.E., Panfil, S.N., Strouse, J.J., and Hubbell, S.P. (1996) Ecology and management of mahogany (*Swietenia macrophylla* King) in the Chimanes Forest, Beni, Bolivia. *Botanical Journal of the Linnean Society* 122, 9-34.

510) Gutiérrez, E. (1993) Effect of glyphosate on different densities of waterhyacinth. *Journal of Aquatic Plant Management* 31, 255-257.

511) Gutiérrez, E., Huerto, R., Saldaña, P., and Arreguín, F. (1996) Strategies for waterhyacinth (*Eichhornia crassipes*) control in Mexico. *Hydrobiologia* 340, 181-185.

512) Gutte, P., Klotz, S., Lahr, C., and Trefflich, A. (1987) *Ailanthus altissima* (Mill.) Swingle — eine vergleichend pflanzengeographische Studie. *Folia Geobotanica et Phytotaxonomica* 22, 241-262.

513) Gutterman, Y. (1996) Effect of day length during plant development and caryopsis maturation on flowering and germination, in addition to temperature during dry storage and light during wetting, of *Schismus arabicus* (Poaceae) in the Negev Desert, Israel. *Journal of Arid Environments* 33, 439-448.

514) Gutterman, Y. (2001) Drought tolerance of the dehydrated root of *Schismus arabicus* seedlings and regrowth after rehydration, affected by caryopsis size and duration of dehydration. *Israel Journal of Plant Sciences* 49, 123-128.

515) Gutterman, Y. and Shem-Tov, S. (1997) Mucilaginous seed coat structure of *Carrichtera annua* and *Anastatica hierochuntica* from the Negev Desert highlands of Israel, and its adhesion to the soil crust. *Journal of Arid Environments* 35, 695-705.

516) Guzikowa, M. and Maycock, P.F. (1986) The invasion and expansion of three North American species of goldenrod (*Solidago canadensis* L. sensu lato, *S. gigantea* Ait. and *S. graminifolia* (L.) Salisb.) in Poland. *Acta Societatis Botanicae Poloniae* 55, 367-384.

517) Haggar, J.P.C. (1988) The structure, composition and status of the cloud forests of the Pico Island in the Azores. *Biological Conservation* 46, 7-22.

518) Hamann, O. (1984) Changes and threats to the vegetation. In: Perry, R. (ed.) *Key environments: Galapagos*. Pergamon Press, Oxford, pp. 115-131.

519) Hamet, A.L. (1980) The *Juncus effusus* aggregate in eastern North America. *Annales Botanici Fennici* 17, 183-191.

520) Hamilton, W.D. and McHenry, W.B. (1982) *Eucalyptus* stump sprout control. *Journal of Arboriculture* 8, 327-328.

521) Hannan-Jones, M.A. and Playford, J. (2002) The biology of Australian weeds. 40. *Bryophyllum* Salisb. species. *Plant Protection Quarterly* 17, 42-57.

522) Hara, T., van der Toorn, J., and Mook, J.H. (1993) Growth dynamics and size structure of shoots of *Phragmites australis*, a clonal plant. *Journal of Ecology* 81, 47-60.

523) Haramoto, T. and Ikusima, I. (1988) Life cycle of *Egeria densa* Planch., an aquatic plant naturalized in Japan. *Aquatic Botany* 30, 389-403.

524) Harding, G.B. (1987) The status of *Prosopis* as a weed. *Applied Plant Science* 1, 43-48.

525) Harley, K.L.S. and Mitchell, D.S. (1981) The biology of Australian weeds. 6. *Salvinia molesta* Mitchell. *Journal of the Australian Institute of Agricultural Science* 47, 67-76.

526) Harley, K.L.S. (1992) *A guide to the management of* Mimosa pigra. CSIRO, Canberra, Australia.

527) Harper, J. (1957) Biological flora of the British Isles. *Ranunculus acris* L. *Journal of Ecology* 45, 289-342.

528) Harradine, A.R. (1980) The biology of African feather grass (*Pennisetum macrourum* Trin.) in Tasmania, I. Seedling establishment. *Weed Research* 20, 165-169.

529) Harradine, A.R. (1980) The biology of African feather grass (*Pennisetum macrourum* Trin.) in Tasmania, II. Rhizome biology. *Weed Research* 20, 171-175.

530) Harradine, A.R. (1985) Dispersal and establishment of slender thistle, *Carduus pycnocephalus*, as affected by ground cover. *Australian Journal of Agricultural Research* 36, 791-798.

531) Harradine, A.R. (1991) The impact of pampas grasses as weeds in southern Australia. *Plant Protection Quarterly* 6, 111-115.

532) Harradine, A.R. and Jones, A.L. (1985) Control of gorse regrowth by Angora goats in the Tasmanian Midlands. *Australian Journal of Experimental Agriculture* 25, 550-556.

533) Harrington, R.A., Brown, B.J., and Reich, P.B. (1989) Ecophysiology of exotic and native shrubs in southern Wisconsin. I. Relationship of leaf characteristics, resource availability, and phenology to seasonal patterns of carbon gain. *Oecologia* 80, 356-367.

534) Harrington, R.A., Brown, B.J., Reich, P.B., and Fownes, J.H. (1989) Ecophysiology of exotic and native shrubs in southern Wisconsin. II. Annual growth and carbon gain. *Oecologia* 80, 368-373.

535) Harris, G.A. (1967) Some competitive relationships between *Agropyron spicatum* and *Bromus tectorum*. *Ecological Monographs* 37, 89-111.

536) Harris, J.A. and Gill, A.M. (1997) History of the introduction and spread of St. John's wort (*Hypericum perforatum* L.) in Australia. *Plant Protection Quarterly* 12, 52-56.

537) Harrod, R.J. and Taylor, R.J. (1995) Reproduction and pollination biology of *Centaurea* and *Acroptilon* species, with emphasis on *C. diffusa*. *Northwest Science* 69, 97-105.

538) Hartemink, A.E. (2001) Biomass and nutrient accumulation of *Piper aduncum* and *Imperata cylindrica* fallows in the humid lowlands of Papua New Guinea. *Forest Ecology and Management* 144, 19-32.

539) Hartemink, A.E. and O'Sullivan, J.N. (2001) Leaf litter decomposition of *Piper aduncum*, *Gliricidia sepium* and *Imperata cylindrica* in the humid lowlands of Papua New Guinea. *Plant and Soil* 230, 115-124.

540) Hartmann, E., Schuldes, H., Kübler, R., and Konold, W. (1995) *Neophyten. Biologie, Verbreitung und Kontrolle ausgewählter Arten.* Ecomed, Landsberg.

541) Harvey, G.J. (1981) Studies on rubber vine (*Cryptostegia grandiflora*) II: field trials using various herbicides. *Australian Weeds* 1, 3-5.

542) Haslam, S.M. (1972) Biological flora of the British Isles. *Phragmites communis* Trin. *Journal of Ecology* 60, 585-610.

543) Hatton, T.J. (1989) Spatial patterning of sweet briar (*Rosa rubiginosa*) by two vertebrate species. *Australian Journal of Ecology* 14, 199-205.

544) Hawkeswood, T.J. (1983) Pollination and fruit production of *Cupaniopsis anacardioides* (A. Rich.) Radlkf. (Sapindaceae) at Townsville, North Queensland. 1. Pollination and floral biology. *Victorian Naturalist* 100, 12-20.

545) Hawkeswood, T.J. (1983) Notes on the fruit production of *Cupaniopsis anacardioides* (A. Rich.) Radlkf. (Sapindaceae) at Townsville, North Queensland. 2. Fruit and seed production. *Victorian Naturalist* 100, 121-124.

546) Heidorn, R. (1991) Vegetation management guideline: exotic buckthorns - common buckthorn (*Rhamnus cathartica* L.), glossy buckthorn (*Rhamnus frangula* L.), Dahurian buckthorn (*Rhamnus davurica* Pall.). *Natural Areas Journal* 11, 216-217.

547) Hellsten, S., Dieme, C., Mbengue, M., Janauer, G.A., den Hollander, N., and Pieterse, A.H. (1999) *Typha* control efficiency of a weed-cutting boat in the Lac de Guiers in Senegal: a preliminary study on mowing speed and re-growth capacity. *Hydrobiologia* 415, 249-255.

548) Henderson, L. (1991) Invasive alien woody plants of the Orange Free State. *Bothalia* 21, 73-89.

549) Henderson L. (2001) *Alien weeds and invasive plants.* Agricultural Research Council, South Africa.

550) Henderson, L. and Musil, K.J. (1984) Exotic woody plant invaders of the Transvaal. *Bothalia* 15, 297-313.

551) Hendrickson, P.E. and Mallory-Smith, C.A. (1999) Response of downy brome (*Bromus tectorum*) and Kentucky bluegrass (*Poa pratensis*) to applications of primisulfuron. *Weed Technology* 13, 461-465.

552) Hendrix, S.D. and Trapp, E.J. (1992) Population demography of *Pastinaca sativa* (Apiaceae): effects of seed mass on emergence, survival, and recruitment. *American Journal of Botany* 79, 365-375.

553) Henson, I.E. (1970) The effects of light, potassium nitrate and temperature on the germination of *Chenopodium album*. *Weed Research* 10, 27-39.

554) Hernandez, H.M. (1981) The reproductive ecology of *Nicotiana glauca*, a cosmopolitan weed. *Boletin de la Sociedad Botanica de Mexico* 41, 47-74. *(Spanish)*

555) Herr, D.G. and Duchesne, L.C. (1995) Jack pine (*Pinus banksiana*) seedling emergence is affected by organic horizon removal, ashes, soil, water and shade. *Water Air and Soil Pollution* 82, 147-154.

556) Herwitz, S.R., Slye, R.E., and Turton, S.M. (1998) Redefining the ecological niche of a tropical rain forest canopy tree species using airborne imagery: long-term crown dynamics of *Toona ciliata*. *Journal of Tropical Ecology* 14, 683-703.

557) Heyligers, P.C. (1983) An appraisal of the beach daisy (*Arctotheca populifolia*) with a view to its possible use for dune stabilization. *Victorian Naturalist* 100, 48-54.

558) Heyligers, P.C. (1984) Beach invaders: sea rockets and beach daisies thrive. *Australian Natural History* 21, 212-214.

559) Heyligers, P.C. (1989) Sea spurge, *Euphorbia paralias*, new record, a strandline pioneer: new to the Perth region (Western Australia, Australia). *Western Australian Naturalist* 18, 1-6.

560) Hickey, B. and Osborne, B. (1998) Effect of *Gunnera tinctoria* (Molina) Mirbel on semi-natural grassland habitats in the west of Ireland. In: Starfinger, U. *et al.* (eds) *Plant invasions - ecological mechanisms and human responses.* Backhuys Publishers, Leiden, pp. 195-208.

561) Hickey, B. and Osborne, B. (2001) Natural seed banks, seedling growth and survival in areas invaded by *Gunnera tinctoria*. In: Brundu, G., Brock, J., Camarda, I., Chiled, L., and Wade, M. (eds) *Plant invasions - species ecology and ecosystem management.* Backhuys Publishers, Leiden, pp. 105-114.

562) Hight, S.D. and Drea, J.J. (1991) Prospects for a classical biological control project against purple loosestrife (*Lythrum salicaria* L.). *Natural Areas Journal* 11, 151-157.

563) Hill, D.J. (1994) A practical strategy for the control of *Fallopia japonica* (Japanese knotweed) in Swansea and the surrounding area, Wales. In: de Waal, L.C., Child, L.E., Wade, P.M., and Brock, J.H. (eds) *Ecology and management of invasive riverside plants.* John Wiley & Sons, Chichester, pp. 195-198.

564) Hill, M.G. (1980) Susceptibility of *Scaevola taccada* bushes to attack by the coccid *Icerya seychellarum*: the effects of leaf loss. *Ecological Entomology* 5, 345-352.

565) Hill, M.P. and Cilliers, C.J. (1999) *Azolla filiculoides* Lamarck (Pteridophyta: Azollaceae), its status in South Africa and control. *Hydrobiologia* 415, 203-206.

566) Hintikka, V. (1990) Germination ecology and survival strategy of *Rumex acetosella* (Polygonaceae) on drought-exposed rock outcrops in South Finland. *Annales Botanici Fennici* 27, 205-215.

567) Hipps, C.B. (1994) Kudzu: a vegetable menace that started out as a good idea. *Horticulture* 72, 36-39.

568) Hirschfeld, J.R., Finn, J.T., and Patterson III, W.A. (1984) Effects of *Robinia pseudoacacia* on leaf litter decomposition and nitrogen mineralization in a northern hardwood stand. *Canadian Journal of Forestry Research* 14, 201-205.

569) Hocking, P.J., Finlayson, C.M., and Chick, A.J. (1983) The biology of Australian weeds. 12. *Phragmites australis* (Cav.) Trin. ex Steud. *Journal of The Australian Institute of Agricultural Science* 49, 123-132.

570) Hoffman, M.T. and Mitchell, D.T. (1986) The root morphology of some legume spp. in the south-western Cape and the relationship of vesicular-arbuscular mycorrhizas with dry mass and phosphorus content of *Acacia saligna* seedlings. *South African Journal of Botany* 52, 316-320.

571) Hoffman, R. and Kearns, K. (1998) *Wisconsin Manual of Control Recommendations for ecologically invasive plants.* Bureau of Endangered Resources, Wisconsin Department of Natural Resources, Madison, Wisconsin, USA.

572) Hoffmann, J.H. and Moran, V.C. (1988) The invasive weed *Sesbania punicea* in South Africa and prospects for its control. *South African Journal of Science* 4, 740-743.

573) Hoffmann, J.H. and Moran, V.C. (1991) Biocontrol of a perennial legume, *Sesbania punicea*, using a florivorous weevil, *Trichapion lativentre*: weed population dynamics with a scarcity of seeds. *Oecologia* 88, 574-576.

574) Hoffmann, J.H. and Moran, V.C. (1998) The population dynamics of an introduced tree, *Sesbania punicea*, in South Africa, in response to long-term damage caused by different combinations of three species of biological control agents. *Oecologia* 114, 343-348.

575) Hoffmann, J.H., Moran, V.C., and Zeller, D.A. (1998) Long-term population studies with the development of an integrated management programme for control of *Opuntia stricta* in Kruger National Park, South Africa. *Journal of Applied Ecology* 35, 156-160.

576) Hoffmann, J.H., Impson, F.A.C., Moran, V.C., and Donnelly, D. (2002) Biological control of invasive golden wattle trees (*Acacia pycnantha*) by a gall wasp, *Trichilogaster* sp. (Hymenoptera: Pteromalidae), in South Africa. *Biological Control* 25, 64-73.

577) Den Hollander, N.G., Schenk, I.W., Diouf, S., Kropff, M.J., and Pieterse, A.H. (1999) Survival strategy of *Pistia stratiotes* L. in the Djoudj National Park in Senegal. *Hydrobiologia* 415, 21-27.

578) Hollingsworth, E.B., Quimby, P.C., and Jaramillo, D.C. (1979) Control of saltcedar (*Tamarix ramosissima*) by subsurface placement of herbicides. *Journal of Range Management* 32, 288-291.

579) Hollingsworth, M. and Bailey, J.B. (2000) Evidence for massive clonal growth in the invasive weed *Fallopia japonica* (Japanese knotweed). *Botanical Journal of the Linnean Society* 133, 463-472.

580) Holm, L.G. *et al.* (1977) *The world's worst weeds*. University of Hawaii Press, Honolulu.

581) Holm, L.G., Doll, J., Holm, E., Pancho J., and Herberger, J. (1997) *World weeds*. John Wiley & Sons, New York.

582) Holmes, P.M. (1988) Implications of alien *Acacia* seed bank viability and germination for clearing. *South African Journal of Botany* 54, 281-284.

583) Holmes, P.M. (1989) Effects of different clearing treatments on the seed dynamics of an invasive Australian shrub, *Acacia cyclops*, in the southwestern Cape, South Africa. *Forest Ecology and Management* 28, 33-46.

584) Holmes, P.M., Dallas, H., and Phillips, T. (1987) Control of *Acacia saligna* in the SW Cape - are the clearing treatments effective? *Veld & Flora* 73, 98-100.

585) Holmes, P.M., Dennill, G.B., and Moll, E.J. (1987) Effects of feeding by native alydid insects on the seed viability of an alien invasive weed, *Acacia cyclops*. *South African Journal of Science* 83, 580-581.

586) Holmes, P.M., Macdonald, I.A.W., and Juritz, J. (1987) Effects of clearing treatment on seed banks of the alien invasive shrubs *Acacia saligna* and *Acacia cyclops* in the southern and south-western Cape, South Africa. *Journal of Applied Ecology* 24, 1045-1051.

587) Holmes, P.M., and Rebelo, A.G. (1988) The occurrence of seed-feeding *Zulubius acaciaphagus* (Hemiptera, Alydidae) and its effects on *Acacia cyclops* seed germination and seed banks in South Africa. *South African Journal of Botany* 54, 319-324.

588) Holthuijzen, A.M.A. and Boerboom, J.H.A. (1982) The *Cecropia* seedbank in the Surinam lowland rain forest. *Biotropica* 14, 62-68.

589) Horn, P. (1997) Seasonal dynamics of aerial biomass of *Fallopia japonica*. In: Brock, J.H., Wade, M., Pysek, P., and Green, D. (eds) *Plant invasions: studies from North America and Europe*. Backhuys Publishers, Leiden, pp. 203-206.

590) Hosking, J.R., Smith, J.M.B., and Sheppard, A.W. (1996) The biology of Australian weeds. 28. *Cytisus scoparius* (L.) Link subsp. *scoparius*. *Plant Protection Quarterly* 11, 102-108.

591) Howard, L.F. and Minnich, R.A. (1989) The introduction and naturalization of *Schinus molle* (pepper tree) in Riverside, California. *Landscape and Urban Planning* 18, 77-95.

592) Howard-Williams, C. and Davies, J. (1988) The invasion of Lake Taupo by the submerged water weed *Lagarosiphon major* and its impact on the native flora. *New Zealand Journal of Ecology* 11, 13-19.

593) Howe, W.H. and Knopf, F.L. (1991) On the imminent decline of Rio Grande cottonwoods in central New Mexico. *The Southwestern Naturalist* 36, 218-224.

594) Howell, J.A. and Blackwell, W.H. (1977) The history of *Rhamnus frangula* (glossy buckthorn) in the Ohio flora. *Castanea* 42, 111-115.

595) Hsiao, A.I. and Huang, W.Z. (1989) Effects of flooding on rooting and sprouting of isolated stem segments and on plant growth of *Paspalum distichum* L. *Weed Research* 29, 335-344.

596) Hu, S.Y. (1979) *Ailanthus. Arnoldia* 39, 29-50.

597) Huang, Z.L., Cao, H.L., Liang, X.D., and Ye, W.H. (2000) The growth and damaging effect of *Mikania micrantha* in different habitats. *Journal of Tropical and Subtropical Botany* 8, 131-138. *(Chinese)*

598) Hubbard, J.C.E. and Partridge, T.R. (1981) Tidal immersion and the growth of *Spartina anglica* marshes in the Waihopai river estuary, New Zealand. *New Zealand Journal of Botany* 19, 115-121.

599) Huenneke, L.F. and Vitousek, P.M. (1990) Seedling and clonal recruitment of the invasive tree *Psidium cattleianum:* implications for management of native Hawaiian forests. *Biological Conservation* 53, 199-211.

600) Hughes, F., Vitousek, P.M., and Tunison, T. (1991) Alien grass invasion and fire in the seasonal submontane zone of Hawaii. *Ecology* 72, 743-746.

601) Huiskes, A.H.L. (1977) The natural establishment of *Ammophila arenaria* from seed. *Oikos* 29, 133-136.

602) Huiskes, A.H.L. (1979) Biological flora of the British Isles. *Ammophila arenaria* (L.) Link (*Psamma arenaria* (L.) Roem. et Schult.; *Calamagrostis arenaria* (L.) Roth). *Journal of Ecology* 67, 363-382.

603) Hulbert, L.C. (1955) Ecological studies of *Bromus tectorum* and other annual brome grasses. *Ecological Monographs* 25, 181-213.

604) Hume, L.J., West, C.J., and Watts, H.M. (1995) Nutritional requirements of *Clematis vitalba* L. (old man's beard). *New Zealand Journal of Botany* 33, 301-313.

605) Humphries, R.N., Jordan, M.A., and Guarino, L. (1982) The effect of water stress on mortality of *Betula pendula* Roth. and *Buddleia davidii* Franch. seedlings. *Plant and Soil* 64, 273-276.

606) Humphries, R.N. and Guarino, L. (1987) Soil nitrogen and the growth of birch and buddleia in abandoned chalk quarries. *Reclamation and Revegetation Research* 6, 55-61.

607) Humphries, S.E., Groves, R.H., and Mitchell, D.S. (1991) *Plant invasions of Australian ecosystems: a status review and management directions.* Kowari 2. Australian National Parks and Wildlife Service. Canberra, Australia.

608) Humphries, S.E., Groves, R.H., and Mitchell, D.S. (1993) Plant invasions: homogenizing Australian ecosystems. In: Moritz, C. and Kikkawa, J. (eds) *Conservation biology in Australia and Oceania.* Surrey Beatty & Sons, Chipping Norton, pp. 149-170.

609) Hunt, W.F. (1979) Effects of treading and defoliation height on the growth of *Paspalum dilatatum. New Zealand Journal of Agricultural Research* 22, 69-76.

610) Hunter, G.G. and Douglas, M.H. (1984) Spread of exotic conifers on South Island rangelands. *New Zealand Journal of Forestry* 29, 78-96.

611) Hunter, R. (1991) *Bromus* invasions on the Nevada test site: present status of *B. rubens* and *B. tectorum* with notes on their relationship to disturbance and altitude. *Great Basin Naturalist* 51, 176-182.

612) Huntley, J.C. (1990) *Robinia pseudacacia* L. Black locust. In: Burns, R.M. and Honkala, B.H. (eds) *Silvics of North America, vol. 2. Hardwoods.* Agriculture Handbook 654, US Department of Agriculture, Washington DC, pp. 755-761.

613) Hussey, B.M.J. (1993) Practical experience with control of pretty watsonia (*Watsonia versfeldii*). *Plant Protection Quarterly* 8, 103-104.

614) Hutchinson, I., Colosi, J., and Lewin, R.A. (1984) The biology of Canadian weeds. 63. *Sonchus asper* (L.) Hill and *S. oleraceus* L. *Canadian Journal of Plant Science* 64, 731-744.

615) Hutchinson, T.F. and Vankat, J.L. (1997) Invasibility and effects of amur honeysuckle in southwestern Ohio forests. *Conservation Biology* 11, 1117-1124.

616) Hutchison, M. (1992) Vegetation management guideline: round-leaved bittersweet (*Celastrus orbiculatus* Thunb.). *Natural Areas Journal* 12, 161-166.

617) Ideker, J. (1996) *Pereskia aculeata* (Cactaceae) in the lower Rio Grande valley of Texas. *SIDA Contributions to Botany* 17, 527-528.

618) Inderjit, I. (1995) Allelopathic potential of an annual weed, *Polypogon monspeliensis*, in crops in India. *Plant and Soil* 173, 251-257.

619) Insausti, P., Grimoldi, A.A., Chaneton, E.J., and Vasellati, V. (2001) Flooding induces a suite of adaptive plastic responses in the grass *Paspalum dilatatum*. *New Phytologist* 152, 291-299.

620) Ipor, I.B. and Price, C.E. (1994) Uptake, translocation and activity of paraquat on *Mikania micrantha* H.B.K. grown in different light conditions. *International Journal of Pest Management* 40, 40-45.

621) Isaacson, D.L. (2000) Impacts of broom (*Cytisus scoparius*) in western North America. *Plant Protection Quarterly* 15, 145-148.

622) Ismail, B.S. (1985) Germination of buffalo grass (*Paspalum conjugatum*) seeds. *Malaysian Applied Biology* 14, 7-11.

623) Ismail, B.S., Shukri, M.S., and Juraimi, A.S. (1994) Studies on the germination of mission grass (*Pennisetum polystachion* (L.) Schultes) seeds. *Plant Protection Quarterly* 9, 122-125.

624) Ismail, B.S., Rosmini, B.I., and Samiah, K. (1996) Factors affecting germination of Siam weed (*Chromolaena odorata* (L.) King & Robinson) seeds. *Plant Protection Quarterly* 11, 2-5.

625) Jacobs, S.W.L., Everett, J., and Torres, M.A. (1998) *Nassella tenuissima* (Gramineae) recorded from Australia, a potential new weed related to serrated tossock. *Telopea* 8, 41-46.

626) Jacobsen, N. and von Bothmer, R. (1995) Taxonomy in the *Hordeum murinum* complex (Poaceae). *Nordic Journal of Botany* 15, 449-458.

627) Jackson, D.I. (1960) A growth study of *Oxalis latifolia* H.B.K. *New Zealand Journal of Science* 3, 600-609.

628) James, C.S., Eaton, J.W., and Hardwick, K. (1999) Competition between three submerged macrophytes, *Elodea canadensis* Michx, *Elodea nuttallii* (Planch.) St. John and *Lagarosiphon major* (Ridl.) Moss. *Hydrobiologia* 415, 35-40.

629) James, E.K. and Sprent, J.I. (1999) Development of N_2-fixing nodules on the wetland legume *Lotus uliginosus* exposed to conditions of flooding. *New Phytologist* 142, 219-231.

630) Jeffrey, P.L. and Bode, R. (1992) Chain pulling of prickly acacia. *Proceedings of the Weed Society of Queensland* 2, 36-40.

631) Jobin, A., Schaffner, U., and Nentwig, W. (1996) The structure of the phytophagous insect fauna on the introduced weed *Solidago altissima* in Switzerland. *Entomologia Experimentalis et Applicata* 79, 33-42.

632) Johnson, S.R. and Allain, L.K. (1998) Observations on insect use of Chinese tallow (*Sapium sebiferum* (L.) Roxb.) in Louisiana and Texas. *Castanea* 63, 188-189.

633) Jones, D.A. and Turkington, R. (1986) Biological flora of the British Isles. *Lotus corniculatus* L. *Journal of Ecology* 74, 1185-1212.

634) Jones, D.T. and Doren, R.F. (1997) The distribution, biology and control of *Schinus terebinthifolius* in southern Florida, with special reference to Everglades National Park. In: Brock, J.H., Wade, M., Pysek, P., and Green, D. (eds) *Plant invasions: studies from North America and Europe*. Backhuys Publishers, Leiden, pp. 81-93.

635) Jones, E.W. (1944) Biological flora of the British Isles. *Acer pseudo-platanus* L. *Journal of Ecology* 32, 215-251.

636) Jones, R.H. and McLeod, K.W. (1989) Shade tolerance in seedlings of Chinese tallow tree, American sycamore, and cherrybark oak. *Bulletin of the Torrey Botanical Club* 116, 371-377.

637) Jones, R.H. and Sharitz, R.R. (1990) Effects of root competition and flooding on growth of Chinese tallow tree seedlings. *Canadian Journal of Forestry Research* 20, 573-578.

638) Jones, R.M. (1963) Preliminary studies of the germination of seed of *Acacia cyclops* and *A. saligna*. *South African Journal of Science* 59, 296-298.

639) Jones, V. and Richards, P.W. (1954) *Juncus acutus* L. *Journal of Ecology* 42, 639-650.

640) Jordano, P. (1982) Migrant birds are the main seed dispersers of blackberries (*Rubus ulmifolius*) in southern Spain. *Oikos* 38, 183-193.

641) Jordano, P. (1984) Seed weight variation and differential avian dispersal in blackberries *Rubus ulmifolius*. *Oikos* 43, 149-153.

642) Joshi, H. and Tewari, K. (1989) An ecological study of population dynamics of *Lolium perenne* and *Poa pratensis*. *Acta Botanica Indica* 17, 255-258.

643) Jubinsky, G. and Anderson, L.C. (1996) The invasive potential of Chinese tallow-tree (*Sapium sebiferum* Roxb.) in the Southeast. *Castanea* 61, 226-231.

644) Julien, M.H. and Bourne, A.S. (1986) Compensatory branching and changes in nitrogen content in the aquatic weed *Salvinia molesta* in response to disbudding. *Oecologia* 70, 250-257.

645) Julien, M.H. and Bourne, A.S. (1988) Alligatorweed is spreading in Australia. *Plant Protection Quarterly* 3, 91-96.

646) Julien, M.H. and Broadbent, J.E. (1980) The biology of Australian weeds. 3. *Alternanthera philoxeroides* (Mart.) Griseb. *Journal of the Australian Institute of Agricultural Science* 46, 150-155.

647) Julien, M.H., Bourne, A.S., and Low, V.H.K. (1992) Growth of the weed *Alternanthera philoxeroides* (Martius) Grisebach, (alligator weed) in aquatic and terrestrial habitats in Australia. *Plant Protection Quarterly* 7, 102-108.

648) Kamaluddin, M. and Grace, J. (1992) Photoinhibition and light acclimation in seedlings of *Bischofia javanica*, a tropical forest tree from Asia. *Annals of Botany* 69, 47-52.

649) Kamaluddin, M. and Grace, J. (1992) Acclimation in seedlings of a tropical tree, *Bischofia javanica*, following a stepwise reduction in light. *Annals of Botany* 69, 557-562.

650) Kammesheidt, L. (2000) Some autecological characteristics of early to late successional tree species in Venezuela. *Acta Oecologica* 21, 37-48.

651) Katznelson, J.S. (1974) Biological flora of Israel 5. The subterranean clover of *Trifolium* subsect. *Calcycomorphum* Katzn. *Trifolium subterraneum* L. (*sensu lato*). *Israel Journal of Botany* 23, 69-108.

652) Kaufmann, M.R. (1977) Soil temperature and drought effects on growth of Monterey pine. *Forest Science* 23, 317-325.

653) Kedzie-Webb, S.A., Sheley, R.L., Borkowski, J.J., and Jacobs, J.S. (2001) Relationships between *Centaurea maculosa* and indigenous plant assemblages. *Western North American Naturalist* 61, 43-49.

654) Keeley, P.E. and Thullen, R.J. (1989) Growth and competition of black nightshade (*Solanum nigrum*) and Palmer amaranth (*Amaranthus palmeri*) with cotton (*Gossypium hirsutum*). *Weed Science* 37, 326-334.

655) Keighery, G. (1993) Distribution, impact and biology of the other 'watsonias', *Chasmanthe* (African corn flag) and *Crocosmia* (Montbretia) in western Australia. *Plant Protection Quarterly* 8, 78-80.

656) Keighery, G.J. (1997) Occurrence and spread of sea spurge (*Euphorbia paralias*) along the west coast of Western Australia. *Nuytsia* 11, 285-286.

657) Kelly, D. (1988) Demography of *Carduus pycnocephalus* and *Carduus tenuiflorus*. *New Zealand Natural Sciences* 15, 17-24.

658) Kelly, D. and Skipworth, J.P. (1984) *Tradescantia fluminensis* in a Manawatu (New Zealand) forest: I. Growth and effects on regeneration. *New Zealand Journal of Botany* 22, 393-397.

659) Kelly, D. and Skipworth, J.P. (1984) *Tradescantia fluminensis* in a Manawatu (New Zealand) forest: II. Management by herbicides. *New Zealand Journal of Botany* 22, 399-402.

660) Kenkel, N.C., Hendrie, M.L., and Bella, I.E. (1997) A long-term study of *Pinus banksiana* population dynamics. *Journal of Vegetation Science* 8, 241-254.

661) Keresztesi, B. (1983) Breeding and cultivation of black locust, *Robinia pseudoacacia*, in Hungary. *Forest Ecology and Management* 6, 217-244.

662) Ketchum, D.E. and Bethune, J.E. (1963) Fire resistance in south Florida slash pine. *Journal of Forestry* 61, 529-530.

663) Khan, M.L. and Tripathi, R.S. (1987) Ecology of forest trees of Meghalaya (India): seed germination, and survival and growth of *Albizia lebbeck* seedlings in nature. *Indian Journal of Forestry* 10, 38-43.

664) Khanna, P.K. (1997) Comparison of growth and nutrition of young monocultures and mixed stands of *Eucalyptus globulus* and *Acacia mearnsii*. *Forest Ecology and Management* 94, 105-113.

665) Kiang, Y.T. (1982) Local differentiation of *Anthoxanthum odoratum* populations on roadsides. *American Midland Naturalist* 107, 340-350.

666) Kibbler, H. and Bahnisch, L.M. (1999) Physiological adaptations of *Hymenachne amplexicaulis* to flooding. *Australian Journal of Experimental Agriculture* 39, 429-435.

667) Kidd, S.B. (1997) A note on *Piper aduncum* in Morobe Province, Papua New Guinea. *Science in New Guinea* 22, 121-123.

668) Kimbel, J.C. and Carpenter, S.R. (1981) Effects of mechanical harvesting on *Myriophyllum spicatum* L. regrowth and carbohydrate allocation to roots and shoots. *Aquatic Botany* 11, 121-127.

669) King, S.E. and Grace, J.B. (2000) The effects of soil flooding on the establishment of cogongrass (*Imperata cylindrica*), a nonindigenous invader of the southeastern United States. *Wetlands* 20, 300-306.

670) King, S.E. and Grace, J.B. (2000) The effects of gap size and disturbance type on invasion of wet pine savanna by cogongrass, *Imperata cylindrica* (Poaceae). *American Journal of Botany* 87, 1279-1286.

671) Kincaid, D.R., Holt, G.A., Dalton, P.D., and Tixier, J.S. (1959) The spread of lehmann lovegrass as affected by mesquite and native perennial grasses. *Ecology* 40, 738-742.

672) Klauck, E.J. (1988) Die *Sambucus nigra-Robinia pseudoacacia*-Gesellschaft und ihre geographische Gliederung. *Tuexenia* 8, 281-286.

673) Klane, M.E., Philman, N.L., Bartuska, C.A., and McConnell, D.B. (1993) Growth regulator effects on in vitro shoot regeneration of *Crassula helmsii*. *Journal of Aquatic Plant Management* 31, 59-64.

674) Klemmedson, J.O. and Smith, J.G. (1964) Cheatgrass (*Bromus tectorum* L.). *Botanical Review* 30, 226-262.

675) Klemow, K.M. and Raynal, D.J. (1981) Population ecology of *Melilotus alba* in a limestone quarry. *Journal of Ecology* 69, 33-44.

676) Klimes, L. and Klimesova, J. (1999) Root sprouting in *Rumex acetosella* under different nutrient levels. *Plant Ecology* 141, 33-39.

677) Klingman, D.L. and Easley, T. (1971) Control of broomsedge. *Weeds Today* 2, 9-10.

678) Klinkhamer, P.G.L. and De Jong, T.J. (1993) Biological flora of the British Isles. *Cirsium vulgare* (Savi) Ten. (*Carduus lanceolatus* L., *Cirsium lanceolatum* (L.) Scop., non Hill). *Journal of Ecology* 81, 177-191.

679) Kloeppel, B.D. and Abrams, M.D. (1995) Ecophysiological attributes of the native *Acer saccharum* and the exotic *Acer platanoides* in urban oak forests in Pennsylvania, USA. *Tree Physiology* 15, 739-746.

680) Kloot, P.M. (1982) The naturalisation of *Echium plantagineum* L. in Australia. *Australian Weeds* 1, 29-31.

681) Kloot, P.M. (1983) The role of common iceplant (*Mesembryanthemum crystallinum*) in the deterioration of mesic pastures. *Australian Journal of Ecology* 8, 301-306.

682) Kluge, R.L. (1991) Biological control of crofton weed, *Ageratina adenophora* (Asteraceae), in South Africa. *Agriculture, Ecosystems and Environment* 37, 187-191.

683) Kluge, R.L. and Neser, S. (1991) Biological control of *Hakea sericea* (Proteaceae) in South Africa. *Agriculture, Ecosystems and Environment* 37, 91-113.

684) Klukas, R.W. and Truesdell, W.G. (1969) *The Australian pine problems in Everglades National Park. Part 1: the problem and some possible solutions. Part 2: Management plan for exotic plant eradication (*Casuarina equisetifolia*).* Everglades National Park, Florida, USA.

685) Knapp, L.B. and Canham, C.D. (2000) Invasion of an old-growth forest in New York by *Ailanthus altissima*: sapling growth and recruitment in canopy gaps. *Journal of the Torrey Botanical Society* 127, 307-315.

686) Knicker, H., Saggar, S., Bäumler, R., McIntosh, P.D., and Koegel, K.I. (2000) Soil organic matter transformations induced by *Hieracium pilosella* L. in tussock grassland of New Zealand. *Biology and Fertility of Soils* 32, 194-201.

687) Knight, R.S. and Macdonald, I.A.W. (1991) Acacias and korhans: an artificially assembled seed dispersal system. *South African Journal of Botany* 57, 220-225.

688) Knopf, F.L. and Olson, T.E. (1984) Naturalization of Russian-olive: implications to Rocky Mountain wildlife. *Wildlife Society Bulletin* 12, 289-298.

689) Knutson, R. (1997) *Hypericum* in National Parks: current control strategies in New South Wales. *Plant Protection Quarterly* 12, 102-103.

690) Kohri, M., Kamada,-M., Yuuki, T., Okabe, T., and Nakagoshi, N. (2002) Expansion of *Elaeagnus umbellata* on a gravel bar in the Naka River, Shikoku, Japan. *Plant Species Biology* 17, 25-36.

691) Kollmann, J. and Reiner, S.A. (1996) Light demands of shrub seedlings and their establishment within scrublands. *Flora* 191, 191-200.

692) Koopowitz, H. and Kaye, H. (1990) *Plant extinction: a global crisis.* Christopher Helm. London.

693) Kourtev, P.S., Ehrenfeld, J.G., and Huang, W.Z. (1998) Effects of exotic plant species on soil properties in hardwood forests of New Jersey. *Water, Air, and Soil Pollution* 105, 493-501.

694) Kourtev, P.S., Huang, W.Z., and Ehrenfeld, J.G. (1999) Differences in earthworm densities and nitrogen dynamics in soils under exotic and native plant species. *Biological Invasions* 1, 237-245.

695) Kowarik, I. (1983) Zur Einbürgerung und zum pflanzensoziologischen Verhalten des Götterbaumes (*Ailanthus altissima* (Mill.) Swingle) im französischen Mittelmeergebiet (Bas-Languedoc). *Phytocoenologia* 11, 389-405. *(German)*

696) Kowarik, I. (1995) Clonal growth in *Ailanthus altissima* on a natural site in West Virginia. *Journal of Vegetation Science* 6, 853-856.

697) Kowarik, I. (1996) Funktionen klonalen Wachstums von Bäumen bei der Brachflächen-Sukzession unter besonderer Beachtung von *Robinia pseudoacacia*. *Verhandlungen der Gesellschaft für Ökologie* 26, 173-181. *(German)*

698) Kowarik, I. and Böcker, R. (1984) Zur Verbreitung, Vergesellschaftung und Einbürgerung des Götterbaumes (*Ailanthus altissima* [Mill.] Swingle) in Mitteleuropa. *Tuexenia* 4, 9-29. *(German)*

699) Krischik, V.A. and Denno, R.F. (1990) Patterns of growth, reproduction, defense, and herbivory in the dioecious shrub *Baccharis halimifolia* (Compositae). *Oecologia* 83, 182-190.

700) Krishnan, P.N. and Rajendraprasad, M. (2000) Changes in growth and physiological attributes in *Adenanthera pavonina* L. saplings grown in normal sunlight and shade. *Indian Journal of Plant Physiology* 5, 47-51.

701) Kriticos, D., Brown, J., Radford, I., and Nicholas, M. (1999) Plant population ecology and biological control: *Acacia nilotica* as a case study. *Biological Control* 16, 230-239.

702) Kruger, F.J. (1977) *Invasive woody plants in the Cape fynbos with special reference to the biology and control of* Pinus pinaster. Proceedings of the second national weeds conference in South Africa, Cape Town, pp. 57-74.

703) Kuehn, K.A. and Suberkropp, K. (1998) Decomposition of standing litter of the freshwater emergent macrophyte *Juncus effusus*. *Freshwater Biology* 40, 717-727.

704) Kumar, A. and Gurumurthi, K. (2000) Genetic divergence studies on clonal performance of *Casuarina equisetifolia*. *Silvae Genetica* 49, 57-60.

705) Küpper, F., Küpper, H., and Spiller, M. (1996) Eine aggressive Wasserpflanze aus Australien und Neuseeland: *Crassula helmsii* (Kirk) Cockayne. *Floristische Rundbriefe* 30, 24-29. *(German)*

706) Kushwaha, S.P.S., Ramakrishnan, P.S., and Tripathi, R.S. (1981) Population dynamics of *Eupatorium odoratum* in successional environments following slash and burn agriculture. *Journal of Applied Ecology* 18, 529-535.

707) Lacey, J.R., Wallander, R., and Olson-Rutz, O. (1992) Recovery, germinability, and viability of leafy spurge (*Euphorbia esula*) seeds ingested by sheep and goats. *Weed Technology* 6, 599-602.

708) Lalonde, R.G. and Roitberg, B.D. (1994) Mating system, life-history, and reproduction in Canada thistle (*Cirsium arvense*; Asteraceae). *American Journal of Botany* 81, 21-28.

709) Lamont, D. (1993) The effect of soil moisture on the distribution of *Watsonia bulbillifera* in Serpentine National Park, Western Australia. *Plant Protection Quarterly* 8, 83-85.

710) Langeland, K.A. (1996) *Hydrilla verticillata* (L.F.) Royle (Hydrocharitaceae), 'the perfect aquatic weed'. *Castanea* 61, 293-304.

711) Lamb, A.F.A. (1968) *Cedrela odorata*. Fast growing timber trees of the lowland tropics, no. 2. Commonwealth Forestry Institute, Oxford.

712) Lambert, J.M. (1947) Biological flora of the British Isles. *Glyceria maxima* (Hartm.) Holmb. *Journal of Ecology* 34, 310-344.

713) Lane, A.M., Williams, R.J., Müller, W.J., and Lonsdale, W.M. (1997) The effects of the herbicide tebuthiuron on seedlings of *Mimosa pigra* and native floodplain vegetation in northern Australia. *Australian Journal of Ecology* 22, 439-447.

714) Lane, D. (1984) Factors affecting the development of populations of *Oxalis pescaprae* L. *Weed Research* 24, 219-225.

715) Langeland, K.A. and Craddock Burks, K. (1998) *Identification and biology of non-native plants in Florida's natural areas.* University of Florida, Gainesville.

716) Laroche, F.B. (ed.) (1999) *Melaleuca management plan. Ten years of successful melaleuca management in Florida 1988-1998.* Florida Exotic Pest Plant Council 1999.

717) LaRosa, A.M. (1984) The biology and ecology of *Passiflora mollissima* in Hawaii. Technical report no. 50, University of Hawaii, Honolulu.

718) LaRosa, A.M. (1987) Note on the idendity of the introduced passion flower vine 'banana poka' in Hawaii. *Pacific Science* 39, 369-371.

719) LaRosa, A.M. (1992) The status of Banana Poka in Hawaii. In: Stone, C.P., Smith, C.W., and Tunison, J.T. (eds) *Alien plant invasions in native ecosystems of Hawaii: management and research.* University of Hawaii, Honolulu, pp. 271-299.

720) LaRosa, A.M., Smith, C.W., and Gardner, D.E. (1985) Role of alien and native birds in the dissemination of firetree (*Myrica faya* Ait. — Myricaceae) and associated plants in Hawaii. *Pacific Science* 39, 372-378.

721) Larsson, C. and Martinsson, K. (1998) Jättebalsamin *Impatiens glandulifera* i Sverige – invasionsart eller harmlös trädgårdsflykting? *Svensk Botanisk Tidskrift* 92, 329-345. (Swedish)

722) Laundon, J.R. (1961) An Australiasian species of *Crassula* introduced into Britain. *Watsonia* 5, 59-63.

723) Lavergne, C. and Shaw, D. (1999) The invasive behaviour and the biological control of *Ligustrum robustum* ssp. *walkeri* on the Mascarene island of La Réunion. *Aliens* 9, 13.

724) Lavergne, C., Rameau, J.-C., and Figier, J. (1999) The invasive woody weed *Ligustrum robustum* subsp. *walkeri* threatens native forests on La Réunion. *Biological Invasions* 1, 377-392.

725) Lawesson, J.E. and Ortiz, L. (1990) Plantas introducidas en las islas Galapagos. *Monographs in Systematic Botany of the Missouri Botanical Gardens* 32, 201-210.

726) Lawrence, J.G., Colwell, A., and Sexton, O.J. (1991) The ecological impact of allelopathy in *Ailanthus altissima* (Simaroubaceae). *American Journal of Botany* 78, 948-958.

727) Lebot, V., Trilles, B., Noyer, J.L., and Modesto, J. (1998) Genetic relationships between *Dioscorea alata* L. cultivars. *Genetic Resources and Crop Evolution* 45, 499-509.

728) Lee, J.M. and Hamrick, J.L. (1983) Demography of two natural populations of musk thistle (*Carduus nutans*). *Journal of Ecology* 71, 923-936.

729) Lee, W.G. and Partridge, T.R. (1983) Rates of spread of *Spartina anglica* and sediment accretion in the New River Estuary, Invercargill, New Zealand. *New Zealand Journal of Botany* 21, 231-236.

730) Lee, W.G., Allen, R.B., and Johnson, P.W. (1986) Succession and dynamics of gorse (*Ulex europaeus* L.) communities in the Dunedin ecological district, South Island, New Zealand. *New Zealand Journal of Botany* 24, 279-292.

731) Leeflang, L. (2000) Response of *Trifolium repens* to a mosaic of bare and vegetated patches. *Plant Species Biology* 15, 59-65.

732) Leege, L.M. and Murphy, P.G. (2000) Growth of the non-native *Pinus nigra* in four habitats on the sand dunes of Lake Michigan. *Forest Ecology and Management* 126, 191-200.

733) Leege, L.M. and Murphy, P.G. (2001) Ecological effects of the non-native *Pinus nigra* on sand dune communities. *Canadian Journal of Botany* 79, 429-437.

734) Le Houérou, H.N. (1996) The role of cacti (*Opuntia* spp.) in erosion control, land reclamation, rehabilitation and agricultural development in the Mediterranean Basin. *Journal of Arid Environments* 33, 135-159.

735) Lejoly, J. and Nyakabwa, M. (1982) The ruderal association with *Paspalum conjugatum* and *Axonopus compressus* in Kisangani (High Zaire). *Bulletin de la Societé Royale de Botanique de Belgique* 114, 228-237. *(French)*

736) Lennox, C.L., Morris, M.J., Samuels, G.A., and Uys, J.L. (2001) Biological control of *Acacia mearnsii* and *Acacia pycnantha* using the mycoherbicide Stumpout®. *South African Journal of Science* 97, xviii (conference abstract).

737) Lesica, P. and Miles, S. (1999) Russian olive invasion into cottonwood forests along a regulated river in north-central Montana. *Canadian Journal of Botany* 77, 1077-1083.

738) Lesica, P. and Miles, S. (2001) Natural history and invasion of russian olive along eastern Montana rivers. *Western North American Naturalist* 61, 1-10.

739) Lester, P.J., Mitchell, S.F., and Scott, D. (1994) Effects of riparian willow tree (*Salix fragilis*) on macroinvertebrate densities in two small Central Otago, New Zealand, streams. *New Zealand Journal of Marine and Freshwater Research* 28, 267-276.

740) Levine, C.M. and Stromberg, J.C. (2001) Effects of flooding on native and exotic plant seedlings: implications for restoring south-western riparian forests by manipulating water and sediment flows. *Journal of Arid Environments* 49, 111-131.

741) Lewin, R.A. (1948) *Sonchus oleraceus* L. emend. Gouan. *Journal of Ecology* 36, 204-216.

742) Leys, A.R., Plater, B., and Lill, W.J. (1991) Response of vulpia (*Vulpia bromoides* (L.) S.F. Gray and *Vulpia myuros* (L.) C.C. Gmelin) and subterranean clover to rate and time of application of simazine. *Australian Journal of Experimental Agriculture* 31, 785-791.

743) Lhotska, M. (1975) Notes on the ecology of germination of *Alliaria petiolata*. *Folia Geobotanica et Phytotaxonomica* 10, 179-183.

744) Lhotska, M. and Kopecky, K. (1966) Zur Verbreitungsbiologie und Phytozoenologie von *Impatiens glandulifera* Royle an den Flusssystemen der Svitava, Svratka und oberen Odra. *Preslia (Praha)* 38, 376-385. *(German)*

745) Liddle, M.J. (1986) Noogoora burr – a successful suite of weeds. In: Kitching, R.L. (ed.) *The ecology of exotic animals and plants*. John Wiley & Sons, Brisbane, pp. 189-220.

746) Lin, W.C., Chen, A.C., Tseng, C.J., and Hwang, S.G. (1958) An investigation and study of Chinese tallow tree in Taiwan (*Sapium sebiferum* Roxb.). *Bulletin of the Taiwan Forestry Research Institute* 57, 1-37.

747) Lindeman, G.V., Enkhsaikhan, D., and Zhalbaa, K. (1994) Natural growth conditions of *Ulmus pumila* L. in Mongolian deserts. *Lesovedenie* 2, 42-53. *(Russian)*

748) Lindeman, G.V., Enkhsaikhan, D., and Zhalbaa, K. (1996) Naturally growing *Ulmus pumila* L. on hillock sands of eastern Mongolia. *Lesovedenie* 4, 68-78. *(Russian)*

749) Lippai, A., Smith, P.A., Price, T.V., Weiss, J., and Lloyd, C.J. (1996) Effects of temperature and water potential on germination of horehound (*Marrubium vulgare*) seeds from two Australian localities. *Weed Science* 44, 91-99.

750) Lippincott, C.L. (2000) Effects of *Imperata cylindrica* (L.) Beauv. (cogongrass) invasion on fire regime in Florida sandhill (USA). *Natural Areas Journal* 20, 140-149.

751) Lisci, M. and Pacini, E. (1994) Germination ecology of drupelets of the fig (*Ficus carica* L.). *Botanical Journal of the Linnean Society* 114, 133-146.

752) Little, E.L. and Skolmen, R.G. (1989) *Common forest trees of Hawaii (native and introduced)*. United States Department of Agriculture.

753) Lockhart, C.S., Austin, D.F., Jones, W.E., and Downey, L.A. (1999) Invasion of Carrotwood (*Cupaniopsis anacardioides*) in Florida natural areas (USA). *Natural Areas Journal* 19, 254-262.

754) Lohman, D.J. and Berenbaum, M.R. (1996) Impact of floral herbivory by parsnip webworm (Oecophoridae: *Depressaria pastinacella* Duponchel) on pollination and fitness of wild parsnip (Apiaceae: *Pastinaca sativa* L.). *American Midland Naturalist* 136, 407-412.

755) Lombardi, T., Fochetti, T., and Onnis, A. (1998) Germination of *Briza maxima* L. seeds: effects of temperature, light, salinity and seed harvesting time. *Seed Science and Technology* 26, 463-470.

756) Longpre, M.H., Bergeron, Y., Pare, D., and Belend, M. (1994) Effect of companion species on the growth of Jack pine (*Pinus banksiana*). *Canadian Journal of Forest Research* 24, 1846-1853.

757) Longstreth, D.J. and Mason, C.B. (1984) The effect of light on growth and dry matter allocation patterns of *Alternanthera philoxeroides* (Mart.) Griseb. *Botanical Gazette* 145, 105-109.

758) Longstreth, D.J. and Strain, B.R. (1975) Effects of salinity and illumination on photosynthesis and water balance of *Spartina alterniflora* Loisel. *Oecologia* 31, 191-199.

759) Lonsdale, W.M. (1988) Litterfall in an Australian population of *Mimosa pigra*, an invasive tropical shrub. *Journal of Tropical Ecology* 4, 381-392.

760) Lonsdale, W.M. (1993) Rates of spread of an invading species - *Mimosa pigra* in northern Australia. *Journal of Ecology* 81, 513-521.

761) Lonsdale, W.M. and Abrecht, D.G. (1989) Seedling mortality in *Mimosa pigra*, an invasive tropical shrub. *Journal of Ecology* 77, 372-385.

762) Lonsdale, W.M. and Miller, I.L. (1993) Fire as a management tool for a tropical woody weed: *Mimosa pigra* in northern Australia. *Journal of Environmental Management* 39, 77-87.

763) Lonsdale, W.M., Harley, K.L.S., and Gillett, J.D. (1988) Seed bank dynamics in *Mimosa pigra*, an invasive tropical shrub. *Journal of Applied Ecology* 25, 963-976.

764) Lonsdale, W.M., Miller, I.L., and Forno, I.W. (1989) The biology of Australian weeds. 20. *Mimosa pigra* L. *Plant Protection Quarterly* 4, 119-131.

765) Loope, L.L. (1992) An overview of problems with introduced plant species in National Parks and biosphere reserves of the United States. In: Stone, C.P., Smith, C.W., and Tunison, J.T. (eds) *Alien plant invasions in native ecosystems of Hawaii*. University of Hawaii Press, Honolulu, pp. 3-28.

766) Loope, L.L., Hamann, O., and Stone, C.P. (1988) Comparative conservation biology of oceanic archipelagos: Hawaii and the Galapagos. *BioScience* 38, 272-282.

767) Loope, L.L., Sanchez, P.G., Tarr, P.W., Loope, W.L., and Anderson, R.L. (1988) Biological invasions of arid land nature reserves. *Biological Conservation* 44, 95-118.

768) Lorence, D.H. (1978) The pteridophytes of Mauritius (Indian Ocean): ecology and distribution. *Botanical Journal of the Linnean Society* 76, 207-247.

769) Lorence, D.H. and Sussman, R.W. (1986) Exotic species invasion into Mauritius wet forest remnants. *Journal of Tropical Ecology* 2, 147-162.

770) Lorence, D.H. and Sussman, R.W. (1988) Diversity, density, and invasion in a Mauritian wet forest. *Monographs of Systematics of the Missouri Botanical Garden* 25, 187-204.

771) Loreti, J. and Oesterheld, M. (1996) Intraspecific variation in the resistance to flooding and drought in populations of *Paspalum dilatatum* from different topographic positions. *Oecologia* 108, 279-284.

772) Loreti, J., Oesterheld, M., and Leon, R.J.C. (1994) Effects of the interaction between grazing and flooding on *Paspalum dilatatum*, a native grass of the Flooding Pampa. *Ecologia Austral* 4, 49-58.

773) Lotan, J.E. and Perry, D.A. (1983) *Ecology and regeneration of lodgepole pine*. US Department of Agriculture, Agriculture Handbook 606. Washington, DC.

774) Lotter, W.D. and Hoffmann, J.H. (1998) An integrated management plan for the control for *Opuntia stricta* (Cactaceae) in the Kruger National Park, South Africa. *Koedoe* 41, 63-68.

775) Lotter, W.D., Thatcher, L., Rossouw, L., and Reinhardt, C.F. (1999) The influence of baboon predation and time in water on germination and early establishment of *Opuntia stricta* (Australian pest pear) in the Kruger National Park. *Koedoe* 42, 43-50.

776) Love, D. and Dansereau, P. (1959) Biosystematic studies on *Xanthium*. Taxonomic appraisal and ecological status. *Canadian Journal of Botany* 37, 173-208.

777) Love, R. and Feigen, M. (1978) Interspecific hybridization between native and naturalized *Crataegus* (Rosaceae) in western Oregon. *Madroño* 25, 211-217.

778) Lovett-Doust, J., Lovett-Doust, L., and Groth, A.T. (1990) The biology of Canadian weeds. 95. *Ranunculus repens. Canadian Journal of Plant Science* 70, 1123-1141.

779) Lowry, J.B., Prinsen, J.H., and Burrows, D.M. (1994) *Albizia lebbeck* – a promising forage tree for semiarid regions. In: Gutteridge, R.C. and Shelton, H.M. (eds) *Forage tree legumes in tropical agriculture*, CAB International, Wallingford, UK, pp. 75-83.

780) Luken, J.O. (1988) Population structure and biomass allocation of the naturalized shrub *Lonicera maackii* (Rupr.) Maxim. in forest and open habitats. *American Midland Naturalist* 119, 258-267.

781) Luken, J.O. (1990) Forest and pasture communities respond differently to cutting of exotic Amur honeysuckle. *Restoration and Management Notes* 8, 122-123.

782) Luken, J.O. and Goessling, N. (1995) Seedling distribution and potential persistence of the exotic shrub *Lonicera maackii* in fragmented forests. *American Midland Naturalist* 133, 124-130.

783) Luken, J.O. and Mattimiro, D.T. (1991) Habitat-specific resilience of the invasive shrub amur honeysuckle (*Lonicera maackii*) during repeated clipping. *Ecological Applications* 1, 104-109.

784) Luken, J.O. and Thieret, J.W. (1996) Amur honeysuckle, its fall from grace. *BioScience* 46, 18-24.

785) Luken, J.O., Kuddes, L.M., Tholemeier, T.C., and Haller, D.M. (1997) Comparative responses of *Lonicera maackii* (amur honeysuckle) and *Lindera benzoin* (spicebush) to increased light. *American Midland Naturalist* 138, 331-343.

786) Lundström, H. and Darby, E. (1994) The *Heracleum mantegazzianum* (giant hogweed) problem in Sweden: suggestions for its management and control. In: de Waal, L.C., Child, L.E., Wade, P.M., and Brock, J.H. (eds) *Ecology and management of invasive riverside plants*. John Wiley & Sons, Chichester, pp. 93-100.

787) Lunt, I.D. and Morgan, J.W. (2000) Can competition from *Themeda triandra* inhibit invasion by the perennial exotic grass *Nassella neesiana* in native grasslands? *Plant Protection Quarterly* 15, 92-94.

788) Lutz, H.J. (1943) Injuries to trees caused by *Celastrus* and *Vitis. Bulletin of the Torrey Botanical Club* 70, 436-439.

789) Lym, R.G. and Messersmith, C.G. (1994) Leafy spurge (*Euphorbia esula*) control, forage production, and economic return with fall-applied herbicides. *Weed Technology* 8, 824-829.

790) Lynn, D.E. and Waldren, S. (2001) Variation in life history characteristics between clones of *Ranunculus repens* grown in experimental garden conditions. *Weed Research* 41, 421-432.

791) Macdicken, K.G., Hairiah, K., Otsamo, A., Duguma, B., and Majid, N.M. (1997) Shade-based control of *Imperata cylindrica*: tree fallows and cover crops. *Agroforestry Systems* 36, 131-149.

792) Macdonald, I.A.W. (1983) Alien trees, shrubs, and creepers invading indigenous vegetation in the Hluhluwe-Umfolozi Game Reserve complex in Natal. *Bothalia* 14, 949-959.

793) Macdonald, I.A.W. (1988) The history, impacts and control of introduced species in the Kruger National Park, South Africa. *Transactions of the Royal Society South Africa* 46, 252-276.

794) Macdonald, I.A.W. and Jarman, M.L. (1985) *Invasive alien plants in the terrestrial ecosystems of Natal, South Africa*. South African National Scientific Programmes report no. 118, CSIR, Pretoria.

795) Macdonald, I.A.W. and Richardson, D.M. (1986) Alien species in terrestrial ecosystems of the fynbos biome. In: Macdonald, I.A.W., Kruger, F.J., and Ferrar, A.A. (eds) *The ecology and management of biological invasions in southern Africa*. Oxford University Press, Cape Town, pp. 77-91.

796) Macdonald, I.A.W. and Wissel, C. (1992) Determining optimal clearing treatments for the alien invasive shrub *Acacia saligna* in the southwestern Cape, South Africa. *Agriculture, Ecosystems and Environment* 39, 169-186.

797) Macdonald, I.A.W., Graber, D.M., DeBenedetti, S., Groves, R.H., and Fuentes, E.R. (1988) Introduced species in nature reserves in Mediterranean-type climatic regions of the world. *Biological Conservation* 44, 37-66.

798) Macdonald, I.A.W., Ortiz, L., Lawesson, J.E. and Nowak, J.B. (1988) The invasion of highlands in Galápagos by the red quinine tree *Cinchona succirubra*. *Environmental Conservation* 15, 215-220.

799) Mack, R.N. (1981) Invasion of *Bromus tectorum* L. into western North America: an ecological chronicle. *Agro-Ecosystems* 7, 145-165.

800) Mack, R.N. and Pyke, D.A. (1984) The demography of *Bromus tectorum*: the role of microclimate, grazing and disease. *Journal of Ecology* 72, 731-748.

801) Mackey, A.P. (1997) The biology of Australian weeds. 29. *Acacia nilotica* ssp. *indica* (Benth.) Brenan. *Plant Protection Quarterly* 12, 7-17.

802) Maddox, D.M., Mayfield, A., and Poritz, N.H. (1985). Distribution of yellow starthistle (*Centaurea solstitialis*) and Russian knapweed (*Centaurea repens*). *Weed Science* 33, 315-327.

803) Maddox, D.M., Joley, D.B., Supkoff, D.M., and Mayfield, A. (1996) Pollination biology of yellow starthistle (*Centaurea solstitialis*) in California. *Canadian Journal of Botany* 74, 262-267.

804) Madhusoodanan, P.V. and Ajit Kumar, K.G. (1993) *Alternanthera philoxeroides* (Mart.) Griseb. - 'alligator weed' - a fast spreading weed in Kerala, south India. *Journal of Economic and Taxonomic Botany* 17, 651-654.

805) Madsen, J.D., Sutherland, J.W., Bloomfield, J.A., Eichler, L.W., and Boylen, C.W. (1991) The decline of native vegetation under dense Eurasian watermilfoil canopies. *Journal of Aquatic Plant Management* 29, 94-99.

806) Makepeace, W. (1981) Polymorphism and the chromosomal number of *Hieracium pilosella* in New Zealand. *New Zealand Journal of Botany* 19, 255-258.

807) Mal, T.K., Lovett-Doust, J., Lovett-Doust, L., and Mulligan, G.A. (1992) The biology of Canadian weeds. 100. *Lythrum salicaria*. *Canadian Journal of Plant Science* 72, 1305-1330.

808) Malan, C., Visser, J.H., and Grobbelaar, N. (1981) Control of problem weeds of maize on the Transvaal Highveld (South Africa). I. *Tagetes minuta* L. *Weed Research* 21, 235-241.

809) Malan, D.E. and Zimmermann, H.G. (1988) Chemical control of *Opuntia imbricata* (Haw.) DC. And *Opuntia rosea* DC. *Applied Plant Science* 2, 13-16.

810) Malecki, R.A. and Rawinski, T.J. (1985) New methods for controlling purple loosestrife. *New York Fish and Game Journal* 32, 9-19.

811) Malecki, R.A., Blossey, B., Hight, S.D., Schroeder, D., Kok, L.T., and Coulson, J.R. (1993) Biological control of Purple Loosestrife. *BioScience* 43, 680-686.

812) Malik, N. and Vanden Born, W.H. (1988) The biology of Canadian weeds. 86. *Galium aparine* L. and *Galium spurium*. *Canadian Journal of Plant Science* 68, 481-499.

813) Marambe, B., Amarasinghe, L., and Dissanayake, S. (2001) Growth and development of *Mimosa pigra*: an alien invasive plant in Sri Lanka. In: Brundu, G., Brock, J., Camarda, I., Child, L., and Wade, M. (eds) Plant invasions: species *ecology and ecosystem management*. Backhuys Publishers, Leiden, pp. 115-122.

814) Marchant, C.J. and Goodman, P.J. (1969) *Spartina alterniflora* Loisel. *Journal of Ecology* 57, 291-295.

815) Marchant, C.J. and Goodman, P.J. (1969) *Spartina maritima* (Curtis) Fernald. *Journal of Ecology* 57, 287-291.

816) Margis, R., Felix, D., Caldas, J.F., Salgueiro, F., de Araujo, D.S.D., Breyne, M.M., de Oliveira, D., and Margis, P.M. (2002) Genetic differentiation among three neighboring Brazil-cherry (*Eugenia uniflora* L.) populations within the Brazilian Atlantic rain forest. *Biodiversity and Conservation* 11, 149-163.

817) Marinucci, A.C. and Bartha, R. (1982) A component model of decomposition of *Spartina alterniflora* in a New Jersey salt marsh. *Canadian Journal of Botany* 60, 1618-1624.

818) Marks, M., Lapin, B., and Randall, J. (1994) *Phragmites australis* (*P. communis*): threats, management, and monitoring. *Natural Areas Journal* 14, 285-294.

819) Marler, R.J., Stromberg, J.C., and Patten, D.T. (2001) Growth response of *Populus fremontii*, *Salix gooddingii*, and *Tamarix ramosissima* seedlings under different nitrogen and phosphorus concentrations. *Journal of Arid Environments* 49, 133-146.

820) Marohasy, J. (1995) Prospects for the biological control of prickly acacia, *Acacia nilotica* (L.) Willd. Ex del. (Mimosaceae) in Australia. *Plant Protection Quarterly* 10, 24-31.

821) Maron, J.L. (1997) Interspecific competition and insect herbivory reduce bush lupine (*Lupinus arboreus*) seedling survival. *Oecologia* 110, 284-290.

822) Maron, J.L. and Connors, P.G. (1996) A native nitrogen-fixing shrub facilitates weed invasion. *Oecologia* 105, 302-312.

823) Marquez, G.J. and Laguna, H.G. (1982) Seed anatomy and germination of *Turbina corymbosa*, Convolvulaceae. *Phyton* (Buenos Aires) 42, 1-8. *(Spanish)*

824) Marquis, D.A. (1990) *Prunus serotina* Ehrh. Black cherry. In: Burns, R.M. and Honkala, B.H. *Silvics of North America, vol. 2. Hardwoods*. Agriculture Handbook 654, US Department of Agriculture, Washington DC, pp. 594-604.

825) Marshall, E.J.P. (1990) Interference between sown grasses and the growth of rhizome of *Elymus repens* (couch grass). *Agriculture, Ecosystems and Environment* 33, 11-22.

826) Marshall, G. (1987) A review of the biology and control of selected weed species in the genus *Oxalis*: *O. stricta* L., *O. latifolia* H.B.K. and *O. pes-caprae* L. *Crop Protection* 6, 355-364.

827) Marshall, G. and Gitari, J.N. (1988) Studies on the growth and development of *Oxalis latifolia*. *Annals of Applied Biology* 112, 143-150.

828) Martin, D.W. and Chambers, J.C. (2001) Effects of water table, clipping, and species interactions on *Carex nebrascensis* and *Poa pratensis* in riparian meadows. *Wetlands* 21, 422-430.

829) Martin, P.H. (1999) Norway maple (*Acer platanoides*) invasion of a natural forest stand: understory consequence and regeneration pattern. *Biological Invasions* 1, 215-222.

830) Martinez-Ropero, E.V., Gomez-Gutierrez, J.M., and Galindo, V.P. (1989) Abiotic factors defining the range of distribution of *Cytisus multiflorus* (L'Hér) Sweet in Spain. *Anales de Edafologia y Agrobiologia* 48, 887-904.

831) Maruta, E. (1983) Growth and survival of current-year seedlings of *Polygonum cuspidatum* at the upper distribution limit on Mt. Fuji. *Oecologia* 60, 316-320.

832) Mather, L.J. and Williams, P.A. (1990) Phenology, seed ecology, and age structure of Spanish heath (*Erica lusitanica*) in Canterbury, New Zealand. *New Zealand Journal of Botany* 28, 207-215.

833) Mathur, G. and Mohan Ram, H.Y. (1986) Floral biology and pollination of *Lantana camara*. *Phytomorphology* 36, 79-100.

834) Mason, R. (1960) Three waterweeds of the family Hydrocharitaceae in New Zealand. *New Zealand Journal of Science* 3, 383-395.

835) Mauchamp, A. (1996) Threats from alien plant species in the Galapagos Islands. *Conservation Biology* 11, 260-263.

836) Maule, H.G., Andrews, M., Morton, J.D., Jones, A.V., and Daly, G.T. (1995) Sun/shade acclimation and nitrogen nutrition of *Tradescantia fluminensis*, a problem weed in New Zealand native forest remnants. *New Zealand Journal of Ecology* 18, 35-46.

837) Mays, W.T. and Kok, L.T. (1988) Seed wasp on multiflora rose, *Rosa multiflora* in Virginia (USA). *Weed Technology* 2, 265-268.

838) Mbalo, B.A. and Witkowski, E.T.F. (1997) Tolerance to soil temperatures experienced during and after the passage of fire in seeds of *Acacia karroo*, *A. tortilis* and *Chromolaena odorata*: a laboratory study. *South African Journal of Botany* 63, 421-425.

839) McCarthy, B.C. (1997) Response of a forest understory community to experimental removal of an invasive nonindigenous plant (*Alliaria petiolata*, Brassicaceae). In: Luken, J.O. and Thieret, J.W. (eds) *Assessment and management of plant invasions.* Springer, New York, pp. 117-130.

840) McCarty, L.B., Colvin, D.L., and Higgins, J.M. (1996) Highbush blackberry (*Rubus argutus*) control in bahiagrass (*Paspalum notatum*). *Weed Technology* 10, 754-761.

841) McCaw, W.L., Smith, R.H., and Neal, J.E. (2000) Post-fire recruitment of red tingle (*Eucalyptus jacksonii*) and karri (*Eucalyptus diversicolor*) following low-moderate intensity prescribed fires near Walpole, south-west Western Australia. *CALMScience* 3, 87-94.

842) McClaran, M. and Anable, M.E. (1992) Spread of introduced Lehmann lovegrass along a grazing intensity gradient. *Journal of Applied Ecology* 29, 92-98.

843) McCormick, L.H. and Hartwig, N.L. (1995) Control of the noxious weed mile-a-minute (*Polygonum perfoliatum*) in reforestation. *Northern Journal of Applied Forestry* 12, 127-132.

844) McDowell, C.R. and Moll, E.J. (1981) Studies of seed germination and seedling competition in *Virgilia oroboides*, *Albizia lophantha* and *Acacia longifolia*. *Journal of South African Botany* 47, 653-686.

845) McFadyen, R.E. and Harvey, G.J. (1990) Distribution and control of rubber vine, *Cryptostegia grandiflora*, a major weed in northern Queensland. *Plant Protection Quarterly* 5, 152-155.

846) McFarland, J.D., Kevan, P.G., and Lane, M.A. (1989) Pollination biology of *Opuntia imbricata* (Cactaceae) in southern Colorado. *Canadian Journal of Botany* 67, 24-28.

847) McIntyre, S. and Ladiges, P.Y. (1985) Aspects of the biology of *Ehrharta erecta* Lam. *Weed Research* 25, 21-32.

848) McKey, D. (1988) *Cecropia peltata*, an introduced neotropical pioneer tree, is replacing *Musanga cecropioides* in southwestern Cameroon. *Biotropica* 20, 262-264.

849) McMahon, A.R.G. (1991) Control of annual grasses with particular reference to *Briza maxima*. *Plant Protection Quarterly* 6, 129.

850) McNab, W.H. and Meeker, M. (1987) Oriental bittersweet: a growing threat to hardwood silviculture in the Appalachians. *Northern Journal of Applied Forestry* 4, 174-177.

851) McNeill, J. (1981) The taxonomy and distribution in eastern Canada of *Polygonum arenastrum* (4x equals 40) and *Polygonum monspeliense* (6x equals 60), introduced members of the *Polygonum aviculare* complex. *Canadian Journal of Botany* 59, 2744-2751.

852) Medd, R.W. (1986) *Carduus nutans* – ingression through indifference towards weeds. In: Kirching, R.L. (ed.) *The ecology of exotic animals and plants*. John Wiley & Sons, pp. 223-239.

853) Medd, R.W. and Smith, R.C.G. (1978) Prediction of the potential distribution of *Carduus nutans* (nodding thistle) in Australia. *Journal of Applied Ecology* 15, 603-612.

854) Medrzycki, P. and Pabjanek, P. (2001) Linking land use and invading species features: a case study of *Acer negundo* in Bialowieza village (NE Poland). In: Brundu, G. *et al.* (eds) Plant invasions: species *ecology and ecosystem management*. Backhuys Publishers, Leiden, pp. 123-132.

855) Meduna, E., Schneller, J., and Holderegger, R. (1999) *Prunus laurocerasus* L., eine sich ausbreitende nichteinheimische Gehölzart: Untersuchungen zu Ausbreitung und Vorkommen in der Nordostschweiz. *Zeitschrift für Ökologie und Naturschutz* 8, 147-155. (German)

856) Medeiros, A.C., Loope, L.L., Conant, P., and McElvaney, S. (1997) Status, ecology, and management of the invasive plant, *Miconia calvescens* DC (Melastomataceae) in the Hawaiian islands. *Bishop Museum Occasional Papers* 48, 23-36.

857) Meerts, P. and Garnier, E. (1996) Variation in relative growth rate and its components in the annual *Polygonum aviculare* in relation to habitat disturbance and seed size. *Oecologia* 108, 438-445.

858) Meerts, P., Baya, T., and Lefebvre, C. (1998) Allozyme variation in the annual weed species complex *Polygonum aviculare* (Polygonaceae) in relation to ploidy level and colonizing ability. *Plant Systematics and Evolution* 211, 239-256.

859) Melgoza, G., Nowak, R.S., and Tausch, R.J. (1990) Soil water exploitation after fire: competition between *Bromus tectorum* (cheatgrass) and two native species. *Oecologia* 83, 7-13.

860) Melville, M.R. and Morton, J.K. (1982) A biosystematic study of the *Solidago canadensis* (Compositae) complex: 1. The Ontario, Canada, populations. *Canadian Journal of Botany* 60, 976-997.

861) Memmott, J., Fowler, S.V., Paynter, Q., Sheppard, A.W., and Syrett, P. (2000) The invertebrate fauna on broom, *Cytisus scoparius*, in two native and two exotic habitats. *Acta Oecologica* 21, 213-222.

862) Mensing, S. and Byrne, R. (1998) Pre-mission invasion of *Erodium cicutarium* in California. *Journal of Biogeography* 25, 757-762.

863) Menwyelet, A., Coppock, D.L., and Detling, J.K. (1994) Fruit production of *Acacia tortilis* and *A. nilotica* in semi-arid Ethiopia. *Agroforestry Systems* 27, 23-30.

864) Merchant, M. (1993) The potential for control of the soft rush (*Juncus effusus*) in grass pasture by grazing goats. *Grass and Forage Science* 48, 395-409.

865) Merchant, M. (1995) The effect of pattern and severity of cutting on the vigour of the soft rush (*Juncus effusus* L.). *Grass and Forage Science* 50, 81-84.

866) Mesleard, F. and Lepart, J. (1989) Continuous basal sprouting from a lignotuber: *Arbutus unedo* L. and *Erica arborea* L., as woody Mediterranean examples. *Oecologia* 80, 127-131.

867) Mesleard, F. and Lepart, J. (1991) Germination and seedling dynamics of *Arbutus unedo* and *Erica arborea* on Corsica (France). *Journal of Vegetation Science* 2, 155-164.

868) Meyer, A.H. and Schmid, B. (1999) Experimental demography of the old-field perennial *Solidago altissima*: the dynamics of the shoot population. *Journal of Ecology* 87, 17-27.

869) Meyer, A.H. and Schmid, B. (1999) Experimental demography of rhizome populations of establishing clones of *Solidago altissima*. *Journal of Ecology* 87, 42-54.

870) Meyer, G.A. (1998) Mechanisms promoting recovery from defoliation in goldenrod (*Solidago altissima*). *Canadian Journal of Botany* 76, 450-459.

871) Meyer, J.-Y. (1996) Status of *Miconia calvescens* (Melastomataceae), a dominant invasive tree in the Society Islands (French Polynesia). *Pacific Science* 50, 66-76.

872) Meyer, J.-Y. (1998) Observations on the reproductive biology of *Miconia calvescens* DC (Melastomataceae), an alien invasive tree on the island of Tahiti (south Pacific Ocean). *Biotropica* 30, 609-624.

873) Meyer, J.-Y. and Florence, J. (1996) Tahiti's native flora endangered by the invasion of *Miconia calvescens* DC. (Melastomataceae). *Journal of Biogeography* 23, 775-781.

874) Mezev-Krichfalishii, G.N., Krichfalushii, V.V., and Komendar, V.I. (1989) Studies on the population biology of *Ornithogalum umbellatum* L. (Liliaceae) for elaborating a strategy of species survival in Transcarpathia. *Tiscia* 24, 3-10.

875) Michael, P.W. (1964) The identity and origin of varieties of *Oxalis pes-caprae* L. naturalized in Australia. *Transactions of the Royal Society of South Australia* 88, 167-173.

876) Milberg, P. and Lamont, B.B. (1995) Fire enhances weed invasion of roadside vegetation in southwestern Australia. *Biological Conservation* 73, 45-49.

877) Miller, G.K., Young, J.A., and Evans, R.A. (1986) Germination of seeds of perennial pepperweed (*Lepidium latifolium*). *Weed Science* 34, 252-255.

878) Miller, J.H. (1985) Testing herbicides for kudzu eradication on a Piedmont site. *Southern Journal of Applied Forestry* 9, 128-132.

879) Miller, J.H. (1990) *Ailanthus altissima* (Mill.) Swingle. Ailanthus. In: Burns, R.M. and Honkala, B.H. (eds) *Silvics of North America, vol. 2. Hardwoods*. Agriculture Handbook 654, US Department of Agriculture, Washington DC, pp. 101-104.

880) Milton, S.J. (1981) Litterfall of the exotic acacias in the south western Cape. *Journal of South African Botany* 47, 147-155.

881) Milton, S.J. and Hall, A.V. (1981) Reproductive biology of Australian acacias in the south-western Cape province, South Africa. *Transactions of the Royal Society of South Africa* 44, 465-487.

882) Mislevy, P., Wilson, R.H., and Hall, D.W. (1987) Sand blackberry (*Rubus cuneifolius*) control and perennial grass recovery. *Soil and Crop Science Society of Florida Proceedings* 46, 64-67.

883) Mislevy, P., Shilling, D.G., Martin, F.G., and Hatch, S.L. (1999) Smutgrass (*Sporobolus indicus*) control in bahiagrass (Paspalum notatum) pastures. *Weed Technology* 13, 571-575.

884) Misra, M.K. and Nisanka, S.K. (1997) Litterfall, decomposition and nutrient release in *Casuarina equisetifolia* plantations on the sandy coast of Orissa, India. *Tropical Ecology* 38, 109-119.

885) Mitchell, D.S. (1980) The water fern *Salvinia molesta* in the Sepick River, Papua New Guinea. *Environmental Conservation* 7, 115-122.

886) Mitchell, D.S. and Tur, N.M. (1975) The rate of growth of *Salvinia molesta* (*S. auriculata* Auct.) in laboratory and natural conditions. *Journal of Applied Ecology* 12, 213-225.

887) Mitchell, R.J. and Ankeny, D.P. (2001) Effects of local conspecific density on reproductive success in *Penstemon digitalis* and *Hesperis matronalis*. *Ohio Journal of Science* 101, 22-27.

888) Mitich, L.W. (1998) Intriguing World of Weeds: Poison-Hemlock (*Conium maculatum* L.). *Weed Technology* 12, 194-197.

889) Mitich, L.W. (2000) Intriguing world of weeds. Kudzu [*Pueraria lobata* (Willd.) Ohwi]. *Weed Technology* 14, 231-235.

890) Mohamed, B.F. and Gimingham, C.F. (1970) The morphology and vegetative regeneration in *Calluna vulgaris*. *New Phytologist* 69, 743-750.

891) Mohan, R.B., Ganeshaiah, K.N., and Uma, S.R. (2001) Paternal parents enhance dispersal ability of their progeny in a wind-dispersed species, *Tecoma stans*. *Current Science* 81, 22-24.

892) Mohnot, K. and Chatterji, U.N. (1965) Chemico-physiological studies on the imbibition and germination of seeds of *Parkinsonia aculeata* Linn. *Österreichische Botanische Zeitschrift* 112, 577-585. *(German)*

893) Molloy, B.P.J., Partridge, T.R., and Thomas, W.P. (1991) Decline of tree lupine (*Lupinus arboreus*) on Kaitorete spit, Canterbury, New Zealand, 1984-1990. *New Zealand Journal of Botany* 29, 349-352.

894) Monasterio, H.E. and Weber, H.E. (1996) Taxonomy and nomenclature of *Rubus ulmifolius* and *Rubus sanctus* (Rosaceae). *Edinburgh Journal of Botany* 53, 311-322.

895) Mooney, H.A., Hamburg, S.P. and Drake, J.A. (1986) The invasions of animals and plants into California. In: Mooney, H.A. and Drake, J.A. (eds) *Ecology of biological invasions of North America and Hawaii.* Springer-Verlag, New York, pp. 250-272.

896) Moore, J.H. and Fletcher, G.E. (1994) Bulbil watsonia (*Watsonia bulbillifera* Mathews & Bolus) control with herbicides in Western Australia. *Plant Protection Quarterly* 9, 82-85.

897) Moore, R.J. (1975) The biology of Canadian weeds. 13. *Cirsium arvense* (L.) Scop. *Canadian Journal of Plant Science* 55, 1033-1048.

898) Moore, R.M. (1953) Studies on perennial weeds: the selctive control of hoary cress (*Cardaria draba* L. Desv.) in wheat. *Journal of the Australian Institute of Agricultural Science* 19, 241-243.

899) Moran, V.C. (1979) Critical reviews of biological pest control in South Africa. 3. The jointed cactus, *Opuntia aurantiaca* Lindley. *Journal of the Entomological Society of South Africa* 42, 299-329.

900) Moran, V.C. and Zimmermann, H.G. (1991) Biological control of jointed cactus, *Opuntia aurantiaca* (Cactaceae), in South Africa. *Agriculture, Ecosystems and Environment* 37, 5-28.

901) Moret, J. (1992) Numerical taxonomy applied to a study of some ploidy levels within the *Ornithogalum umbellatum* complex (Hyacinthaceae) in France. *Nordic Journal of Botany* 12, 183-195.

902) Morison, J.I.L., Piedade, M.T.F., Mueller, E., Long, S.P., Junk, W.J., and Jones, M.B. (2000) Very high productivity of the C_4 aquatic grass *Echinochloa polystachya* in the Amazon floodplain confirmed by net ecosystem CO_2 flux measurements. *Oecologia* 125, 400-411.

903) Morrison, S.L. and Molofsky, J. (1998) Effects of genotypes, soil moisture, and competition on the growth of an invasive grass, *Phalaris arundinacea* (reed canary grass). *Canadian Journal of Botany* 76, 1939-1946.

904) Morton, C.V. (1966) The Mexican species of *Tectaria*. *American Fern Journal* 56, 120-137.

905) Morton, J.F. (1978) Brazilian pepper — its impact on people, animals and the environment. *Economic Botany* 32, 353-359.

906) Morton, J.F. (1994) Lantana, or red sage (*Lantana camara* L., [Verbenaceae]), notorious weed and popular garden flower; some cases of poisoning in Florida. *Economic Botany* 48, 259-270.

907) Morton, J.K. (1989) The clustered dock, *Rumex conglomeratus* (Polygonaceae), in Canada. *Canadian Field-Naturalist* 103, 86-88.

908) Mueller-Dombois, D. (1973) A non-adapted vegetation interferes with water removal in a tropical rainforest area in Hawaii. *Tropical Ecology* 14, 1-16.

909) Mueller-Dombois, D. and Whiteaker, L.D. (1990) Plants associated with *Myrica faya* and two other pioneer trees on a recent volcanic surface in Hawaii Volcanoes National Park. *Phytocoenologia* 19, 29-41.

910) Mullahey, J.J. (1996) Tropical soda apple (*Solanum viarum* Dunal), a biological pollutant threatening Florida. *Castanea* 61, 255-260.

911) Mullahey, J.J. and Cornell, J. (1994) Biology of tropical soda apple (*Solanum viarum*) an introduced weed in Florida. *Weed Technology* 8, 465-469.

912) Mullahey, J.J., Cornell, J.A., and Colvin, D.L. (1993) Tropical soda apple (*Solanum viarum*) control. *Weed Technology* 7, 723-727.

913) Mullahey, J.J., Nee, M., Wunderlin, R.P., and Delaney, K.R. (1993) Tropical soda apple (*Solanum viarum*): a new weed threat in subtropical regions. *Weed Technology* 7, 783-786.

914) Mullahey, J.J., Shilling, D.G., Mislevy, P., and Akanda, R.A. (1998) Invasion of tropical soda apple (*Solanum viarum*) into the US: lessons learned. *Weed Technology* 12, 733-736.

915) Mullett, T.L. (2001) Effects of the native environmental weed *Pittosporum undulatum* Vent. (sweet pittosporum) on plant biodiversity. *Plant Protection Quarterly* 16, 117-121.

916) Mullett, T. and Simmons, D. (1995) Ecological impacts of the environmental weed sweet pittosporum (*Pittosporum undulatum* Vent.) in dry sclerophyll forest communities, Victoria. *Plant Protection Quarterly* 10, 131-138.

917) Mulligan, G.A. and Frankton, C.E. (1962) Taxonomy of the genus *Cardaria* with particular reference to the species introduced into North America. *Canadian Journal of Botany* 40, 1411-1425.

918) Mulligan, G.A. and Finflay, J.N. (1974) The biology of Canadian weeds. 3. *Cardaria draba*, *C. chalepensis* and *C. pubescens*. *Canadian Journal of Plant Science* 54, 149-160.

919) Mulligan, G.A. and Munro, D.B. (1981) The biology of Canadian weeds. 51. *Prunus virginiana* L. and *P. serotina* Ehrh. *Canadian Journal of Plant Science* 61, 977-992.

920) Muniappan, R. (1996) Biological control of *Chromolaena odorata*. In: Caligari, P.D.S. and Hind, D.J.N. (eds) *Compositae: Biology & Utilization. Proceedings of the International Compositae Conference*. Royal Botanic Gardens, Kew, pp. 333-337.

921) Munir, A.A. (2002) A taxonomic revision of the genus Verbena L. (Verbenaceae) in Australia. *Journal of the Adelaide Botanic Gardens* 20, 21-103.

922) Musil, C.F. (1993) Effect of invasive Australian acacias on the regeneration, growth and nutrient chemistry of South African lowland fynbos. *Journal of Applied Ecology* 30, 361-372.

923) Musil, C.F. and Midgley, G.F. (1990) The relative impact of Australian invasive acacias, fire and season on the soil chemical status of a sand plain lowland fynbos community. *South African Journal of Botany* 56, 419-427.

924) Muyt, A. (2001) *Bush invaders of south-east Australia.* R.G. and F.J. Richardson, Meredith.

925) Myers, R.L. (1983) Site susceptibility to invasion by the exotic tree *Melaleuca quinquenervia* in southern Florida. *Journal of Applied Ecology* 20, 645-658.

926) Mytinger, L. and Williamson, G.B. (1987) The invasion of *Schinus* into saline communities of Everglades National Park. *Florida Scientist* 50, 7-12.

927) National Research Council (1984) *Casuarinas: nitrogen-fixing trees for adverse sites.* National Academy Press, Washington DC.

928) Nauman, C.E. and Austin, D.F. (1978) Spread of the exotic fern *Lygodium microphyllum* in Florida. *American Fern Journal* 68, 65-66.

929) Navarro, L. (1998) Effect of pollen limitation, additional nutrients, flower position and flowering phenology on fruit and seed production in *Salvia verbenaca* (Lamiaceae). *Nordic Journal of Botany* 18, 441-446.

930) Newbury, D.M. (1980) Interactions between the coccid, *Icerya seychellarum* and its host tree species on Aldabra Atoll, Indian Ocean. 2. *Scaevola taccada. Oecologia* 46, 180-185.

931) Newman, I.V. (1934) Studies in the Australian acacias. IV. Life history of *Acacia baileyana. Proceedings of the Linnean Society of New South Wales* 59, 277-313.

932) Nieschalk, C. (1989) Contributions to the knowledge of the rose flora of northern Hesse (West Germany): VI. The genus *Rosa* L. section *caninae* DC. subsection *eucaninae* Crepin, the formenkreis of the *Rosa canina* group. *Philippia* 6, 155-199. *(German)*

933) Nijs, J.C. (1984) Biosystematic studies of the *Rumex acetosella* complex: (Polygonaceae) 8. A taxonomic revision. *Feddes Repertorium* 25, 43-66.

934) Nilsen, E.T. and Muller, W.H. (1980) A comparison of the relative naturalization ability of two *Schinus* species in southern California. I. Seed germination. *Bulletin of the Torrey Botanical Club* 107, 51-56.

935) Nilsen, E.T. and Muller, W.H. (1980) A comparison of the relative naturalization ability of two *Schinus* species in southern California. II. Seedling establishment. *Bulletin of the Torrey Botanical Club* 107, 232-237.

936) Nilsen, E.T., Rundel, P.W., and Sharifi, M.R. (1981) Summer water relations of the desert phreatophyte *Prosopis glandulosa* in the Sonoran desert of southern California. *Oecologia* 50, 271-276.

937) Nilsen, E.T., Karpa, D., Mooney, H.A., and Field, C. (1993) Patterns of stem photosynthesis in two invasive legumes (*Spartium junceum, Cytisus scoparius*) of the California coastal region. *American Journal of Botany* 80, 1126-1136.

938) Nishimoto, R.K. and Murdoch, C.L. (1994) Smutgrass (*Sporobolus indicus*) control in bermudagrass (Cynodon dactylon) turf with triazine-MSMA applications. *Weed Technology* 8, 836-839.

939) Northcroft, E.F. (1927) The blackberry pest. *New Zealand Journal of Agriculture* 34, 376-388.

940) Novak, S.J., Mack, R.N., and Soltis, P.S. (1993) Genetic variation in *Bromus tectorum* (Poaceae): introduction dynamics in North America. *Canadian Journal of Botany* 71, 1441-1448.

941) Nowak, D.J. and Rowntree, R.A. (1990) History and range of Norway maple. *Journal of Arboriculture* 16, 291-296.

942) Nuzzo, V.A. (1991) Experimental control of garlic mustard (*Alliaria petiolata* (Bieb.) Cavara & Grande) in northern Illinois using fire, herbicide, and cutting. *Natural Areas Journal* 11, 158-167.

943) Nuzzo, V.A. (1993) Distribution and spread of the invasive biennial *Alliaria petiolata* (garlic mustard) in North America. In: McKnight, B.N. (ed.) *Biological pollution: the control and impact of invasive exotic species.* Indiana Academy of Science, Indianapolis, pp. 137-145.

944) Nuzzo, V.A. (1999) Invasion pattern of the herb garlic mustard (*Alliaria petiolata*) in high quality forests. *Biological Invasions* 1, 169-179.

945) Nyboer, R. (1992) Vegetation management guideline: bush honeysuckles: Tatarian, Morrow's, belle, and amur honeysuckle (*Lonicera tatarica* L., *Lonicera morrowii* Gray, *Lonicera* x *bella* Zabel, and *Lonicera maackii* (Rupr.) Maxim.). *Natural Areas Journal* 12, 218-219.

946) Obeso, J.R. (1997) Costs of reproduction in *Ilex aquifolium*: effects at tree, branch and leaf levels. *Journal of Ecology* 85, 159-166.

947) Obeso, J.R. (1998) Patterns of variation in *Ilex aquifolium* fruit traits related to fruit consumption by birds and seeds predation by rodents. *Ecoscience* 5, 463-469.

948) Obeso, J.R. (1998) Effects of defoliation and girdling on fruit production in *Ilex aquifolium. Functional Ecology* 12, 486-491.

949) Obeso, J.R., Alvarez, S.M., and Retuerto, R. (1998) Sex ratios, size distributions, and sexual dimorphism in the dioecious tree *Ilex aquifolium* (Aquifoliaceae). *American Journal of Botany* 85, 1602-1608.

950) O'Connell, A.M. (1988) Nutrient dynamics in decomposing litter in Karri (*Eucalyptus diversicolor* F. Muell.) forests of south-western Australia. *Journal of Ecology* 76, 1186-1203.

951) O'Connell, A.M. (1989) Nutrient accumulation in and release from the litter layer of karri (*Eucalyptus diversicolor*) forests of southwestern Australia. *Forest Ecology and Management* 26, 95-112.

952) Oertli, B. and Lachavanne, J. (1995) The effects of shoot age on colonization of an emergent macrophyte (*Typha latifolia*) by macroinvertebrates. *Freshwater Biology* 34, 421-431.

953) Ogg, A.G. and Rogers, B.S. (1989) Taxonomy, distribution, biology, and control of black nightshade (*Solanum nigrum*) and related species in the United States and Canada. *Reviews in Weed Science* 4, 25-28.

954) Ogg, A.G., Rogers, B.S., and Schilling, E.E. (1981) Characterization of black nightshade (*Solanum nigrum*) and related species in the United States. *Weed Science* 29, 27-32.

955) Ogle, C.C., La Cock, G.D., Arnold, G., and Mickleson, N. (2000) Impact of an exotic vine *Clematis vitalba* (F. Ranunculaceae) and of control measures on plant biodiversity in indigenous forest, Taihape, New Zealand. *Austral Ecology* 25, 539-551.

956) Olckers, T. and Hulley, P.E. (1989) Seasonality and biology of common insect herbivores attacking *Solanum* plants in the eastern Cape Province (South Africa). *Journal of the Entomological Society of Southern Africa* 52, 109-118.

957) Olckers, T. and Zimmermann, H.G. (1991) Biological control of silverleaf nightshade, *Solanum elaeagnifolium*, and bugweed, *Solanum mauritianum*, (Solanaceae) in South Africa. *Agriculture Ecosystems and Environment* 37, 137-156.

958) Olckers, T. (2000) Biology, host specifity and risk assessment of *Gargaphia decoris*, the first agent to be released in South Africa for the biological control of the invasive tree *Solanum mauritianum. BioControl* 45, 373-388.

959) Oliver, J.D. (1992) Carrotwood: an invasive plant new to Florida. *Aquatics* 14, 4-9.

960) Oliver, J.D. (1996) Mile-a-minute weed (*Polygonum perfoliatum* L.), an invasive vine in natural and disturbed sites. *Castanea* 61, 244-251.

961) Olson, T.E. and Knopf, F.L. (1986) Naturalization of Russian-olive in the western United States. *Western Journal of Applied Forestry* 1, 65-69.

962) Opravil, E. (1983) *Xanthium strumarium*: an European archaeophyte? *Flora* 173, 71-79.

963) Osborne, B., Doris, F., Cullen, A., McDonald, R., Campbell, G., and Steer, M. (1991) *Gunnera tinctoria*: an unusual nitrogen-fixing invader. *BioScience* 41, 224-234.

964) Ostendorp, W. (1989) 'Die-back' of reeds in Europe – a critical review of literature. *Aquatic Botany* 35, 5-26.

965) Otte, A. and Franke, R. (1998) The ecology of the Caucasian herbaceous perennial *Heracleum mantegazzianum* Somm. et Lev. (giant hogweed) in cultural ecosystems of Central Europe. *Phytocoenologia* 28, 205-232.

966) Overton, R.P. (1990) *Acer negundo* L. Boxelder. In: Burns, R.M. and Honkala, B.H. (eds) *Silvics of North America, vol. 2. Hardwoods.* Agriculture Handbook 654, US Department of Agriculture, Washington DC, pp. 41-45.

967) Owen, D.F. (1977) Insect fauna of *Buddleia davidii*. *Entomological Records* 89, 344.

968) Owen, D.F. and Whiteway, R. (1980) *Buddleia davidii* in Britain: history and development of an associated fauna. *Biological Conservation* 17, 149-155.

969) Paine, R.W. (1934) The control of Koster's Curse (*Clidemia hirta*) on Taveuni. *Fiji Agricultural Journal* 7, 10-21.

970) Palaniappan, V.M., Marrs, R.H., and Bradshaw, A.D. (1979) The effect of *Lupinus arboreus* on the nitrogen status of China clay wastes. *Journal of Applied Ecology* 16, 825-831.

971) Palmer, J.H. and Sagar, G.R. (1963) Biological flora of the British Isles. *Agropyron repens* (L.) Beauv. *Journal of Ecology* 51, 783-794.

972) Panda, T. and Mohanty, R.B. (1998) Litter production by *Casuarina equisetifolia* L. in coastal sandy belt of Orissa. *Tropical Ecology* 39, 149-150.

973) Pandey, A.K., Singh, G., and Mishra, O.P. (2000) Effect of flooding on the control of *Oxalis latifolia* H.B.K. *Agricultural and Biological Research* 16, 30-36.

974) Panetta, F.D. (1979) Shade tolerance as reflected in population structures of the woody weed, groundsel bush (*Baccharis halimifolia* L.). *Australian Journal of Botany* 27, 609-615.

975) Panetta, F.D. (1979) The effects of vegetation development upon achene production in the woody weed, groundsel bush (*Baccharis halimifolia* L.). *Australian Journal of Agricultural Research* 30, 1053-1065.

976) Panetta, F.D. (1985) Population studies on pennyroyal mint (*Mentha pulegium*). 1. Germination and seedling establishment. *Weed Research* 25, 301-310.

977) Panetta, F.D. (1985) Population studies on pennyroyal mint (*Mentha pulegium*). 2. Seed banks. *Weed Research* 25, 311-316.

978) Panetta, F.D. (1988) Studies on the seed biology of arum lily (*Zantedeschia aethiopica* (L.) Spreng.). *Plant Protection Quarterly* 3, 169-171.

979) Panetta, F.D. (2001) Seedling emergence and seed longevity of the tree weeds *Celtis sinensis* and *Cinnamomum camphora*. *Weed Research* 41, 83-95.

980) Panetta, F.D. and McKee, J. (1997) Recruitment of the invasive ornamental, *Schinus terebinthifolius*, is dependent upon frugivores. *Australian Journal of Ecology* 22, 432-438.

981) Panetta, F.D. and Mitchell, N.D. (1991) Homoclime analysis and the prediction of weediness. *Weed Research* 31, 273-284.

982) Parker, I.M. (1997) Pollinator limitation of *Cytisus scoparius* (Scotch broom), an invasive exotic shrub. *Ecology* 78, 1457-1470.

983) Parker, I.M. (2001) Safe sites and seed limitation in *Cytisus scoparius* (Scotch broom): invasibility, disturbance, and the role of cryptogams in a glacial outwash prairie. *Biological Invasions* 3, 323-332.

984) Parsons, J.J. (1972) Spread of African pasture grasses to the American tropics. *Journal of Range Management* 25, 12-17.

985) Parsons, W.T. (1973) *Noxious weeds of Victoria*. Inkata Press, Melbourne.

986) Parsons, W.T. and Cuthbertson, E.G. (2001) *Noxious weeds of Australia*. CSIRO Publishing, Collingwood, 698 pp.

987) Pathak, P.S., Debroy, R., and Rai, P. (1974) Autecology of *Leucaena leucocephala* (Lam.) de Wit. 1. Seed polymorphism and germination. *Tropical Ecology* 15, 1-10.

988) Patil, P.K. and Patil, V.K. (1983) Effect of soil salinity levels on growth and chemical composition of *Syzygium cumini*. *Journal of Horticultural Science* 58, 141-146.

989) Patnaik, S. (1976) Autecology of *Ipomoea aquatica* Forssk. *Journal of the Inland Fisheries Society India* 8, 77-82.

990) Paton, D.C., Tucker, J.R., Paton, J.B., and Paton, P.A. (1988) Avian vectors of the seeds of the European Olive, *Olea europaea*. *South Australian Ornithologist* 30, 158-159.

991) Patterson, D.T. (1994) Temperature responses and potential range of the grass weed, serrated tussock (*Nassella trichotoma*), in the United States. *Weed Technology* 8, 703-712.

992) Patterson, D.T., McGowan, M., Mullahey, J.J., and Westbrooks, R.G. (1997) Effects of temperature and photoperiod on tropical soda apple (*Solanum viarum* Dunal) and its potential range in the US. *Weed Science* 45, 404-408.

993) Paul, S. and Gill, H.S. (1980) Ecology of *Phalaris minor* in wheat crop ecosystem. *Tropical Ecology* 20, 186-191.

994) Pavlik, B.M. (1983) Nutrient and productivity relations of the dune grasses *Ammophila arenaria* and *Elymus mollis*. I. Blade photosynthesis and nitrogen use efficiency in the laboratory and field. *Oecologia* 57, 227-232.

995) Pavlik, B.M. (1983) Nutrient and productivity relations of the dune grasses *Ammophila arenaria* and *Elymus mollis*. III. Spatial aspects of the clonal expansion with reference to rhizome growth and the dispersal of buds. *Bulletin of the Torrey Botanical Club* 110, 271-279.

996) Paynter, Q., Fowler, S.V., Memmott, J., and Sheppard, A.W. (1998) Factors affecting the establishment of *Cytisus scoparius* in southern France: implications for managing both native and exotic populations. *Journal of Applied Ecology* 35, 582-595.

997) Paynter, Q., Fowler, S.V., Memmott, J., Shaw, R.H., and Sheppard, A.W. (2000) Determinants of broom (*Cytisus scoparius* (L.) Link) abundance in Europe. *Plant Protection Quarterly* 15, 149-155.

998) Pedroni, F. and Sanchez, M. (1997) Seed dispersal of *Pereskia aculeata* Muller (Cactaceae) in a forest fragment in southeast Brazil. *Revista Brasileira de Biologia* 57, 479-486.

999) Peet, N.B., Watkinson, A.R., Bell, D.J., and Sharma, U.R. (1999) The conservation management of *Imperata cylindrica* grassland in Nepal with fire and cutting: an experimental approach. *Journal of Applied Ecology* 36, 374-387.

1000) Peirce, J.R. (1993) Watsonia: chemicals and their problems. *Plant Protection Quarterly* 8, 81-82.

1001) Peirce, J.R. (1997) The biology of Australian weeds. 31. *Oxalis pes-caprae* L. *Plant Protection Quarterly* 12, 110-119.

1002) Pemberton, R.W. (1998) The potential of biological control to manage old world climbing fern (*Lygodium microphyllum*), an invasive weed in Florida. *American Fern Journal* 88, 176-182.

1003) Pemberton, R.W. (2000) Waterblommetjie (*Aponogeton distachyos*, Aponogetonaceae), a recently domesticated aquatic food crop in Cape South Africa with unusual origins. *Economic Botany* 54, 144-149.

1004) Pemberton, R.W. and Ferriter, A.P. (1998) Old world climbing fern (*Lygodium microphyllum*), a dangerous invasive weed in Florida. *American Fern Journal* 88, 165-175.

1005) Penfound, W.T. (1948) The biology of the water hyacinth. *Ecological Monographs* 18, 447-472.

1006) Peng, C.-I. and Yang, K.-C. (1998) Unwelcome naturalization of *Chromolaena odorata* (Asteraceae) in Taiwan. *Taiwania* 43, 289-294.

1007) Pepper, T.F. (1984) Chemical and biological control of downy brome (*Bromus tectorum*) in wheat and alfalfa in North America. *Weed Science* 32, 18-25.

1008) Perrins, J., Fitter, A., and Williamson, M. (1993) Population biology and rates of invasion of three introduced *Impatiens* species in the British Isles. *Journal of Biogeography* 20, 33-44.

1009) Peterken, G.F. and LLoyd, P.S. (1967) Biological flora of the British Isles. *Ilex aquifolium* L. *Journal of Ecology* 55, 841-858.

1010) Peters, E.J. and Lowance, S.A. (1974) Fertility and management treatments to control broomsedge in pastures. *Weed Science* 22, 201-205.

1011) Peters, H.A. (2001) *Clidemia hirta* invasion at the Pasoh Forest Reserve: an unexpected plant invasion in an undisturbed tropical forest. *Biotropica* 33, 60-68.

1012) Peterson, D.J. and Prasad, R. (1998) The biology of Canadian weeds. 109. *Cytisus scoparius* (L.) Link. *Canadian Journal of Plant Science* 78, 497-504.

1013) Pfitzenmeyer, C.D.C. (1962) Biological flora of the British Isles. *Arrhenatherum elatius* (L.) J. & C. Presl (*A. avenaceum* Beauv.). *Journal of Ecology* 50, 235-245.

1014) Pheloung, P.C. and Scott, J.K. (1996) Climate-based prediction of *Asparagus asparagoides* and *A. declinatus* distribution in Western Australia. *Plant Protection Quarterly* 11, 51-53.

1015) Philip, L.J., Posluszny, U., and Klironomos, J.N. (2001) The influence of mycorrhizal colonization on the vegetative growth and sexual reproductive potential of *Lythrum salicaria* L. *Canadian Journal of Botany* 79, 381-388.

1016) Phillips, J.F.V. (1928) The behaviour of *Acacia melanoxylon* R.Br. in the Knysna forests: an ecological study. *Transactions of the Royal Society of South Africa* 16, 31-43.

1017) Pickart, A.J., Miller, L.M., and Duebendorfer, T.E. (1998) Yellow bush lupine invasion in northern California coastal dunes. I. Ecological impacts and manual restoration techniques. *Restoration Ecology* 6, 59-68.

1018) Pickart, A.J., Theiss, K.C., Stauffer, H.B., and Olsen, G.T. (1998) Yellow bush lupine invasion in northern California coastal dunes. II. Mechanical restoration techniques. *Restoration Ecology* 6, 69-74.

1019) Piedade, M.F.G., Junk, W.J., and Long, S.P. (1997) Nutrient dynamics of the highly productive C-4 macrophyte *Echinochloa polystachya* on the Amazon floodplain. *Functional Ecology* 11, 60-65.

1020) Pienaar, K. (1977) *Sesbania punicea* (Cav.) Benth. The handsome plant terrorist. *Veld & Flora* 63, 17-18.

1021) Pierce, S.M. (1983) Estimation of the nonseasonal production of *Spartina maritima* in a South African estuary. *Estuarine Coastal and Shelf Science* 16, 241-254.

1022) Pierson, E.A. and Mack, R.N. (1990) The population biology of *Bromus tectorum* in forests: distinguishing the opportunity for dispersal from environmental restriction. *Oecologia* 84, 519-525.

1023) Pierson, E.A. and Mack, R.N. (1990) The population biology of *Bromus tectorum* in forests: effect of disturbance, grazing, and litter on seedling establishment and reproduction. *Oecologia* 84, 526-533.

1024) Pierson, E.A., Mack, R.N., and Black, R.A. (1990) The effect of shading on photosynthesis, growth, and regrowth following defoliation for *Bromus tectorum*. *Oecologia* 84, 534-543.

1025) Pieterse, P.J. and Cairns, A.L.P. (1986) The effect of fire on an *Acacia longifolia* seed bank in the south-western Cape. *South African Journal of Botany* 52, 233-236.

1026) Pieterse, P.J. and Cairns, A.L.P. (1987) The effect of fire on an *Acacia longifolia* seed bank and the growth, mortality and reproduction of seedlings establishment after a fire in the South West Cape. *Applied Plant Science* 1, 34-38.

1027) Pieterse, P.J. and Cairns, A.L.P. (1988) Factors affecting the reproductive success of *Acacia longifolia* (Andr.) Willd. in the Banhoek Valley, south-western Cape, Republic of South Africa. *South African Journal of Botany* 54, 461-464.

1028) Pickard, J. (1984) Exotic plants on Lord Howe Island: distribution in space and time, 1853-1981. *Journal of Biogeography* 11, 181-208.

1029) Piggin, C.M. (1976) Factors affecting seed germination of *Echium plantagineum* L. and *Trifolium subterraneum* L. *Weed Research* 16, 337-344.

1030) Piggin, C.M. (1976) Factors affecting seedling establishment and survival of *Echium plantagineum* L., *Trifolium subterraneum* L. and *Lolium rigidum* Gaud. *Weed Research* 16, 267-272.

1031) Piggin, C.M. (19782) The biology of Australian weeds. 8. *Echium plantagineum* L. *Journal of the Australian Institute of Agricultural Science* 48, 3-16.

1032) Pigott, J.P. and Farrell, P. (1996) Factors affecting the distribution of bridal creeper (*Asparagus asparagoides*) in the lower south-west of Western Australia. *Plant Protection Quarterly* 11, 54-56.

1033) Pivello, V.R., Shida, C.N., and Meirelles, S.T. (1999) Alien grasses in Brazilian savannas: a threat to the biodiversity. *Biodiversity and Conservation* 8, 1281-1294.

1034) Pizzaro, H. (1999) Periphyton biomass of *Echinochloa polystachya* (H.B.K.) Hitch. of a lake of the lower Parana River floodplain, Argentina. *Hydrobiologia* 397, 227-239.

1035) Platt, W.J. and Gottschalk, R.M. (2001) Effects of exotic grasses on potential fine fuel loads in the groundcover of south Florida slash pine savannas. *International Journal of Wildland Fire* 10, 155-159.

1036) Playford, J., Bell, J.C., and Moran, G.F. (1993) A major disjunction in genetic diversity over the geographic range of *Acacia melanoxylon* R. Br. *Australian Journal of Botany* 41, 355-368.

1037) Plucknett, D.L. and Stone, B.C. (1961) The principal weedy Melastomataceae in Hawaii. *Pacific Science* 15, 301-303.

1038) Plyler, D.B. and Proseus, T.E. (1996) A comparison of the seed dormancy characteristics of *Spartina patens* and *Spartina alterniflora* (Poaceae). *American Journal of Botany* 83, 11-14.

1039) Pompeo, M.L.M., Henry, R., and Moschini, C.V. (2001) The water level influence on biomass of *Echinochloa polystachya* (Poaceae) in the Jurumirim Reservoir (Sao Paulo, Brazil). *Brazilian Journal of Biology* 61, 19-26.

1040) Popay, A.I. and Medd, R.W. (1990) The biology of Australian weeds 21. *Carduus nutans* L. ssp. *nutans*. *Plant Protection Quarterly* 5, 3-13.

1041) Porembski, S. (2000) The invasibility of tropical granite outcrops ('inselbergs') by exotic weeds. *Journal of the Royal Society of Western Australia* 83, 131-137.

1042) Post, T.W., McCloskey, E., and Klick, K.F. (1989) Two-year study of fire effects on *Rhamnus frangula* L. *Natural Areas Journal* 9, 175-176.

1043) Pozo, J. and Colino, R. (1992) Decomposition processes of *Spartina maritima* in a salt marsh of the Basque Country. *Hydrobiologia* 231, 165-175.

1044) Pratt, P.D. and Pemberton, R.W. (2001) Geographic expansion of the invasive weed *Paederia foetida* into tropical south Florida. *Castanea* 66, 307.

1045) Prior, S.L. and Armstrong, T.R. (2001) A comparison of the effects of foliar applications of glyphosate and fluoroxypyr on Madeira vine, *Anredera cordifolia* (Ten.) van Steenis. *Plant Protection Quarterly* 16, 33-36.

1046) Pritchard, G.H. (1993) Evaluation of herbicides for the control of common prickly pear (*Opuntia stricta* var. *stricta*) in Victoria. *Plant Protection Quarterly* 8, 40-43.

1047) Pritchard, G.H. (2002) Evaluation of herbicides for the control of the environmental weed bridal creeper (*Asparagus asparagoides*). *Plant Protection Quarterly* 17, 17-26.

1048) Prowse, A. (1998) Patterns of early growth and mortality in *Impatiens glandulifera*. In: Starfinger, U., Edwards, K., Kowarik, I., and Williamson, M. (eds) *Plant invasions. Ecological mechanisms and human responses*. Backhuys Publishers, Leiden, pp. 245-252.

1049) Puetz, N. (1994) Vegetative spreading of *Oxalis pes-caprae*. *Plant Systematics and Evolution* 191, 57-67.

1050) Puff, C. (1991) The genus *Paederia* L. (Rubiaceae – Paederieae): a multidisciplinary study. *Opera Botanica Belgica* 3. National Botanic Garden of Belgium, Meise.

1051) Puntieri, J.G. (1991) Vegetation response on a forest slope cleared for a ski-run with special reference to the herb *Alstroemeria aurea* Graham (Alstroemeriaceae), Argentina. *Biological Conservation* 56, 207-221.

1052) Puntieri, J.G. and Gómez, I.A. (1992) Growth analysis of the reproductive shoots of *Alstroemeria aurea* in two contrasting populations. *Preslia, Praha* 64, 343-355.

1053) Putz, F.E. and Holbrook, N.M. (1988) Further observations on the distribution of mutualism between *Cecropia* and its ants: the Malaysian case. *Oikos* 53, 121-125.

1054) Pysek, P. (1991) *Heracleum mantegazzianum* in the Czech Republic: dynamics of spreading from the historical perspective. *Folia Geobotanica et Phytotaxonomica Praha* 26, 439-454.

1055) Pysek, P. (1994) Ecological aspects of invasion by *Heracleum mantegazzianum* in the Czech Republic. In: de Waal, L.C., Child, L.E., Wade, P.M., and Brock, J.H. (eds) *Ecology and management of invasive riverside plants*. John Wiley & Sons, Chichester, pp. 45-54.

1056) Pysek, P. (1995) Invasion dynamics of *Impatiens glandulifera* - a century of spreading reconstructed. *Biological Conservation* 74, 41-48.

1057) Pysek, P. and Pysek, A. (1995) Invasion by *Heracleum mantegazzianum* in different habitats in the Czech Republic. *Journal of Vegetation Science* 6, 711-718.

1058) Pysek, P., Kucera, T., Puntieri, J., and Mandak, B. (1995) Regeneration in *Heracleum mantegazzianum* - response to removal of vegetative and generative parts. *Preslia* 67, 161-171.

1059) Rabeler, R.K. (1985) *Petrorhagia* (Caryophyllaceae) of North America. *SIDA Contributions to Botany* 11, 6-44.

1060) Radcliffe, J.E. (1985) Grazing management of goats and sheep for gorse control. *New Zealand Journal of Experimental Agriculture* 13, 181-190.

1061) Radford, I.J., Nicholas, D.M., and Brown, J.R. (2001) Assessment of the biological control impact of seed predators on the invasive shrub *Acacia nilotica* (prickly acacia) in Australia. *Biological Control* 20, 261-268.

1062) Radho-Toly, S., Majer, J.D., and Yates, C. (2001) Impact of fire on leaf nutrients, arthropod fauna and herbivory of native and exotic eucalypts in Kings Park, Perth, Western Australia. *Austral Ecology* 26, 500-506.

1063) Randall, J.M. (1993) Exotic weeds in North American and Hawaiian natural areas: the Nature Conservancy's plan of attack. In: *Biological pollution: the control and impact of invasive exotic species.* Indiana Academy of Science, Indianapolis, pp. 159-171.

1064) Randall, J.M. (1993) Interference of bull thistle (*Cirsium vulgare*) with growth of ponderosa pine (*Pinus ponderosa*) seedlings in a forest plantation. *Canadian Journal of Forest Research* 23, 1507-1513.

1065) Randall, J.M. and Marinelli, J. (eds) (1996) *Invasive plants. Weeds of the global garden.* Brooklyn Botanic Gardens, Brooklyn, New York.

1066) Rapson, G.L. and Wilson, J.B. (1992) Genecology of *Agrostis capillaris* L. (Poaceae) – an invader into New Zealand. 1. Floral phenology. *New Zealand Journal of Botany* 30, 1-11.

1067) Rapson, G.L. and Wilson, J.B. (1992) Genecology of *Agrostis capillaris* L. (Poaceae) – an invader into New Zealand. 2. Responses to light, soil fertility, and water availability. *New Zealand Journal of Botany* 30, 13-24.

1068) Rasmussen, G.A., Smith, R.P., and Scifres, C.J. (1985) Seedling growth responses of buffelgrass (*Pennisetum ciliare*) to tebuthiuron and honey mesquite (*Prosopis glandulosa*). *Weed Science* 34, 88-93.

1069) Reader, R.J. and Bonser, S.P. (1993) Control of plant frequency on an environmental gradient: effects of abiotic variables, neighbours, and predators on *Poa pratensis* and *Poa compressa* (Gramineae). *Canadian Journal of Botany* 71, 592-597.

1070) Reddy, K.N. and Singh, M. (1992) Germination and emergence of hairy beggarticks (*Bidens pilosa*). *Weed Science* 40, 195-199.

1071) Redman, D.E. (1995) Distribution and habitat types for Nepal microstegium [*Microstegium vimineum* (Trin.) Camus] in Maryland and the district of Columbia. *Castanea* 60, 270-275.

1072) Reece, P.E. and Wilson, R.G. (1983) Effect of Canada thistle (*Cirsium arvense*) and musk thistle (*Carduus nutans*) control on grass herbage. *Weed Science* 31, 488-492.

1073) Reigosa, M.J., Casal, J.F., and Carballeira, A. (1984) Allelopathic effect of *Acacia dealbata* Link. during flowering. *Studies in Ecology* 3, 135-150.

1074) Reinhardt, C.F. and Rossouw, L. (2000) Ecological adaptation of an alien invader plant (*Opuntia stricta*) determines management strategies in the Kruger National Park. *Zeitschrift für Pflanzenkrankheiten und Pflanzenschutz* 17, 77-84.

1075) Reinhardt, C.F., Rossouw, L., Thatcher, L., and Lotter, W.D. (1999) Seed germination of *Opuntia stricta:* implications for management strategies in the Kruger National Park. *South African Journal of Botany* 65, 295-298.

1076) Renne, I.J. and Gauthreaux, S.A. (2000) Seed dispersal of the Chinese tallow tree (*Sapium sebiferum* (L.) Roxb.) by birds in coastal South Carolina. *American Midland Naturalist* 144, 202-215.

1077) Rey, P.J. (1995) Spatio-temporal variation in fruit and frugivorous bird abundance in olive orchards. *Ecology* 76, 1625-1635.

1078) Riba, M. (1997) Effects of cutting and rainfall pattern on resprouting vigour and growth of *Erica arborea* L. *Journal of Vegetation Science* 8, 401-404.

1079) Riba, M. (1998) Effects of intensity and frequency of crown damage on resprouting of *Erica arborea* L. (Ericaceae). *Acta Oecologica* 19, 9-16.

1080) Rice, E.L. (1972) Allelopathic effects of *Andropogon virginicus* and its persistence in old fields. *American Journal of Botany* 59, 752-755.

1081) Rice, K.J. and Mack, R.N. (1991) Ecological genetics of *Bromus tectorum*. III. The demography of reciprocally sown populations. *Oecologia* 88, 91-101.

1082) Rice, P.M. (1997) Plant community diversity and growth form responses to herbicide applications for control of *Centaurea maculosa*. *Journal of Applied Ecology* 34, 1397-1412.

1083) Richards, P.W. and Clapham, A.R. (1941) Biological Flora of the British Isles. *Juncus effusus* L. (*Juncus communis* & *effusus* E. Mey.). *Journal of Ecology* 29, 375-380.

1084) Richardson, D.M. and Brink, M.P. (1985) Notes on *Pittosporum undulatum* in the south-western Cape. *Veld & Flora* 71, 75-77.

1085) Richardson, D.M. and Manders, P.T. (1985) Predicting pathogen-induced mortality in *Hakea sericea* (Proteaceae), an aggressive alien plant invader in South Africa. *Annals of Applied Biology* 106, 243-254.

1086) Richardson, D.M. and Brown, P.J. (1986) Invasion of mesic mountain fynbos by *Pinus radiata*. *South African Journal of Botany* 52, 529-536.

1087) Richardson, D.M. and van Wilgen, B.W. (1986) Effects of thirty-five years of afforestation with *Pinus radiata* on the composition of mesic mountain fynbos near Stellenbosch. *South African Journal of Botany* 52, 309-315.

1088) Richardson, D.M., Williams, P.M., and Hobbs, R.J. (1994) Pine invasions in the southern hemisphere: determinants of spread and invadability. *Journal of Biogeography* 21, 511-527.

1089) Richardson, D.M., Macdonald, I.A.W., Hoffmann, J.H., and Henderson, L. (1997) Alien plant invasions. In: Cowling, R.M. *et al.* (eds) *Vegetation of Southern Africa*. Cambridge University Press, Cambridge, pp. 535-570.

1090) Richardson, R.G. and Hill, R.L. (1998) The biology of Australian weeds. 34. *Ulex europaeus* L. *Plant Protection Quarterly* 13, 46-58.

1091) Rickson, F.R. (1977) Progressive loss of ant related traits of *Cecropia peltata* on selected Caribbean islands. *American Journal of Botany* 64, 585-592.

1092) Rivera, L.W. and Aide, T.M. (1998) Forest recovery in the karst region of Puerto Rico. *Forest Ecology and Management* 108, 63-75.

1093) Rivera, L.W., Zimmerman, J.K., and Aide, T.M. (2000) Forest recovery in abandoned agricultural lands in a karst region of the Dominican Republic. *Plant Ecology* 148, 115-125.

1094) Roberts, H.A. and Lockett, P.M. (1978) Seed dormancy and field emergence in *Solanum nigrum* L. *Weed Research* 18, 231-241.

1095) Robertson, D.J., Robertson, M.C., and Tague, T. (1994) Colonization dynamics of four exotic plants in a northern Piedmont natural area. *Bulletin of the Torrey Botanical Club* 121, 107-118.

1096) Roche, B.F. (1991) Achene dispersal in yellow starthistle (*Centaurea solstitialis* L.). *Northwest Science* 66, 62-65.

1097) Rodriguez-Riaño, T., Ortega-Olivencia, A., and Devesa, J.A. (1999) Reproductive phenology in three Genisteae (Fabaceae) shrub species of the W Mediterranean region. *Nordic Journal of Botany* 19, 345-354.

1098) Rogers, H.M. and Hartemink, A.E. (2000) Soil seed bank and growth rates of an invasive species, *Piper aduncum*, in the lowlands of Papua New Guinea. *Journal of Tropical Ecology* 16, 243-251.

1099) Rolston, M.P. (1981) Wild oats in New Zealand: a review. *New Zealand Journal of Experimental Agriculture* 9, 115-121.

1100) Rolston, M.P. and Talbot, J. (1980) Soil temperatures and regrowth of gorse after treatment with herbicides. *New Zealand Journal of Experimental Agriculture* 8, 55-61.

1101) Romo, J.T. and Grilz, P.L. (1990) Invasion of the Canadian prairies by an exotic perennial. *Blue Jay* 48, 130-135.

1102) Room, P.M. *et al.* (1981) Successful biological control of the floating weed *Salvinia*. *Nature* 294, 78-80.

1103) Room, P.M. and Thomas, P.A. (1985) Nitrogen and establishment of a beetle for biological control of the floating weed *Salvinia* in Papua New Guinea. *Journal of Applied Ecology* 22, 139-156.

1104) Room, P.M. and Thomas, P.A. (1986) Population growth of the floating weed *Salvinia molesta*: field observations and a global model based on temperature and nitrogen. *Journal of Applied Ecology* 23, 1013-1028.

1105) Room, P.M. (1983) 'Falling apart' as a lifestyle: the rhizome architecture and population growth of *Salvinia molesta*. *Journal of Ecology* 71, 349-365.

1106) Room, P.M. (1990) Ecology of a simple plant-herbivore system: biological control of *Salvinia*. *Trends in Ecology and Evolution* 5, 74-79.

1107) Rosa, M.L. and Corbineau, F. (1986) Some aspects of the germination of caryopses of *Leersia oryzoides*. *Weed Research* 26, 99-104.

1108) Rose, S. and Fairweather, P.G. (1997) Changes in floristic composition of urban bushland invaded by *Pittosporum undulatum* in northern Sydney, Australia. *Australian Journal of Botany* 45, 123-149.

1109) Rose, S. (1997) Integrating management of *Pittosporum undulatum* with other environmental weeds in Sydney's urban bushland. *Pacific Conservation Biology* 3, 350-365.

1110) Rose, S. (1997) Influences of suburban edges on invasion of *Pittosporum undulatum* into the bushland of northern Sydney, Australia. *Australian Journal of Ecology* 22, 89-99.

1111) Rothfels, C.J., Beaton, L.L., and Dudley, S. (2002) The effects of salt, manganese, and density on life history traits in *Hesperis matronalis* L. from oldfield and roadside populations. *Canadian Journal of Botany* 80, 131-139.

1112) Rouget, M., Richardson, D.M., Milton, S.J., and Polakow, D. (2001) Predicting invasion dynamics of four alien *Pinus* species in a highly fragmented semi-arid shrubland in South Africa. *Plant Ecology* 152, 79-92.

1113) Rouw, de A. (1991) The invasion of *Chromolaena odorata* (L.) King & Robinson (ex *Eupatorium odoratum*), and competition with the native flora, in a rain forest zone, south-west Côte d'Ivoire. *Journal of Biogeography* 18, 13-23.

1114) Royer, T.V., Monaghan, M.T., and Minshall, G.W. (1999) Processing of native and exotic leaf litter in two Idaho (USA) streams. *Hydrobiologia* 40, 123-128.

1115) Rudolf, P.O. (1965) Jack pine (*Pinus banksiana* Lamb.). In: Fowells, H.A. (ed.) *Silvics of forest trees of the United States. Agriculture Handbook 271.* US Department of Agriculture, Washington, DC, pp. 338-354.

1116) Ruyle, G.B., Roundy, B.A., and Cox, J.R. (1988) Effects of burning on germinability of Lehmann lovegrass. *Journal of Range Management* 41, 404-406.

1117) Ryser, P. and Aeschlimann, U. (1999) Proportional dry-mass content as an underlying trait for the variation in relative growth rate among 22 Eurasian populations of *Dactylis glomerata* s.l. *Functional Ecology* 13, 473-482.

1118) Sachse, U. (1992) Invasion patterns of boxelder on sites with different levels of disturbance. *Verhandlungen der Gesellschaft für Ökologie* 21, 103-111.

1119) Saenz, R.C. and Guries, R.P. (2002) Landscape genetic structure of *Pinus banksiana*: seedling traits. *Silvae Genetica* 51, 26-35.

1120) Sagar, G.R. and Harper, J.L. (1964) *Plantago lanceolata* L. *Journal of Ecology* 52, 211-221.

1121) Sahid, I.B., Ibrahim, R.B., and Kadri, S. (1996) Effects of watering frequency, shade and glyphosate application on *Paspalum conjugatum* Berg (sour grass). *Crop Protection* 15, 15-19.

1122) Sale, P.J.M., Orr, P.T., Shell, G.S., and Erskine, D.J.C. (1985) Photosynthesis and growth rates in *Salvinia molesta* and *Eichhornia crassipes*. *Journal of Applied Ecology* 22, 125-137.

1123) Salgado, I. (1997) Contribution to ecology research about *Swietenia macrophylla* King (Meliaceae) in Brazilian Amazonia. *Acta Botanica Gallica* 144, 231-242. *(French)*

1124) Sallabanks, R. (1992) Fruit fate, frugivory, and fruit characteristics: a study of the hawthorn, *Crataegus monogyna* (Rosaceae). *Oecologia* 91, 296-304.

1125) Sallabanks, R. (1993) Fruiting plant attractiveness to avian seed dispersers: native vs. invasive *Crataegus* in western Oregon. *Madroño* 40, 108-116.

1126) Samarakoon, S.P., Wilson, J.R., and Shelton, H.M. (1990) Growth, morphology and nutritive quality of shaded *Stenotaphrum secundatum*, *Axonopus compressus* and *Pennisetum clandestinum*. *Journal of Agricultural Science* 114, 161-170.

1127) Sammul, M., Kull, K., Oksanen, L., and Veromann, P. (2000) Competition intensity and its importance: results of field experiments with *Anthoxanthum odoratum*. *Oecologia* 125, 18-25.

1128) Sampson, C. (1994) Cost and impact of current control methods used against *Heracleum mantegazzianum* (giant hogweed) and the case for instigating a biological control programme. In: de Waal, L.C., Child, L.E., Wade, P.M., and Brock, J.H. (eds) *Ecology and management of invasive riverside plants.* John Wiley & Sons, Chichester, pp. 55-65.

1129) Samways, M.J., Caldwell, P.M., and Osborn, R. (1996) Ground-living invertebrate assemblages in native, planted and invasive vegetation in South Africa. *Agriculture, Ecosystems and Environment* 58, 19-32.

1130) Sanchez, J.M., Otero, X.L., Izco, J., and Macias, F. (1997) Growth form and population density of *Spartina maritima* (Curtis) Fernald in northwest Spain. *Wetlands* 17, 368-374.

1131) Sanchez, J.M., San Leon, D.G., and Izco, J. (2001) Primary colonisation of mudflat estuaries by *Spartina maritima* (Curtis) Fernald in Northwest Spain: vegetation structure and sediment accretion. *Aquatic Botany* 69, 15-25.

1132) Santos, I. (1998) *Wedelia trilobata* on Pohnpei. *Aliens* 7, 3.

1133) Santra, S.C., Adhya, T.K., and Desarkar, D.K. (1981) Ecological studies on *Bidens pilosa*: effect of light, temperature, salt and different extracts on seed germination and seedling growth. *Tropical Ecology* 22, 162-169.

1134) Sasek, T.W. and Strain, B.R. (1991) Effects of CO_2 enrichment on the growth and morphology of a native and an introduced honeysuckle vine. *American Journal of Botany* 78, 69-75.

1135) Saxena, K.G. and Ramakrishnan, P.S. (1984) Growth and patterns of resource allocation in *Eupatorium odoratum* L. in the secondary successional environments following slash and burn agriculture (Jhum). *Weed Research* 24, 127-134.

1136) Schatz, T.J. (2001) The effect of cutting on the survival of *Mimosa pigra* and its application to the use of blade ploughing as a control method. *Plant Protection Quarterly* 16, 50-54.

1137) Schierenbeck, K.A., Mack, R.N., and Sharitz, R.R. (1994) Effects of herbivory on growth and biomass allocation in native and introduced species of *Lonicera*. *Ecology* 75, 1661-1672.

1138) Schiffman, P.M. (1994) Promotion of exotic weed establishment by endangered giant kangaroo rats (*Dipodomys ingens*) in a California grassland. *Biodiversity and Conservation* 3, 524-537.

1139) Schimming, W.K. and Messersmith, C.G. (1988) Freezing resistance of overwintering buds of four perennial weeds. *Weed Science* 36, 568-573.

1140) Schmid, B. and Weiner, J. (1993) Plastic relationships between reproductive and vegetative mass in *Solidago altissima*. *Evolution* 47, 61-74.

1141) Schnitzler, A. (1995) Community ecology of arboreal lianas in gallery forests of the Rhine valley, France. *Acta Oecologica* 16, 219-236.

1142) Schofield, E.K. (1989) Effects of introduced plants and animals on island vegetation: examples from the Galapagos archipelago. *Conservation Biology* 3, 227-238.

1143) Schreiner, M., Bauer, E.-M., and Kollmann, J. (2000) Reducing predation of conifer seeds by clear-cutting *Rubus fruticosus* agg. in two montane forest stands. *Forest Ecology and Management* 126, 281-290.

1144) Schroder, M. and Howard, C. (2000) Controlling broom (*Cytisus scoparius* (L.) Link) in natural ecosystems in Barrington Tops National Park. *Plant Protection Quarterly* 15, 169-172.

1145) Schroeer, A.E., Hendrick, R.L., and Harrington, T.B. (1999) Root, ground cover, and litterfall dynamics within canopy gaps in a slash pine (*Pinus elliottii* Engelm.) dominated forest. *Ecoscience* 6, 548-555.

1146) Schweitzer, J.A. and Larson, K.C. (1999) Greater morphological plasticity of exotic honeysuckle species may make them better invaders than native species. *Journal of the Torrey Botanical Society* 126, 15-23.

1147) Scott, D., Robertson, J.S., and Archie, W.J. (1990) Plant dynamics of New Zealand tussock grassland infested with *Hieracium pilosella*. I. Effects of seasonal grazing, fertilizer and overdrilling. *Journal of Applied Ecology* 27, 224-234.

1148) Scott, D., Robertson, J.S., and Archie, W.J. (1990) Plant dynamics of New Zealand tussock grassland infested with *Hieracium pilosella*. II. Transition matrices of vegetation changes. *Journal of Applied Ecology* 27, 235-241.

1149) Scott, J.K. and Wykes, B.J. (eds) (1997) *Proceedings of a workshop on arum lily (*Zantedeschia aethiopica*)*. CSIRO Entomology, Wembly, Australia.

1150) Scurfield, G. (1962) Biological Flora of the British Isles. *Cardaria draba* (L.) Desv. (*Lepidium draba* L.). *Journal of Ecology* 50, 489-499.

1151) Searle, S.D., Bell, J.C., and Moran, G.F. (2000) Genetic diversity in natural populations of *Acacia mearnsii*. *Australian Journal of Botany* 48, 279-286.

1152) Seiger, L.A. (1997) The status of *Fallopia japonica* (*Reynoutria japonica*; *Polygonum cuspidatum*) in North America. In: Brock, J.H., Wade, M., Pysek, P., and Green, D. (eds) *Plant invasions: studies from North America and Europe*. Backhuys Publishers, Leiden, pp. 95-102.

1153) Seiger, L.A. and Merchant, H.C. (1997) Mechanical control of Japanese knotweed *Fallopia japonica* (Houtt. Ronse Decraene): effects of cutting regime on rhizomatous reserves. *Natural Areas Journal* 17, 341-345.

1154) Semenza, R.J., Young, J.A., and Evans, R.A. (1978) Influence of light and temperature on the germination and seedbed ecology of common mullein (*Verbascum thapsus*). *Weed Science* 26, 577-581.

1155) Semer, C.R. and Charudattan, R. (1997) First report of *Rhizoctonia solani* causing a foliar leaf spot on Brazilian pepper-tree (*Schinus terebinthifolius*) in Florida. *Plant Disease* 81, 424.

1156) Sena Gomes, A.R. and Kozlowski, T.T. (1980) Responses of *Pinus halepensis* seedlings to flooding. *Canadian Journal of Research* 10, 308-311.

1157) Sessions, L.A. and Kelly, D. (2000) The effects of browntop (*Agrostis capillaris*) dominance after fire on native shrub germination and survival. *New Zealand Natural Sciences* 25, 1-9.

1158) Sessions, L.A. and Kelly, D. (2002) Predator-mediated apparent competition between an introduced grass, *Agrostis capillaris*, and a native fern, *Botrychium australe* (Ophioglossaceae), in New Zealand. *Oikos* 96, 102-109.

1159) Shad, R.A. and Siddiqui, S.U. (1996) Problems associated with *Phalaris minor* and other grass weeds in India and Pakistan. *Experimental Agriculture* 32, 151-160.

1160) Shafroth, P.B., Auble, G.T., and Scott, M.L. (1995) Germination and establishment of the native plains cottonwood (*Populus deltoides* Marshall subsp. *monilifera*) and the exotic russian-olive (*Elaeagnus angustifolia* L.). *Conservation Biology* 9, 1169-1175.

1161) Shafroth, P.B., Friedman, J.M., and Ischinger, L.S. (1995) Effects of salinity on establishment of *Populus fremontii* (cottonwood) and *Tamarix ramosissima* (saltcedar) in southwestern United States. *Great Basin Naturalist* 55, 58-65.

1162) Sharma, B.M. and Sridhar, M.K.C. (1989) Growth characteristics of water lettuce (*Pistia stratiotes* L.) in south-west Nigeria. *Archiv für Hydrobiologie* 115, 305-312.

1163) Sharma, M.P., McBeath, D.K., and Vanden Born, W.H. (1977) Studies on the biology of wild oats. II. Growth. *Canadian Journal of Plant Science* 57, 811-817.

1164) Sharma, M.P. and Vanden Born, W.H. (1978) The biology of Canadian weeds. 27. *Avena fatua* L. *Canadian Journal of Plant Science* 58, 141-157.

1165) Sharma, R. and Dakshini, K.M.M. (1991) A comparative assessment of the ecological effects of *Prosopis cineraria* and *P. juliflora* on the soil of revegetated spaces. *Vegetatio* 96, 87-96.

1166) Sharma, R. and Dakshini, K.M.M. (1996) Ecological implications of seed characteristics of the native *Prosopis cineraria* and the alien *P. juliflora*. *Vegetatio* 124, 101-105.

1167) Shaw, M.W. (1984) *Rhododendron ponticum* - ecological reasons for the success of an alien species in Britain and features that may assist its control. *Aspects of Applied Biology* 5, 231-242.

1168) Sheil, D. (1994) Naturalized and invasive plant species in the evergreen forests of the East Usambara Mountains, Tanzania. *African Journal of Ecology* 32, 66-71.

1169) Sheley, R.L., Larson, L.L., and Johnson, D.E. (1993) Germination and root dynamics of range weeds and forage species. *Weed Technology* 7, 234-237.

1170) Sheley, R.L. and Larson, L.L. (1994) Comparative growth and interference between cheatgrass and yellow starthistle seedlings. *Journal of Range Management* 47, 470-474.

1171) Shelton, H.M. and Brewbaker, J.L. (1994) *Leucaena leucocephala* – the most widely used forage tree legume. In: Gutteridge, R.C. and Shelton, H.M. (eds) *Forage tree legumes in tropical agriculture*. CAB International, Wallingford, UK, pp. 15-29.

1172) Shem-Tov, S., Zaady, E., and Gutterman, Y. (2002) Germination of *Carrichtera annua* (Brassicaceae) seeds on soil samples collected along a rainfall gradient in the Negev Desert of Israel. *Israel Journal of Plant Sciences* 50, 113-118.

1173) Sheppard, A.W., Hosking, J.R., and Leys, A.R. (2000) Broom management. *Plant Protection Quarterly* 15, 134-138.

1174) Sher, A.A., Marshall, D.L., and Gilbert, S.A. (2000) Competition between native *Populus deltoides* and invasive *Tamarix ramosissima* and the implications for reestablishing flooding disturbance. *Conservation Biology* 14, 1744-1754.

1175) Shoulders, E. and Tiarks, A.E. (1980) Predicting height and relative performance of major southern pines from rainfall, slope, and available moisture. *Forest Science* 26, 437-447.

1176) Silander, J.A. Jr. and Klepeis, D.M. (1999) The invasion ecology of Japanese barberry (*Berberis thunbergii*) in the New England landscape. *Biological Invasions* 1, 189-201.

1177) Silander, S.R. and Lugo, A.E. (1990) *Cecropia peltata* L. Yagrumo hembra, trumpet-tree. In: Burns, R.M. and Honkala, B.H. *Silvics of North America, vol. 2. Hardwoods.* Agriculture Handbook 654, US Department of Agriculture, Washington DC, 244-249.

1178) Silveri, A., Dunwiddie, P.W., and Michaels, H.J. (2001) Logging and edaphic factors in the invasion of an Asian woody vine in a mesic North American forest. *Biological Invasions* 3, 379-389.

1179) Simmons, D.M. and Flint, P.W. (1986) Variation in *Chrysanthemoides monilifera* (Compositae) in eastern Australia. *Weed Research* 26, 427-432.

1180) Simpson, C.E. and Bashaw, E.C. (1969) Cytology and reproductive characteristics in *Pennisetum setaceum*. *American Journal of Botany* 56, 31-36.

1181) Simpson, D.A. (1984) A short history of the introduction and spread of *Elodea* Michx. in the British Isles. *Watsonia* 15, 1-9.

1182) Simpson, D.A. (1990) Displacement of *Elodea canadensis* Michx by *Elodea nuttallii* (Planch.) H. St. John in the British Isles. *Watsonia* 18, 173-177.

1183) Sindel, B.M. (1991) A review of the ecology and control of thistles in Australia. *Weed Research* 31, 189-201.

1184) Singh, D. and Dhaliwals, H.S. (1984) Control of *Phalaris minor* and broad-leaved weeds in wheat with selective herbicides. *Pesticides* (Bombay) 18, 45-47.

1185) Singh, S., Kirkwood, R.C., and Marshall, G. (1999) Biology and control of *Phalaris minor* Retz. (littleseed canarygrass) in wheat. *Crop Protection* 18, 1-16.

1186) Singh, T.P. and Paliwal, G.S. (1988) Comparative analysis of the performance of seedlings of some tree species under the influence of fertilizers: I. Field conditions. *Indian Forester* 114, 417-428.

1187) Singh, Y., Van Wyk, A.E., and Baijnath, H. (1996) Floral biology of *Zantedeschia aethiopica* (L.) Spreng. (Araceae). *South African Journal of Botany* 62, 146-150.

1188) Singhurst, J.R., Ledbetter, W.J., and Holmes, W.C. (1997) *Ardisia crenata* (Myrsinaceae): new to Texas. *Southwestern Naturalist* 42, 503-504.

1189) Sinha, S. and Sharma, A. (1984) *Lantana camara* L. – a review. *Feddes Repertorium* 95, 621-633.

1190) Skerman, P.J. and Riveros, F. (1989) *Tropical grasses*. FAO Plant Production and Protection Series, no. 23, Rome, Italy.

1191) Skolmen, R.G. and Ledig, F.T. (1990) *Eucalyptus globulus* Labill. Bluegum eucalyptus. In: Burns, R.M. and Honkala, B.H. *Silvics of North America, vol. 2. Hardwoods.* Agriculture Handbook 654, US Department of Agriculture, Washington DC, pp. 299-304.

1192) Skowno, A.L., Midgley, J.J., Bond, W.J., and Balfour, D. (1999) Secondary succession in *Acacia nilotica* (L.) savanna in the Hluhluwe game reserve, South Africa. *Plant Ecology* 145, 1-9.

1193) Slavik, B. (1995) *Erodium botrys* - a new species in the flora of the Czech Republic. *Preslia, Prague* 67, 305-309.

1194) Slobodchikoff, C.N. and Doyen, J.T. (1977) Effects of *Ammophila arenaria* on sand dune arthropod communities. *Ecology* 58, 1171-1175.

1195) Smale, M.C. (1990) Ecological role of Buddleia (*Buddleja davidii*) in streambeds in the Urewera National Park. *New Zealand Journal of Ecology* 14, 1-6.

1196) Smathers, G.A. and Gardner, D.E. (1979) Stand analysis of an invading firetree (*Myrica faya* Aiton) population, Hawaii. *Pacific Science* 33, 239-255.

1197) Smekens, M.J. and van Tienderen, P.H. (2001) Genetic variation and plasticity of *Plantago coronopus* under saline conditions. *Acta Oecologica* 22, 187-200.

1198) Smirnov, I.A. (1981) Salt tolerance of Siberian elm, *Ulmus pumila*. *Ekologiya* (Moscow) 4, 82-85. *(Russian)*

1199) Smith, C.W. (1985) Impact of alien plants on Hawaii's native biota. In: Stone, C.P. and Scott, J.M. (eds) *Hawaii's terrestrial ecosystems: preservation and management.* University of Hawaii, Honolulu, pp. 180-250.

1200) Smith, C.W. (1989) Non-native plants. In: Stone, C.P. and Stone, D.B. (eds) *Conservation biology in Hawaii*. University of Hawaii Cooperative National Park Resources Studies Unit, Honolulu, pp. 60-69.

1201) Smith, C.W. (1992) Distribution, status, phenology, rate of spread, and management of *Clidemia* in Hawaii. In: Stone, C.P., Smith, C.W., and Tunison, J.T. (eds) *Alien plant invasions in native ecosystems of Hawaii: management and research.* University of Hawaii, Honolulu, pp. 241-253.

1202) Smith, J.M.B. (1994) The changing ecological impact of broom (*Cytisus scoparius*) at Barrington Tops, New South Wales. *Plant Protection Quarterly* 9, 6-11.

1203) Smith, L.L. and Vankat, J.L. (1991) Communities and tree seedling distribution in *Quercus rubra*- and *Prunus serotina*-dominated forests in southwestern Pennsylvania. *American Midland Naturalist* 126, 294-307.

1204) Smith, L.M. and Kok, L.T. (1984) Dispersal of musk thistle (*Carduus nutans*) seeds. *Weed Science* 32, 120-125.

1205) Smith, M.A., Bell, D.T., and Loneragan, W.A. (1999) Comparative seed germination of *Austrostipa compressa* and *Ehrharta calycina* (Poaceae) in a western Australian Banksia woodland. *Australian Journal of Ecology* 24, 35-42.

1206) Smith, R.G.B. and Brock, M.A. (1996) Coexistence of *Juncus articulatus* L. and *Glyceria australis* C.E.Hubb. in a temporary shallow wetland in Australia. *Hydrobiologia* 340, 147-151.

1207) Soldaat, L.L. and Auge, H. (1998) Interactions between an invasive plant, *Mahonia aquifolium*, and a native phytophagous insect, *Rhagoletis meigenii*. In: Starfinger, U., Edwards, K., Kowarik, I., and Williamson, M. (eds) *Plant invasions. Ecological mechanisms and human responses.* Backhuys Publishers, Leiden, pp. 347-360.

1208) Solecki, M.K. (1989) The viability of cut-leaved teasel (*Dipsacus laciniatus* L.) seed harvested from flowering sterms: management implications. *Natural Areas Journal* 9, 102-105.

1209) Solecki, M.K. (1993) Cut-leaved and common teasel (*Dipsacus laciniatus* L. and *D. sylvestris* Huds.): profile of two invasive aliens. In: McKnight, B.N. (ed.) *Biological pollution: the control and impact of invasive exotic species.* Indiana Academy of Science, Indianapolis, pp. 85-92.

1210) Song, S.D., Park, T.G., An-Chung, S., and Kim, J.H. (1994) Effects of environmental factors on growth and nitrogen fixation activity on autumn olive (*Elaeagnus umbellata*) seedlings. *Journal of Plant Biology* 37, 387-393. *(Korean)*

1211) Soros, C.L. and Dengler, N.G. (1996) Leaf morphogenesis and growth in *Cyperus eragrostis* (Cyperaceae). *Canadian Journal of Botany* 74, 1753-1765.

1212) Sparkes, E.C., Grace, S., and Panetta, F.D. (2002) The effects of various herbicides on *Bryophyllum pinnatum* (Lam.) Pers in Nudgee Wetlands Reserve, Queensland. *Plant Protection Quarterly* 17, 77-80.

1213) Spears, B.M., Rose, S.T., and Belles, W.S. (1980) Effect of canopy cover, seeding depth, and soil moisture on emergence of *Centaurea maculosa* and *C. diffusa*. *Weed Research* 20, 87-90.

1214) Spencer, N.R. and Coulson, J.R. (1976) The biological control of alligatorweed, *Alternanthera philoxeroides*, in the United States of America. *Aquatic Botany* 2, 177-190.

1215) Spencer-Jones, D. (1994) Some observations on the use of herbicides for control of *Crassula helmsii*. In: de Waal, L.C., Child, L.E., Wade, P.M., and Brock, J.H. (eds) *Ecology and management of invasive riverside plants*. John Wiley & Sons, Chichester, pp. 15-18.

1216) Spennemann, D.H.R. and Allen, L.R. (2000) Feral olives (*Olea europaea*) as future woody weeds in Australia: a review. *Australian Journal of Experimental Agriculture* 40, 889-901.

1217) Speroni, F.C. and De Viana, M.L. (1998) Fruit and seed production in *Gleditsia triacanthos*. In: Starfinger, U., Edwards, K., Kowarik, I., and Williamson, M. (eds) *Plant invasions. Ecological mechanisms and human responses*. Backhuys Publishers, Leiden, pp. 155-160.

1218) Speroni, F.C. and De Viana, M.L. (2000) Seed scarification requirements in native and alien species. *Ecologia Austral* 10, 123-131. *(Spanish)*

1219) Speroni, F.C. and De Viana, M.L. (2001) Community characteristics in a mountain forest invaded by *Gleditsia triacanthos*. In: Brundu, G., Brock, J., Camarda, I., Chiled, L., and Wade, M. (eds) Plant invasions: species *ecology and ecosystem management*. Backhuys Publishers, Leiden, pp. 75-82.

1220) Spicer, K.W. and Catling, P.M. (1988) The biology of Canadian weeds. 88. *Elodea canadensis* Michx. *Canadian Journal of Plant Science* 68, 1035-1051.

1221) Spicher, D. and Josselyn, M. (1985) *Spartina* (Gramineae) in northern California: distribution and taxonomic notes. *Madroño* 32, 158-167.

1222) Spies, T.A. and Barnes, B.V. (1981) A morphological analysis of *Populus alba*, *Populus grandidentata* and their natural hybrids in southeastern Michigan, USA. *Silvae Genetica* 30, 102-106.

1223) Spies, T.A. and Barnes, B.V. (1982) Natural hybridization between *Populus alba* and the native aspens in southeastern Michigan, USA. *Canadian Journal of Forest Research* 12, 653-660.

1224) Spongberg, S.A. and Burch, I.H. (1979) Lardizabalaceae hardy in temperate North America. *Journal of the Arnold Arboretum Harvard University* 60, 302-315.

1225) Springett, J.A. (1976) The effect of planting *Pinus pinaster* Ait. On populations of soil microarthropods and on litter decomposition at Gnangara, Western Australia. *Australian Journal of Ecology* 1, 83-87.

1226) Spyreas, G., Gibson, D.J., and Basinger, M. (2001) Endophyte infection levels of native and naturalized fescues in Illinois and England. *Journal of the Torrey Botanical Society* 128, 25-34.

1227) Srivastava, R.C. (1992) Taxonomic revision of the genus *Hiptage* Gaertn. (Malpighiaceae) in India. *Candollea* 47, 601-612.

1228) Starfinger, U. (1991) Population biology of an invading tree species - *Prunus serotina*. In: Seitz, A. and Loeschke, V. (eds) *Species conservation: a population-biological approach*. Birkhäuser, Basel, Switzerland, pp. 171-184.

1229) Starfinger, U. (1997) Introduction and naturalization of *Prunus serotina* in central Europe. In: Brock, J.H., Wade, M., Pysek, P., and Green, D. (eds) *Plant invasions: studies from North America and Europe*. Backhuys Publishers, Leiden, pp. 161-171.

1230) Stary, P. and Laska, P. (1999) Adaptation of native syrphid flies to new exotic plants (*Impatiens* spp.) – aphid-ant associations in central Europe (Dipt., Syrphidae; Hom., Aphididae; Hym., Formicidae). *Anzeiger für Schädlingskunde* 72, 72-75.

1231) Stasiak, J. (1990) Structure and dynamics of population *Juncus articulatus* ssp. *litoralis* (Buch.) Lemke in deflation fields of the Leba Bar. *Ekologia Polska* 38, 413-441.

1232) Steenkamp, H.E. and Chown, S.L. (1996) Influence of dense stands of an exotic tree, *Prosopis glandulosa* Benson, on a savanna dung beetle (Coleoptera: Scarabaeinae) assemblage in southern Africa. *Biological Conservation* 78, 305-311.

1233) Steward, K.K. and Van, T.K. (1987) Comparative studies of monoecious and dioecious hydrilla (*Hydrilla verticillata*) biotypes. *Weed Science* 35, 204-210.

1234) Stewart, C.A., Chapman, R.B., and Frampton, C.M.A. (2000) Growth of alligator weed (*Alternanthera philoxeroides* (Mart.) Griseb. (Amaranthaceae)) and population development of *Agasicles hygrophila* Selman & Vogt (Coleoptera: Chrysomelidae) in northern New Zealand. *Plant Protection Quarterly* 15, 95-101.

1235) Stewart, G. and Hull, A.C. (1949) Cheatgrass (*Bromus tectorum* L.) – an ecological intruder in southern Idaho. *Ecology* 30, 58-74.

1236) Stoll, P., Egli, P., and Schmid, B. (1998) Plant foraging and rhizome growth patterns of *Solidago altissima* in response to mowing and fertilizer application. *Journal of Ecology* 86, 341-354.

1237) Stougaard, R.N., Stivers, J.I., and Holen, D.L. (1999) Hoary cress (*Cardaria draba*) management with imazethapyr. *Weed Technology* 13, 581-585.

1238) Stoutemyer, V.T., O'Rourke, F.L., and Steiner, W.W. (1944) Some observations on the vegetative propagation of honey locust. *Journal of Forestry* 42, 32-36.

1239) Strahm, W. (1999) Invasive species in Mauritius: examining the past and charting the future. In: Sandlund, O.T. *et al.* (eds) *Invasive species and biodiversity management.* Kluwer Academic Publishers, Dordrecht, pp. 325-347.

1240) Stransky, J.J. (1984) Forage yield of Japanese honeysuckle after repeated burning or mowing. *Journal of Range Management* 37, 237-238.

1241) Style, B.T. (1962) The taxonomy of *Polygonum aviculare* and its allies in Britain. *Watsonia* 5, 177-214.

1242) Suehs, C.M., Médail, F., and Afre, L. (2001) Ecological and genetic features of the invasion by the alien *Carpobrotus* plants in mediterranean island habitats. In: Brundu, G., Brock, J., Camarda, I., Child, L., and Wade, M. (eds) *Plant invasions: species ecology and ecosystem management.* Backhuys Publishers, Leiden, pp. 145-158.

1243) Sukopp, H. and Sukopp, U. (1988) *Reynoutria japonica* Houtt. in Japan and in Europe. *Veröffentlichungen des Geobotanischen Institutes ETH Stiftung Rübel Zürich* 98, 354-372.

1244) Sumrall, L.B., Roundy, B.A., Cox, J.R., and Winkel, V.K. (1991) Influence of canopy removal by burning or clipping on emergence of *Eragrostis lehmanniana* seedlings. *International Journal of Wildland Fire* 1, 35-40.

1245) Sun, G.C. (1992) The responses of photosynthesis on water stress in leaves of *Ardisia quinquegona* and *Rhodomyrtus tomentosa* of subtropical monsoon broad-leaves forest. *Acta Botanica Yunnanica* 14, 307-313. *(Chinese)*

1246) Sundblad, K. (1990) The effects of cutting frequency on natural *Glyceria maxima* stands. *Aquatic Botany* 37, 27-38.

1247) Sundblad, K. and Robertson, K. (1988) Harvesting reed sweetgrass (*Glyceria maxima*, Poaceae): effects on growth and rhizome storage of carbohydrates. *Economic Botany* 42, 495-502.

1248) Susko, D.J. and Lovett-Doust, L. (1998) Variable patterns of seed maturation and abortion in *Alliaria petiolata* (Brassicaceae). *Canadian Journal of Botany* 76, 1677-1686.

1249) Susko, D.J., Mueller, J.P., and Spears, J.F. (1999) Influence of environmental factors on germination and emergence of *Pueraria lobata*. *Weed Science* 47, 585-588.

1250) Sutherland, W.J. (1990) Biological flora of the British Isles. *Iris pseudacorus* L. *Journal of Ecology* 78, 833-848.

1251) Sutherland, W.J. and Walton, D. (1990) The changes in morphology and demography of *Iris pseudacorus* L. at different heights on a saltmarsh. *Functional Ecology* 4, 655-660.

1252) Suzuki, N. (2000) Pollinator limitation and resource limitation of seed production in the Scotch broom, *Cytisus scoparius* (Leguminosae). *Plant Species Biology* 15, 187-193.

1253) Suzuki, J.-I. (1994) Growth dynamics of shoot height and foliage structure of a rhizomatous perennial herb, *Polygonum cuspidatum*. *Annals of Botany* 73, 629-638.

1254) Swamy, P.S. and Ramakrishnan, P.S. (1987) Effect of fire on population dynamics of *Mikania micrantha* H.B.K. during early succession after slash and burn agriculture (jhum) in northeastern India. *Weed Research* 27, 397-404.

1255) Swamy, P.S. and Ramakrishnan, P.S. (1987) Weed potential of *Mikania micrantha* H.B.K., and its control in fallows after shifting agriculture (Jhum) in northeast India. *Agriculture, Ecosystems and Environment* 18, 195-204.

1256) Swamy, P.S. and Ramakrishnan, P.S. (1988) Effect of fire on growth and allocation strategies of *Mikania micrantha* under early successional environments. *Journal of Applied Ecology* 25, 653-658.

1257) Swamy, P.S. and Ramakrishnan, P.S. (1988) Growth and allocation patterns of *Mikania micrantha* in successional environments after slash and burn agriculture. *Canadian Journal of Botany* 66, 1465-1469.

1258) Swanton, C.J., Cavers, P.B., Clements, D.R., and Moore, M.J. (1992) The biology of Canadian weeds. 101. *Helianthus tuberosus* L. *Canadian Journal of Plant Science* 72, 1367-1382.

1259) Swarbrick, J.T. (1986) History of the lantanas in Australia and origins of the weedy biotypes. *Plant Protection Quarterly* 1, 115-121.

1260) Swarbrick, J.T. (1999) Seedling production by Madeira vine (*Anredera cordifolia*). *Plant Protection Quarterly* 14, 38-39.

1261) Swarbrick, J.T. and Hart, R. (2001) Environmental weeds of Christmas Island (Indian Ocean) and their management. *Plant Protection Quarterly* 16, 54-57.

1262) Swarbrick, J.T., Finlayson, C.M., and Cauldwell, A.J. (1981) The biology of Australian weeds 7. *Hydrilla verticillata* (L.f.) Royle. *The Journal of the Australian Institute of Agricultural Science* 47, 183-190.

1263) Swarbrick, J.T., Timmins, S.M., and Bullen, K.M. (1999) The biology of Australian weeds 36. *Ligustrum lucidum* Aiton and *Ligustrum sinense* Lour. *Plant Protection Quarterly* 14, 122-130.

1264) Swarbrick, J.T., Willson, B.W., and Hannan-Jones, M.A. (1995) The biology of Australian weeds 25. *Lantana camara* L. *Plant Protection Quarterly* 10, 82-95.

1265) Symon, D.E. (1981) The solanaceous genera, *Browallia*, *Capsicum*, *Cestrum*, *Cyphomandra*, *Hyoscyamus*, *Lycopersicon*, *Nierembergia*, *Physalis*, *Petunia*, *Salpichroa* and *Withania*, naturalized in Australia. *Journal of the Adelaide Botanic Gardens* 3, 133-166.

1266) Syrett, P. (1990) Prospects for the biological control of *Rosa rubiginosa* (sweet brier) in New Zealand. *Plant Protection Quarterly* 5, 18-22.

1267) Szafoni, R.E. (1991) Vegetation management guideline: Autumn olive, *Elaeagnus umbellata* Thunb. *Natural Areas Journal* 11, 121-122.

1268) Szentesi, Á. (1999) Predispersal seed predation of the introduced false indigo, *Amorpha fruticosa* L. in Hungary. *Acta Zoologica Academiae Scientiarum Hungaricae* 45, 125-141.

1269) Tabbush, P.M. and Sale, J.S.P. (1984) Experiments on the chemical control of *Rhododendron ponticum* L. *Aspects of Applied Biology* 5, 243-263.

1270) Tabbush, P.M. and Williamson, D.R. (1987) *Rhododendron ponticum* as a forest weed. *Forestry Commission Bulletin* 73, 1-7.

1271) Takahashi, S. (1997) Population dynamics of an invasive plant, *Cirsium vulgare* Ten., in the grassland of Japan. In: Yano, E., Matsuo, K., Shiyomi, M., and Andow, D.A. (eds) *Biological invasions of ecosystem by pests and beneficial organisms.* National Institute of Agro-Environmental Sciences (NIAES) Series 3, Tsukuba, Japan, pp. 122-127.

1272) Talbot, E. (2000) Cutting and mulching broom (*Cytisus scoparius* (L.) Link): a Tasmanian perspective. *Plant Protection Quarterly* 15, 183-185.

1273) Tanaka, H. (1993) Pollination and pre-germination of pollen of *Stachytarpheta urticifolia* (Verbenaceae). *Journal of Japanese Botany* 68, 174-178. *(Japanese)*

1274) Tanner, C.C., Clayton, J.S., and Coffey, B.T. (1990) Submerged-vegetation changes in Lake Rotoroa (Hamilton, New Zealand) related to herbicide treatment and invasion by *Egeria densa. New Zealand Journal of Marine and Freshwater Research* 24, 45-57.

1275) Tanner, C.C., Clayton, J.S., and Wells, R.D.S. (1993) Effects of suspended solids on the establishment and growth of *Egeria densa. Aquatic Botany* 45, 299-310.

1276) Tanner, G.W., Wood, J.M., and Jones, S.A. (1992) Cogongrass (*Imperata cylindrica*) control with glyphosate. *Florida Scientist* 55, 112-115.

1277) Tanphiphat, K. and Appleby, A.P. (1990) Growth and development of bulbous oatgrass (*Arrhenatherum elatius* var. *bulbosum*). *Weed Technology* 4, 843-848.

1278) Tassin, J. (1999) Plant invaders in Réunion Island (French overseas territory, Indian Ocean). *Aliens* 9, 10.

1279) Terry, P.J., Adjers, G., Akobundu., I.O., Anoka, A.U., Drilling, M.E., Tjitrosemito, S., and Utomo, M. (1997) Herbicides and mechanical control of *Imperata cylindrica* as a first step in grassland rehabilitation. *Agroforestry Systems* 36, 151-179.

1280) Thaman, R.R. (1974) *Lantana camara*: its introduction, dispersal, and impact on islands of the tropical Pacific Ocean. *Micronesica* 10, 17-39.

1281) Thieret, J.W. (1986) *Scaevola* (Goodeniaceae) in southeastern United States. *SIDA* 11, 445-453.

1282) Thomas, K.J. (1975) Biological control of *Salvinia* by the snail *Pila globosa* Sw. *Biological Journal of the Linnean Society* 7, 243-247.

1283) Thomas, K.J. (1979) The extent of *Salvinia* infestation in Kerala (South India): its impact and suggested methods of control. *Environmental Conservation* 6, 63-69.

1284) Thomas, K.J. (1981) The role of aquatic weeds in changing the pattern of ecosystems in Kerala. *Environmental Conservation* 8, 63-66.

1285) Thomas, L. and Anderson, L. (1984) Waterhyacinth control in California. *Aquatics* 6, 11-16.

1286) Thomas, P.B., Possingham, H., and Roush, R. (2000) Effects of soil disturbance and weed removal on germination within woodlands infested by boneseed (*Chrysanthemoides monilifera* ssp. *monilifera*). *Plant Protection Quarterly* 15, 6-13.

1287) Thompson, D.J. and Shay, J.M. (1985) The effects of fire on *Phragmites australis* in the Delta Marsh, Manitoba. *Canadian Journal of Botany* 63, 1964-1969.

1288) Thompson, D.Q. (1991) History of purple loosestrife (*Lythrum salicaria* L.) biological control efforts. *Natural Areas Journal* 11, 148-157.

1289) Thompson, J.D. (1991) The biology of an invasive plant. What makes *Spartina anglica* so successful? *BioScience* 41, 393-401.

1290) Thompson, J.D. and Turkington, R. (1988) The biology of Canadian weeds. 82. *Holcus lanatus* L. *Canadian Journal of Plant Science* 68, 131-147.

1291) Tibbetts, T.J. and Ewers, F.W. (2000) Root pressure and specific conductivity in temperate lianas: exotic *Celastrus orbiculatus* (Celastraceae) vs. native *Vitis riparia* (Vitaceae). *American Journal of Botany* 87, 1272-1278.

1292) Tickner, D.P., Angold, P.G., Gurnell, A.M., Mountford, J.O., and Sparks, T. (2001) Hydrology as an influence on invasion: experimental investigations into competition between the alien *Impatiens glandulifera* and the native *Urtica dioica* in the UK. In: Brundu, G., Brock, J., Camarda, I., Child, L., and Wade, M. (eds) *Plant invasions: species ecology and ecosystem management.* Backhuys Publishers, Leiden, pp. 159-168.

1293) Tiley, G.E.D. and Philp, B. (1994) *Heracleum mantegazzianum* (giant hogweed) and its control in Scotland. In: de Waal, L.C., Child, L.E., Wade, P.M., and Brock, J.H. (eds) *Ecology and management of invasive riverside plants.* John Wiley & Sons, Chichester, pp. 101-109.

1294) Tiley, G.E.D. and Philp, B. (1997) Observations on flowering and seed production in *Heracleum mantegazzianum* in relation to control. In: Brock, J.H., Wade, M., Pysek, P., and Green, D. (eds) *Plant invasions: studies from North America and Europe.* Backhuys Publishers, Leiden, pp. 115-121.

1295) Tiley, G.E.D., Dodd, F.S., and Wade, P.M. (1996) Biological flora of the British Isles. *Heracleum mantegazzianum* Sommier & Levier. *Journal of Ecology* 84, 297-319.

1296) Toft, R.J., Harris, R.J., and Williams, P.A. (2001) Impacts of the weed *Tradescantia fluminensis* on insect communities in fragmented forests in New Zealand. *Biological Conservation* 102, 31-46.

1297) Tomley, A.J. (1995) The biology of Australian weeds. 26. *Cryptostegia grandiflora* R. Br. *Plant Protection Quarterly* 10, 122-130.

1298) Tourn, G.M., Menvielle, M.F., Scopel, A.L., and Pidal, B. (1999) Clonal strategies of a woody weed: *Melia azedarach. Plant and Soil* 217, 111-117.

1299) Traveset, A., Willson, M.F., and Sabag, C. (1998) Effect of nectar-robbing birds on fruit set of *Fuchsia magellanica* in Tierra del Fuego: a disrupted mutualism. *Functional Ecology* 12, 459-464.

1300) Trémolières, M., Carbiener, R., Exinger, A., and Turlot, J.C. (1988) Un exemple d'interaction non compétitive entre espèces ligneuses: le cas du lierre arborescent (*Hedera helix* L.) dans la forêt alluviale. *Acta Oecologica Oecologica Plantarum* 9, 187-209. *(French)*

1301) Tripathi, R.S. and Yadav, A.S. (1987) Population dynamics of *Eupatorium adenophorum* Spreng. and *Eupatorium riparium* Regel in relation to burning. *Weed Research* 27, 229-236.

1302) Tsuchiya, T., Shinozuka, A., and Ikusima, I. (1993) Population dynamics, productivity and biomass allocation of *Zizania latifolia* in an aquatic-terrestrial ecotone. *Ecological Research* 8, 193-198.

1303) Tunison, J.T. (1992) Fountain grass control in Hawaii Volcanoes National Park: management considerations and strategies. In: Stone, C.P., Smith, C.W., and Tunison, J.T. (eds) *Alien plant invasions in native ecosystems of Hawaii: management and research.* University of Hawaii Cooperative National Park Resources Studies Unit, Honolulu, pp. 376-393.

1304) Turkington, R. and Aarssen, L.W. (1983) Biological flora of the British Isles. *Hypochoeris radicata* L. *Journal of Ecology* 71, 999-1022.

1305) Turkington, R. and Burdon, J.J. (1983) The biology of Canadian weeds. 57. *Trifolium repens* L. *Canadian Journal of Plant Science* 63, 243-266.

1306) Turkington, R. and Franko, G.D. (1980) The biology of Canadian weeds. 41. *Lotus corniculatus* L. *Canadian Journal of Plant Science* 60, 965-979.

1307) Turkington, R., Cavers, P.B., and Rempel, E. (1978) The biology of Canadian weeds. 29. *Melilotus alba* Desr. and *M. officinalis* (L.) Lam. *Canadian Journal of Plant Science* 58, 523-537.

1308) Turner, B.L. (1995) Taxonomy and nomenclature of *Schkuhria pinnata* (Asteraceae, Helenieae). *Phytologia* 79, 364-368.

1309) Turner, D.R. and Vitousek, P.M. (1987) Nodule biomass of the nitrogen-fixing alien *Myrica faya* Ait. in Hawaii Volcanoes National Park. *Pacific Science* 41, 186-190.

1310) Turrill, W.B. (1951) Wild and cultivated olives. *Kew Bulletin* 6, 437-442.

1311) Twyford, K.L. and Baxter, G.S. (1999) Chemical control of blue periwinkle (*Vinca major* L.) in Croajingolong National Park, Victoria. *Plant Protection Quarterly* 14, 47-50.

1312) Tybirk, K. (1989) Flowering, pollination and seed production of *Acacia nilotica*. *Nordic Journal of Botany* 9, 375-381.

1313) Tye, A. (1999) Invasive plant problems and requirements for weed risk assessment in the Galapagos islands. - 1st international workshop on weed risk assessment, Adelaide, Australia. www.hear.org/iwraw/index

1314) Tyser, R.W. (1992) Vegetation associated with two alien plant species in a fescue grassland in Glacier National Park, Montana. *Great Basin Naturalist* 52, 189-193.

1315) Tyser, R.W. and Key, C.H. (1988) Spotted knapweed in natural area fescue grasslands: an ecological assessment. *Northwest Science* 62, 151-160.

1316) Udensi, U.E., Akobundu, I.O., Ayeni, A.O., and Chikoye, D. (1999) Management of cogongrass (*Imperata cylindrica*) with velvetbean (*Mucuna pruriens* var. *utilis*) and herbicides. *Weed Technology* 13, 201-208.

1317) Udvardy, L. (1998) Spreading and coenological circumstances of the tree of heaven (*Ailanthus altissima*) in Hungary. *Acta Botanica Hungarica* 41, 299-314.

1318) Ullah, E., Shepherd, R.C.H., Baxter, J.T., and Peterson, J.A. (1989) Mapping flowering Paterson's curse (*Echium plantagineum*) around Lake Hume, north eastern Victoria, using Landsat TM data. *Plant Protection Quarterly* 4, 155-157.

1319) Ultsch, G.R. (1973) The effects of water hyacinths (*Eichhornia crassipes*) on the microenvironment of aquatic communities. *Archiv für Hydrobiologie* 72, 460-473.

1320) Upadhyaya, M.K., Turkington, R., and McIlvride, D. (1986) The biology of Canadian weeds. 75. *Bromus tectorum* L. *Canadian Journal of Plant Science* 66, 689-709.

1321) Uriarte, M. (2000) Interactions between goldenrod (*Solidago altissima* L.) and its insect herbivore (*Trirhabda virgata*) over the course of succession. *Oecologia* 122, 521-528.

1322) Van, T.K. and Steward, K.K. (1986) The use of controlled-release fluridone fibers for control of hydrilla (*Hydrilla verticillata*). *Weed Science* 34, 70-76.

1323) Van, T.K. and Steward, K.K. (1990) Longevity of monoecious *Hydrilla* propagules. *Journal of Aquatic Plant Management* 28, 74-76.

1324) Van den Tweel, P.A. and Eijsackers, H. (1987) Black cherry, a pioneer species or 'forest pest'. *Proceedings of the Koninklijke Nederlandse Akademie van Wetenschappen* C90, 59-66.

1325) Van der Sommen, F.J. (1986) Colonisation of forest and woodland communities by exotic plants. In: Wallace, H.R. (ed.) *The ecology of the forests and woodlands of South Australia*. D.J. Woolman, Government Printer, Australia, pp. 248-267.

1326) Van der Toorn, J. (1980) The ecology of *Cotula coronopifolia* and *Ranunculus sceleratus*. 1. Geographic distribution, habitat and field observations. *Acta Botanica Neerlandica* 29, 385-396.

1327) Van der Toorn, J. and Mock, J.H. (1982) The influence of environmental factors and management on stands of *Phragmites australis*. I. Effects of burning, frost and insect damage on shoot density and shoot size. *Journal of Applied Ecology* 19, 477-499.

1328) Van der Toorn, J. and Ten-Hove, H.J. (1982) The ecology of *Cotula coronopifolia* and *Ranunculus sceleratus*. 2. Experiments on germination, seed longevity, and seedling survival. *Acta Oecologica Oecologia Plantarum* 3, 409-418.

1329) Van de Venter, H.A., Hosten, L., Lubke, R.A., and Palmer, A.R. (1984) Morphology of *Opuntia aurantiaca* (jointed cactus) biotypes and its close relatives, *O. discolor* and *O. salmiana* (Cactaceae). *South African Journal of Botany* 3, 331-339.

1330) Van Dijk, G.M., Thayer, D.D., and Haller, W.T. (1986) Growth of *Hygrophila* and *Hydrilla* in flowing water. *Journal of Aquatic Plant Management* 24, 85-87.

1331) Van Haaren, P. and Vitelli, J. (1997) Chemical control of thunbergia (*Thunbergia grandiflora*). *Plant Protection Quarterly* 12, 29-32.

1332) Van Klinken, R.D. and Campbell, S.D. (2001) The biology of Australian weeds. 37. *Prosopis* L. species. *Plant Protection Quarterly* 16, 2-20.

1333) Van Wilgen, B.W. and Richardson, D.M. (1985) The effects of alien shrub invasions on vegetation structure and fire behaviour in South African fynbos shrublands: a simulation study. *Journal of Applied Ecology* 22, 955-966.

1334) Van Wilgen, B.W. and Siegfried, W.R. (1986) Seed dispersal properties of three pine species as a determinant of invasive potential. *South African Journal of Botany* 52, 546-548.

1335) Vandiver, V.V. (1980) *Hygrophila*. *Aquatics* 2, 4-11.

1336) Vanstone, V.A. and Paton, D.C. (1988) Extrafloral nectaries and pollination of *Acacia pycnantha* Benth. by brids. *Australian Journal of Botany* 36, 519-531.

1337) Vasellati, V., Oesterheld, M., Medan, D., and Loreti, J. (2001) Effects of flooding and drought on the anatomy of *Paspalum dilatatum*. *Annals of Botany* 88, 355-360.

1338) Vilà, M. and D'Antonio, C.M. (1998) Hybrid vigor for clonal growth in *Carpobrotus* (Aizoaceae) in coastal California. *Ecological Applications* 8, 1196-1205.

1339) Vilà, M. and D'Antonio, C.M. (1998) Fitness of invasive *Carpobrotus* (Aizoaceae) hybrids in coastal California. *Ecoscience* 5, 191-199.

1340) Vilà, M. and D'Antonio, C.M. (1998) Fruit choice and seed dispersal of invasive vs. noninvasive *Carpobrotus* (Aizoaceae) in coastal California. *Ecology* 79, 1053-1060.

1341) Vilà, M. and Gimeno, I. (2001) Patterns of invasion of *Opuntia* sp.PL. in abandoned olive groves in Catalona (Spain). In: Brundu, G. *et al.* (eds) *Plant invasions: species ecology and ecosystem management*. Backhuys Publishers, Leiden, pp. 169-174.

1342) Vilà, M., Weber, E., and D'Antonio, C.M. (1998) Flowering and mating system in hybridizing *Carpobrotus* (Aizoaceae) in coastal California. *Canadian Journal of Botany* 76, 1165-1169.

1343) Viljoen, B.D. (1987) Pasture recovery after nassella tussock control with tetrapion. *Applied Plant Science* 1, 18-22.

1344) Vincente, M. (1972) Germinacao de sementes de *Solanum viarum* Dunal. III. Luz. *Revista Brasileira de Biologia* 32, 585-591.

1345) Vitelli, J., Mayer, R.J., and Jeffrey, P.J. (1994) Foliar application of 2,4-D/picloram, imazapyr, metsulfuron, triclopyr/picloram, and dicamba kills individual rubber vine (*Cryptostegia grandiflora*) plants. *Tropical Grasslands* 28, 120-126.

1346) Vitelli, J. and van Haaren, P. (2001) Chemical control of harungana (*Harungana madagascariensis*) shrubs in Queensland. *Plant Protection Quarterly* 16, 41-43.

1347) Vitousek, P.M. and Walker, L.R. (1989) Biological invasion by *Myrica faya* in Hawai'i: plant demography, nitrogen fixation, ecosystem effects. *Ecological Monographs* 59, 247-265.

1348) Vitousek, P., Walker, L.R., Whiteaker, L.D., Mueller-Dombois, D., and Matson, P.A. (1987) Biological invasion by *Myrica faya* alters ecosystem development in Hawaii. *Science* 238, 802-804.

1349) Vivrette, N.J. and Muller, C.H. (1977) Mechanism of invasion and dominance of coastal grassland by *Mesembryanthemum crystallinum*. *Ecological Monographs* 47, 301-318.

1350) Von Bothmer, R., Flink, J., Jacobsen, N., and Jørgensen, R.B. (1989) Variation and differentiation in *Hordeum marinum* (Poaceae). *Nordic Journal of Botany* 9, 1-10.

1351) Voser-Huber, M.L. (1983) Studien an eingebürgerten Arten der Gattung *Solidago* L. *Dissertationes Botanicae* 68, 1-97.

1352) Vranjic, J.A., Woods, M.J., and Barnard, J. (2000) Soil-mediated effects on germination and seedling growth of coastal wattle (*Acacia sophorae*) by the environmental weed, bitou bush (*Chrysanthemoides monilifera* ssp. *rotundata*). *Austral Ecology* 25, 445-453.

1353) Wade, M., Darby, E.J., Courtney, A.D., and Caffrey, J.M. (1997) *Heracleum mantegazzianum*: a problem for river managers in the Republic of Ireland and the United Kingdom. In: Brock, J.H., Wade, M., Pysek, P., and Green, D. (eds) *Plant invasions: studies from North America and Europe*. Backhuys Publishers, Leiden, pp. 129-151.

1354) Wager, V.A. (1927) The structure and life history of the South African Lagarosiphons. *Transactions of the Royal Society of South Africa* 16, 191-212.

1355) Wagner, G.M. (1997) *Azolla*: a review of its biology and utilization. *The Botanical Review* 63, 1-26.

1356) Waite, S. (1984) Changes in the demography of *Plantago coronopus* at 2 coastal sites. *Journal of Ecology* 72, 809-826.

1357) Walker, L.R. and Vitousek, P.M. (1991) An invader alters germination and growth of a native dominant tree in Hawaii. *Ecology* 72, 1449-1455.

1358) Wallace, A. (1997) The biology of Australian weeds. 30. *Vulpia bromoides* ((L.) S.F. Gray) and *V. myuros* ((L.) C.C. Gmelin). *Plant Protection Quarterly* 12, 18-28.

1359) Walther, G.-R. (1999) Distribution and limits of evergreen broad-leaved (laurophyllous) species in Switzerland. *Botanica Helvetica* 109, 153-167.

1360) Warren, D.K. and Turner, R.M. (1975) Saltcedar seed production, seedling establishment, and response to inundation. *Arizona Academy of Science Journal* 10, 131-144.

1361) Warren, J.M. (2000) The role of white clover in the loss of diversity in grassland habitat restoration. *Restoration Ecology* 8, 318-323.

1362) Watson, A.K. (1985) Introduction. The leafy spurge problem. In: Watson, A.K. (ed.) *Leafy Spurge*. Weed Science Society of America, Champaign, Illinois, pp. 1-6.

1363) Watson, A.K. and Renney, A.J. (1974) The biology of Canadian weeds. 6. *Centaurea diffusa* and *C. maculosa*. *Canadian Journal of Plant Science* 54, 687-701.

1364) Weaver, S.E. and Warwick, S.I. (1984) The biology of Canadian weeds. 64. *Datura stramonium* L. *Canadian Journal of Plant Science* 64, 979-991.

1365) Weaver, P.L. (1992) An ecological comparison of canopy trees in the montane rain forest of Puerto Rico's Luquillo Mountains. *Caribbean Journal of Science* 28, 62-69.

1366) Webb, C.E., Oliver, I., and Pik, A.J. (2000) Does coastal foredune stabilization with *Ammophila arenaria* restore plant and arthropod communities in southeastern Australia? *Restoration Ecology* 8, 283-288.

1367) Webb, S.L. and Kaunzinger, C.K. (1993) Biological invasion of the Drew University (New Jersey) forest preserve by norway maple (*Acer platanoides* L.). *Bulletin of the Torrey Botanical Club* 120, 343-349.

1368) Webb, S.L., Dwyer, M., Kaunzinger, C.K., and Wyckoff, P.H. (2000) The myth of the resilient forest: case study of the invasive Norway maple (*Acer platanoides*). *Rhodora* 102, 332-354.

1369) Webb, S.L., Pendergast, T.H., and Dwyer, M.E. (2001) Response of native and exotic maple seedling banks to removal of the exotic, invasive Norway maple (*Acer platanoides*). *Journal of the Torrey Botanical Society* 128, 141-149.

1370) Weber, E. (1997) Morphological variation of the introduced perennial *Solidago canadensis* L. sensu lato in Europe. *Botanical Journal of the Linnean Society* 123, 197-210.

1371) Weber, E. (1997) Phenotypic variation of the introduced perennial *Solidago gigantea* Ait. in Europe. *Nordic Journal of Botany* 17, 631-638.

1372) Weber, E. (1998) The dynamics of plant invasions: a case study of three exotic goldenrod species (*Solidago* L.) in Europe. *Journal of Biogeography* 25, 147-154.

1373) Weber, E. (2000). Biological flora of Central Europe: *Solidago altissima* L. *Flora* 195, 123-134.

1374) Weber, E. and D'Antonio, C.M. (1999) Germination and growth responses of hybridizing *Carpobrotus* ssp. (Aizoaceae) from coastal California to soil salinity. *American Journal of Botany* 86, 1257-1263.

1375) Weber, E. and D'Antonio, C.M. (1999) Phenotypic plasticity in hybridizing *Carpobrotus* ssp. from coastal California and its role for plant invasion. *Canadian Journal of Botany* 77, 1411-1418.

1376) Weber, E., Vilà, M., Albert, M., and D'Antonio, C.M. (1998) Invasion by hybridization: *Carpobrotus* in coastal California. In: Starfinger, U., Edwards, K., Kowarik, I., and Williamson, M. (eds) *Plant invasions: ecological mechanisms and human responses.* Backhuys Publishers, Leiden, pp. 275-284.

1377) Wedderburn, M.E. and Gwynne, D.C. (1981) Seasonality of rhizome and shoot production and nitrogen fixation in *Lotus uliginosus* under upland conditions in southwest Scotland, UK. *Annals of Botany* 48, 5-14.

1378) Weisner, S.E.B. (1993) Long-term competitive displacement of *Typha latifolia* by *Typha angustifolia* in a eutrophic lake. *Oecologia* 94, 451-456.

1379) Weiss, J. and Sagliocco, J.L. (2000) Horehound (*Marrubium vulgare*): a comparison between European and Australian populations. *Plant Protection Quarterly* 15, 18-20.

1380) Weiss, J. and Wills, E. (2000) Integrated management of horehound (*Marrubium vulgare* L.) in Wyperfeld National Park. *Plant Protection Quarterly* 15, 40-42.

1381) Weiss, P.W. (1984) Seed characteristics and regeneration of some species in invaded coastal communities. *Australian Journal of Ecology* 9, 99-106.

1382) Weiss, P.W. (1986) The biology of Australian weeds. 14. *Chrysanthemoides monilifera* (L.) T. Norl. *Journal of the Australian Institute of Agricultural Science* 52, 127-134.

1383) Weiss, P.W. and Noble, I.R. (1984) Status of coastal dune communities invaded by *Chrysanthemoides monilifera*. *Australian Journal of Ecology* 9, 93-98.

1384) Weller, R.F. and Phipps, R.H. (1979) A review of black nightshade (*Solanum nigrum* L.). *Protection Ecology* 1, 121-139.

1385) Wells, R.D.S. and Clayton, J.S. (1991) Submerged vegetation and spread of *Egeria densa* Planchon in Lake Rotorua, central North Island, New Zealand. *New Zealand Journal of Marine and Freshwater Research* 25, 63-70.

1386) Werner, P.A. (1975) The biology of Canadian weeds. 12. *Dipsacus sylvestris* Huds. *Canadian Journal of Plant Science* 55, 783-794.

1387) Werner, P.A. and Rioux, R. (1977) The biology of Canadian weeds. 24. *Agropyron repens* (L.) Beauv. *Canadian Journal of Plant Science* 57, 905-919.

1388) Werner, P.A., Bradbury, I.K., and Gross, R.S. (1980) The biology of Canadian weeds. 45. *Solidago canadensis* L. *Canadian Journal of Plant Science* 60, 1393-1409.

1389) Westbrooks, R.G. and Cross, G. (1993) Serrated tussock (*Nassella trichotoma*) in the United States. *Weed Technology* 7, 525-528.

1390) Wester, L.L. and Wood, H.B. (1977) Koster's curse (*Clidemia hirta*), a weed pest in Hawaiian forests. *Environmental Conservation* 4, 35-41.

1391) Westman, W.E., Panetta, F.D., and Stanley, T.D. (1975) Ecological studies on reproduction and establishment of the woody weed, groundsel bush (*Baccharis halimifolia* L.: Asteraceae). *Australian Journal of Agricultural Research* 26, 855-870.

1392) Wheeler, A.G. and Henry, T.J. (1976) Biology of the honeylocust plant bug, *Diaphnocoris chlorionis*, and other Mirids associated with ornamental honeylocust. *Annals of the Entomological Society of America* 69, 1095-1104.

1393) Wheeler, A.G. and Mengel, S.A. (1984) Phytophagous insect fauna of *Polygonum perfoliatum*, an asiatic weed recently introduced to Pennsylvania (USA). *Annals of the Entomological Society of America* 77, 197-202.

1394) Wheeler, C.T., Helgerson, O.T., Perry, D.A., and Gordon, J.C. (1987) Nitrogen fixation and biomass accumulation in plant communities dominated by *Cytisus scoparius* L. in Oregon and Scotland. *Journal of Applied Ecology* 24, 231-237.

1395) Whelan, R.J. and York, J. (1998) Post-fire germination of *Hakea sericea* and *Petrophile sessilis* after spring burning. *Australian Journal of Botany* 46, 367-376.

1396) White, D.J., Haber, E., and Keddy, C. (1993) *Invasive plants of natural habitats in Canada*. Canadian Museum of Nature, Ottawa, Ontario, Canada.

1397) White, O.E. and Bowden, W.M. (1947) Oriental and American bittersweet hybrids. *Journal of Heredity* 38, 125-127.

1398) Whiteaker, L.D. and Gardner, D.E. (1985) *The distribution of* Myrica faya *Ait. in the state of Hawaii*. Technical report no. 55, University of Hawaii, Honolulu.

1399) Whiteaker, L.D. and Gardner, D.E. (1992) Firetree (*Myrica faya*) distribution in Hawaii. In: Stone C.P., Smith C.W., and Tunison J.T. (eds) *Alien plant invasions in native ecosystems of Hawaii: management and research*. Cooperative National Park Resources Studies Unit, University of Hawaii. University of Hawaii Press, Honolulu, pp. 225-240.

1400) Whitson, T.D. and Koch, D.W. (1998) Control of downy brome (*Bromus tectorum*) with herbicides and perennial grass competition. *Weed Technology* 12, 391-396.

1401) Whittaker, E. and Gimingham, C.H. (1962) The effects of fire on the regeneration of *Calluna vulgaris* (L.) Hull from seed. *Journal of Ecology* 50, 815-822.

1402) Whittle, C.A., Duchesne, L.C., and Needham, T. (1998) Soil seed bank of a jack pine (*Pinus banksiana*) ecosystem. *International Journal of Wildland Fire* 8, 67-71.

1403) Wicks, G.A. (1984) Integrated systems for control and management of downy brome (*Bromus tectorum*) in cropland. *Weed Science* 32, 26-31.

1404) Wiese, A.F., Salisbury, C.D., and Bean, B.W. (1995) Downy brome (*Bromus tectorum*), jointed goatgrass (*Aegilops cylindrica*) and horseweed (*Conyza canadensis*) control in fallow. *Weed Technology* 9, 249-254.

1405) Wilcut, J.W., Dute, R.R., Truelove, B., and Davis, D.E. (1988) Factors limiting the distribution of cogongrass, *Imperata cylindrica*, and torpedograss, *Panicum repens*. *Weed Science* 36, 577-582.

1406) Wilcut, J.W., Truelove, B., Davis, D.E., and Williams, J.C. (1988) Temperature factors limiting the spread of cogongrass (*Imperata cylindrica*) and torpedograss (*Panicum repens*). *Weed Science* 36, 49-55.

1407) Willard, T.R., Shilling, D.G., Gaffney, J.F., and Currey, W.L. (1996) Mechanical and chemical control of cogongrass (*Imperata cylindrica*). *Weed Technology* 10, 722-726.

1408) Willard, T.R., Gaffney, J.F., and Shilling, D.G. (1997) Influence of herbicide combinations and application technology on cogongrass (*Imperata cylindrica*) control. *Weed Technology* 11, 76-80.

1409) Williams, C.E., Ralley, J.J., and Taylor, D.H. (1992) Consumption of seeds in the invasive amur honeysuckle, *Lonicera maackii* (Rupr.) Maxim., by small mammals. *Natural Areas Journal* 12, 86-89.

1410) Williams, D.G. and Black, R.A. (1993) Phenotypic variation in contrasting temperature environments: growth and photosynthesis in *Pennisetum setaceum* from different altitudes on Hawaii. *Functional Ecology* 7, 623-633.

1411) Williams, H.E. (2002) Life history and laboratory host range of *Charidotis auroguttata* (Boheman) (Coleoptera: Chrysomelidae), the first natural enemy released against *Macfadyena unguis-cati* (L.) Gentry (Bignoniaceae) in South Africa. *Coleopterists Bulletin* 56, 299-307.

1412) Williams, J.T. (1963) Biological flora of the British Isles. *Chenopodium album* L. *Journal of Ecology* 51, 711-725.

1413) Williams, P.A. (1981) Aspects of the ecology of broom (*Cytisus scoparius*) in Canterbury, New Zealand. *New Zealand Journal of Botany* 19, 31-43.

1414) Williams, P.A. (1992) *Hakea salicifolia*: biology and role in succession in Abel Tasman National Park, New Zealand. *Journal of the Royal Society of New Zealand* 22, 1-18.

1415) Williams, P.A. (1997) *Ecology and management of invasive weeds*. Conservation Sciences Publication no. 7. Department of Conservation, Wellington, New Zealand.

1416) Williams, P.A. and Buxton, R.P. (1986) Hawthorn (*Crataegus monogyna*) populations in Mid-Canterbury. *New Zealand Journal of Ecology* 9, 11-17.

1417) Williams, P.A. and Buxton, R.P. (1995) Aspects of the ecology of two species of *Passiflora* (*P. mollissima* (Kunth) L. Bailey and *P. pinnatistipula* Cav.) as weeds in South Islands, New Zealand. *New Zealand Journal of Botany* 33, 315-323.

1418) Williams, P.A. and Timmins, S.M. (1990) *Weeds in New Zealand protected natural areas: a review for the department of conservation*. Science and Research Series, report no. 14, Dept. of Conservation, Wellington.

1419) Williams, P.A. and Timmins, S.M. (1999) Biology and ecology of Japanese honeysuckle (*Lonicera japonica*) and its impacts in New Zealand. *Science for Conservation* no. 99. New Zealand Department of Conservation, Wellington.

1420) Williams, P.A., Timmins, S.M., Smith, J.M.B., and Downey, P.O. (2001) The biology of Australian weeds. 38. *Lonicera japonica* Thunb. *Plant Protection Quarterly* 16, 90-100.

1421) Williamson, J.A. and Forbes, J.C. (1982) Giant hogweed (*Heracleum mantegazzianum*): its spread and control with glyphosate in amenity areas. *Proceedings of the British Crop Protection Conference* 1982, 967-971.

1422) Williamson, M. (1998) Measuring the impact of plant invaders in Britain. In: Starfinger, U., Edwards, K., Kowarik, I., and Williamson, M. (eds) *Plant invasions. Ecological mechanisms and human responses*. Backhuys Publishers, Leiden, pp. 57-68.

1423) Willson, G.D. and Stubbendieck, J. (1997) Fire effects on four growth stages of smooth brome (*Bromus inermis* Leyss.). *Natural Areas Journal* 17, 306-312.

1424) Wilson, M.V. and Clark, D.L. (2001) Controlling invasive *Arrhenatherium elatius* and promoting native prairie grasses through mowing. *Applied Vegetation Science* 4, 129-138.

1425) Wilson, P.A. and Conran, J.G. (1993) The effect of slashing on the growth of *Watsonia meriana* (L.) Mill. cv *bulbillifera* in the Adelaide Hills. *Plant Protection Quarterly* 8, 85-90.

1426) Wilson, R.G. (1979) Germination and seedling development of Canada thistle (*Cirsium arvense*). *Weed Science* 27, 146-151.

1427) Winkler, E. and Stöcklin, J. (2002) Sexual and vegetative reproduction of *Hieracium pilosella* L. under competition and disturbance: a grid-based simulation model. *Annals of Botany* 89, 525-536.

1428) Witkowski, E.T.F. (1991) Effects of invasive alien acacias on nutrient cycling in the coastal lowlands of the Cape fynbos. *Journal of Applied Ecology* 28, 1-15.

1429) Witkowski, E.T.F. (1991) Growth and competition between seedlings of *Protea repens* (L.) L. and the alien invasive, *Acacia saligna* (Labill.) Wendl. in relation to nutrient availability. *Functional Ecology* 5, 101-110.

1430) Witkowski, E.T.F. and Wilson, M. (2001) Changes in density, biomass, seed production and soil seed banks of the non-native invasive plant, *Chromolaena odorata*, along a 15 year chronosequence. *Plant Ecology* 152, 13-27.

1431) Wohl, N. (1994) Density and distribution of Japanese barberry (*Berberis thunbergii*), an exotic shrub species naturalized in the Morristown National Historical Park, Morris County, New Jersey. *Bulletin of the New Jersey Academy of Science* 39, 1-5.

1432) Woirnarski, J.C.Z. (1992) *A survey of the wildlife and vegetation of Purnululu (Bungle Bungle) National Park and adjacent area*. Department of Conservation and Land Management, Research Bulletin 6, Western Australia.

1433) Wolf, J.J. and Rohrs, J. (2001) The influence of physical soil conditions on the formation of root nodules of *Melilotus officinalis* in the montane zone of Rocky Mountain National Park. *European Journal of Soil Biology* 37, 51-57.

1434) Wood, H. (1994) The introduction and spread of capeweed, *Arctotheca calendula* (L.) Levyns (Asteraceae) in Australia. *Plant Protection Quarterly* 9, 94-100.

1435) Wood, H. and Degabriele, R. (1985) Genetic variation and phenotypic plasticity in populations of Paterson's Curse (*Echium plantagineum* L.) in south-eastern Australia. *Australian Journal of Botany* 33, 677-685.

1436) Woodall, S.L. (1982) Seed dispersal in *Melaleuca quinquenervia*. *Florida Scientist* 45, 81-93.

1437) Woods, K.D. (1993) Effects of invasion by *Lonicera tatarica* L. on herbs and tree seedlings in four New England forests. *American Midland Naturalist* 130, 62-74.

1438) Wunderlin, R.P., Hansen, B.F., Delaney, K.R., Nee, M., and Mullahey, J.J. (1993) *Solanum viarum* and *S. tampicense* (Solanaceae): two weedy species new to Florida and the United States. *SIDA Contributions to Botany* 15, 605-611.

1439) Wyckoff, P.H. and Webb, S.L. (1996) Understory influence of the invasive Norway maple (*Acer platanoides*). *Bulletin of the Torrey Botanical Club* 123, 197-205.

1440) Yadav, A.S. and Tripathi, R.S. (1985) Effect of soil moisture and sowing density on population growth of *Eupatorium adenophorum* and *Eupatorium riparium*. *Plant and Soil* 88, 441-447.

1441) Yamasaki, S. (1997) Rhizome formation and survival of *Zizania latifolia* (Griseb.) Stapf, under limited oxygen supply in deep water. *Japanese Journal of Limnology* 58, 205-214.

1442) Yamasaki, S. and Tange, I. (1981) Growth responses of *Zizania latifolia*, *Phragmites australis* and *Miscanthus sacchariflorus* to varying inundation. *Aquatic Botany* 10, 229-240.

1443) Yamashita, N., Ishida, A., Kushima, H., and Tanaka, N. (2000) Acclimation to sudden increase in light favoring an invasive over native trees in subtropical islands, Japan. *Oecologia* 125, 412-419.

1444) Yang, S.-Z. (2001) A new record and invasive species in Taiwan – *Clidemia hirta* (L.) D. Don. *Taiwania* 46, 232-237.

1445) Yannitsaros, A. (1979) *Tagetes minuta* L. in Greece. *Candollea* 34, 99-107.

1446) Yatazawa, M., Hambali, G.G., and Uchino, F. (1983) Nitrogen fixing activity in warty lenticellate tree barks. *Soil Science and Plant Nutrition* 29, 285-294.

1447) Young, J.A. and Evans, R.A. (1986) Germination of white horehound (*Marrubium vulgare*) seeds. *Weed Science* 34, 266-270.

1448) Young, J.A., Palmquist, D.E., and Wotring, S.O. (1997) The invasive nature of *Lepidium latifolium*: a review. In: Brock, J.H., Wade, M., Pysek, P., and Green, D. (eds) *Plant invasions: studies from North America and Europe*. Backhuys Publishers, Leiden, pp. 59-68.

1449) Young, J.A., Palmquist, D.E., and Blank, R.R. (1998) The ecology and control of perennial pepperweed (*Lepidium latifolium* L.). *Weed Technology* 12, 402-405.

1450) Zaady, E., Gutterman, Y., and Boeken, B. (1997) The germination of mucilaginous seeds of *Plantago coronopus*, *Reboudia pinnata* and *Carrichtera annua* on cyanobacterial soil crust from the Negev Desert. *Plant and Soil* 190, 247-252.

1451) Zabkiewicz, J.A. and Balneaves, J.M. (1984) Gorse control in New Zealand forestry - the biology and the benefits. *Aspects of Applied Biology* 5, 255-264.

1452) Zancola, B.J., Wild, C., and Hero, J.-M. (2000) Inhibition of *Ageratina riparia* (Asteraceae) by native Australian flora and fauna. *Austral Ecology* 25, 563-569.

1453) Zavagno, F. and D'Auria, G. (2001) Synecology and dynamics of *Amorpha fruticosa* communities in the Po plain (Italy). In: Brundu, G., Brock, J., Camarda, I., Child, L., and Wade, M. (eds) Plant invasions: species *ecology and ecosystem management*. Backhuys Publishers, Leiden, pp. 175-182.

1454) Zedler, J. and Loucks, O.L. (1969) Differential burning response of *Poa pratensis* fields and *Andropogon scoparius* prairies in central Wisconsin. *American Midland Naturalist* 81, 341-352.

1455) Zedler, P.H. and Scheid, G.A. (1988) Invasion of *Carpobrotus edulis* and *Salix lasiolepis* after fire in a coastal chaparral site in Santa Barbara county, California. *Madroño* 35, 196-201.

1456) Zika, P.F. (2002) *Cotoneaster divaricatus* (Rosaceae) naturalized in Massachusetts. *Rhodora* 104, 302-303.

1457) Zika, P.F., Alverson, E.R., and Wilson, L. (2000) *Berberis darwinii* Hook. (Berberidaceae). *Madroño* 47, 214-216.

1458) Zimmerman, J.K. and Weis, I.M. (1984) Factors affecting survivorship, growth, and fruit production in a beach population of *Xanthium strumarium*. *Canadian Journal of Botany* 62, 2122-2127.

1459) Zimmermann, H.G. (1991) Biological control of mesquite, *Prosopis* spp. (Fabaceae), in South Africa. *Agriculture, Ecosystems and Environment* 37, 175-186.

1460) Zimmermann, H.G. and Malan, D.E. (1980) A modified technique for the herbicidal control of jointed cactus, *Opuntia aurantiaca* Lindley, in South Africa. *Agroplantae* 12, 65-67.

1461) Zobolo, A.M. and van Staden, J. (1999) The effects of deflowering and defruiting on growth and senescence of *Bidens pilosa* L. *South African Journal of Botany* 65, 86-88.

1462) Zotz, G., Bermejo, P., and Dietz, H. (1999) The epiphyte vegetation of *Annona glabra* on Barro Colorado Island, Panama. *Journal of Biogeography* 26, 761-776.

List of synonyms

The following list contains common synonyms of the species treated in the book. Synonyms are listed in the species entries together with authorship.

Name	Treated as
Abrus abrus	*Abrus precatorius*
Acacia arabica	*Acacia nilotica*
Acacia cyanophylla	*Acacia saligna*
Acacia cyclopsis	*Acacia cyclops*
Acacia decurrens var. *dealbata*	*Acacia dealbata*
Acacia decurrens var. *mollis*	*Acacia mearnsii*
Acacia lebbeck	*Albizia lebbeck*
Acetosa sagittata	*Rumex sagittatus*
Acetosella vulgaris	*Rumex acetosella*
Achyranthes repens	*Alternanthera pungens*
Aegilops incurva	*Parapholis incurva*
Agapanthus orientalis	*Agapanthus praecox* spp. *orientalis*
Agave rasconensis	*Agave americana*
Agropyron junceum	*Elytrigia juncea* subsp. *juncea*
Agropyron repens	*Elytrigia repens*
Agrostis indica	*Sporobolus indicus*
Agrostis tenuis	*Agrostis capillaris*
Agrostis vulgaris	*Agrostis capillaris*
Ailanthus glandulosa	*Ailanthus altissima*
Albizia lophantha	*Paraserianthes lophantha*
Alliaria officinalis	*Alliaria petiolata*
Alopecurus monspeliensis	*Polypogon monspeliensis*
Alstroemeria aurantiaca	*Alstroemeria aurea*
Alternanthera achyrantha	*Alternanthera pungens*
Alternanthera repens	*Alternanthera pungens*
Amorpha virgata	*Amorpha fruticosa*
Ampelopsis heterophylla	*Ampelopsis brevipedunculata*
Anacharis canadensis	*Elodea canadensis*
Anacharis densa	*Egeria densa*
Andrachne trifoliata	*Bischofia javanica*
Andropogon condensatus	*Schizachyrium condensatum*
Andropogon glomeratus	*Schizachyrium condensatum*
Andropogon rufus	*Hyparrhenia rufa*
Androsaemum officinale	*Hypericum androsaemum*
Anisantha tectorum	*Bromus tectorum*
Annona palustris	*Annona glabra*
Antholyza floribunda	*Chasmanthe floribunda*
Arctotis calendula	*Arctotheca calendula*
Ardisia crenulata	*Ardisia crenata*
Ardisia humilis	*Ardisia elliptica*
Argemone mexicana var. *ochroleuca*	*Argemone ochroleuca*
Argemone subfusiformis	*Argemone ochroleuca*
Arundo phragmites	*Phragmites australis*
Arundo reynaudiana	*Neyraudia reynaudiana*

Name	Treated as
Arundo selloana	*Cortaderia selloana*
Asclepias procera	*Calotropis procera*
Asclepias procera	*Calotropis procera*
Asparagus aethiopicus	*Asparagus densiflorus*
Asparagus longifolius	*Asparagus officinalis*
Asparagus sprengeri	*Asparagus densiflorus*
Aspidium macrophyllum	*Tectaria incisa*
Atropa rhomboidea	*Salpichroa origanifolia*
Azolla rubra	*Azolla filiculoides*
Banisteria benghalensis	*Hiptage benghalensis*
Banksia gibbosa	*Hakea gibbosa*
Berberis aquifolium	*Mahonia aquifolium*
Berberis chitria	*Berberis glaucocarpa*
Berberis diversifolia	*Mahonia aquifolium*
Bidens chinensis	*Bidens pilosa*
Bidens leucantha	*Bidens pilosa*
Bignonia pallida	*Tabebuia pallida*
Bignonia stans	*Tecoma stans*
Bignonia tweedieana	*Macfadyena unguis-cati*
Bignonia unguis-cati	*Macfadyena unguis-cati*
Bischofia trifoliata	*Bischofia javanica*
Boussingaultia cordifolia	*Anredera cordifolia*
Brachiaria mutica	*Urochloa mutica*
Brachiaria purpurascens	*Urochloa mutica*
Bromus madritensis subsp. *rubens*	*Bromus rubens*
Bromus purpurascens	*Bromus rubens*
Bryophyllum calycinum	*Bryophyllum pinnatum*
Buchholzia philoxeroides	*Alternanthera philoxeroides*
Buddleja variabilis	*Buddleja davidii*
Cactus pereskia	*Pereskia aculeata*
Caesalpinia japonica	*Caesalpinia decapetala*
Caesalpinia sepiaria	*Caesalpinia decapetala*
Caladium esculentum	*Colocasia esculenta*
Calamagrostis arenaria	*Ammophila arenaria*
Calla aethiopica	*Zantedeschia aethiopica*
Calophyllum calaba	*Calophyllum antillanum*
Cardaria draba	*Lepidium draba*
Cardiospermum barbicule	*Cardiospermum grandiflorum*
Cardiospermum hirsutum	*Cardiospermum grandiflorum*
Carduus lanceolatus	*Cirsium vulgare*
Carduus marianus	*Silybum marianum*
Cassia alata	*Senna alata*
Cassia bicapsularis	*Senna bicapsularis*
Cassia didymobotrya	*Senna didymobotrya*
Cassia emarginata	*Senna bicapsularis*
Cassia obtusifolia	*Senna obtusifolia*
Cassia tora	*Senna obtusifolia*
Ceanothus asiaticus	*Colubrina asiatica*
Cedrela australis	*Toona ciliata*
Cedrela mexicana	*Cedrela odorata*

Name	Treated as
Cedrela toona	*Toona ciliata*
Celastrus articulatus	*Celastrus orbiculatus*
Celtis japonica	*Celtis sinensis*
Cenchrus asperifolius	*Pennisetum setaceum*
Centaurea maculosa	*Centaurea stoebe* subsp. *stoebe*
Centaurea myacantha	*Centaurea calcitrapa*
Cerasus laurocerasus	*Prunus laurocerasus*
Chloris compressa	*Chloris virgata*
Chloris elegans	*Chloris virgata*
Cinchona succirubra	*Cinchona pubescens*
Cinnamomum zeylanicum	*Cinnamomum verum*
Cirsium lanceolatum	*Cirsium vulgare*
Cochlearia draba	*Lepidium draba*
Colocasia antiquorum	*Colocasia esculenta*
Conchium drupaceum	*Hakea drupacea*
Convolvulus acuminatus	*Ipomoea indica*
Coprosma baueri	*Coprosma repens*
Coronilla varia	*Securigera varia*
Cortaderia argentea	*Cortaderia selloana*
Cotoneaster angustifolius	*Pyracantha angustifolia*
Cotoneaster crenulatus	*Pyracantha crenulata*
Cotoneaster serotinus	*Cotoneaster glaucophyllus*
Critesion marinum	*Hordeum marinum* subsp. *marinum*
Cryptostemma calendulaceum	*Arctotheca calendula*
Cupania anacardiodes	*Cupaniopsis anacardioides*
Cymbopogon rufus	*Hyparrhenia rufa*
Cytisus albus	*Cytisus multiflorus*
Cytisus linifolius	*Genista linifolia*
Cytisus lusitanicus	*Cytisus multiflorus*
Cytisus monspessulanus	*Genista monspessulana*
Cytisus palmensis	*Chamaecytisus prolifer*
Cytisus prolifer	*Chamaecytisus prolifer*
Dactylis glaucescens	*Dactylis glomerata*
Dactylis maritima	*Spartina maritima*
Datura inermis	*Datura stramonium*
Datura tatula	*Datura stramonium*
Daubentonia punicea	*Sesbania punicea*
Digitaria paspalodes	*Paspalum distichum*
Dioscorea rubella	*Dioscorea alata*
Dipsacus sylvestris	*Dipsacus fullonum*
Dolichos gibbosus	*Dipogon lignosus*
Doxantha unguis-cati	*Macfadyena unguis-cati*
Echinochloa spectabilis	*Echinochloa polystachya*
Echium lycopsis	*Echium plantagineum*
Ehrharta panicea	*Ehrharta erecta*
Elaeagnus crispa	*Elaeagnus umbellata*
Elaeagnus hortensis	*Elaeagnus angustifolia*
Elodea densa	*Egeria densa*
Elymus repens	*Elytrigia repens*
Eragrostis robusta	*Eragrostis curvula*

Name	Treated as
Eragrostis subulata	*Eragrostis curvula*
Erigeron mucronatus	*Erigeron karvinskianus*
Eugenia brasiliana	*Eugenia uniflora*
Eugenia cumini	*Syzygium cumini*
Eugenia jambolana	*Syzygium cumini*
Eugenia jambos	*Syzygium jambos*
Eugenia michelii	*Eugenia uniflora*
Eulalia viminea	*Microstegium vimineum*
Eupatorium adenophorum	*Ageratina adenophora*
Eupatorium odoratum	*Chromolaena odorata*
Eupatorium riparium	*Ageratina riparia*
Festuca bromoides	*Vulpia bromoides*
Festuca elatior	*Festuca arundinacea*
Ficus aggregata	*Ficus microcarpa*
Fraxinus americana var. *uhdei*	*Fraxinus uhdei*
Fraxinus angustifolia	*Fraxinus rotundifolia*
Fuchsia gracilis	*Fuchsia magellanica*
Fuchsia macrostemma	*Fuchsia magellanica*
Geranium botrys	*Erodium botrys*
Geranium cicutarium	*Erodium cicutarium*
Glycine abrus	*Abrus precatorius*
Gnaphalium candidissimum	*Vellereophyton dealbatum*
Gunnera chilensis	*Gunnera tinctoria*
Hakea acicularis	*Hakea sericea*
Hakea saligna	*Hakea salicifolia*
Hakea suaveolens	*Hakea drupacea*
Hakea tenuifolia	*Hakea sericea*
Harongana madagascariensis	*Harungana madagascariensis*
Helianthus tomentosus	*Helianthus tuberosus*
Herpetica alata	*Senna alata*
Hiptage madablota	*Hiptage benghalensis*
Holcus argenteus	*Holcus lanatus*
Homalocenchrus oryzoides	*Leersia oryzoides*
Homeria flaccida	*Moraea flaccida*
Hottonia serrata	*Hydrilla verticillata*
Impatiens roylei	*Impatiens glandulifera*
Imperata arundinacea	*Imperata cylindrica*
Ipomoea acuminata	*Ipomoea indica*
Ipomoea burmannii	*Turbina corymbosa*
Ipomoea congesta	*Ipomoea indica*
Ipomoea palmata	*Ipomoea cairica*
Ipomoea reptans	*Ipomoea aquatica*
Ischaemum secundatum	*Stenotaphrum secundatum*
Ixia alba	*Sparaxis bulbifera*
Ixia bulbifera	*Sparaxis bulbifera*
Ixia rosea	*Romulea rosea*
Jacaranda ovalifolia	*Jacaranda mimosifolia*
Jasminum revolutum	*Jasminum humile*
Juncus compressus	*Juncus articulatus*
Juncus lampocarpus	*Juncus articulatus*

Name	Treated as
Justicia polysperma	*Hygrophila polysperma*
Kalanchoe pinnata	*Bryophyllum pinnatum*
Lantana aculeata	*Lantana camara*
Lantana nivea	*Lantana camara*
Laurocerasus officinalis	*Prunus laurocerasus*
Laurus camphora	*Cinnamomum camphora*
Laurus cinnamomum	*Cinnamomum verum*
Leucaena glauca	*Leucaena leucocephala*
Ligustrum indicum	*Ligustrum sinense*
Ligustrum insulare	*Ligustrum vulgare*
Ligustrum microcarpum	*Ligustrum sinense*
Litsea laurifolia	*Litsea glutinosa*
Lolium boucheanum	*Lolium perenne*
Lonicera insularis	*Lonicera morrowii*
Lotus ambiguus	*Lotus corniculatus*
Lotus balticus	*Lotus corniculatus*
Lotus decumbens	*Lotus uliginosus*
Lycium macrocalyx	*Lycium ferocissimum*
Lygodium scandens	*Lygodium microphyllum*
Melastoma hirsutum	*Clidemia hirta*
Melastoma hirta	*Clidemia hirta*
Melia toosendan	*Melia azedarach*
Melianthus minor	*Melianthus comosus*
Melilotus arvensis	*Melilotus officinalis*
Melilotus officinale	*Melilotus officinalis*
Melinis tenuinervis	*Melinis minutiflora*
Memecylon caeruleum	*Memecylon floribundum*
Mesembryanthemum edule	*Carpobrotus edulis*
Mesembryanthemum pugioniforme	*Conicosia pugioniformis*
Mimosa asperata	*Mimosa pigra*
Mimosa lebbeck	*Albizia lebbeck*
Mimosa saligna	*Acacia saligna*
Morus indica	*Morus alba*
Myosotis oblongata	*Myosotis sylvatica*
Myrica faya	*Morella faya*
Myriophyllum brasiliense	*Myriophyllum aquaticum*
Myrsiphyllum asparagoides	*Asparagus asparagoides*
Myrsiphyllum scandens	*Asparagus asparagoides*
Myrtus tomentosa	*Rhodomyrtus tomentosa*
Nephrolepis auriculata	*Nephrolepis cordifolia*
Ophioglossum japonicum	*Lygodium japonicum*
Oplismenus polystachyus	*Echinochloa polystachya*
Opuntia arborescens	*Cylindropuntia imbricata*
Opuntia gymnocarpa	*Opuntia ficus-indica*
Opuntia imbricata	*Cylindropuntia imbricata*
Opuntia inermis	*Opuntia stricta*
Opuntia stricta var. *dillenii*	*Opuntia dillenii*
Oryza oryzoides	*Leersia oryzoides*
Oxalis cernua	*Oxalis pes-caprae*
Oxalis humilis	*Oxalis purpurea*

Name	Treated as
Paederia scandens	*Paederia foetida*
Paederia tomentosa	*Paederia foetida*
Panicularia aquatica	*Glyceria maxima*
Panicularia fluitans	*Glyceria fluitans*
Panicum amplexicaulis	*Hymenachne amplexicaulis*
Panicum gouinii	*Panicum repens*
Panicum hirsutissimum	*Panicum maximum*
Panicum littorale	*Panicum repens*
Panicum melinis	*Melinis minutiflora*
Panicum polystachion	*Pennisetum polystachion*
Panicum purpurascens	*Urochloa mutica*
Paspalum ciliatifolium	*Paspalum conjugatum*
Paspalum paspalodes	*Paspalum distichum*
Passiflora mollissima	*Passiflora tripartita* var. *mollissima*
Passiflora tomentosa	*Passiflora mixta*
Pectis pinnata	*Schkuhria pinnata*
Pennisetum angolense	*Pennisetum macrourum*
Pennisetum benthamii	*Pennisetum purpureum*
Pennisetum cenchroides	*Cenchrus ciliaris*
Pennisetum ciliare	*Cenchrus ciliaris*
Pennisetum giganteum	*Pennisetum macrourum*
Pennisetum ruppelii	*Pennisetum setaceum*
Persicaria perfoliata	*Polygonum perfoliatum*
Phalaris oryzoides	*Leersia oryzoides*
Phalaris stenoptera	*Phalaris aquatica*
Phalaris tuberosa	*Phalaris aquatica*
Phleum nodosum	*Phleum pratense*
Phragmites communis	*Phragmites australis*
Physalis edulis	*Physalis peruviana*
Pilosella officinarum	*Hieracium pilosella*
Piper celtidifolium	*Piper aduncum*
Piper elongatum	*Piper aduncum*
Poa aquatica	*Glyceria maxima*
Pollinia imberbis	*Microstegium vimineum*
Polygala speciosa	*Polygala virgata*
Polygonum cuspidatum	*Fallopia japonica*
Polygonum heterophyllum	*Polygonum aviculare*
Polypodium cordifolium	*Nephrolepis cordifolia*
Pontederia crassipes	*Eichhornia crassipes*
Protasparagus densiflorus	*Asparagus densiflorus*
Prunus grandifolia	*Prunus laurocerasus*
Psamma arenaria	*Ammophila arenaria*
Psidium littorale	*Psidium cattleianum*
Psidium pomiferum	*Psidium guajava*
Psidium pumilum	*Psidium guajava*
Pueraria hirsuta	*Pueraria montana*
Pueraria lobata	*Pueraria montana*
Pueraria triloba	*Pueraria montana*
Pulegium vulgare	*Mentha pulegium*
Racosperma baileyanum	*Acacia baileyana*

Name	Treated as
Racosperma dealbatum	*Acacia dealbata*
Racosperma longifolium	*Acacia longifolia*
Racosperma mearnsii	*Acacia mearnsii*
Racosperma melanoxylon	*Acacia melanoxylon*
Reynoutria japonica	*Fallopia japonica*
Rhamnus frangula	*Frangula alnus*
Rhamnus jujuba	*Ziziphus mauritiana*
Rhododendron lancifolium	*Rhododendron ponticum*
Rhodomyrtus parviflora	*Rhodomyrtus tomentosa*
Rhoeo spathacea	*Tradescantia spathacea*
Rhus cacodendron	*Ailanthus altissima*
Richardia africana	*Zantedeschia aethiopica*
Rorippa islandica	*Rorippa palustris*
Rosa eglanteria	*Rosa rubiginosa*
Rosa lutetiana	*Rosa canina*
Rosa polyantha	*Rosa multiflora*
Rubus albescens	*Rubus niveus*
Rubus discolor	*Rubus ulmifolius*
Rubus flavus	*Rubus ellipticus*
Rubus inermis	*Rubus ulmifolius*
Rubus lasiocarpus	*Rubus niveus*
Rubus micranthus	*Rubus niveus*
Salix acuminata	*Salix cinerea*
Salix aquatica	*Salix cinerea*
Salix elegantissima	*Salix babylonica*
Salpichroa rhomboidea	*Salpichroa origanifolia*
Salsola australis	*Salsola kali*
Salvia clandestina	*Salvia verbenaca*
Salvia cleistogama	*Salvia verbenaca*
Salvinia auriculata	*Salvinia molesta*
Sapium sebiferum	*Triadica sebifera*
Sarothamnus scoparius	*Cytisus scoparius*
Scaevola frutescens	*Scaevola taccada*
Scaevola sericea	*Scaevola taccada*
Schinus areira	*Schinus molle*
Senecio mikanioides	*Delairea odorata*
Solanum auriculatum	*Solanum mauritianum*
Solanum capsicastrum	*Solanum pseudocapsicum*
Solanum diflorum	*Solanum pseudocapsicum*
Solanum hermannii	*Solanum linnaeanum*
Solanum jasminoides	*Solanum laxum*
Solidago serotina	*Solidago gigantea*
Spartium scoparium	*Cytisus scoparius*
Spathodea nilotica	*Spathodea campanulata*
Sporobolus africanus	*Sporobolus indicus*
Stachytarpheta urticifolia	*Stachytarpheta cayennensis*
Stenolobium stans	*Tecoma stans*
Stipa neesiana	*Nassella neesiana*
Stipa tenuissima	*Nassella tenuissima*
Stipa trichotoma	*Nassella trichotoma*

Name	Treated as
Swietenia candollei	*Swietenia macrophylla*
Syzygium jambolanum	*Syzygium cumini*
Tabebuia pentaphylla	*Tabebuia pallida*
Tacsonia quitensis	*Passiflora mixta*
Tagetes glandulifera	*Tagetes minuta*
Tagetes glandulosa	*Tagetes minuta*
Tamarix orientalis	*Tamarix aphylla*
Tamarix pentandra	*Tamarix ramosissima*
Tectaria martinicensis	*Tectaria incisa*
Teline linifolia	*Genista linifolia*
Teline monspessulana	*Genista monspessulana*
Tetranthera laurifolia	*Litsea glutinosa*
Thelechitonia trilobata	*Sphagneticola trilobata*
Thespesia macrophylla	*Thespesia populnea*
Thinopyrum junceum	*Elytrigia juncea* subsp. *juncea*
Tillaea helmsii	*Crassula helmsii*
Tracaulon perfoliatum	*Polygonum perfoliatum*
Tradescantia discolor	*Tradescantia spathacea*
Trichonema purpurascens	*Romulea rosea*
Tunica prolifera	*Petrorhagia prolifera*
Ugena microphylla	*Lygodium microphyllum*
Uhdea bipinnatifida	*Montanoa bipinnatifida*
Verbena cayennensis	*Stachytarpheta cayennensis*
Verdcourtia lignosa	*Dipogon lignosus*
Vitis brevipedunculata	*Ampelopsis brevipedunculata*
Vulpia dertonensis	*Vulpia bromoides*
Watsonia bulbillifera	*Watsonia meriana*
Wedelia trilobata	*Sphagneticola trilobata*
Xanthium echinatum	*Xanthium strumarium*
Xanthium orientale	*Xanthium strumarium*
Xanthium pungens	*Xanthium strumarium*
Xanthium vulgare	*Xanthium strumarium*

Glossary

achene: a dry one-seeded fruit, with the seed not fused to the fruit wall, as in Ranunculaceae and Asteraceae.

acuminate: gradually tapering to a point.

acute: having a short sharp point.

anther: the pollen-bearing part of the stamen.

apex: the tip or end.

apomixis: the production of viable seeds without fertilization.

areole: a cluster of hairs and/or spines borne at the node of a leafless stem, e.g. in Cactaceae.

aril: a fleshy or membranous appendage, sometimes partially or wholly covering the surface of the seed, often brightly coloured.

auricle: an ear-shaped outgrowth at the base of the sheath of some grasses; an ear-shaped lobe at the base of a leaf.

bipinnate: a compound leaf with the leaf blade divided twice pinnately, i.e. the pinnae themselves are divided pinnately.

bract: a more or less modified leaf, especially a smaller one associated with a flower or part of an inflorescence.

bracteole: a bract-like structure borne singly or in pairs on the pedicel or calyx of a flower.

caespitose: growing in tufts.

calyx: the sepals of a single flower.

capsule: a dry dehiscent fruit.

caryopsis: a dry, indehiscent one-seed fruit in which the seed is fused to the fruit wall, as in the Poaceae.

ciliate: a margin that is fringed with hairs.

cleistogamous: of flowers remaining closed and self-pollinating, producing fertile seeds.

connate: fusion of similar parts, e.g. petals into a corolla.

cordate: heart-shaped with a basal notch.

corolla: the petals of a flower.

corymb: an inflorescence in which all the flowers are at the same level but the pedicels arise at different levels and are of different length.

cyathium: an inflorescence of reduced unisexual flowers, surrounded by involucral bracts, e.g. in *Euphorbia*.

decumbent: horizontally spreading stems with the ends growing upwards.

decurrent: extending downwards beyond the point of insertion, often in petioles.

dehiscent: a fruit opening at maturity to release the seeds.

dioecious: a plant with male and female flowers borne on different individuals.

drupe: an indehiscent fleshy fruit with the seeds enclosed in an inner stony layer.

fertile: 1) capable of reproduction, 2) portions of a plant producing reproductive structures.

filament: the stalk of a stamen.

filiform: thread-like.

floret: a small flower, one of a dense cluster, flowerhead, or spikelet. In grasses a flower together with the lemma and palea that enclose it.

follicle: a dry fruit derived from a single carpel.

frond: the leaf of a fern or cycad.

glandular: with glands.

glume: one of the two bracts at the base of a grass spikelet.

hastate: spear-shaped.

hermaphrodite: a plant with all flowers bisexual.

hypanthium: a cup-like or tubular structure formed above the base of the ovary, with the stamens and perianth parts inserted on the rim.

imbricate: closely packed and overlapping.

indehiscent: a fruit not opening at maturity to release the seeds.

indusium: the tissue covering the sorus of a fern.

inflorescence: the flower-bearing part or parts of a plant.

interjugary: of glands, present on the rachis of a bipinnate leaf between the junction of pairs of pinnae or of pinnules.

involucre: a whorl or several whorls of bracts surrounding a flowerhead, as in many Asteraceae.

lanceolate: lance-shaped, i.e. 3-6 times as long as wide and broadest below the middle.

lemma: the lower of two bracts enclosing the flower of a grass.

lenticel: a small raised corky spot or line appearing on young bark, through which gas exchange occurs.

ligule: 1) grasses: a variously shaped appendage facing towards the base of a leaf. 2) Asteraceae: the corolla lobe or limb in ray florets.

node: the level of a stem at which the leaves arise.

nutlet: a dry indehiscent one-seeded fruit.

oblong: rectangular in shape with the length greater than the width.

obovate: ovate in shape, but broadest above the middle.

obtuse: blunt or broadly rounded.

palea: the upper of two bracts enclosing the flower of a grass.

palmate: a compound leaf with three or more leaflets arising from the same point at the end of the petiole.

panicle: a compound inflorescence with a main axis and lateral branches which are further branched, each axis ending in a flower or flower bud.

pappus: the group of appendages above the ovary and outside the corolla in Asteraceae. Usually hairs or scales, often persisting on the fruit.

pedicel: the stalk of a flower.

peduncle: the stalk of an inflorescence, i.e. the axis between the last foliage leaf and the first flower bearing branch.

peltate: a leaf with the stalk or point of attachment on its lower surface not at the margin but on the surface.

perianth: the calyx and corolla together.

pericarp: the wall of the fruit, developed from the ovary wall.

petal: a free segment of the corolla.

petiole: the stalk of a leaf.

phyllary: a single involucral bract of the Asteraceae.

phyllode: a flattened leaf-like petiole, replacing the leaf blade.

pilose: hairy with long soft and weak hairs.

pinna: a primary segment of the blade of a compound leaf.

pinnate: the leaf blade divided into pinnae in two rows along a rachis.

pome: a fleshy false fruit.

procumbent: with the stems trailing or spreading over the ground.

prostrate: lying flat on the ground.

pubescent: with a somewhat dense covering of short and weak hairs.

punctate: marked with dots.

raceme: a simple inflorescence ending in a non-floral bud and in which the flowers are stalked.

rachis: the axis of a pinnate leaf, or a pinna, or an inflorescence.

receptacle: 1) the top of the stalk on which a flower or flowerheads arises, 2) an axis on which sporangia arise in ferns.

rhizome: a belowground stem, usually growing horizontally.

rugose: covered with coarse lines or furrows.

sagittate: shaped like an arrow-head, with the two lobes at the base directed backwards.

samara: a dry indehiscent fruit with its wall expanded into a wing.

scandent: climbing without special climbing organs.

sepal: one of the outer leaf-like structures surrounding the corolla and fertile organs of a flower, usually green.

silique: a dry dehiscent fruit derived from a superior ovary of two carpels, as in Brassicaceae.

sinuate: with a deeply wavy margin.

sorus: a discrete aggregate of sporangia in ferns.

spadix: a spike-like inflorescence with a thickened, often succulent axis, the whole often being surrounded by a spathe, as in many Araceae.

spathe: a large bract at the base of a spadix, partially or wholly enclosing it.

spike: a simple inflorescence, ending in a non-floral bud, in which the flowers are sessile.

spikelet: the small partial inflorescence in Poaceae and Cyperaceae, consisting of an axis bearing glumes and florets.

spinose: with spines.

sporangium: a structure in which spores are formed, e.g. in ferns.

stamen: one of the male organs of a flower, consisting of a pollen-bearing anther and a filament.

stellate: star-shaped.

sterile: 1) without reproductive structures, not producing any seed, spores or pollen, 2) seeds, spores or pollen not capable of germination.

stipule: one or a pair of appendages sometimes developed at the base of a leaf and of various shape.

stolon: a horizontal stem growing aboveground and rooting at the nodes.

subulate: narrow and gradually tapering to a fine apex.

syncarp: a multiple fruit consisting of several united single fruits, usually fleshy.

tendril: a long slender organ derived from an axis or leaf, or from part of one of these.

terete: cylindric and elongated.

terminal: at the end.

ternate: in threes, e.g. a leaf with leaflets arranged in groups of three.

thyrse: a compound inflorescence ending in a vegetative bud and with mixed types of branching, the main axis bearing lateral cymes.

tomentose: covered with dense intertwined hairs.

triangular: three-angled and three-sided in shape.

umbel: an inflorescence in which all the flowers or flower-stalks arise from one point at the end of the peduncle.

undulate: leaf blades that are wavy and not flat.

villous: covered with long shaggy hairs.

zygomorphic: a bilaterally symmetrical flower, e.g. the lower parts differing in shape or size from the upper parts.